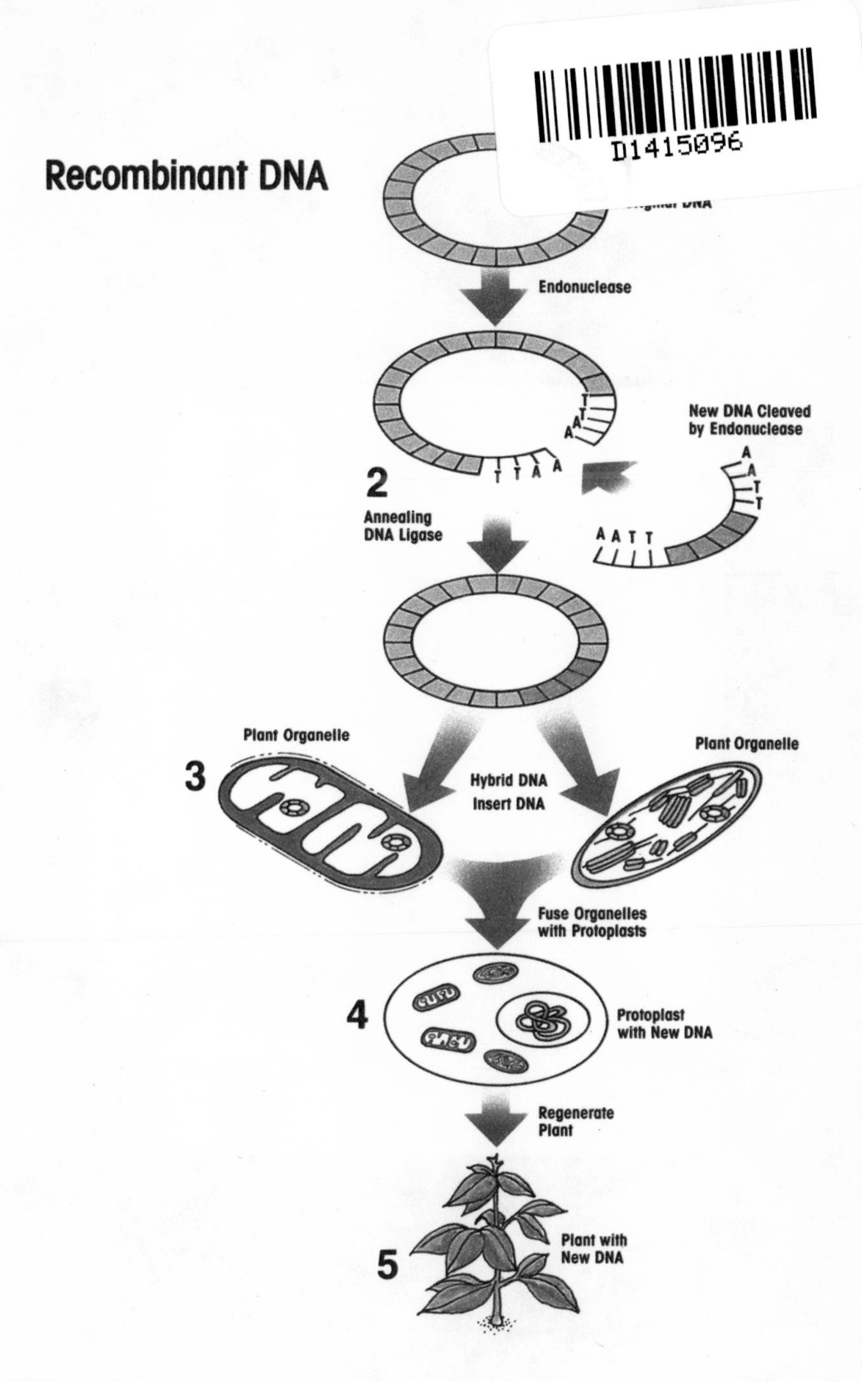
Recombinant DNA
Endonuclease
T T A A
New DNA Cleaved
by Endonuclease
A A T T
2
Annealing
DNA Ligase
Plant Organelle
3
Hybrid DNA
Insert DNA
Plant Organelle
Fuse Organelles
with Protoplasts
4
Protoplast
with New DNA
Regenerate
Plant
Plant with
New DNA
5

UMASS
University of
Massachusetts
Amherst

HANDBOOK OF PLANT CELL CULTURE

Published

Volume 1, Techniques for Propagation and Breeding
Editors: David A. Evans, William R. Sharp, Philip V. Ammirato, Yasuyuki Yamada

Volume 2, Crop Species
Editors: William R. Sharp, David A. Evans, Philip V. Ammirato, Yasuyuki Yamada

Volume 3, Crop Species
Editors: Philip V. Ammirato, David A. Evans, William R. Sharp, Yasuyuki Yamada

Forthcoming volumes will be published biennially.

HANDBOOK OF PLANT CELL CULTURE,

Volume 3

Crop Species

Editors

Philip V. Ammirato
Barnard College, Columbia University
DNA Plant Technology Corporation

David A. Evans
DNA Plant Technology Corporation

William R. Sharp
DNA Plant Technology Corporation

Yasuyuki Yamada
Kyoto University

Macmillan Publishing Company
A Division of Macmillan, Inc.
NEW YORK

Collier Macmillan Publishers
LONDON

Macmillan Publishing Co.
866 Third Avenue, New York, NY 10022

Collier Macmillan Canada, Inc.

Printed in the United States of America

printing number
2 3 4 5 6 7 8 9 10

Library of Congress Cataloging in Publication Data

(Revised for volume 3)

Main entry under title:

Handbook of plant cell culture.

Vol. 3- published: New York : Macmillan ; London : Collier Macmillan.
Includes bibliographies and indexes.
Contents: v. 1. Techniques for propagation and breeding / David Evans . . . [et al.] — v. 2. Crop species / William R. Sharp . . . [et al.] — v. 3. Crop species / Philip V. Ammirato . . . [et al.] editors.
1. Plant cell culture—Collected works. I. Evans, David, 1952- . II. Sharp, William R., 1936- . III. Ammirato, Philip V., 1943- .
SB123.H317 1983 631.5'2 82-73774
ISBN 0-02-949230-0 (v. 1)
0-02-949780-9 (v. 2)
0-02-949010-3 (v. 3)

Contents

Section V. Root and Tuber Crops

Section VI. Tropical and Subtropical Fruits

Section VII. Temperate Fruits

Section VIII. Fiber and Wood

Section IX. Extractable Products

Preface

The tools of plant cell culture are increasingly being applied to a wide range of biotechnology ventures and in particular to the clonal propagation and genetic improvement of crop plants. For this, the approaches and methodologies must be specifically adapted to the differing problems and potentialities of each crop and to the differing responses of plants that may be herbaceous or woody, dicotyledonous or monocotyledonous, annual or perennial, inbred or highly heterozygous. It is the application of plant cell culture techniques to the improvement of specific crop plants that is the subject of this volume, as it was in Volume 2.

The list of plants that play an important part in agribusiness is longer than one might initially surmise. The selection of those to be included in these volumes reflects a necessary amalgam of several factors—the plants chosen must be of recognized economic importance, they need to have been successfully employed in cell culture research, and a key investigator had to be available and willing to contribute. Some important crops, then, are not here because these elements did not come together. And, it is to be expected in any area where technology is only just being applied that there will be varying degrees of experience and success. This can be seen in the varying lengths of the presentations. The plants finally selected do represent major crop plants where cell culture technology has been demonstrably applied.

In addition, in each volume, several general chapters provide overviews of topics that are particularly relevant to the subject at hand. In this volume, we have included a discussion of plant germplasm resources, emphasizing the problems of rapidly diminishing natural and cultivated populations, and the requirements for germplasm collection, maintenance, evaluation, and distribution. These resources serve as both source material for introduction into cell culture programs and as a repository for the plants these programs generate. A second chapter discusses a number of crop plants that are not now fully exploited but offer the potential for increased production of food, oil, fiber, and other plant products, especially with the utilization of cell culture techniques. The last overview summarizes the application of anther culture techniques to the development of new varieties in China, a country where this technology has been advanced and applied with notable success.

As in our previous volumes, we are pleased to have an introductory essay by a distinguished scientist. E. C. Cocking is a pioneer in the development and application of protoplast technology and was recently elected to the Royal Society of London, joining a long list of distinguished scholars and scientists past and present, including F. C. Steward, who contributed the introductory chapter for the first volume of this treatise. In his essay, Dr. Cocking discusses aspects of the history of plant hybridization and in particular the contributions of Luther Burbank, a pioneer in the development of hybrids between distantly related species. Dr. Cocking's essay serves to illustrate the essential relationship between traditional sexual hybridization and the newer techniques of somatic hybridization.

The centerpiece for this volume, as for the entire series, is practical methodology, and we have again included for each chapter a major section with actual protocols, whether as recipes, tables, charts, or narratives. To introduce the specific crop, the history and economic importance are presented, culminating in a discussion of important breeding and propagation problems, areas in which cell culture methods may be particularly applied. A critical review of the literature summarizes past and current cell culture work, and a discussion of future prospects details where we may expect the technology to go. Again, key references have been highlighted, and references are given with full citation.

As before, our goal is to provide a comprehensive and practical compilation that students and scientists, academicians and businessmen alike will find useful in both understanding current strategies and in extending scientific frontiers.

The publication of Volume 3 completes the initial phase of this treatise. However, plant cell culture continues to progress at an astonishing pace, with the refinement of old techniques, the development of new ones, and their successful application to an ever greater number of economically important plants. We are delighted that this series is being continued and that additional volumes are already being readied to document and foster these advances.

At this juncture, after having witnessed the metamorphosis of piles of manuscripts into three finished volumes, we again thank those that helped in this endeavor but do so with an extra measure of gratitude.

We are indebted to our editors at Macmillan, Sarah Greene and Frances Tindall, for the necessary support in bridging the gap from concept to book. Especial thanks go to Susan Dale, who saw to the translation of words into text with more than a full measure of care. Janis Bravo again served as editorial assistant, the crucial link from us to both authors and editors, for which she deserves our sincere appreciation. We also wish to thank our typists, Karen Selover, Flora Loose, Cheryl Baskin, and Joann Morrison, and Mary Ellen Curtin for preparation of the index. Lastly, we offer our deep thanks to our many authors who provided manuscripts on time and ushered them through the various publication stages with despatch and good cheer.

CONTRIBUTORS

B.S. Ahloowalia	Plant Breeding Department, Agricultural Institute, Oak Park Research Center, Carlow, Ireland
P.V. Ammirato	Department of Biological Sciences, Barnard College, Columbia University, New York, NY and DNA Plant Technology Corporation, Cinnaminson, NJ
Y.P.S. Bajaj	Punjab Agricultural University, Ludhiana, Punjab, India
P. Boxus	Station des Cultures Fruitières et Maraichères, Gembloux, Belgium
E. Brasseur	Station des Cultures Fruitières et Maraichères, Gembloux, Belgium
J.E. Bravo	DNA Plant Technology Corporation, Cinnaminson, NJ
R.A. Buchanan	Soil and Land Use Technology, Inc., Columbia, MD
A. Carvalho	Department of Genetics, Institute of Agronomy, Campinas, S.P., Brazil
E.C. Cocking	Department of Botany, The University of Nottingham, University Park, Nottingham, U.K.
W.M. Costa	Department of Genetics, Institute of Agronomy, Campinas, S.P., Brazil
C. Damiano	Istituto Sperimentale per la Frutticoltura, Rome, Italy
P. Dublin	Laboratoire de Culture In Vitro, G.E.R.D.A.T., Centre de Recherches de Montpellier, Montpellier, France
D.A. Evans	DNA Plant Technology Corporation, Cinnaminson, NJ
L.C. Fazuoli	Department of Genetics, Institute of Agronomy, Campinas, S.P., Brazil
H. Hu	Institute of Genetics, Academia Sinica, Beijing, China
B.B. Johnson	Department of Botany, Oklahoma State University, Stillwater, OK
E.G. Knox	Soil and Land Use Technology, Inc., Columbia, MD
S.A. Kut	DNA Plant Technology Corporation, Cinnaminson, NJ
G.L. Laidig	5467 Hildebrand Court, Columbia, MD
L. Lipschutz	DNA Plant Technology Corporation, Cinnaminson, NJ
W.H. Loh	International Plant Research Institute, San Carlos, CA
P.M. Lyrene	Fruit Crops Department, University of Florida, Gainesville, FL
T. McCoy	USDA/ARS, College of Agriculture, University of Nevada, Reno, NV
H.P. Medina-Filho	Department of Genetics, Institute of Agronomy, Campinas, S.P., Brazil, and DNA Plant Technology Corporation, Cinnaminson, NJ

S.A. Miller	DNA Plant Technology Corporation, Cinnaminson, NJ
T. Nakamura	Department of Genetics, Institute of Agronomy, Campinas, S.P., Brazil
K. Paranjothy	Palm Oil Research Institute of Malaysia, Kuala Lumpur, Malaysia
H.J. Price	Department of Plant Sciences, Texas A & M University, College Station, TX
T.S. Rangan	Phytogen Corporation, Pasadena, CA
C.M. Rick	Department of Vegetable Crops, University of California, Davis, CA
R.M. Skirvin	Department of Horticulture, University of Illinois, Urbana, IL
J.M. Smagula	Department of Plant and Soil Sciences, University of Maine, Orono, ME
R.H. Smith	Department of Plant Sciences, Texas A & M University, College Station, TX
H.E. Sommer	School of Forest Resources, University of Georgia, Athens, GA
M.R. Sondahl	Department of Genetics, Institute of Agronomy, Campinas, S.P., Brazil, and DNA Plant Technology Corporation, Cinnaminson, NJ
P. Spiegel-Roy	Division of Fruit Breeding and Genetics, Volcani Center, Bet Dagan, Israel
A. Vardi	Division of Fruit Breeding and Genetics, Volcani Center, Bet Dagan, Israel
I.K. Vasil	Department of Botany, University of Florida, Gainesville, FL
K. Walker	Plant Genetics, Inc., Davis, CA
H.Y. Wetzstein	Department of Horticulture, University of Georgia, Athens, GA
Y. Yamada	Research Center for Cell and Tissue Culture, Kyoto University, Kyoto, Japan
S.Y. Zee	Botany Department, University of Hong Kong
J.Z. Zeng	Institute of Genetics, Academia Sinica, Beijing, China

"Sorry, he's all tied up at the moment."

Reprinted courtesy of *Parade Magazine* and Bill Hoest.

HANDBOOK OF PLANT CELL CULTURE

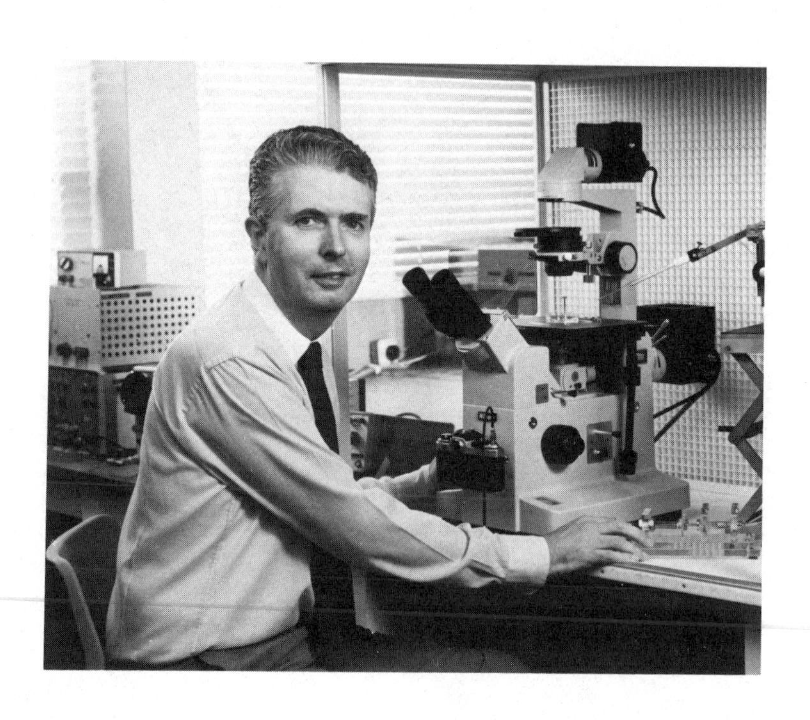

SPECIAL ESSAY:

Hybridizations Past and Present

E. C. Cocking

It was through the demonstration of the possibility of hybridization that the sexuality of plants, for a long time doubted, was indisputably proven; it was with this objective in view that hybrids were raised in great numbers by Koelreuter as early as 1761. Occasionally the sexual cells of different varieties, species, or even genera were shown to be able to unite and produce descendants capable of development (for detailed references to this early work see Strasburger et al., 1908). As early as 1676 Nehemiah Grew in an address to the Royal Society in London suggested that the stamens are the male organs of the plant. In 1694 the German botanist Camerarius in his classic work *De Sexu Plantarum* provided experimental proof that pollen was necessary in a number of plants for seed development. In 1717 the English gardener Thomas Fairchild obtained the first sexual hybrid of which there is an authentic record, a cross between carnation (*Dianthus caryophyllus*) and the Sweet William (*D. barbatus*). He noted that the progeny of the cross resembled both parents (Dreyer, 1975).

These studies, of course, predated the enumeration of Mendel's laws in relation to intervarietal sexual crosses. They laid the foundation for ongoing work on hybridization, which found its expression in the extensive, but at present largely forgotten, work of Luther Burbank on plant improvement by sexual hybridization. Currently a reappraisal of Luther Burbank's work is required because the resurgence of interest in extending the range of hybridizations by somatic hybridization is highlighting the need for a comprehensive knowledge of the limitations of sexual procedures.

The most recent comprehensive survey of interspecific hybridization in plant breeding (Sanchez-Monge and Garcia-Olmedo, 1977) does not mention the work of Luther Burbank. As pointed out by Lacadena (1977), what is considered as the first international congress of genetics was the International Conference on Hybridization held in London in 1899, and this intense interest in the possible usefulness of hybridization led to the postulation by Winge (1917) that hybridization and chromosome doubling could generate new species. The work of Karpechenko in the 1920s, which showed that it was indeed often possible to restore fertility in distant hybrids by doubling the chromosome number, was being undertaken at a time when Luther Burbank was approaching the end of his working life (he died in 1926). It was also Karpechenko (1927) who succeeded in developing an amphidiploid between radish and cabbage (Raphanobrassica), but as will be described later Luther Burbank was paralleling this with the development of crosses between various fruit species, including the cross between plum and apricot (the plumcot), and between a range of horticultural species, including the cross between Nicotiana and Petunia (the Nicotunia). Also paralleling this work was the production by others of experimental hybrids between wheat and rye (Triticale) in the early 1900s. Only in recent years however has this man-made cereal become adequately perfected (Cauderon, 1977). As recently pointed out by Dreyer (1975) in his penetrating reassessment of the life of Luther Burbank, it was not until 1943 that Jones in a review in the *Journal of the New York Botanical Garden* observed that "Burbank's most valuable contribution to science, his *Rubus* hybrids that bred true, have been largely overlooked by geneticists. When they were first announced, Mendelian segregation was expected in all hybrids and his statements were not accepted. Now that amphidiploids are known to transmit without segregation, Burbank is not given the priority he deserves for producing and putting these on record." As pointed out by Dreyer, Burbank's priority lies in having produced such hybrids and clearly distinguishing them as "Another Mode of Species Forming" (Burbank, 1909).

Luther Burbank's name is associated with the Burbank potato, and it is interesting to recall that Burbank developed this variety when he was a young man of 24 in 1873. Even in 1906 the U.S. Department of Agriculture estimated that not less than 17 million dollars worth of Burbank potatoes had been grown in the United States. What Luther Burbank found in his Massachusetts garden was a seed ball on one of his Early Rose potato vines. It required an imaginative mind to conceive that there was importance in this exceptional phenomenon. There were 23 seeds in the cluster and these were planted the next spring, each one by itself; his plan was nearly frustrated by the loss of the seed ball which was broken off by the wind just as it ripened. As noted by his biographer (Williams, 1915), after patient search the treasure was recovered and carefully preserved over the winter. In the fall, when the potatoes were dug up it was obvious that they represented 23 different varieties. Two of these were of altogether exceptional size and quality and from these were developed the Burbank potato. Williams remarked that the remarkable Burbank potato was thus developed in a single season merely by potato seeds that developed quite independently of human effort: "Forty years of subsequent effort in which vast numbers of hybridizing experiments have been performed have failed to produce another variety of potato superior to the one that was virtually a gift of Nature. In this field of endeavor as in so many others there is an element of uncertainty that adds to its charm."

Luther Burbank sold this new potato variety for $150 to a practical gardener, who gave it the name of the Burbank potato. This sale helped him to migrate to the better climate of California and set up experimental work as a plant developer on 4 acres in Santa Rosa, together with a tract of 18 acres at Sebastopol 7 miles away. Within this relatively small area more than 100,000 distinct experiments were carried out.

Williams (1915) has emphasized that the fundamental principles of plant development through which Burbank sought to develop new and improved varieties were not in themselves novel or revolutionary. They consisted essentially of the careful selection among a mass of plants of any individual that showed exceptional qualities of a desirable type, the saving of seed of this exceptional individual, and the carrying out of the same process of selection of the progeny through successive generations. Couple this method of selection with the method of cross-pollination of different varieties of species, to produce hybrid forms showing a tendency to greater variation of desired characters, and we have, as so succinctly stated by Williams, the outline of the fundamental principles of plant breeding as known to horticulturalists for generations. What Luther Burbank did was to add the extra dimension of screening very large numbers, including a wide range of varieties and species. Through the astute use of his highly trained senses, backed by an amazing fund of practical knowledge, the entire procedure took on a mystifying aspect of wizardry (Williams, 1915).

As we have seen, while others have explored the theoretical reasons why an increase in the number of chromosome sets should play an important role in the processes of speciation in plants, Luther Burbank was par excellence an experimenter who saw a closer parallel between plant breeding and evolution; he was less interested in theories and more interested in producing better plants by extensive assessments of hybridizations, coupled with rigorous selection of the best. As his biographer (Williams, 1915) commented, his work was not conducted to prove or test any particular scientific theories or make scientific discoveries, but it had for its sole aim the production of more and better varieties of cultivated plants. "I shall be contented if because of me there shall be better fruits and fairer flowers," Burbank said.

The printed records of Burbank's life and work are found almost exclusively in newspapers, magazines, and books. Written under Burbank's immediate direction are twelve beautifully illustrated volumes entitled "Luther Burbank, His Methods and Discoveries and their Practical Application, prepared from his original notes covering more than 100,000 experiments made during forty years devoted to plant improvement" (Whitson et al., 1914-1915). As far as I can discover Burbank only published one or two scientific papers—somewhat refreshing for these days! Burbank suffered from the absurdly exaggerated character of certain newspaper and magazine accounts of his work (Howard, 1945), a happening not totally unfamiliar to some of us today.

In 1980 I was invited to give a series of lectures on plant genetic manipulations at Ames, Iowa, to celebrate the opening of their new Horticulture Department. In one of these lectures I had reason to survey the use of somatic fusions, as contrasted with gametic fusions, for the production of hybrids. I also discussed the recent success in producing somatic hybrids between potato and tomato by the fusion of protoplasts (Melchers et al., 1978). After my lecture Dr. C.V. Hall drew my attention to Burbank's work on the attempted sexual hybridization of potato and tomato described in his biography "New Creations in Plant Life" by W.S. Harwood (1906). Burbank thought that a new and important vegetable might be produced by hybridizing the flowers of the potato and tomato. Burbank was unable to effect this hybridization, but was foremost to proclaim that such negative experiments are never final. Burbank carried out extensive grafting of potato on tomato root stock, and vice versa, and from among these chimeras he obtained a new fruit that grew on the potato which he named the pomato. "Looking to the common origin of tomato and potato and considering the general appearance of the new fruit, he has happily combined the two names in designating this new creation" (Harwood, 1906). As mentioned by Howard (1945), Burbank was not looked upon as a scientist by many of his contemporaries because he did not have what was then, and perhaps is also now, regarded as the correct attitude for a scientist; the basic urge he had for experimentation was utilitarian, to produce something that was saleable rather than to make contributions to knowledge. The main reason, however, that he was not accepted as a scientist, as has also been pointed out by many of his biographers, was that he did not keep proper records of his crosses.

This was my first aquaintance with Burbank's work on hybridizations. Fortunately while visiting York in 1981 I was able to purchase a secondhand copy of the biography of Luther Burbank by Williams (1915). Reproduced from one of its plates is a photograph of "Mr. Burbank Planting Choice Seeds" (Fig. 1). This biography succinctly summarizes the extensive range of hybridizations undertaken, including intensive work with plums, berries, and lilies. He originated numerous new varieties of plums, some of which are among the best known and most successful kinds now grown. Most of the new plum varieties were the result of multiple crossings in which Japanese plums played an important part. One of Luther Burbank's most celebrated experiments was that in which he hybridized the plum and apricot, producing a wonderful new fruit, the plumcot (a color illustration appears in Williams, 1915). The hybridization was "effected with difficulty because the two strains are so distantly related, but it was finally accomplished" (Williams, 1915).

Unequivocal proof that hybridization had in fact been achieved was not in fact presented; and even in current hybridization assessments problems can arise in this respect! Most workers had accepted, based largely on biochemical characterization, that potato and tomato fusions had produced somatic hybrid plants (Melchers et al., 1978). Krikorian (1982), however, has noted that, since it was not possible to perform sophisticated cytological techniques to demonstrate unequivocally that

Figure 1. Mr. Burbank planting choice seeds (from Williams, 1915).

true hybridization had been achieved, the possibility still remains that the plants were chimeras. More recent work has now removed this possibility of doubt (Schiller et al., 1982).

His work with berries for 35 years resulted in the introduction commercially of ten or more varieties obtained through various hybridizations of dewberries, blackberries, and raspberries. Sixteen years of work, and the use of more than fifty varieties of lily, resulted in a brilliant array of new forms. The famous Shasta daisy was the result of multiple crossing between an American and a European species of field daisy, and then between the hybrids and a Japanese form.

My second aquaintance with Burbank's hybridizations was when visiting Washington State University at Pullman to attend the recent NATO Meeting on Genetic Manipulation of Eukaryotic Systems in 1982. In the library I found the twelve volumes of Luther Burbank's work (Whitson et al., 1914-1915) together with a copy of his descriptive catalogue entitled "New Creations in Fruits and Flowers" (1893-1901). This publication is of particular interest; his biographer in the Dictionary of American Biography (1929) describes it as of "peculiar interest." It was not for public distribution. He urged correspondents to be as brief as possible and to ask no questions that could be answered elsewhere. The fruits and flowers mentioned were described as the best of millions of cross-bred, hybrid, and seedling plants. He stressed that "visitors to my grounds who expect to find long, uniform rows of beautiful plants will not see what they expect; but, instead, will see chaos, utter chaos; a workshop, the birthplace of new vegetable forms." Among the numerous fruits and flowers described are the stoneless plums and prunes: "Since the plum was first grown nothing more wonderful in its history has occurred than the production of these stoneless fruits—it is a prune having the general appearance of the common Californian prune, but growing much larger, not yet educated up to the production of sufficient sugar." Also described is a new plant the Nicotunia:

> This is the first time this word was ever printed, and the plant is one which has never before been seen; some botanist asks "What is it?" It is the name which I have given to the new race of plants produced by crossing the large flowering Nicotianas with the Petunias. If one thinks he can take right hold and produce Nicotunias as he would hybrid Petunias or cross-bred Primroses, let him try; there is no patent on their manufacture; but if the five hundredth crossing succeeds, or even the five thousandth, under the best conditions obtainable he will surely be very successful; I do not fear any immediate competition.
>
> The plants have slender, drooping or trailing tomentose green, red and purple stalks and leaves twice or three times as large as the Petunia; the flowers are handsome, white, pink, carmine or striped and borne in bounteous profusion. No seed is ever produced, but they are very readily multiplied by cuttings.

This description of the intergeneric hybrid and the naming of it by Luther Burbank, is of particular interest in view of the attempts currently being undertaken almost a century later to obtain this hybrid by protoplast fusion. Burbank also documented some of his failures at other disparate hybridizations, and indeed the Nicotunias proved difficult to propagate (Whitson et al., 1914-1915). Burbank suggested that a more extended series of experiments with Petunias "might result in hybridizing more satisfactorily with *Salpiglossis*, for the plants are botanically related pretty closely" (Whitson et al., 1914-1915).

Luther Burbank, of course, did not have available the special refinements in sexual hybridization currently available to the plant breeder by using in vitro techniques. If these had been available to him no doubt his Catalogue would have been even more extensive!

The present scenario of plant hybridizations is of some interest when viewed in the context of the work of Burbank and others, who in the early years of this century laid the foundation for the current range of sexual hybridizations. The development of an additional method of hybridization by the fusion of somatic cells (Cocking, 1977), coupled with in vitro culture techniques applied to sexual hybridization, has brought fresh enthusiasm and many new investigations. Yet the genetical problems have not essentially changed. At this stage in the development of the subject it is important to approach hybridizations on as broad a front as possible with a wide range of species wild and cultivated; and to extend the range of sexual hybridizations by in vitro techniques of embryo culture using carefully selected media. Such procedures have recently been successfully utilized in wide hybridizations in the *Glycine* genus between soybean and a wild perennial relative (Newell and Hymowitz, 1982). Only in this way can we begin to obtain a better basic knowledge of the factors influencing gene flow between plant species. Even the contemporary biographers of Burbank foresaw with characteristic prescience the opportunities in this respect, extending even as far as protoplasmic fusions between plants and animals. Writing in 1906, Howard, in discussing the origin of new species in the light of Luther Burbank's work on hybridizations, added "looking to the future, will it be possible, granting the common protoplasmic basis of plant and animal life, eventually to interblend the two? Such a union should it come must be scarcely more marvelous than the union here recorded, effecting creations which Nature in the very amplitude of her powers, never could have achieved alone." The first experimental steps in this direction were taken in 1975 (Ahkong et al., 1975).

In the present competitive world of the commercialization of hybrids it is interesting to note that the struggles of Luther Burbank, who often had to sell his stock material at highly inflated prices to maintain financial solvency, led to the introduction of plant breeders' rights in the United States (Dreyer, 1975).

REFERENCES

Ahkong, Q.F., Howell, J.I., Lucy, J.A., Safwat, F., Davey, M.R., and Cocking, E.C. 1975. Fusion of hen erythryocytes with yeast protoplasts induced by polyethylene glycol. Nature London 255:66-67.

Burbank, L. 1909. Another Mode of Species Forming. Popular Science Monthly, September. [Read at the annual meeting of the American Breeders' Association at Columbia, Missouri, Jan. 5-8, 1909 (reproduced in full in Dreyer)].

Cauderon, Y. 1977. Alloploidy. In: Interspecific Hybridization in Plant Breeding. Proceedings of the 8th Congress of Eucarpia, 1977 (E. Sanchez-Monge and F. Garcia-Olmedo, eds.) p. 131. Escuela Technica Superior de Ingenieros Agromonas Universidad Politecnica de Madrid, Madrid, Spain.

Cocking, E.C. 1977. The outlook for the interspecies somatic hybridization of sexually incompatible species. In: Interspecific Hybridization in Plant Breeding. Proceedings of the 8th Congress of Eucarpia, 1977 (E. Sanchez-Monge and F. Garcia-Olmedo, eds.) pp. 229-235. Escuela Technica Superior de Ingenieros Agromonas Universidad Politecnica de Madrid, Madrid, Spain.

Dictionary of American Biography (A. Johnson, ed.) 1929. Burbank, Luther, Mar. 7, 1849-Apr. 11, 1926. Charles Scribener's & Sons, New York.

Dreyer, P. 1975. A Gardener Touched with Genius: The Life of Luther Burbank. Coward, McCann and Geoghegan, New York.

Harwood, W.S. 1906. New Creations in Plant Life: An Authoritative Account of the Life and Work of Luther Burbank. Macmillan, New York.

Howard, W.L. 1945. Luther Burbank: A victim of hero worship. Chron. Bot. 9:300-506.

Karpechenko, G.D. 1927. On the problem of experimental species. Bull. App. Bot. Plant Breed. 17:305.

Krikorian, A.D. 1982. Cloning higher plants from aseptically cultured tissues and cells. Biol. Rev. 57:151-218.

Lacadena, J.R. 1977. Interspecific gene transfer in plant breeding. In: Interspecific Hybridization in Plant Breeding. Proceedings of the 8th Congress of Eucarpia, 1977 (E. Sanchez-Monge and F. Garcia-Olmedo, eds.) p. 45. Escuela Tecnica Superior de Ingenieros Agromonos Universidad Politecnica de Madrid, Madrid, Spain.

Melchers, G., Sacristan, M.D., and Holder, A.A. 1978. Somatic hybrid plants of potato and tomato regenerated from fused protoplasts. Carlsberg Res. Commun. 43:203-218.

Newell, C.A. and Hymowitz, T. 1982. Successful wide hybridization between the soybean and a wild perennial relative, *G. tomentella* Hayata. Crop Sci. 22:1062-1065.

Sanchez-Monge, E. and Garcia-Olmedo, F. (eds.) 1977. Interspecific Hybridization in Plant Breeding, Proceedings of the 8th Congress of Eucarpia, 1977. Escuela Technica Superior de Ingenieros Agromonos Universidad Politecnica de Madrid, Madrid, Spain.

Schiller, B., Herrmann, R.G., and Melchers, G. 1982. Restriction endonuclease analysis of plastid DNA from tomato, potato and some of their somatic hybrids. Molec. Gen. Genet. 186:453-459.

Strasburger, E., Noll, F., Schenck, H., and Karsten, G. 1908. A Text-Book of Botany, 3rd ed., English. Macmillan, New York.

Whitson, J., John, R., and Williams, H.S. 1914-1915. Luther Burbank: His Methods and Discoveries and their Practical Application, 12 vol. Luther Burbank Press, New York and London.

Williams, H.S. 1915. Luther Burbank: His Life and Work. Hearst's International Library, New York.

Winge, O. 1917. The chromosomes: Their numbers and general importance. C. R. Trav. Lab. Carlsberg Ed. Fsc. 13:131.

SECTION I
Overview

CHAPTER 1
Plant Germplasm Resources

C. M. Rick

When primitive man learned the art of farming and saved selected seed from one season to the next, he was actually practicing a type of germplasm management. In effect, a single bag of seeds is a limited germplasm bank, although that term tends to be applied today to massive, modern units. Whatever their dimensions, such resources are of vital importance to the breeding of improved cultivars, which serve human needs by providing larger yields of more nutritious and varied foods, condiments, narcotics, timber, fiber, and other products.

In this cursory treatment I shall attempt to present a broad picture of the various aspects of plant germplasm management and their role in the development of improved cultivars. Yet applications for other purposes should not be overlooked. Defective mutants have been used to prove the nature of such fundamental processes as photosynthesis, mineral translocation, stress tolerance, action of growth substances, and host-parasite interactions. The improved understanding of these basic processes often contributes to better agricultural production. Thus many advantages that accrue from investigating diverse germplasm lines only emphasize the need to acquire and preserve them.

The United States has pioneered in the husbandry of plant germplasm resources; as a matter of fact, the federal Department of Agriculture (USDA) traces its beginnings to plant collection activities. The consulates were in the practice of collecting plant materials and sending them to Washington. Plant introduction was thus a charter function of the USDA when it was established during Lincoln's administration. Subsequently, plant germplasm resource activities were innovated in the United States, and our system has served as the model for units in many other countries. For many years the United States led the way in plant collecting expeditions, in the establishment of substations for major crops, in cataloging systems, in worldwide seed distribution, and in constructing facilities for long-term storage of germplasm resources.

Contributions by other countries should not be overlooked. Thus leadership in plant collecting was assumed for several decades from the 1920s to the 1940s by the Soviet Union under the direction of N.I. Vavilov. Within the last decade the International Board for Plant Genetic Resources (IBPGR), headquartered in Rome, has organized, directed, and supported important advances in this field and also aims to coordinate worldwide plant germplasm activities.

In recent years public interest in plant germplasm activities has been stimulated by advances made by plant breeders in the solution of various agricultural problems and the risk of extinction of certain plant materials. The response to this great volume of publicity has evoked more words than action. Many critical needs have yet to be met.

KINDS OF PLANT GERMPLASM

In the interests of efficiency and saving time, plant breeders logically choose sources of germplasm genetically most closely related to the best cultivars in use. Their choice of plant resources is therefore as follows, in which the lower the rank, the more distantly related is the material from modern cultivars: (a) modern cultivars, (b) outmoded or obsolete cultivars, (c) breeding stocks, (d) landraces (primitive cultivars), (e) wild forms of cultivated species, and (f) closely related wild species.

In this ranking, the closer an accession is to the beginning of the list, generally the less difficult is the task of combining desired genes and the shorter the time to achieve objectives. In descending the list, the proportion of desired genes decreases, until, with strictly wild, allospecific accessions, the investigator usually faces the task of ridding his populations of all wild traits save the few desired ones.

In very fortunate situations, the investigator may find the needed variation within optimal cultivars. This situation arises more often in allogamous plants, in which the outbreeding system and intolerance of inbreeding maintain at least a moderate level of genetic variability in open-pollinated cultivars. In the naturally self-pollinated crops, the needed variation can arise as a result of spontaneous or induced mutation.

Modern Cultivars

This group comprises the most productive and best adapted genotypes for a particular region. Characterized by the least extent of genetic variation, it is consequently most vulnerable to epidemics, plagues, and environmental vicissitudes. It provides the standard against which new cultivars are compared and recurrent parents into which new genes are incorporated for amelioration.

Obsolete Cultivars

The "modern" cultivars of the recent past, though outdistanced by new introductions, are held in reserve as a valuable gene pool for various breeding objectives. As a whole they comprise a much wider range of variation than the preceding group and, on a worldwide basis, constitute a massive number of stocks of wide adaptation. They are frequently called into action to recover valuable genes that have been inadvertently lost in breeding modern cultivars, as in the case of brown blight resistance of lettuce in the Salinas Valley, California.

Breeding Stocks

In breeding investigations countless selected lines are developed, which, although highly meritorious in most respects, suffer one or more defects that render them unacceptable as commercial cultivars. Although unquestionable value is to be found in this group, its rate of extinction is unfortunately high. Countless lines are culled during selection programs. Decisions concerning which lines, aside from the selected cultivars, should be saved, and which discarded, are difficult and often arbitrary; obviously it is impossible to save all useful lines, and the future utility of each line is often impossible to predict. Other types of stocks that are logically classified here are populations developed via recurrent selection and the composites via bulk selection and other methods.

Landraces

These are the nearest approximations to the original domesticates. Their range of genetic variation is still larger than the preceding groups, but average performance is inferior. This group is usually the least well represented, as many valuable accessions have been lost through neglect, accidents, or other causes. Landraces may originate in the region of origin or from secondary centers of variability. Their adaptations to these regions often imply selection for valuable genes for disease, insect, and stress resistance not available in the preceding categories.

Wild Forms of Cultivated Species

The limitations on availability of this category are more stringent than those for any others yet considered. Wild forms of many crop species are extinct. For the remainder, some accessions are usually available, yet the full range of variation is seldom adequately sampled in living collections. In the case of small grains and other groups, "weedy races" are also known, and products of hybridization between cultivars and wild races have been identified. This class of germplasm is probably most utilized for various kinds of resistances that account for survival in nature.

Closely Related Wild Species

Defined as those species that can be hybridized with the crop species, this group is usually the last resort of the investigator, for whom all previous categories have failed to provide the desired variation. Hybridization with this group is usually faced with less enthusiasm, because it can be fraught with such problems as noncrossability, hybrid sterility and degeneration, epistasis, and linkage restrictions; furthermore, the bulk of the wild genome is generally undesirable, and the investigator is faced with the task of ridding hybrid progenies of unwanted alleles while retaining the desirable ones. This group is seldom satisfactorily represented in germplasm banks; for certain species only a single accession is available, or none. In view of the usual high level of polymorphy in wild species, an adequate representation requires many collections, the quantity of each mandated by the extent of intrapopulational variability.

Spontaneous or Induced Mutation

Failing to attain the desired characteristic in the above categories, the investigator may have to resort to induced mutation, an approach that might also be justified by other considerations. Although mutant lines are stocked in germplasm banks, the need for such variants is seldom satisfied by existing lines and requires new mutation breeding programs. Examples of problems solved this way are enumerated by Sigurbjörnsson and Micke (1974). A well-stocked bank should have an assortment of M_2 seeds of many lines in the best cultivars. The accessibility of such stocks would save time and facilitate the work of investigators, who could systematically screen them with efficient selection procedures.

In tapping these resources, an expert will normally seek sources in the order indicated above, not proceeding to any level lower than necessary to secure the desired trait(s). While plausible, this rationale leads to the neglect of the lower, usually vastly richer, sources of genetic variation. Success with the latter categories depends to a large extent on the degree of selection efficiency. Examples of such utilization will be presented later in this chapter.

SOURCES OF GERMPLASM

The agencies or areas where germplasm can be obtained are as diverse as the categories of the resources themselves. Again, in the interests of efficiency, the investigator will usually follow the path of least resistance in selecting sources. The following group of sources is listed more or less in this order: (a) Seed companies, generally a good source of modern and standard cultivars. (b) National germplasm banks—for example, the U.S. National Germplasm System (facilities and organization elaborated later). (c) Curatorial collections, like the Maize Genetics Collection and the Tomato Genetics Stock Center, usually devoted to a single crop and related species. (d) Contributions from other experts, generally consisting of advanced breeding stocks. (e) Botanical gardens, many of which maintain supplies of seeds for exchange purposes, mostly of related species and other exotics. (f) Local expeditions in regions of considerable crop antiquity. The collecting of "heirloom beans" in New Hampshire (Yeager, 1951), local cultivars of melons in Spain (Esquinas, 1981), and various crops in The Netherlands (Zeven, 1979) were unexpectedly productive. As a variation on this theme of "plant hunting at home," large commercial plantings may be searched successfully for desired characters if appropriate screening devices are exploited; examples are the discovery of spontaneous mutants for brown blight resistance in lettuce (Jagger, 1940), for monogerm seeds in beets (Savitsky, 1950), and for male sterility in tomatoes (Rick, 1945). (g) Distant exploration to primary or secondary centers of variability. This is an activity of inestimable value. Such enterprises can be of a private, institutional, or international character. They have been popularized by the writings of such plant explorers as Fairchild (1938), Goodspeed (1961), Kingdon-Ward (1960), Wilson (1927), and many others. It is beyond the scope of this treatment to consider the many aspects of plant collecting. The volumes edited by Frankel and Bennett (1970) and by Frankel and Hawkes (1975) provide invaluable recommendations.

MAINTENANCE

If certain plant lines are worth acquiring, it follows that they deserve to be properly maintained. Obvious as that statement might seem, it is remarkable how often it is not heeded, and the number of valuable collections lost through such neglect is too revolting to enumerate. The objectives of proper maintenance are adequate supply, viability, and preservation of the original genetic constitution.

Frequent monitoring is needed for the first two of these objectives, and their requirements (length of storage life, quantity of seeds, etc.) will vary with the species and sometimes with the genotype. Similarly, optimal storage conditions will depend upon the species and the plant part preserved, although low humidity and low temperature are best for seeds of the great majority of crop plants.

The third objective may present the greatest difficulties. If an accession is in the form of a pure line, i.e., a single genotype, the situation is simple and requires monitoring only to prevent contamination via cross-pollination or seed mixture. But in accessions comprising mixed genotypes, as they usually are in allogamous species, the problem of genotype preservation is much more challenging, and preventing changes is virtually impossible. Space limitation prevents a full consideration of this topic. Suffice it to state that an arbitrary decision must often be made to maintain the identity as near as possible to the original within the economic constraints of the program. Of course, other limitations may exist, such as a limited number of viable seeds in the stocks of an accession. For a thorough treatment of this problem, refer to Marshall and Brown (1975).

The writer has faced this problem with the maintenance of the allogamous tomato species, of which *Lycopersicon peruvianum* (L.) Mill. is a good example. Accessions of this species usually consist of the descendents of a single wild population. In a sample of 50-100 plants of a single accession no two plants will have the same genotype (as assayed by isozymes), even for a limited number of loci; as many as 8-10 alleles may be detected at one locus in one accession.

The complications engendered by attempts to preserve the original variability of a population during replenishment increase are legion (Allard, 1970). To mention some, seeds of different genotypes may have different longevities or different germination requirements; plants of different genotypes may differ in their time of flowering, relative pollen production, and seed-production capacity; attractivity to insect pollen vectors, or efficiency of wind cross-pollination, may differ significantly from plant to plant; also, restrictions that limit population size—often difficult to avoid—inevitably lead to erratic changes via genetic drift, including loss ("erosion") of genetic resources (Allard and Hansche, 1964). A casual consideration of these processes reveals how the compounding of the effects in repeated generations of propagation can lead to radical changes in the genetic composition. Thus, it is well-nigh impossible to preserve genotypes or even gene frequencies of a polymorphic population in a constant, predictable fashion during replenishments. It follows that, at best, only an approximation can be effected. Another conclusion reached is that the less frequently stocks are renewed, the less the original composition will be altered. An example of genetic changes that may occur in the course of frequent propagations is found in the work of Allard and Jain (1962) on a composite population of barley. Here, even though barley is highly self-pollinating, and in spite of multiplication without intentional selection in very large populations, this accession lost 50-70% of its original genetic variation in 19 generations.

Possible alternatives to the maintenance of mixed stocks under artificial, cultural conditions have been considered by several authorities. One such system utilizes bulk or composite populations as conceived and originated for barley germplasm by Harlan and Martini (1938) for selection and plant improvement. Not intended by their originators for germplasm conservation, these "mass reservoirs" were advocated for this purpose by Simmonds (1962). These populations originate by systematic intercrossing of many genotypes and are maintained thereafter in large mass groups without intentional selection. The very changes that are expected to eliminate unadapted genotypes and favor the more productive ones would per se vitiate their use for preserving all possible genetic variability; in fact, Harlan and Martini (1938) themselves and, as previously mentioned, Allard and Jain (1962) detected radical genetic changes in such composites. Other shortcomings of this alternative have been discussed by Frankel (1970a) and Jain (1975); for example, the fact that culture of a composite in many different locations as a measure to diversify its composition might be more costly than the separate maintenance of its parental components. Another objection is the loss of gene combinations of the component cultivars—the products of many generations of careful breeding. Altogether, bulks or composites suffer too many shortcomings to be satisfactory repositories of plant germplasm.

Another concept of maintenance is the use of genetic or natural reserves. Frankel (1970b, 1974) advocated them for preservation of certain kinds of plant germplasm. The rationale is that a wild species will maintain itself best in its native habitat, where, in fact, the only hope for long-term preservation exists. Other species of unknown but potential value will also be protected in this arrangement, as well as the ecological and other scientific and esthetic values of the habitat. Such areas obviously require vigilant control to protect the species in their natural state free from human disturbances. The pros and cons of natural reserves have been weighed by Frankel (1974) and Jain (1975). One difficulty that must be faced in the case of wild tomato species would be the number of habitats necessary to preserve the various subspecific taxa. The species *Lycopersicon peruvianum*, for example, already the source of invaluable genes for fruit quality and disease resistance, has at least 30 races adapted to widely differing habitats, many in isolated areas of the Andes. The logistics of acquiring and patrolling reserves adequately to maintain the main segments of variability within this single species is bewildering. In the same country a distinct series of reserves would be required for the other tomato species, the potato species, the *Capsicum* species, the cotton relatives, and others. The same set of considerations would apply to Ecuador and other Andean countries. The requirements of land, manpower, and administration would be enormous, perhaps inconceivable in these times when conservation of plant germplasm evokes more oratory than action.

As acknowledged by Frankel (1970a), natural reserves are visualized for preservation only of wild species. No doubt, as emphasized by Jain (1975), such areas will have to be adopted eventually, but in the meanwhile, we cannot neglect alternative procedures for maintaining and distributing seeds of wild relatives of crop plants.

The regulation of seed viability by ambient conditions has been reviewed by Harrington (1970). For seeds of most angiosperms low temperatures and low humidity are optimal for long-term storage, the latter being more critical. Exceptions are found in certain large-seeded trees and aquatics that will not tolerate low moisture content. Although expensive and sometimes fastidious to maintain in working order, the equipment necessary to regulate relative humidity is indispensible for proper germplasm storage. Long-term storage in liquid nitrogen is an alternative that is being investigated. Anything short of optimal seed storage conditions should not be tolerated in germplasm banks.

As mentioned above, silent extinction via death of seeds that have been stored past their term of viability is one of the risks in germplasm maintenance. Little is known about the effect of genotype on seed longevity. An exception is the research of Harrison (1966) demonstrating a threefold difference in storage life of seeds of different lettuce cultivars. Another risk of long-term storage is the increased rate of gene mutation and chromosome breakage, the literature for which has been summarized by Roberts (1975). Fortunately, since these studies have determined that the frequency of such changes is inversely related to germination percentages, the risk is minimized by renewal before germination falls below a certain level, 80% being often advocated.

The foregoing considerations are directed at seed-propagated crops; the vegetatively maintained cultigens require special considerations. Repositories for the preservation of valuable clonal stocks are being established in the United States, but considerations of the available resources and the expense of maintenance at the current state of technology limit the number of accessions that can be stocked. Storage of cell or tissue cultures at low temperature by lyophilization is an alternative that requires vigorous investigation (see Chapter 28, Volume 1). Another alternative for clonal, as well as seed-propagated crops, is pollen storage, the viability and requirements of which are similar to those of seeds (Harrington, 1970). The small space requirement is an attractive feature of pollen storage; however, pollen of the many trinucleate species (for example Cruciferae and Gramineae) is short lived and usually does not tolerate low humidity. Clearly much research must be invested to solve some of these problems of long-term pollen storage. From a genetic standpoint, pollen storage by itself suffers several severe shortcomings. Although genes could thereby be preserved, the genotypes of superior cultivars resulting from centuries of hybridization and selection would be lost. Also, unless techniques are perfected for regeneration of plants from stored pollen, it would be necessary to maintain clones of some stocks to serve as pistillate parents.

A final consideration of storage is sanitation. Great care must be observed to keep seed vaults free of seed-eating insects. Prevention is better than cure because at least certain fumigants are known to reduce seed viability.

BASIC STUDIES

Plant germplasm programs cannot be properly planned and executed without a knowledge of the genetic structure of cultivated plants and of their genetic relationships with closely allied species. Basic studies in genetics, cytogenetics, biochemistry, morphology, distribution, and ecology contribute to an understanding of interspecific relationships. Information thus acquired is also essential for the sound and workable taxonomy needed to catalog and use the vast number of accessions in germplasm collections. The need for such basic studies was recognized by the National Plant Genetic Resources Board (Cutler et al., 1979).

A thorough treatment of this subject is not warranted here; instead, a brief synopsis of activities in these fields will be presented in order to highlight some aspects of their relevance to germplasm resources and utilization.

Genetics

Linkage maps constitute the basic genetic framework of a species. The clearly defined marker genes used for such mapping provide the tools for genetic analysis of economic traits that are derived from germplasm collections. Such mutants also have great potential for solving problems in other areas of research. In exploiting exotics, genetic techniques are used to analyze the feasibility of gene transfers to the cultivated taxa. They ascertain the limits of hybridization as well as fertility and viability of F_1 hybrids and later generations, and the extent of genetic and cytoplasmic differences between parents. Much of the information thus acquired also aids in biosystematic studies.

Cytogenetics

Genetic analysis of a species is inseparable from a study of its chromosomes. Cytological deviations provide the tools for cytogenetic analysis of a species, hence the need to maintain stocks of polyploid and structural variants. A key role in interpreting interspecific relationships is played by investigations of chromosome number and morphology and of meiotic-pairing behavior in diploid and polyploid hybrids. These studies clarify the nature of hybrid sterility and lead the way to the most efficient utilization of exotic germplasm.

Biochemistry

The analysis of biochemical constituents has become increasingly important in unraveling systematic snarls. Thus, qualitative and quantitative differences in terpene content are of systematic importance in pines, as are differences in storage proteins in legumes. Clues to relationships between closely affiliated taxa in many groups are often provided by studies of enzymatic variation. Recent developments in assays of genic and cytoplasmic DNA promise to shed much light on phylogenetic relationships. Besides aiding taxonomists, such determination may reveal new sources of compounds with nutritional or industrial significance. They are also routinely used in studies of pest resistance and food safety.

Morphology

The data for classical taxonomy and often the only criteria for identification of herbarium specimens and for field determination are provided by morphological characters. In plant germplasm collections, the acquisition of such data is usually limited to characters of systematic and economic significance; furthermore, the integration of morphology and genetics is mutually beneficial.

Geographic Distribution

The usefulness of distributional information in directing collectors to areas of critical importance and in aiding the biosystematist is obvious.

The area of cultivation versus the natural distribution of a species is often significant in relation to the presence or absence of various disease and insect pests. Geographic isolation frequently promotes the differentiation of new biotypes and taxa and may therefore affect plant collection tactics. Weedy races may accompany the cultivated forms and play a significant role in the evolution and use of plant species.

Autecology

The distribution of a plant taxon is determined by its ecological preferences. Reproductive isolation may be vested in autecological factors, and hence is of evolutionary and biosystematic importance. Observations on the responses of plants to temperature, light intensity, photoperiod, soil characteristics, precipitation, and other factors in their native habitats and in the first trial plantings may provide important clues for the effective use of plant germplasm resources (Bunting and Kuckuck, 1970; Rick, 1973).

Documentation

A system of documentation is essential to handle effectively the immense amount of data that is required for a plant germplasm collection. Information that must be available for each accession includes reference numbers, taxonomic and agronomic classification, source, date of acquisition, age and any available generation data of the stock, and any other information that can be provided concerning features of the accession. For collections of a wild or feral nature, additional details should be provided concerning geographic location, elevation, size of the source population, developmental stage, plant associations and other pertinent ecological information, and local uses. Evaluations are expected to yield data on adaptation to different areas and morphological, biochemical, and physiological traits of genetic, biosystematic, and agronomic significance. Methods for systematizing all of these data and storing them in memory banks have been and are being developed. Rapid retrieval of data from such an information source is clearly essential to full utilization of a germplasm resource. Reference is made later in this chapter to the current status and development of information systems in the U.S. National Germplasm System.

EVALUATION

Collections of plant germplasm cannot be fully utilized unless the accessions have been evaluated for useful traits. As self-evident as this statement might seem, most collections have not been satisfactorily evaluated; consequently, their real potential cannot be realized. The reasons for this deficiency are various, but lack of adequate resource support is mainly responsible. Satisfactory evaluation of collections for important economic traits is costly; furthermore, the data amassed is no more valuable than the accuracy with which it has been collected. Organization of a successful evaluation program demands much dedication from competent experts in several fields because the labor is tedious and unrewarding—another factor that accounts for the lag in such efforts. Also, today's requirements of the various crop industries may not be tomorrow's: the appearance of new disease and insect pests or strains of them, or changes in crop practices, may require reevaluation for traits not previously assayed, and new accessions are being continually added to collections. It follows from these and other reasons that evaluation is a never-ending task.

The resources available determine the extent of evaluation programs; therefore, the list of descriptors for which a collection is screened will have to be arbitrarily limited. Whereas all information concerning every morphological character, every physiological character pertinent to culture and harvest, reactions to different climates, reactions to various pests, and stress tolerances might ultimately be useful, it is inevitably necessary to restrict the number of descriptors to a list deemed the most important by experts in pertinent fields.

Another factor that complicates evaluation is genetic variability within accessions. Whereas many crop accessions are highly inbred, and consequently uniform, accessions of allogamous crops and related wild and feral taxa must be maintained in a genetically polymorphic condition. Evaluation of such variable lines requires special considerations and much additional effort. Thus, to ascertain the presence or absence of a qualitative character or to characterize a quantitative character, a larger population sample is obviously needed than for homozygous lines. New questions are raised such as how large the samples must be and what statistics should be computed.

A meaningful evaluation of performance in many characters requires repeated tests in different years and in different regions because the adaptation features of an accession can be of critical importance. Clearly, as this very brief consideration reveals, proper execution of an evaluation program is no simple affair.

It also follows from these considerations that the many demands of adequate evaluation cannot ordinarily be done by maintainers of germplasm collections; rather, it must be done by users. The wide range of expertise demanded also requires that evaluating be accomplished by teams of specialists in such fields as plant breeding, entomology, pathology, soil science, and crop physiology. Currently these tasks are being planned and administrated in the United States by committees consisting of such experts for each major crop. The validity of the team approach is manifest by the long history of success in discovering sources of resistance to downy mildew in maize, the maize viruses, the several smuts and rusts of small grains, late blight of potatoes, spotted alfalfa aphid, and many others (Cutler et al., 1979). The systematic screening of wild tomato species in the Plant Introduction collection for reaction to 16 major diseases by a large team of workers is a good example of such cooperative evaluation (Alexander and Hoover, 1955).

DISTRIBUTION

An important function of germplasm banks is the distribution of stocks to users. The guidelines for this activity fall more in the purview of common sense than science, yet, without adequate distribution of these resources, the progress of science would be impeded. In the past, distributional policies of certain institutions were inadequate or nonexistent. Not only was there little or no opportunity for scientists elsewhere to develop their programs with these stocks, but also a large portion of the stocks lost viability after the programs ended. Were it not for the special efforts of other investigators with broader vision to acquire, replenish, and properly store the items that they could salvage, we might not have any viable stocks from these programs for future investigation.

Stocks are normally provided only to workers with a genuine scientific objective. Samples are generally sent free, since fees would lead to cumbersome bookkeeping and tend to discourage deserving investigators. The quantity sent per sample varies according to several factors. Supply, particularly of gene or chromosome stocks, often limits the number of propagules per sample. If accessions are requested for the search of some desired trait, it is reasonable to provide more seeds per sample of the heterogeneous collections (bulks, composites, allogamous crop lines, or wild species) than of pure-line material.

Records are kept of each shipment as required by supporting agencies and also for various reference purposes. Problems may be reported by users in respect to germinability of seeds or identity of the accession. Occasionally a report will be made by the user to the distributor as to important traits of the accessions, which should be duly noted in the records of the germplasm bank and transmitted to the information system. Such feedback is relatively rare but should be encouraged for mutual benefit.

CORRELATED ACTIVITIES

Plant breeding generates new, improved genotypes that become part of the total germplasm resources, and hence needs to be considered here along with the correlated activities of genetic analysis of economic traits, prebreeding, and seed production. It is not the purpose of this article to present an extensive treatment of these areas; rather, they will be discussed briefly and described within the total picture of plant germplasm resource management.

Genetic Mechanisms Controlling Desired Traits

Analysis of the inheritance of specific traits is a prerequisite for the development of an effective strategy for incorporating them in new, improved cultivars. Such knowledge will assist in various ways, including the selection of appropriate breeding methods among the multitude available to plant breeders. The most efficient strategy may differ radically, depending on whether a desired trait is determined by a single dominant gene or many polygenes of equal, small, additive effect. In many situations, success of these endeavors may depend on effective cooperation with experts in such fields as soil science, plant pathology, entomology, nematology, food science, and with farmers, seed trade experts, processors, and consumers.

Prebreeding

The search for desired traits may lead the investigator to exotic germplasm types. If necessary to travel this far afield, the process of transferring genes by conventional methods may be prolonged and beset by various cytogenetic problems. In some situations, the exotic accessions may be so poorly adapted to a given region that they cannot be properly evaluated for the presence of useful traits. Another complexity is the frequent phenomenon of modified expression of a gene from an exotic source by the genetic background of the cultigen. Faced by these impediments, the plant breeder understandably may avoid the use of exotic stocks and seek other avenues of approach or even switch to other projects. Clearly, exotic sources would be better exploited if their genes were bred into adapted lines of the cultigen to provide more utilizable material—a process commonly termed prebreeding or developmental breeding.

The process of prebreeding may be conducted by different agencies. The plant breeder may be forced by necessity to do the job himself. In other situations, a botanist or cytogeneticist, whose major objective might be different, may accomplish the prebreeding. Again, special cooperative projects may be organized for these purposes. The conversion of exotic sorghums to adapt them to U.S. growing conditions is a fine example of such organized effort. The USDA, cooperating with the Texas Agricultural Experiment Station and the Federal Station in Mayagüez, Puerto Rico, undertook this project to adapt some 509 alien lines (Stephens et al., 1967). Their goal was to incorporate in these exotics the genes necessary to permit them to flower under long daylengths and to convert them to dwarf stature, with a separate battery of 3-4 genes determining each character. Since the program envisaged backcrossing to the exotic accessions, the initial and subsequent hybridizations had to be made under short-day conditions obtaining at Mayagüez. Progenies from such hybridizations and the next, segregating generation were grown in Texas, where dwarf, adapted individuals could be selected, the three generations of each cycle being grown within one year. Many benefits have already accrued from this program. Useful characters that are being utilized are high lysine content (Singh and Axtell, 1973), a new cytoplasmic-genic sterility system (Miller and Edminster, 1976), resistance to sorghum midge (Johnson et al., 1973), Banks grass mite (Foster et al., 1977), a series of diseases (Miller, 1976), lodging (Esechie et al., 1977), and tolerance of acid soils (Duncan, 1981). The future will undoubtedly witness continued and expanded utilization of these converted lines. Certainly this intelligently conceived and executed program should lead the way to badly needed prebreeding projects in other crops.

A similar, though less extensive conversion effort has been and is being conducted in maize by Gerrish (1981). Races from Mexico, the Caribbean, and South America are hybridized with corn-belt maize, and, via backcrossing and selection, adapted lines have been bred in the form of reconstituted races, inbreds, or composites. Tests with these newly constituted lines have already revealed promising sources of high combining ability and resistance to the southwestern corn borer and to aflatoxin problems.

Breeding Improved Cultivars

All efforts in plant germplasm resource management culminate in the breeding of improved cultivars. Success in this enterprise depends on many factors, but proper marshalling of the available genetic resources is of prime importance. The application of traditional methods of hybridization and selection has resulted in sensational yield increases of most U.S. crops over the past 40 years. Improvement of field maize yields during this period is a familiar, and, in terms of monetary gain, deservedly the most impressive example. Part of the gain in yield must have resulted from improved cultivation practices, but by comparing yields of the leading cultivars of each decade in replicated plots in different areas at different spacings, Russell (1974) and Duvick (1977) demonstrated that at least 60% of the gain can be attributed to genotype.

Successful plant breeders are ready to employ any method that can aid in reaching their goals quickly and efficiently. Many innovations have already been utilized to various extents, including mutation breeding, meristem culture, and such cytogenetic procedures as analytical breeding, aneuploidy, euploidy, and chromosome structural change. Undoubtedly cell and tissue culture methods will be exploited to a greater extent in the future.

A necessary, though less glamorous, part of plant breeding is line testing. A large share of the resources supporting such experimental programs must be invested in careful, systematic, routine performance testing of advanced breeding lines. It is logical and necessary that, no matter how such lines are produced, they will not be employed by the industry unless they are superior to existing cultivars. The only way to determine their superiority is by such evaluation.

Seed Production and Distribution

Production and distribution of seed to growers in the United States have become largely functions of the commercial seed industry. It is their responsibility to supply growers with an uninterrupted source of improved, high-quality planting stock. These high standards are usually but not always stimulated by competition.

Many seed firms also participate in other germplasm activities, notably breeding improved cultivars. Such research is usually directed at the more practical and less technical areas designed to produce the maximum number of commercially acceptable cultivars in a minimum of time. Other areas of germplasm managment for which the seed industry is not in a position to assume responsibility are usually investigated by public agencies. The interplay between these two groups is best described by the following statement of the National Plant Genetic Resources Board (Cutler et al., 1979): "The cooperation between the public institutions engaged in the genetic manipulation of plant germplasm and the private seed industry is unique to the United States. Over the years the two groups of organizations have arrived voluntarily at a logical division of labor that includes minimal duplication of effort. The complementary nature of the relationship has served American agriculture well. The need for this kind of cooperation is as great today as at any time in the past. Each crop improvement program of industry and the public agencies does not need all seven phases [of germplasm management]; the nation itself needs all phases. This program can be a model for the national sharing of the workload among State and Federal agencies and private industry."

ORGANIZATION OF GERMPLASM ACTIVITIES IN THE U.S.

The U.S. plant germplasm activities are coordinated through the National Plant Germplasm System (NPGS). Involvement of the USDA in collecting and maintaining plant germplasm can be traced back to the USDA Section of Seed and Plant Introduction formalized in 1898. The current NPGS has the following mission: "In order to provide genetic diversity to increase crop production and reduce genetic vulnerability in future food and agriculture development, not only in the U.S., but for the entire world, it is the mission of the NPGS to acquire, maintain, evaluate, and make readily accessible to crop breeders and other plant scientists as wide as possible a range of genetic diversity in the form of seed and clonal germplasm of our crops and potential new crops" (Anonymous, 1977).

The NPGS, coordinated by the Agricultural Research Service (ARS) of the USDA, consists of a national network of scientists from the federal, state, and private sectors of the agricultural research community. The NPGS maintains a total of approximately 400,000 accessions of germplasm in the form of seed and vegetatively propagated stocks. These accessions consist of cultivars and unimproved germplasm from foreign sources as well as some domestic breeding lines and cultivars that are also maintained in certain subunits of the system. These items are available to any bona fide plant scientist in the U.S.; additionally, material in the NPGS is given to other countries in exchange for germplasm needed by U.S. scientists. New accessions of germplasm are added to the system at the rate of 7,000-15,000 per year. Accessions are also acquired through formal explorations sponsored by the International Board for Plant Genetic Resources (IBPGR) and the NPGS, which account for 20-30% of new introductions. Proposals for explorations are received from various sources, channeled through the Regional Plant Introduction Stations (RPIS) coordinators, ranked by the USDA Plant Germplasm Coordinating Committee, and those of highest priority are funded by ARS.

The NPGS coordinates its efforts with the IBPGR. One of the administrators of the NPGS currently serves as a member of the IBPGR.

The diffuse NPGS is managed by the USDA through the Assistant to the Deputy Administrator for Germplasm within ARS. Many state or private workers who participate as curators or users are associated with the land-grant system and private industry.

The NPGS is served by several advisory components. The National Plant Genetic Resources Board (NPGRB) is a policy advisory group appointed by the Secretary of Agriculture and is composed of representatives from federal, state, university, and private groups. The NPGRB advises the Secretary of Agriculture on a wide scope of matters relating to plant germplasm. The National Plant Germplasm Committee provides advice and coordination directly to the NPGS. The ARS Plant Germplasm Coordinating Committee advises the Administrator on operational matters, particularly those pertaining to ARS-funded explorations. Individual Crop Advisory Committees provide technical advice on the management of collections for ten major crops.

The NPGS consists of several major working units. The Germplasm Resources Laboratory, which is a part of the ARS Plant Genetics and Germplasm Institute at the Beltsville Agricultural Research Center in Maryland, includes the Plant Introduction Office, the USDA Small Grains Collection, the USDA Rice Collection, and the USDA Plant Introduction Station at Glenn Dale, Maryland. The Plant Introduction Office monitors acquisition and exchange of plant germplasm, assigns PI (Plant Inventory) numbers, provides initial documentation, and distributes germplasm. No collections are maintained there. Plant materials provided to foreign scientists are obtained and forwarded by this Office. The Station at Glenn Dale, besides maintaining collections of pome and stone fruits and woody ornamentals, also serves as the USDA/Animal and Plant Health Inspection Service (APHIS) plant quarantine facility and enforces phytosanitary regulation. Also included in the Plant Genetics and Germplasm Institute (PGGI) are the Plant Taxonomy Laboratory and the Economic Botany Laboratory, which are important components of the NPGS. The other major working units of NPGS are listed in Table 1.

Seeds held in the National Seed Storage Laboratory (NSSL) constitute the base collection of the NPGS. The seeds stored there are not intended to meet day-to-day needs of plant scientists, but rather to serve as reserve stocks to prevent loss of germplasm. Samples are held in this laboratory for long-term storage and are grown no more often than necessary to ensure seed viability and to prevent genetic changes resulting from differential deterioration. Optimal storage conditions, especially of humidity and temperature, are maintained, and the unit is effectively fireproof. Presently more than 117,000 accessions are cataloged and stored in the NSSL, and an approximately equal number are stored there but not yet cataloged.

The NPGS also includes a series of working collections of several crops that are maintained by curators at appropriate U.S. locations having the requisite climate and facilities. These working collections play an important role in the NPGS by maintaining and supplying stocks primarily of genetic or chromosomal type. Such collections are maintained for barley, maize, cotton, oats, peas, tomatoes, and wheat.

Information concerning accession numbers, location, supply, and known characteristics of each germplasm collection is stored in a

Table 1. Regional Stations of the National Germplasm System

UNIT NAME	LOCATION	RESPONSIBILITIES
SEA Plant Introduction Station	Miami, FL	Collections of tropical and subtropical species including coffee, mango, cacao
Regional Plant Introduction Stations (RPIS)		Maintain the working collections of seed-propagated crops; distribute and administrate evaluation of collections
Northeastern RPIS	Geneva, NY	Perennial clover, onion, pea, broccoli, timothy
Southern RPIS	Experiment, GA	Cantaloupe, cowpea, millet, peanut, sorghum, pepper
North Central RPIS	Ames, IA	Alfalfa, maize, sweet clover, beet, tomato, cucumber
Western RPIS	Pullman, WA	Bean, cabbage, fescue, wheat grasses, lentils, lettuce, safflower, chickpea
Interregional Potato Introduction Laboratory	Sturgeon Bay, WI	Potato and related species
Northwest Clonal Repository	Corvallis, OR	Pear, filbert, small nuts, hops, mints
Fruit and Nut Germplasm Repository	Davis, CA	Grapes, stonefruits, nuts
National Seed Storage Laboratory	Fort Collins, CO	Long-term storage of seed collections and research on seed storage methodology

centralized, computerized information system so that the facts about germplasm accessions will be readily retrievable by investigators nationwide. The Information System of the NPGS works directly with the Crop Advisory Committees to establish mutually acceptable lists of descriptors and to coordinate evaluations and gain needed accession data.

The units of NPGS are financed by federal, state, and private sources in the U.S. and by such international sources as the IBPGR.

UTILIZATION

The Role of Exotics

By definition, each new cultivar that is bred originates from some type of germplasm utilization. Undoubtedly large, the proportion of examples that stems from accessions secured from genebanks would be difficult to estimate. According to the common, more restricted viewpoint, germplasm signifies exotic sources—i.e., "all germplasm that does not have immediate usefulness without selection for adaptation to a given area" (Hallauer and Miranda, 1981). It is this vast reservoir of genetic resources that is probed by plant breeders when they fail to find the needed variation in adapted lines. It follows that the extent to which exotic sources are tapped depends largely on the range of genetic variation in adapted germplasm for the crop species in question. At one extreme are the native allogamous crops like maize, which embrace such a wealth of genetic variation that professionals scarcely ever need to seek exotic sources. At the other extreme we find introduced crops like the tomato and potato, the principal cultivars of which are so depleted of genetic variation for various reasons that progress in improvement has depended upon extensive exploitation of exotics for the past 30-40 years. Between these two extremes lie many cultigens with intermediate degrees of resource variability in their adapted genotypes.

According to the aforementioned definition, exotics comprise any unadapted crossable source, whether modern, older, or primitive cultivar, or wild form of the cultigen, or related species. In the following part of this section, some cases of the utilization of this class of germplasm are presented, not with the intention of completely covering the territory, rather to exemplify the opportunities and limitations. Certain aspects of this subject have already been considered in the section on prebreeding, in which noteworthy instances are presented of the wealth of useful traits already appearing in converted lines of sorghum and maize exotics.

As stated by Brown (1982) and others, most crop species have evolved via introgression between taxa of various levels of relationship. It is therefore logical that these ancestral taxa possess germplasm of potential value for further improvement of the cultigens. It is hoped that awareness of the spectacular achievements already realized will whet the appetite of other investigators to probe the vast amount of germplasm that awaits exploitation. If any doubts linger concerning the magnitude of genetic variation in exotics, recent comparisons (CSIRO, 1980) between allozyme variation cultivars and wild forms of barley should dispel them. This survey revealed 4.5 times as many alleles per locus in *Hordeum spontaneum* Koch—the presumed ancestor of barley (*H. vulgare* L.)—than in an assortment of barley cultivars. Much future utilization of these resources can be accomplished by conventional breeding methods, but much will require special techniques to circumvent crossability barriers and solve other problems in interspecific transfers.

Potato

Germplasm resources in wild species and primitive cultivars have come to the rescue of the cultivated potato on many occasions. As in the case of many other crops, these reserves have been tapped to the greatest extent as sources of disease resistance. The best known situation is that of the battle against late blight. According to the admirable review of the role of exotic lines in potato improvement by Ross (1979), *Solanum demissum* Lindl. and its derivative *S. edense* (*S. tuberosum* L. x *S. demissum*) have been the principal sources of resistance. The first category of resistance derived from these accessions was the vertical, highly specific, monogenic type, but the subsequent outbreak of new pathogenic biotypes of the parasite in 1936 rendered it useless. Fortunately, highly useful horizontal resistance was detected in the same exotics and, when bred into horticultural lines, proved more stable and effective than the vertical type. By retarding reproduction and spread of the infective organism, the horizontal resistance delays infection sufficiently to prevent severe losses and, by acting in a more generalized fashion against the many different pathogenic races, it provides a longer lasting solution to the problem. Ross (1979) also enumerates an impressive list of resistances to other diseases derived from *S. demissum* and other exotics, many of which have already been bred into established cultivars. This list includes resistance to Potato X Virus (*S. tuberosum* ssp. *andigena* Juz and Buk. and ssp. *tuberosum* Juz and Buk., *S. acaule* Bitt.), Potato Y Virus (*S. tuberosum* ssp. *andigena*, ssp. *tuberosum*, *S. stoloniferum* Schlechtd.), Potato A Virus (*S. stoloniferum*), Potato S Virus (*S. tuberosum* ssp. *andigena*), Potato Leaf Curl Virus (*S. acaule*, *S. demissum*), scab and fusarium (*S. spegazzinii* Bitt.), several species of nematodes (*S. tuberosum* ssp. *andigena*, *S. spegazzinii*), and *Pseudomonas* wilt (*S. phureja* Juz. & Buk.).

Aside from disease resistance, other useful traits have been extracted from exotic lines. Frost resistance was detected by Ross and Rowe (1969) in 13 related species and, from certain F_1 hybrid combinations, resistant derivatives were selected. Ross (1979) reports the unexpected appearance of high starch content in derivatives of *S. tuberosum* x *S. vernei* Bitt. & Wittm. Significant heterosis for yield has been reported in different interspecific hybrids by Muñoz and Plaisted (1981) and by Cubillos and Plaisted (1976); additional references are provided by Ross (1979). It is noteworthy that significant transgressive variation was encountered in the segregation for all three of these examples. As a matter of fact, hybridization with *S. vernei* was not made for the purpose of improving starch content, but this improvement appeared as a bonus in the hybrid progenies.

A measure of the importance of exotics in current potato breeding activity is epitomized by Ross (1979) in his statement, "The present situation in the Federal Republic of Germany is illustrated by the fact that at least 90% of the more than one million seedlings produced in a year have wild species or primitive cultivars in their ancestry."

Tobacco

Disease control is an especially urgent matter in tobacco production. Even the slightest blemishes render the leaves unfit for such uses as cigar wrappers. Fungicide applications are often prohibited for similar reasons and due to possible toxicity of residues. An ideal solution is provided by inherited resistance, often the best of which must be extracted from wild sources. A remarkable wealth of germplasm possessing potentially valuable resistances exists in the *Nicotiana* species: Burk and Heggestad (1966) enumerate resistance to 16 tobacco diseases among 65 species. Although *N. tabacum* L. is not crossable with all other species, the desired genes might eventually be transferred to tobacco by use of bridging species or other devices.

Examples of the successful transfer of monogenic resistance from other species to cultivated tobacco resulting in improved cultivars are to be found in the case of resistance to Tobacco Mosaic Virus derived from *N. glutinosa* L. (Sand and Taylor, 1961), to wildfire from *N. longiflora* Cav. (Heggestad et al., 1960), and to black root rot from *N. debneyi* Domin. (Heggestad, 1966). Substantial progress has been made in the transfers of resistance to blue mold from *N. debneyi* (Clayton, 1958) and to black shank from *N. longiflora* (Valleau et al., 1960). In most of these situations, the reduced pairing between chromosomes of the parent species necessitated breeding by the sesquidiploid method--i.e., by first producing the interspecific amphidiploid between them, which was successively backcrossed to *N. tabacum* with selection for resistance, culminating in the incorporation in the latter of a wild chromosome segment containing the resistant gene.

Another achievement in tobacco breeding is the development of cytoplasmic male sterility via interspecific hybridization. These crosses constitute classic examples of the many and varied morphological types of such sterility. The combination of cytoplasm of the wild species and genome of cultivated tobacco, secured through successive backcrossing to the latter and by protoplast fusion (Gerstel, 1980), often results in cytoplasmic male sterility. Examples are presented by Frankel and Galun (1977) for the following: pistilloid phenotype (*N. suaveolens* Lehm. plasmatype), petalloid anther phenotype [*N. bigelovii* (Torr.) Wats. and *N. undulata* Ruiz & Pav. plasmatypes], carpelloid functional phenotype (*N. megalosiphon* Huerck & Mueller plasmatype), and stigmoid, split corolla phenotype (*N. debneyi* plasmatype). As summarized by Gerstel (1980), various workers have demonstrated how fertility of such biotypes may be restored by incorporating genes or a chromosome fragment (which has nucleolar organizer properties) from the parent that contributes the cytoplasm. By appropriate manipulation of these cytoplasms and genotypes it might be possible to derive a workable system of male sterility for the production of hybrid tobacco cultivars.

Tomato

The gene pool that was available to tomato breeders in temperate North America until early in the twentieth century consisted largely of a few cultivars imported from Europe, where tomatoes had been grown and selected for more than three centuries. The European stocks, in turn, trace to imports from the area of domestication, probably Mesoamerica. The Mesoamerican sources presumably trace to migrations of the semiwild *Lycopersicon esculentum* Mill. var. cerasiforme Alef.

through northern South America, Panama, and Central America from the native Andean region. Throughout all of these wanderings, the ancestral forms were exposed to natural and artificial selection and were doubtlessly subject to frequent reductions in population size. The inevitable result of such selection and bottlenecking is rapid fixation of genes and depletion of variability. It is therefore not unexpected that the total variation in the European stocks and early North American cultivars, as assayed by allozymes, is nearly nil in contrast to var. cerasiforme and related taxa in the native region (Rick and Fobes, 1975). Since the 1940s, however, when workers started exploiting exotics, detectable allozyme variation has increased, and the tomato industry has been benefiting from the new characters introgressed from these sources. In contrast to the other genera discussed in this part, chromosomes of the tomato species are identical in number and homology; thus except for minor pairing irregularities and reduced recombination, gene exchange between *L. esculentum* and the wild species is less difficult than in many other plant genera.

The advances realized in various areas of research from recent hybridization with exotics have been summarized (Rick, 1982). The most spectacular gains have been achieved in the area of disease resistance. The first utilization was fusarium wilt resistance derived from *L. pimpinellifolium* Mill. by Bohn and Tucker (1940). To date resistance to no less than 27 different serious diseases has been discovered in wild species and other exotics, and 14 of these resistances have been incorporated in well-adapted and widely grown cultivars; moreover, resistance to up to 8 diseases has been incorporated in a number of these cultivars. Promising resistance to arthropod pests and tolerance to various environmental stresses have also been detected, but progress in breeding these traits into acceptable cultivars has been slow for various technical reasons. Among other desired characters, the visual and nutritional quality of tomatoes has also been improved by such utilization of exotics.

Wheat

The transfers of disease resistance from other species to wheat are classic examples of the utilization of exotic germplasm. These instances are noteworthy because they entail the crossing of species with different euploid chromosome numbers. Bread wheat (*Triticum aestivum* L.) is an allohexaploid; the contributing species are diploid and allotetraploid. According to Knott and Dvořák (1976) the *Triticum-Aegilops* complex is composed of about 30 species, and the cultivated species can also be hybridized with rye (*Secale cereale* L.) and certain species of *Agropyron*. The even more distant cross with barley (*Hordeum vulgare* L.) has been made recently, although problems with flowering and sterility have obstructed utilization. Considerable advantage has been taken of this vast reserve of germplasm to breed resistances and other useful economic traits into cultivated wheat.

The pioneer interspecific transfers were made with the object of gaining stem rust resistance from tetraploid cultivated wheats. Despite considerable sterility in the F_1 hybrid, Hayes et al. (1920) succeeded in transfers from the Iumillo *T. durum* Desf. to derive cv. Marquillo, the cultivar destined to be a parent of the very important cv. Thatcher. In similar fashion, McFadden (1930) derived resistance from Yaroslav emmer to breed cv. Hope, also an important parent in later wheat improvement. Subsequently the battle against the highly polymorphic stem rust organism was aided by the discovery and transfer of more genes for resistance from *T. monococcum* L. and *T. timopheevi* Zhuk. by other investigators.

Different sources of resistance have likewise been exploited by several investigators in the fight against leaf rust. In a classic experiment, Sears (1956) ingeniously manipulated hybridizations, ionizing radiation, and selection to effect a transfer of leaf rust resistance from *Ae. umbellulata* Zhuk. to bread wheat. The complete failure of pairing between wheat and *Ae. umbellulata* chromosomes and the situation of the resistance gene in its chromosome required chromosomal translocation of a specific type. Other genes for leaf rust resistance were transferred by Kerber and Dyck (1969, 1973) from *Ae. squarrosa* L. by a technique requiring the synthesis of a new alloploid between it and a tetraploid parent.

The discovery of genetic control of homoeologous chromosome pairing in the wheat complex facilitates transfers without requiring the fastidious procedure of inducing and manipulating chromosome rearrangements (Riley, 1958; Sears and Okamoto, 1958). Deletion of the wheat chromosome 5B or the substitution of a different *Ph* allele allows pairing and exchange by normal crossingover between the wheat chromosomes and homoeologues of other genomes. One of the first successful uses of this technique resulted in the transfer of stripe rust resistance from *Ae. comosa* Sibth. & Sm. (Riley et al., 1968a,b).

The wide cross of wheat x *Agropyron elongatum* (Host) Beauv. was effected and, via the induced translocation method, Sharma and Knott (1966) succeeded in breeding a stem-rust resistant wheat, although resistance was tightly linked with an undesirable yellow flour color. Another cytogenetic technique that transfers exotic alleles is substitution of an alien chromosome for its wheat homoeologue. Such substitution is not satisfactory if the introduced chromosome is derived from a wild relative, and hence carries undesirable genes; it can succeed if the introduced chromosome is derived from a cultivated species, as in the case of mildew and stripe rust resistance in the substitution of a rye chromosome (Mettin et al., 1973; Zeller, 1973).

Cytoplasmic male sterility was discovered in descendents of interspecific hybrids of wheat by Kihara (1951). This report was followed by intensive research by many investigators, who, with Kihara, demonstrated that this phenomenon is a consequence of the substitution of wheat nuclei in the cytoplasms of *Ae. caudata* L., *Ae. ovata* L., *T. timopheevi*, and *T. zhukovskyi* Meu. & Er. A difficult problem encountered in the utilization of such male sterility is restoration of complete fertility—a sine qua non of successful F_1 hybrids.

Oats

Oat breeding has consisted mainly of selection within progenies derived from matings of pure-line cultivars. According to Brown

(1982), "the lack of significant genetic variability within and between cultivars has restricted progress in oat development, although it has permitted the development of improved straw strength and better resistance to fungal pathogens." Brown also suggests that a lack of understanding of the evolution of cultivated hexaploid oats has retarded the use of diploid and tetraploid exotics.

Oat breeding advanced in the 1920s and 1930s when the cultivated *Avena sativa* L. was crossed with *A. byzantina* Koch for the purpose of deriving disease resistance from the latter. Comparative yield tests revealed that cultivars developed from this mating outyielded standard pure-line cultivars by 9 to 14% (Browning et al., 1964; Langer et al., 1978). This yield improvement presumably resulted from interaction of genes of the two species—i.e., a form of transgressive variation.

Langer et al. (1978) maintain that oat yields did not increase to any significant extent during the 40-year period since the boost from *A. byzantina* introgression was realized. The next major improvement can be traced to crosses made between *A. sativa* and the wild hexaploid *A. sterilis* L., which were made for the purpose of breeding crown rust resistance from the latter. By backcrossing to *A. sativa* and selecting for resistance it was possible to derive resistant oat lines, which, surprisingly, outyielded modern cultivars. The yield improvement of 11 such lines amounted to 20% or more. These experiments were repeated by Lawrence and Frey (1975), who verified that lines recovered from BC_2 to BC_5 of the same parentage manifested transgressive segregation for yields above those of either parent. As Brown (1982) aptly concludes, these experiences in the use of exotic germplasm ". . . carry an important message for the breeders of oats and perhaps other crops as well."

Novel Variation

As manifest in the foregoing examples, studies of progenies from wide crosses frequently display characters that were not known in either parent—a phenomenon I have labeled "novel variation" (Rick, 1967; Rick and Smith, 1953). Such variation can consist of a new qualitative trait or extreme quantitative variants that exceed the known range of one or both parents. Although not always useful for a particular breeding objective, novel variation can be of appreciable consequences for various purposes. Qualitative novel variation can be as random in nature as spontaneous or induced mutation; therefore only a tiny fraction of such types might be useful for plant improvement to meet a specific requirement. A degree of predictability is nevertheless manifest in cytoplasmic male sterility, since this phenotype commonly appears when the nucleus of one species is associated with the cytoplasm of another. If manifest as a quantitative variant, the phenomenon is equivalent to transgressive variation, which clearly can be of considerable importance for such selected characters as earliness, yield, nutritional content, etc. Some examples of novel variation in interspecific plant hybrids are described in the following sections.

QUALITATIVE NOVEL VARIATION. (a) The B gene, coding for β-carotene, is not manifest in the source parent, *Lycopersicon hirsutum* Humb. & Bonpl., but affects carotenoid synthesis in the background of

L. esculentum (Lincoln and Porter, 1950); (b) the fruit pigment intensifier *Ip* is derived from *L. chmielewskii* Rick et al. (Rick, 1974); (c) there is genic male sterility in hybrids of *L. esculentum* x *L. parviflorum* (Kesicki, 1980); and (d) cytoplasmic male sterility is found in the aforementioned examples in tobacco and wheat.

QUANTITATIVE NOVEL VARIATION. (a) Ability to flower in short-day, low-light intensity conditions and earlier flowering in hybrids of *L. esculentum* x *L. hirsutum* (Maxon-Smith, 1978); (b) increased nitrogen fixation activity in beans (McFerson et al., 1982); (c) increased yields of interspecific oat derivatives in the two aforementioned examples; and (d) increased starch content and heterosis for yield in interspecific progenies of potatoes cited earlier.

It should be appreciated from this series of examples that novel variation is not an exceptional phenomenon. Its occurrence is sufficiently frequent, in fact, that investigators can ill afford to neglect the opportunities that it provides.

CONCLUSION

In spite of great public interest and devoted efforts of investigators, much valuable germplasm has been lost. Many examples could be cited. In the area with which I am familiar—tomato germplasm—I have witnessed the complete disappearance of primitive cultivars from the Peruvian coast, chiefly as a result of their replacement by introduced cultivars with higher yields and other features of greater appeal to the industry. Many sites of related wild species have been lost in the same region as well as in the Peruvian Sierra, as a reult of human development activities, particularly in the widespread grazing of goats, being responsible. The same pressures are exacting their toll elsewhere in the Andes. For certain taxa, all we have to show is a pitifully small assortment of accessions that were collected before these changes and fortunately have been maintained in viable form.

A similar, dismal picture is presented by Ochoa (1975) for the decimation of primitive potato cultivars from the native Andean area. In Chiloe, Chile, and elsewhere, recently introduced, highly productive cultivars have rapidly replaced the multitudinous, highly variable assortment of native primitive cultivars. The preservation of many of these heirlooms has been obstructed by the spread of virus diseases and other problems incurred in clonal propagation. Tubers of the aboriginal material were notable for their astounding diversity of colors, shapes, flavors, nutritional content, and uses. It is frightening and lamentable that much of this irreplaceable treasure has been lost forever.

The same sad story is repeated in the situation of the primitive races of maize in Latin America, and, to various degrees of seriousness, in other crops. The situation is desperate and urgent. For the sake of ourselves, the rest of the world, and posterity, the remnants must be saved and plant germplasm systems otherwise strengthened.

The enormous gains already realized by utilization of germplasm resources stress the critical importance of proper management of all phases of the system. The considerations of this chapter have been completely limited to plant breeding as practiced according to conventional methods. As the field becomes facilitated by genetic engineering techniques, the role of germplasm systems can only become more important.

ACKNOWLEDGMENTS

Bibliographic search partly supported by NSF Grant DEB 80-05542. The manuscript benefitted from Dora Hunt's masterful editorial talents. Valuable references were provided by Drs. F.A. Bliss, W.L. Brown, D.U. Gerstel, and P. McGuire. Dr. Q. Jones verified statements in the section describing germplasm activities in the U.S. To all of these colleagues, as well as students and staff members who assisted in other ways, I am most grateful.

REFERENCES

Allard, R.W. 1970. Problems of maintenance. In: Genetic Resources in Plants: Their Exploration and Conservation (O.H. Frankel and E. Bennett, eds.) pp. 491-494. Blackwell, London.

______ and Hansche, P.E. 1964. Some parameters of population variability and their implications in plant breeding. Adv. Agron. 15:281-325.

______ and Jain, S.K. 1962. Population studies in predominantly self-pollinated species. II. Analysis of quantitative changes in a bulk hybrid population of barley. Evolution 16:90-101.

Alexander, L.J. and Hoover, M.M. 1955. Disease Resistance in the Wild Species of Tomato. North Central Reg. Pub. 51. Ohio Agricultural Experiment Station, Research Bulletin 752. Wooster, Ohio.

Anonymous. 1977. The National Plant Germplasm System. USDA Program Aid No. 1188, USDA, Beltsville, Maryland.

Bohn, G.W. and Tucker, C.M. 1940. Studies on *Fusarium* wilt of the tomato. I. Immunity in *Lycopersicon pimpinellifolium* Mill. and its inheritance in hybrids. Missouri Agriculture Experiment Station Research Bulletin 311:1-82. Missouri State, Missouri.

Brown, W.L. 1982. Exotic germplasm in cereal crop improvement. In: Frontiers in Plant Breeding (I.K. Vasil et al., eds.) pp. 29-42. Academic Press, New York.

Browning, J.A., Frey, K.J., and Grindeland, R.L. 1964. Breeding of multiline oat varieties for Iowa. Iowa Farm Sci. 18:629-632.

Bunting, A.H. and Kuckuck, H. 1970. Ecological and agronomic studies related to plant exploration. In: Genetic Resources in Plants: Their Exploration and Conservation (O.H. Frankel and E. Bennett, eds.) pp. 181-188. Blackwell, London.

Burk, L.G. and Heggestad, H.E. 1966. The genus *Nicotiana*: A source of resistance to diseases of cultivated tobacco. Econ. Bot. 20:76-88.

Clayton, E.E. 1958. The genetics and breeding progress in tobacco during the last 50 years. Agron. J. 50:352-356.

Commonwealth Scientific and Industrial Research Organization. 1980. Division of Plant Industry Report 1979-80, pp. 75-76. Canberra, Australia.

Cubillos, A.G. and Plaisted, R.H. 1976. Heterosis for yield in hybrids between *S. tuberosum* ssp. *tuberosum* and *S. tuberosum* ssp. *andigena*. Am. Potato J. 53:143-150.

Cutler, M., Brown, W.L., Lewis, C.F., Beard, D.F., Fitzgerald, P.J., Gabelman, W.H., Gardner, C.O., Loden, H.D., Peters, D.C., Rick, C.M., Robinson, H.F., Sprague, G.F., and Wortman, L.S. 1979. Plant Genetic Resources: Conservation and Use. USDA, Washington, D.C.

Duncan, R.R. 1981. Registration of GPIR acid soil tolerant sorghum germplasm population. Crop Sci. 21:637.

Duvick, D.N. 1977. Genetic gains in hybrid maize yields during the past 40 years. Maydica 22:187-196.

Esechie, H.A., Maranville, J.W., and Ross, W.M. 1977. Relationship of stalk morphology and chemical composition to lodging resistance in sorghum. Crop Sci. 17:609-612.

Esquinas, J.T. 1981. Alloenzyme variation and relationships among Spanish landraces of *Cucumis melo* L. Kulturpfl. 29:337-353.

Fairchild, D.G. 1938. The World Was My Garden. Scribners, New York.

Foster, D.G., Teetes, G.L., Johnson, J.W., Rosenow, D.T., and Ward, C.R. 1977. Field evaluation of resistance in sorghums to Bank's grass mite. Crop Sci. 17:821-823.

Frankel, O.H. 1970a. Genetic conservation in perspective. In: Genetic Resources in Plants: Their Exploration and Conservation (O.H. Frankel and E. Bennett, eds.) pp. 469-489. Blackwell, London.

______ 1970b. Genetic conservation of plants useful to man. Biol. Conserv. 2:162-169.

______ 1974. Genetic conservation: Our evolutionary responsibility. Proceedings XIII International Genetics Congress, Genetics 78:53-65.

______ and Bennett, E. (eds.) 1970. Genetic Resources in Plants: Their Exploration and Conservation. Blackwell, London.

______ and Hawkes, J.G. (eds.) 1975. Crop Genetic Resources for Today and Tomorrow. Cambridge Univ. Press, Cambridge, England.

Frankel, R. and Galun, E. 1977. Pollination Mechanisms, Reproduction, and Plant Breeding. Springer-Verlag, Berlin.

Gerrish, E.E. 1981. Germplasm collection, preservation and use. In: Plant Breeding II (K.F. Frey, ed.) pp. 71-73. Iowa State Univ. Press, Ames.

Gerstel, D. 1980. Cytoplasmic Male Sterility in *Nicotiana* (a review). North Carolina Agricultural Research Station Technical Bulletin 263. Raleigh, North Carolina.

Goodspeed, T.H. 1961. Plant Hunters in the Andes. University California Press, Berkeley.

Hallauer, A.R. and Miranda, J.B. 1981. Quantitative Genetics and Maize Breeding. Iowa State Univ. Press, Ames.

Harlan, H.V. and Martini, M.L. 1938. The effect of natural selection in a mixture of barley varieties. J. Agric. Res. 57:189-199.

Harrington, J.F. 1970. Seed and pollen storage for conservation of plant gene resources. In: Genetic Resources in Plants: Their Exploration and Conservation (O.H. Frankel and E. Bennett, eds.) pp. 501-521. Blackwell, London.

Harrison, B.J. 1966. Seed deterioration in relation to storage conditions and its influence upon germination, chromosomal damage, and plant performance. J. Nat. Inst. Agric. Bot. 10:644-663.

Hayes, H.K., Parker, J.H., and Kurtzweil, C. 1920. Genetics of rust resistance in crosses of varieties of *Triticum vulgare* with varieties of *T. durum* and *T. dicoccum*. J. Agric. Res. 19:523-542.

Heggestad, H.E. 1966. Registration of Burley 1, Burley 2, Burley 11A, Burley 11B, Burley 21, Burley 37, and Burley 49 tobaccos. Crop Sci. 6:612-613.

______, Clayton, E.E., Neas, M.O., and Skoog, H.A. 1960. Development of Burley 21, the first wildfire-resistant tobacco variety, including results of variety trials. Tennessee Agricultural Experiment Station Bulletin 321. Knoxville, Tennessee.

Jagger, I.C. 1940. Brown blight of lettuce. Phytopathology 30:53-64.

Jain, S.K. 1975. Genetic reserves. In: Crop Genetic Resources for Today and Tomorrow (O.H. Frankel and J.G. Hawkes, eds.) pp. 379-396. Cambridge Univ. Press, Cambridge, England.

Johnson, J.W., Rosenow, D.T., and Teetes, G.L. 1973. Resistance to the sorghum midge in converted exotic sorghum cultivars. Crop Sci. 13:754-755.

Kerber, E.R. and Dyck, P.L. 1969. Inheritance in hexaploid wheat of leaf rust resistance and other characters derived from *Aegilops squarrosa*. Can. J. Genet. Cytol. 11:639-647.

______ 1973. Inheritance of stem rust resistance tranferred from diploid wheat (*Triticum monococcum*) to tetraploid and hexaploid wheat and chromosome location of the gene involved. Can. J. Genet. Cytol. 15:397-409.

Kesicki, E. 1980. A new type of functional male sterility derived from interspecific hybrid. Acta Horticul. 100:365-369.

Kihara, H. 1951. Substitution of nucleus and its effects on genome manifestations. Cytologia 16:177-193.

Kingdon-Ward, F. 1960. Pilgrimage for Plants. G.G. Harrap, London.

Knott, D.R. and Dvořák, J. 1976. Alien germplasm as a source of resistance to disease. Annu. Rev. Phytopathol. 14:211-235.

Langer, I., Frey, K.J., and Bailey, T.B. 1978. Production response and stability characteristics of oat cultivars developed in different eras. Crop Sci. 18:938-942.

Lawrence P. and Frey, K.J. 1975. Backcross variability for grain yield in oat species crosses (*Avena sativa* L. x *Avena sterilis* L.). Euphytica 24:77-85.

Lincoln, R.E. and Porter, J.W. 1950. Inheritance of Beta-carotene in tomatoes. Genetics 35:206-211.

Marshall, D.R. and Brown, A.H.D. 1975. Optimum sampling strategies in genetic conservation. In: Crop Genetic Resources for Today and Tomorrow (O.H. Frankel and J.G. Hawkes, eds.) pp. 53-80. Cambridge Univ. Press, Cambridge, England.

Maxon-Smith, J.W. 1978. *Lycopersicon hirsutum* as a source of genetic variation for the cultivated tomato. In: Interspecific Hybridization in Plant Breeding, Proceedings 8th Congress Eucarpia (E. Sanchez-Monge and F. Garcia-Olmedo, eds.) pp. 119-128. Madrid.

McFadden, E.S. 1930. A successful transfer of emmer characteristics to *vulgare* wheat. J. Am. Soc. Agron. 22:1020-1034.

McFerson, J., Bliss, F.A., and Rosas, J.C. 1982. Selection for enhanced N_2 fixation in common bean, *Phaseolus vulgaris* L. Niftal, Cali, Colombia (in press).

Mettin, D., Blüthner, W.D., and Schelge, G. 1973. Additional evidence on spontaneous 1B/1R wheat-rye substitutions and translocations. Proceedings 4th International Wheat Genetics Symposium, pp. 179-184. Columbia, Missouri.

Miller, F.R. 1976. Release of Tx430, a disease resistant inbred male. Texas Agricultural Experiment Station Communication, Feb. 20, 1976.

Miller, J.E. and Edminster, T.W. 1976. Germplasm release of A and B pair of sorghum lines with a new cytoplasmic-genic sterility system. Texas Agricultural Experiment Station/USDA/ARS Communication Dec. 22, 1976. College Station, Texas.

Muñoz, F.J. and Plaisted, R.L. 1981. Yield and combining abilities in *andigena* potatoes after six cycles of recurrent phenotypic selection to long-day conditions. Am. Potato J. 58:469-480.

Ochoa, C. 1975. Potato collecting expeditions in Chile, Bolivia, Peru and the genetic erosion of indigenous cultivars. In: Crop Genetic Resources for Today and Tomorrow (O.H. Frankel and J.G. Hawkes, eds.) pp. 167-173. Cambridge Univ. Press, Cambridge, England.

Rick, C.M. 1945. A survey of cytogenetic causes of unfruitfulness in the tomato. Genetics 30:347-362.

______ 1967. Exploiting species hybrids for vegetable improvement. Proceedings XVII International Horticulture Congress 3:217-229. College Park, Maryland.

______ 1973. Potential genetic resources in tomato species: Clues from observations in native habitats. In: Genes, Enzymes, and Populations (A. Srb, ed.) pp. 255-269. Plenum, New York.

______ 1974. High soluble-solids content in large-fruited tomato lines derived from a wild green-fruited species. Hilgardia 42:493-510.

______ 1982. The potential of exotic germplasm for tomato improvement. In: Frontiers in Plant Breeding (I.K. Vasil et al., eds.) pp. 1-28. Academic Press, New York.

______ and Fobes, J.F. 1975. Allozyme variation in the cultivated tomato and closely related species. Bul. Torrey Bot. Club 102:376-384.

______ and Smith, P.G. 1953. Novel variation in tomato species hybrids. Am. Nat. 87:359-373.

Riley, R. 1958. Chromosome pairing and haploids in wheat. Proceedings X International Congress Genetics 2:234-235. Montreal, Canada.

______, Chapman, V., and Johnson, R. 1968a. Introduction of yellow rust resistance of *Aegilops comosa* into wheat by genetically induced homoeologous recombination. Nature 217:383-384.

______, Chapman, V., and Johnson, R. 1968b. The incorporation of alien disease resistance in wheat by genetic interference with the regulation of meiotic chromosome synapsis. Genet. Res. 12:199-219.

Roberts, E.H. 1975. Problems of long-term storage of seeds and pollen for resources conservation. In: Crop Genetic Resources for Today and Tomorrow (O.H. Frankel and J.G. Hawkes, eds.) pp. 269-295. Cambridge Univ. Press, Cambridge, England.

Ross, H. 1979. Wild species and primitive cultivars as ancestors of potato varieties. In: Broadening the Genetic Base of Crops (A.C. Zeven and A.M. van Harten, eds.) pp. 237-245. PUDOC, Wageningen, The Netherlands.

Ross, R.W. and Rowe, P.R. 1969. Utilizing frost resistance of diploid *Solanum* species. Am. Potato J. 46:5-13.

Russell, W.A. 1974. Comparative performance for maize hybrids representing different eras of maize breeding. Proceedings Corn Sorghum Research Conference 29:81-101.

Sand, S.A. and Taylor, G.S. 1961. C2, a new mosaic-resistant Connecticut broadleaf tobacco. Conn. Agric. Exp. Stn. Bull. 636. New Haven, Connecticut.

Savitsky, V.F. 1950. Monogerm sugar beets in the United States. Proc. Am. Soc. Sugar Beet Techn. 1:156-159.

Sears, E.R. 1956. The transfer of leaf rust resistance from *Aegilops umbellulata* to wheat. Brookhaven Symp. Biol. 9:1-22.

______ and Okamoto, M. 1958. Intergenomic chromosome relationships in hexaploid wheat. Proceedings X International Congress Genetics 2:258-259. Montreal, Canada.

Sharma, D. and Knott, D.R. 1966. The transfer of leaf-rust resistance from *Agropyron* to *Triticum* by irradiation. Can. J. Genet. Cytol. 8: 137-143.

Sigurbjörnsson, B. and Micke, A. 1974. Philosophy and accomplishments of mutation breeding. In: Polyploidy and Induced Mutations in Plant Breeding, pp. 303-343. International Energy Agency, Vienna.

Simmonds, N.W. 1962. Variability in crop plants, its use, and conservation. Biol. Rev. 37:422-465.

Singh, R. and Axtell, J.D. 1973. High lysine mutant gene (*hl*) that improves protein quality and biological value of grain sorghum. Crop Sci. 13:535-539.

Stephens, J.C., Miller, F.R., and Rosenow, D.T. 1967. Conversion of alien sorghums to early combine genotypes. Crop Sci. 7:396.

Valleau, W.D., Stokes, W.D., and Johnson, E.M. 1960. Nine years experience with the *Nicotiana longiflora* factor for resistance to *Phytophthora parasitica* var. nicotianae in the control of black shank. Tob. Sci. 4:92-94.

Wilson, E.H. 1927. Plant Hunting. Stratford, Boston.

Yeager, A.F. 1951. Adventures in horticulture. Proc. Am. Soc. Hortic. Sci. 58:377-383.

Zeller, F.J. 1973. 1B/1R wheat-rye chromosome substitutions and translocations. Proceedings 4th International Wheat Genetics Symposium pp. 209-221. Columbia, Missouri.

Zeven, A.C. 1979. Collecting genetic resources in highly industrialized Europe, especially in The Netherlands. In: Broadening the Genetic Base of Crops (A.C. Zeven and A.M. van Harten, eds.) pp. 49-58. PUDOC, Wageningen, The Netherlands.

CHAPTER 2
Underexploited Crops

G. L. Laidig, E. G. Knox, and *R. A. Buchanan*

Underexploited crops are plants that have potential for generation of a significant new flow of one or more crop products from production areas to consumers. These plants offer a special challenge for the use of cell culture because most are unfamiliar to plant breeders, nearly all require extensive efforts to select and improve plant materials, and at least some promise major changes in United States and world agriculture.

The flow of crop products takes place by the operation of a production-marketing-consumption (PMC) system. A PMC system, portrayed in Fig. 1, "ties together the individuals and institutions which plant, nurture, and harvest the crop; those who collect, process, store, transport, distribute, and sell the crop products; and those who buy and consume the products. In most cases, there are a large number of producers, a much smaller number of marketing channels, and a large number of consumers. The consumers may be individuals at the retail level, or they may be industrial concerns which utilize the crop within a compound product or as an agent in some industrial process." (Theisen et al., 1978.)

The initiation and expansion of production of an underexploited crop is made possible by domestication of a plant previously harvested from a natural stand, domestication and utilization of a previously unused plant, importation of an exotic crop, change to a new use for an existing crop or other cultivated plant, commercialization of a crop previously grown primarily for home use, or some combination of these coupled with an appropriate increase in area of production, increase in yield, or both. The area of production can be increased by displace-

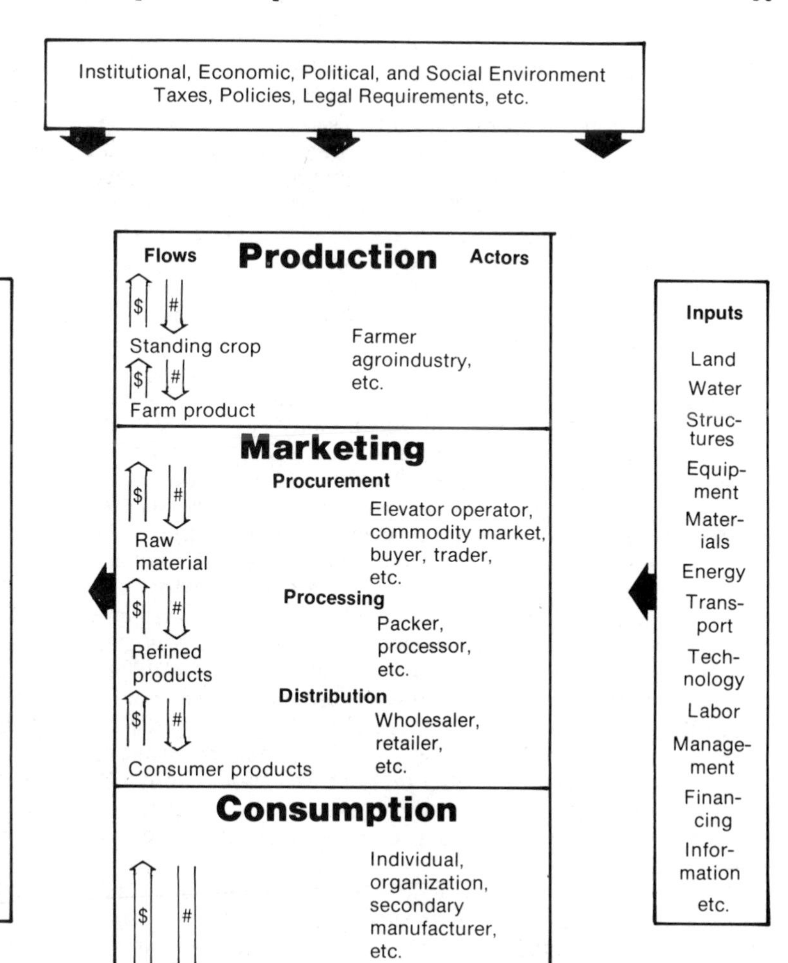

Figure 1. General production-marketing-consumption (PMC) system.

ment of other crops, change to multiple cropping on existing land, or conversion of other land to crop use.

Initiation and expansion of consumption is made possible by substitution of the new crop product for some other domestic product, import substitution, export of the new crop product, satisfaction of previously unsatisfied demand, expansion of demand, expansion of geographic market area, or development of demand for higher quality or new products.

Initiation and expansion of marketing (procurement, processing, and distribution) is made possible by reassignment of existing facilities and institutions or construction and development of new facilities and institutions. New facilities may be needed because of special characteristics and requirements of the new crop products or because the areas of production, consumption, or both are distant from existing facilities. Of the three marketing phases, processing is most likely and distribution least likely to require new linkages and organizations.

The core feature of a functioning PMC system is a flow of goods in response to economic demand for the end product that is great enough to activate all the necessary functions of the system. Consumer demand for end products is of crucial importance for the profitability of the system and the volume of flow. At any given volume, the price paid by consumers must produce a satisfactory economic return in all levels of the production and marketing subsystems. The price consumers are willing to pay tends to decrease with increasing volume; thus the volume of flow can increase only up to a price level that is acceptable to the production and marketing subsystems.

At a given price level, profitability is controlled by costs of production and marketing. Reductions in cost that make lower consumer prices acceptable allow a greater volume of flow. Thus, both reductions in cost and increases in demand permit increases in volume of flow. Increases in demand result from restriction of imports, replacement of competitive products, technological developments that call for new products, changes in consumer tastes and preferences, development of export markets, expansion of economic activity, population growth, or some combination of these. Reductions in cost may result from more efficient processing, utilization of by-products, more efficient farm machinery, improved cultivars (with higher or more reliable yields or both, pest resistance, characteristics that facilitate harvest, etc.), improved storage and handling technology, etc. Relatively minor changes in demand or costs can lead to major changes in volume of flow.

The development of an underexploited crop is a difficult and complex process that requires new technology, institutions, linkages, regulations, and services; recruitment of personnel; reassignment of land, other facilities, materials, and financial resources; etc. In short, it requires the creation of a new physical, technological, social, economic, and institutional infrastructure.

The process, problems, and requirements of new crop development have been analyzed and described (Knox and Theisen, 1981). The selection and development of appropriate plant materials is only one of the many essential activities, but it is important because it is one of the very first. Much of the work must be done before there are clear indications of financial returns, and it may take many years to provide materials that allow other activities to proceed. Furthermore, it is never finished.

Plant selection and breeding must adapt a new crop to a new environment. In a new geographic area, the introduced crop is likely to face different temperature, day length, or moisture regimes. A newly domesticated crop within its native range faces changes in stand composition and density, tillage, altered nutrient and water regimes, and controlled propagation.

Because most of the cost of production depends directly on area, genetic work must strive for higher yields to reduce cost per unit of crop product. Also, it must strive for dependable yields in the face of variations in weather conditions. Resistance to or tolerance of weeds, insects, fungi, etc., is especially important for new crops because the clearance by federal and state agencies of pesticide chemicals for application on a new crop takes several years. Moreover, work toward clearance is not likely to be initiated until commercial experience with the crop justifies the expense.

Adaptations to soil conditions are important. These include the capacity to withstand periods of water deficiency, the capacity to utilize soil moisture fully including that less readily available, resistance to water saturation and low oxygen levels in the soil, the capacity to utilize phosphorus and potassium in forms not readily available to most plants, symbiotic relationships with microorganisms that facilitate extraction of soil nutrients or that fix nitrogen, tolerance of high levels of aluminum, manganese, or sodium, tolerance of acid or alkaline reaction, tolerance of high levels of calcium and magnesium carbonates, gypsum, and more soluble salts, etc.

Uniformity of plant material in time of maturity, size, and growth form is necessary to facilitate harvest. Efficiency of harvest may also require modification of the growth form, change in the fruit (e.g., to eliminate a shattering tendency), reduction in the length of the fruiting period, etc.

Genetic work must treat those characteristics that influence procurement and processing and determine quality of the consumer products as well as those that relate to production. Resistance to deterioration during transportation and storage and high content of valuable chemicals (oil, latex, protein, etc.) are examples of such characteristics.

Some underexploited crops are of interest as sources of human food or livestock feed, but the more dramatic possibilities lie with those potential new crops that promise to supply oils, latex, fiber, cellulose, and other industrial feedstocks for production of fuels, paper, rubber, chemicals, and other materials. The demands of processors in the complicated PMC system of such industrial crops, for high volume, uniform high quality, and dependable year round supply present a special challenge to genetic workers.

Several potentially new crops—guayule, jojoba, kenaf, crambe, milkweed, and grain amaranth—are briefly discussed below. Selected references for each crop at the end of the chapter provide additional information.

GUAYULE

Guayule (*Parthenium argentatum* A. Gray) is a latex-producing shrub, native to the Chihuahuan Desert of north central Mexico and the Trans-Pecos region of southwestern Texas. Though the rubber tree (*Hevea brasiliensis* Willd. ex A. Joss)—grown in Southeast Asia, South America, and Africa—is our major source of natural rubber, guayule has been an important source of commercial quantities of high quality natural rubber.

History

In the early 1900s, wild guayule was a source of natural rubber for the United States and Mexico. In response to reckless exploitation of the wild shrubs, a group of American industrialists formed the Intercontinental Rubber Company in Salinas, California. This company planted more than 3200 ha of guayule and produced about 1.2 million kg of rubber in the 1930s, but could not meet competition resulting from favorable trade policies with Southeast Asian rubber producers.

In 1942, the U.S. government initiated a $45 million Emergency Rubber Project (ERP) to offset the wartime loss of natural rubber from Southeast Asia. Technology, plantings, and processing facilities were ready by the end of the war, but with a national policy to favor Southeast Asian countries and the rapid development of synthetic rubber, the guayule project was destroyed. All 13,000 ha of guayule at 13 sites in three southwestern states (California, Arizona, and Texas) were burned or plowed under. Limited agronomic and processing research continued at the USDA Natural Rubber Research Station in Salinas but by 1953 even this effort had been terminated.

In the early 1970s, Mexico started a long-term guayule development program to provide employment for its rural poor. Mexico planned first to use wild guayule stands and later to incorporate regenerative techniques and plantation agriculture. A 450 kg/day guayule pilot processing plant was constructed in Saltillo, Mexico, in 1976.

The United States also has revived its interest in guayule. In 1978, Congress passed the Native Latex Commercialization and Economic Development Act. This act authorized $30 million over 4 years to conduct research and to encourage commercial growing and processing of guayule and other rubber-producing plants in the Southwest. The Federal Emergency Management Agency (FEMA) is interested in guayule as a domestic source of natural rubber to help meet a 730,000 metric ton deficit in the U.S. strategic stockpile. Firestone and Goodyear Tire and Rubber companies both have been investigating guayule. Agronomic and plant breeding research programs are being conducted at Texas A & M University, Arizona State University, New Mexico State University, the University of California at Riverside, the California Department of Food and Agriculture, and the Los Angeles State and County Arboretum.

Description

Guayule, a member of the sunflower family, Compositae, is a symmetrical, closely branched, perennial shrub that reaches a height of 60-90 cm and a circumference of 90-120 cm. The leaves are narrow, about 5 cm long, silvery and grayish in color, and are covered by a drought-protective white wax. Its tap root may penetrate the soil more than 6 m; lateral fiberous roots can extend 9-12 m. The individual flowers are inconspicuous, wind- or insect-pollinated, and are formed in heads on long stems.

Production

Wild guayule grows on upland plateaus with subtropical, temperate climates, where rainfall is low and erratic. It survives drought up to 7 months, but if annual rainfall is less than 35 cm, the rubber yield diminishes. Annual moisture greater than about 65 cm promotes excessive vegetative growth rather than rubber formation. Guayule survives high temperatures but becomes dormant at low temperatures and can be killed by frost. Approximately 3 million ha have been identified as agroclimatically suitable for commercial guayule production in the southern border regions of Texas, Arizona, New Mexico, and California (Foster et al., 1980a).

Direct seeding guayule is the simplest and least costly method of propagation. However, direct seeding is not always successful because of poor seed germination, inadequate moisture at planting, low temperatures, and a variety of other factors. Transplanting seedling or root stock is the most successful but also the most expensive method of propagation.

Guayule is susceptible to a number of plant diseases common in the southwestern U.S.: *Verticillium dahliae*, *Phymototrichum omnivorium*, and *Sclerotium bataticola*. Weed control is important, since guayule seedlings are intolerant of shade and competition.

Two-thirds of the rubber in guayule is in the stems and branches and the remainder is in the roots. Thus, a partial harvest by clipping the plant at about 3 years of age, regeneration from the roots, and a final harvest by digging the whole plant at 5 years of age may allow the maximum yield of rubber. Alternatively, the whole plant can be harvested at 3-5 years.

Yields of guayule rubber depend on total dry matter and rubber content and vary by strain or variety, the method of field establishment, plant spacing, climate, irrigation, and age at harvest. Researchers expect that present varieties could produce about 2300 kg/ha of rubber after four years under irrigated conditions. For dryland production, yields would be lower, about 1100 kg/ha after 5 years (Foster et al., 1980a).

At harvest, guayule's hydrocarbon content begins to deteriorate by oxidation. The fresh field product must be delivered directly to the processing plant to prevent loss of rubber. Unless a means to control deterioration is found, harvested guayule will be difficult to store. Scheduling of year-round harvests and delivery to the processing plant could present logistical difficulties given a large scale of production and processing.

Processing

There are no commercial-scale guayule processing facilities in the United States at this time, nor is any existing facility easily converted to guayule processing. Little actual processing of guayule has occurred in the United States since World War II, although Firestone and Goodyear Tire and Rubber companies have undertaken some laboratory-scale research.

Firestone's processing approach uses either a solvent-extraction or a flotation process to extract all the resins from the shrub and then to recover the rubber (Weihe and Nivert, 1980). Since the flotation process is water intensive, in semi-arid regions the solvent extraction process seems more feasible.

Consumption

Researchers have concluded that guayule rubber is an acceptable substitute for hevea rubber and synthetic cis-polyisoprene rubber in many types of tires (National Academy of Science, 1977). Radial tires produced by Goodyear Tire and Rubber Company containing 30-40% guayule rubber have passed rigorous Department of Transportation high speed and endurance tests. Goodyear is also encouraged by the results of tests on 100% guayule tires for heavy equipment. U.S. Navy tests of 100% guayule in retread aircraft tires indicate that guayule is an acceptable substitute for such high stress applications.

Although rubber is considered the primary product of guayule, various by-products—resins and bagasse—also have potential industrial value. Thirty-seven percent of the resin obtained during processing for rubber is linoleic acid, which has an established position in the paint and varnish industry. Cinnamic acid, also found in guayule resins, is used by the cosmetic and pharmaceutical industries. Terpenes from guayule resin, notably cyclic di- and tri-terpenes, probably can substitute for naval store products used in paint and paper-sizing industries (Foster et al., 1980a).

Guayule bagasse could be burned for direct energy generation. Alternatively, guayule bagasse could be combined with longer fiber woodpulp for papermaking. Cork obtained in guayule processing can be burned as fuel. It may have a higher value, however, as a filler for plastics and rubber or for making structural board and insulation products.

Guayule's PMC System Status

The commercial development of guayule as a domestic source of natural rubber in the United States, short of a major change in processing technology, will require a decision by either public agencies or private rubber companies or a combination of both to invest large capital resources in the construction of a guayule processing plant in the Southwest. Alternatively, given technological improvements that permit economical processing at smaller scale, other investors could initiate commercial development. Unfortunately, private rubber companies are facing a depressed tire market and general economic uncertainty. Funds for the Native Latex Act have not been forthcoming to support a publicly funded guayule development project. The government's special interest in guayule as a strategic material has not yet produced funding in support of commercialization.

The prospects for commercialization are limited by the need for better high-yielding varieties, improved farm machinery, development and clearance of pesticides, and intensive pilot research on processing technologies. Coordination between production and processing phases is crucial. Growing areas must be close to the processing plant and production must be matched with plant capacity. Because guayule is a perennial crop, production will require relatively long-term (and risky) investments that are unlikely to be made by producers alone. Careful planning and scheduling of harvest and delivery to the processing plant will be needed to support year-round operation of the processing plant (Taylor et al., 1981a).

JOJOBA

A liquid wax with properties remarkably similar to sperm whale oil can be extracted from the seed of jojoba (*Simmondsia chinensis* Link Schneider), a desert shrub native to northwestern Mexico and southwestern United States. During the past few years, a great deal of attention has focused on the commercial potential of this low-water-use shrub as an agricultural crop for the warm, arid and semiarid regions of the United States and other countries.

History

Although American Indians and early settlers in the Southwest used the oil from native jojoba seeds for medical, cosmetic, and other domestic purposes, the commercial potential of the oil was ignored until the 1920s. Chemists then found that jojoba seeds contained a liquid wax, similar to sperm whale oil and unique among plants.

Research on jojoba was a series of starts and stops from 1930 until 1970. When submarine and other warship activity during World War II severely limited the hunting of sperm whales and drastically reduced the availability of sperm oil, interest in jojoba ran high. In 1946, the Research and Marketing Act ordered the USDA Division of Plant Introduction and Exploration to identify new or underutilized sources of industrial raw materials. Jojoba research under this act continued into the 1960s but no commercial activity resulted.

Following the enactment of the Endangered Species Conservation Act in 1969, sperm whales were placed on the protected species list. To discourage the killing of sperm whales, the United States banned the importation of sperm whale oil in 1971. Industry was then forced to search for substitutes for the 25,000 metric tons of sperm oil annually used in the United States (Foster et al., 1980b). Small amounts of jojoba oil began to enter the market from small-scale, low-technology processing of seed collected from wild plants.

In the early 1970s, the Office of Economic Opportunity funded a project to identify the commercial potential of jojoba on Indian reservations in California and Arizona. The program encompassed agronomic research at the University of Arizona and the University of California at Riverside. It also involved seed collection from indigenous jojoba stands on Indian reservations. The seed was processed into hydrogenated wax and sent to more than 75 corporations and research institutions in the United States and overseas for evaluation.

The National Academy of Science (NAS) reported on the results of this evaluation in 1975 (NAS, 1975b). The report stimulated further interest in jojoba in the United States and abroad. By 1977, the international demand for jojoba seed and oil had increased approximately 700% and the price of jojoba oil had increased from $11 to nearly $22/kg (Miller, 1978). Initial plantings of jojoba were established in 1977 and 1978. However, a freeze in the winter of 1978 killed many of the young plants (Foster et al., 1980b).

Since 1979, about 6800 ha of jojoba have been planted in the United States: about 3200 ha in Arizona, 3200 ha in California, and 400 ha in Texas. Major research has been conducted in Israel and Mexico since the early 1960s and, more recently, in Costa Rica, South Africa, and Australia. Preliminary investigations are being conducted in Sudan, India, Jamaica, Haiti, Kenya, Paraguay, and Brazil.

Description

Jojoba is a long-lived, multi-stemmed, dioecious, evergreen shrub which is from 0.6 to 2 m in height and about 3 m in diameter. It develops several tap roots that may penetrate to a depth of 10 m. Jojoba may be propagated by direct seeding, seedling transplant, or rooted cuttings. Flowering begins on wild plants from three to seven years after germination. Pollination typically occurs in March and seeds reach maturity in July and August.

Jojoba produces fruit at alternate nodes on new shoots, commonly with a single seed, 1-2 cm in diameter. In native stands, mature female plants yield from 0.2 to 2.2 kg of dry seed per year. Yields, however, vary greatly from year to year for any given locality, depending on seasonal variations in temperature and rainfall (Gentry, 1958; Yermanos, 1974).

Production

Natural populations of jojoba grow in the Sonoran Desert of Mexico and the United States between longitudes 101° and 117° west, at elevations from sea level on the Sonoran and Baja Coast to over 1500 m in southern Arizona. Mature plants can withstand a low temperature of -9.5 C and a daily summer high of 46 C (Gentry, 1958; Yermanos, 1974; Hogan, 1979). Annual precipitations in these areas range from 7 to 46 cm. Jojoba tolerates saline and alkaline soils.

Although jojoba grows wild in these harsh environments, researchers agree that commercial production requires more favorable conditions (Hogan, 1979; Yermanos, 1979). For example, jojoba should not be planted in areas where temperatures go below freezing. Frost can destroy young seedlings and at -4 C new growth on mature plants is retarded. During the first 3 years of growth, jojoba thrives on about

75 cm of water per year. Productive mature plants need about 50 cm per year. Soils suitable for jojoba must have good drainage, good percolation, and good aeration.

Jojoba can be planted either by direct seeding or by transplanting seedlings or rooted cuttings. Because jojoba is dioecious and because plant sex cannot be determined readily before flowering (about 3 years), plantations have been over planted and then thinned to a 1:6 male to female ratio.

According to American Jojoba Industries of Bakersfield, California, jojoba can be harvested with a blueberry harvester. Two annual harvests are required because seeds do not mature evenly.

Reliable yield data for mature cultivated jojoba shrubs are not yet available. Most researchers believe that yields of 1.4 kg per shrub after ten years of growth are reasonable. This is equivalent to 2800 kg/ha, at 2000 trees/ha.

Processing

Jojoba seeds contain about 50% oil. The oil can be obtained from seed by mechanical pressing and solvent extraction. Mechanical extraction yields a slightly yellow oil and a press cake meal containing about 25% oil. Solvent extraction of the press cake removes the remaining oil.

Oil-free meal contains between 26 and 30% crude protein. The meal is a potential ingredient for animal feed, but because of the presence of simmondsin and other structurally related glycosides it is toxic. Anver Bioscience Design of Sierra Madre, California has devised a method to detoxify the meal, making rations with 10% jojoba meal palatable to mice, chickens, and sheep (Verbiscar et al., 1980).

Consumption

In chemical structure, jojoba seed oil is not a fat but a liquid wax. Sulfurized jojoba oil compares with sulfurized sperm whale oil (a) as an extreme pressure additive in motor oils, gear lubricants, and automatic transmission oils, (b) in hydraulic fluids that need a low coefficient of friction, and (c) for metal cutting and machining and metal forming (NAS, 1975b; Miwa and Rothfus, 1978).

Hydrogenated or bleached jojoba oil is virtually odorless, nongreasy, and immune to rancidity. It is used by the cosmetic industry as a base for shampoo, hair conditioners, lipstick, and creams. Fully hydrogenated jojoba oil has a melting point of 70 C and is very hard. It has applications in car and furniture waxes and polishes (Foster et al., 1980b).

Jojoba may also be beneficial in penicillin and other drug fermentation processes (NAS, 1975a,b). Derivatives of jojoba oil yield products such as surfactants, emulsifiers, disinfectants, and detergents with potential application in polyethylene extrusion, water evaporation retardation, contiguous steel and brass castings, and a variety of other uses (Miwa and Rothfus, 1978).

Jojoba's PMC System Status

Although a small PMC system for jojoba seeds hand-harvested from wild plants, processed in small-scale mechanical extraction facilities, and used by specialty cosmetic companies, is functioning today, as commercial plantings expand and come into production, this system must be greatly enlarged and transformed.

A growing demand for jojoba oil, primarily for the cosmetics industry, has exceeded the supply readily available from harvest of wild plants. The resulting high prices have been the major stimulus for establishment of plantations. Production from plantations already growing in the United States, Mexico, Israel, and elsewhere can be anticipated to exceed the current supply several-fold. Greater competition will drive prices down unless consumption expands significantly.

Jojoba oil is known to be a versatile chemical with many potential uses. Some of these have been translated into markets; many have not. Product versatility and market development are not synonymous. Potential consumers must be identified and sold on the new products and uses developed in the laboratory. Industrial consumers demand large quantities of feedstock of uniform quality and low cost. It is not known whether jojoba oil can be produced and sold at a suitable, high-volume price.

If a high-volume market does develop, processing will probably move to larger scale plants that use solvent extraction, perhaps in combination with mechanical extraction.

Because of economic uncertainties in jojoba's PMC system, conventional sources of financing for producers have been lacking or very limited. Speculative investors may encounter unanticipated capital requirements to purchase expensive harvestors, plant new higher-yielding cultivars, construct seed storage facilities, or to survive periods of market glut, low prices, or freezing temperatures. Producers will find that the registration of herbicides and other pesticides is a time-consuming and expensive process, particularly if end uses for the oil involve the cosmetic industry. Producers, processors, and consumers will need to develop more formal marketing arrangements with a procurement infrastructure and establish grades and standards to maintain high quality.

KENAF

Kenaf (*Hisbiscus cannabinus* L.) is a high-yielding annual fiber crop, adapted to the warmer parts of the United States. Although kenaf is grown for cordage in Thailand, Taiwan, and other countries, current interest in kenaf in the United States centers on its potential for pulping and papermaking.

History

Most authorities agree that kenaf, despite its Persian name, originated in Africa (Murdock, 1959). For most of its early history, kenaf was a subsistence crop, supplying crude fiber for household purposes and leaves and shoots for food. Presently, it is grown commercially as a cordage crop in more than 20 countries, mostly by small growers using hand methods.

Interest in kenaf in the United States has progressed through several stages. In 1943, the USDA and the Cooperative Fiber Commission initiated a program to assess the feasibility of producing, harvesting, and processing the bast fiber of kenaf in southern Florida. This research continued into the late 1960s.

Beginning in 1956, the USDA Agricultural Research Service identified the potential of kenaf as feedstock for the manufacture of paper and as a profitable crop for farmers. By 1967, the studies on pulping kenaf had gained such momentum that the Technical Association of the Pulp and Paper Industry (TAPPI) organized the Ad-hoc Committee on Kenaf and Related Raw Materials, which sponsored the First Conference on Kenaf for Pulp in Gainesville, Florida. In addition, several companies and other research institutions conducted laboratory and pilot-plant pulping investigations (Atchinson and Collins, 1976).

In the late 1970s, the USDA emphasis for kenaf shifted from pulp to biomass as a renewable energy source. Interest in kenaf as a pulp crop continues through support of the American Newspaper Publishers Association (ANPA). ANPA has sponsored a comprehensive study on the feasibility of using kenaf pulp for newsprint, in anticipation that kenaf can help stabilize the rising price of newsprint.

Description

Kenaf is a short-day, annual, herbaceous member of the Malvaceae or mallow family. Some familiar relatives are cotton (*Gossypium* sp.), okra (*H. esculentus* L.), and hollyhock (*Althea* sp.). Roselle (*H. sabdariffa* L.) is closely related (White et al., 1970).

Kenaf plants are mostly unbranched when grown in dense stands and range in height from 2 to 4 m. The bark contains long bast fibers, which are used in some countries for cordage. A small core of weakly dispersed pith cells is surrounded by a thick cylinder of short, woody fibers. The bast fiber constitutes 20–25% by weight (dry basis) of the stem.

Production

Kenaf has been grown at latitudes up to 45°, but satisfactory production requires a long period of high day and night temperatures. Flowering, initiated by short days, curtails vegetative growth so that in temperate regions the effective growing season may be terminated in the fall by the photo-period sensitivity even before the plant is killed by frost. Best yields require a growing season of at least four or five months in which the mean temperature is about 20 C or higher. Satisfactory seed germination and seedling growth require soil temperatures of about 13 C or higher. In the continental United States, these conditions are not satisfied north of about 35°.

Kenaf is adapted to humid areas; it transpires freely and is not drought resistant. Rainfall or irrigation is necessary to balance evapotranspiration. Depending on temperature, this may be 100 mm per month or more. Large, established plants seem to extract moisture efficiently from a thick rooting zone.

Relatively good soil drainage is required. Kenaf does not tolerate wet soil conditions, especially in early stages of growth. Traditionally, it has been grown on the acid soils of humid regions. Recently in the United States, it has been grown, with irrigation, on alkaline soils of arid regions. It has been grown on relatively infertile soils but responds to higher nutrient levels. Fertilizer requirements are comparable to those of corn (*Zea mays* L.). The use of farm machinery limits kenaf production to nearly level or gently sloping soils.

The most serious pest problems for kenaf production is its susceptibility to nematodes, particularly to two root-knot species, *Meloidogyne incognita* and *M. incognita acrita*. Two relatives of kenaf, *H. sabdariffa* L. and *H. radiatus* Cav., possess some tolerance to nematodes and could be used in breeding programs to resolve this problem. In the United States, nematodes have not been identified as a problem in the Southwest. In the Southeast, kenaf would be restricted to nematode-free fields or would require soil treatments to control nematodes.

Kenaf is planted in rows, like corn and cotton, and production costs are comparable. It can be cut with a heavy duty silage cutter or sugarcane harvester. For storage and transportation, compaction of the bulky stalks can be accomplished by baling or cubing.

Yields on a dry weight basis of 15-30 tons/ha and higher can be expected in favorable areas.

Kenaf seed was produced satisfactorily in the United States in 1981.

Processing

Kenaf has substituted for wood as feedstock for existing pulp mills. Kenaf newsprint has been produced using thermomechanical (TMP) and chemical thermomechanical (CTMP) pulping processes.

Kenaf pulp can be used alone or blended with other pulps to produce newsprint or other types of paper. No significant plant or process modifications are anticipated. Press runs on kenaf paper have shown it to be fully as satisfactory as conventional newsprint. Experience to date suggests that in addition to producing high quality newsprint, kenaf may offer significant savings in energy and labor (Hodgson et al., 1980).

For the traditional use of kenaf as a source of cordage fiber, the bast fibers are separated and cleaned by a labor-intensive retting process. Mechanization of this process could open new market possibilities. Bast fibers separated without retting have been used as pulping feedstock for cigarette and other specialty papers.

Consumption

The United States consumption of newsprint in 1980 was about 10.1 million metric tons. About 6 million metric tons were imported, nearly all from Canada. None of this newsprint was produced from kenaf. By the end of 1980, the average price of newsprint was $470/ton, about 183% higher than the average price of $166/ton in 1970 (Taylor, 1981b). With even a slight price advantage over pulpwood and with only a fraction of the total demand for newsprint, kenaf newsprint could achieve a substantial volume of consumption.

Kenaf's PMC System Status

A study of the feasibility of a kenaf-newsprint system in the United States, sponsored by ANPA, identifies three regions where kenaf could be produced and delivered to nearby mills at costs no higher than current prices for pulpwood. Of the 13 pulp and paper mills that produce newsprint in these regions, 6 are located to take advantage of potential kenaf producing areas. Production technology, including seed supply, is ready for the production of as much as 400 ha (perhaps 8000 metric tons) even in 1982. Printing trials indicate that kenaf paper will be accepted readily by publishers. A decision to accept kenaf feedstock for pulping at one or more newsprint mills would initiate a commercial PMC system. New pulping and papermaking facilities especially designed for kenaf can be anticipated, but probably after existing mills have proven that kenaf is satisfactory.

Use of kenaf newsprint will require no modification on the part of publishers. Producers will be motivated to shift from other crops by predictions of higher net incomes. Production practices are comparable to those for corn or cotton. Continuing work is needed to develop better varieties, especially with resistance to damage by nematodes, and varieties that are adapted to more northern regions of the country. Improvements in harvesting and handling equipment can be anticipated when kenaf comes into production. Likewise, herbicides and other pesticides will be approved for use on kenaf. No significant changes in papermaking are required but pulping will require some modifications of equipment to handle the feedstock, temperature, and amount of chemicals used. The greatest need for innovation is in procurement, the link between production areas and pulp mills. A new procurement system to handle a seasonal nonwood raw material and to deliver it to the mill in an appropriate form in a timely fashion will require adjustments by farmers and the pulp industry.

MILKWEED

The two most widely distributed milkweeds in the United States and Canada are the common or eastern milkweed, *Asclepias syriaca* L., and the showy or western milkweed, *A. speciosa* Torr. Both are perennial herbs. Milkweeds are widely adapted to the agroclimatic conditions of the United States and harvest of the whole plant, above ground level, offers a potential source of biomass, fiber, protein animal feed components, crude petroleum substitute, natural rubber, and chemical intermediates.

History

Early interest in milkweed focused on the use of its fibers. North American Indians used the long bast fibers of the stem for cordage, fishing lines, and sewing threads and the short floss fiber from the seed pod for padding. In the 1930s, several attempts were made to commercialize milkweed fiber in Europe under trade names such as ozone fiber, artificial cotton, and cotine (Whiting, 1943).

During World War II, school children throughout the Midwest collected more than 5000 metric tons of milkweed pods from wild plants. The pods were processed at Petosky, Michigan, where the floss was ginned from the seed and hulls. The floss played a vital role as a substitute for kapok in the manufacture of naval life jackets. The by-product seeds were processed for oil and feed meal. The results were so successful that Berkman (1949) suggested the value of the floss, bast fiber, and oilseed justified commercialization of milkweed and advocated research toward that end. He was not successful in obtaining USDA or other governmental support.

The Arab oil embargo revitalized U.S. interest in latex-producing plants. USDA studies on oil- and hydrocarbon-producing plants in 1975 called attention to milkweed as a potential multiuse crop (Buchanan et al., 1978). The Plant Resources Institute in Salt Lake City, Utah, undertook research based on the earlier Utah studies on showy milkweed. Batelle Columbus, USDA, and other groups began studies on common milkweed and other *Asclepias* species.

Description

The common and showy milkweed species, family Asclepiadaceae, are similar in appearance and morphology, but showy milkweed has more stem branching and has fewer, larger flowers in a cluster than common milkweed.

Milkweeds are herbaceous perennials with stout upright stems that grow rapidly to a height of 0.9–1.5 m. Leaves of both species are opposite, broadly lanceolate, or ovate with conspicuous midribs and veins and have gray down or velvet on the underside.

Milkweeds have characteristic and unusual flowers, pollinated by insects. The fruit of both species are plump, rough-textured pods that split open after a frost. Usually only one or two pods are produced per flower cluster. Pods are 1-2 cm long and are filled with about 80 precisely placed, overlapping flat seeds. Each seed is covered with a tuft of over 900 silky hairs.

Production

Both milkweed species are native to North America. They grow rapidly, persist as perennials, and can reach maturity even at high latitudes in subarctic Canada. Milkweeds shed their leaves as the pods ripen in the fall. After frost the stems die; the roots remain dormant until late the following spring, when new plants grow from root buds. The roots enable the plant to survive cold winters and prolonged periods of drought.

Showy milkweed adapts to semiarid conditions but it is not a strict xerophyte. It responds to irrigation, growing rapidly while moisture is available; it also endures prolonged dry spells. It is found near the shores of the Great Salt Lake and apparently tolerates high soil salinity, at least where moisture is adequate.

Common milkweed tolerates summer droughts but apparently adapts poorly to semiarid conditions. For optimum growth, it has about the same moisture requirements as conventional midwestern crops.

Milkweeds do not have exacting soil requirements; they grow on soils ranging from dry and sandy to wet and clayey, including marginal soils unsuited to conventional crops. Optimum growth of common milkweed occurs in fertile well-drained soils suited for corn. Showy milkweed grows well under irrigation on land previously used for sugar beets (*Beta vulgaris* L.).

Both milkweed species are easily propagated by seed or by root cuttings. Once established, milkweed may be harvested from the same roots for many years—up to 15 years for common milkweed in Russian experiments. Along roadsides, it survives late summer mowings and produces flowers and fruits when cut late in the season. The plant density of milkweed increases, particularly during the second and third years, as new shoots grow from adventitious roots.

Processing and Consumption

Although milkweeds have received attention for products such as cellulose, triglyceride seed oil, and natural rubber, current interest centers on the organic-soluble extractives of the whole plant (Buchanan and Duke, 1981). The extractives customarily have been separated into two crude fractions termed polyphenol and whole-plant oil, respectively. The polyphenol fraction is at least sparingly soluble in acetone and freely soluble in 87.5% aqueous ethanol. The whole-plant oil is very soluble in hexane. Polyphenol is considered a potential chemical feedstock; its most likely use is as an extender in the manufacture of phenolic adhesives and resins. Whole-plant oil is considered as a potential substitute for crude petroleum.

If recovery of polyphenol does not prove to be worthwhile, the most promising method to obtain whole-plant oils is by solvent extraction with hexane, using equipment similar to that used for soybean extraction. With this process, common milkweed yields from 5 to 10% oil on a dry weight basis or from 45-90 kg of crude oil per ton of processed milkweed. The crude, hexane soluble extract contains about 30% natural rubber and only a very small portion of polyphenol. Calvin (1980) suggests that this type of product might be 25% more valuable than crude petroleum, because its cracking products are similar to those from naphtha.

Pure rubber can be separated from the crude hexane extract by precipitation in methanol or by other methods. The rubber from milkweed is too low in molecular weight to substitute for natural rubber from other sources for conventional mixing and processing. It is potentially of interest as a plasticizing additive (processing aid) to rubber mixes, for liquid rubber processing methods, for making cements (adhesives), and, if low enough in cost, as a hydrocarbon feedstock.

We foresee little demand for polyphenols from milkweed. Natural rubber (polyisoprenes) is apt to be too expensive to separate. Fortunately, it is not necessary to separate low-molecular-weight rubber from whole-plant oils intended for use as a petroleum refinery or chemical feedstock. Thus, perhaps the plant breeder's goal should be to develop crop varieties high in total content of oil plus low-molecular-weight polyisoprenes (Buchanan et al., 1980a,b).

Milkweed's PMC System Status

Of all the crops discussed in this chapter, milkweed probably is the least advanced in its development continuum. However, it is also one of the most promising potential crops. Research and development programs need to be established to pursue the full range of the PMC systems components, including plant breeding and genetics; agronomic, harvesting, and delivery research; economic feasibility analysis; processing studies; and product development.

CRAMBE

The seed of crambe (*Crambe abyssinica* L.) contains a glyceride oil rich in erucic acid. Industries using erucic acid rely on imported rapeseed oil. However, many producing countries are shifting from industrial to edible oil markets by selecting rapeseed varieties low in erucic acid. This has increased interest in crambe as a potential annual crop for the extraction of an industrial oil high in erucic acid.

History

The first investigations into the use of crambe as an industrial oilseed began in 1932 in Russia. Since then, crambe has been evaluated in many areas of the world, including Germany, Canada, Venezuela, Denmark, Sweden, and Poland. Serious evaluation of crambe in the United States began in 1958 in Texas.

Throughout the 1960s and early 1970s, the USDA Northern Regional Research Center (NRRC) in Peoria, Illinois, and Purdue University conducted research on crambe production, processing, and utilization. Results indicated that crambe could be produced using conventional crop machinery and processed with conventional prepress-solvent extraction processes.

Although research on crambe at NRRC and Purdue University was terminated in the late 1970s, two small research projects are currently underway at Murray State University in Kentucky and at New Mexico State University. These investigations consider crambe as a potential alternative crop for western Kentucky and evaluate crambe's saline and alkaline tolerance and low-water-use potential.

Description

Crambe belongs in the Cruciferae (mustard) family. The plant is an erect, annual herb, which grows to a height of 60–90 cm and matures in 90–100 days (Whiteley and Rinn, 1963). The crambe plant branches abundantly and produces a number of tiny (0.8–2.6 mm diameter) seeds.

Flowering is indeterminate, but pods formed early usually adhere until later ones mature. The pod remains on the seed at harvest and is considered part of the harvested product. As harvested, seed is 30% oil. Dehulling leaves a seed with oil content ranging from 35 to 60% and protein content from 30 to 50% (Papathanassiou and Lessman, 1966).

Production

There are few climatic restrictions to crambe production in the United States. Crambe can be grown as an early season crop in wheat-growing areas of the Pacific Northwest and the Corn Belt. It tolerates cold better than most conventional crops and can be planted in mid- to late-April in the midwest. It has potential for double-crop sequences with soybeans in the lower Mississippi valley and also as a winter crop in Texas, New Mexico, and Arizona. Although crambe responds to irrigation in New Mexico, it is able to withstand moderate drought conditions.

Weed competition can be a devastating production problem. Conventional herbicides have been suggested for weed control, although none have been registered with the EPA for use on crambe.

In experimental plantings, crambe has been relatively free from disease. The most potentially serious disease problems are *Alternaria circinans* and *Fusarium* spp. Crambe also is susceptible to *Marmor brassicae* H. (Thornberry and Phillippe, 1965).

Several insect pests pose severe economic hazards to crambe production, including *Lygus rugulipennis*, *Phyllotera* spp., *Camapanutus caryae*, and *Hylemya brassicae* (White and Higgins, 1966). No specific insecticide has been labeled for use on crambe.

Crambe can be harvested with combines like those used for small grains.

Seed yields on experimental plots have been extremely variable, ranging from 1000 to 2600 kg/ha. Researchers believe that consistent yields of 1600 to 1900 kg/ha can be achieved with existing varieties.

Processing and Consumption

Crambe seed can be processed efficiently by prepress-solvent extraction or direct solvent extraction techniques using conventional equipment (Mustakas et al., 1965).

The resulting crambe oil resembles rapeseed oil in composition but contains higher and more consistent levels of esterified erucic acid (cis-13-docosenoic). Rapeseed oil contains 30-45% erucic acid, while crambe oil contains 55-60% erucic acid. Erucic acid is a long-chain fatty acid with 22 carbon atoms in its molecule. A valuable commercial product, this acid also imparts useful characteristics to the oil. The viscosity, density, and smoke point of crambe oil is higher than that of other domestic vegetable oils.

Refined crambe oil may be used without further processing, or the erucic acid may be extracted by fractional distillation of the free fatty acids obtained by hydrolysis of the glycerides. Alternatively, erucic acid may be obtained by converting the oil to methyl esters then separating methyl erucate by distillation and hydrolyzing it to the free acid (Green et al., 1967).

Refined crambe oil has excellent lubricating qualities and can be used in the continuous casting of steel, as a spinning lubricant in the textile industry, as a rolling oil in the processing of light gauge steel, and as a component in lubricants for metal forming operations. The oil can be blended with natural and synthetic rubbers to yield soft, highly elastic products resistant to light and ozone. Hydrogenation of crambe oil yields a glossy wax with a melting point slightly higher than beeswax.

Derivatives of erucic acid also have commercial applications. Erucamide is considered one of the best additives for extruded polyethylene and polypropylene film. It prevents sticking and improves the finished surface of wax-coated paper. Other useful derivatives of erucic acid are: eruconitrile, used in the manufacture of nylon-13; behenamide, used in water repellants for textiles; and behenyl amine, used in saltwater corrosion prevention.

Brassylic acid is derived by oxidative ozonolysis of erucic acid. It is used in the formulation of nylon-13/13 and in the perfume industry, where it substitutes for natural musk. A by-product of the conversion of erucic acid to brassylic acid is pelargonic acid. It is used in making jet aircraft lubricants, vinyl stabilizers, plasticizers, hydrotropic salts, pharmaceuticals, and insect repellents (Nieschlag and Wolff, 1971).

Crambe meal contains 45-55% protein. The feed value of crambe meal, however, is limited due to the presence of toxic glucosinolates. Detoxified meal could be used as a source of supplemental protein for finishing beef cattle. In early 1981, its use in animal feeds was approved by the Food and Drug Administration. Research has shown no adverse response when cattle are fed 8.5% crambe meal.

Crambe's PMC System Status

The future of crambe in the United States is tied closely with the production status of rapeseed high in erucic acid. The major producers of high-erucic-acid rapeseed are Poland, Sweden, and several other northern European countries. Canada was formerly the major United States supplier of high-erucic rapeseed, but in the past several years has shifted to production of low-erucic rapeseed for use in the edible oil market. There is concern among United States users of erucic acid that northern European countries also will shift to low-erucic-acid rapeseed production and that political turmoil in Poland will disrupt the supply to the United States.

Given the proper price stimulus, crambe commercialization appears feasible in the United States. Twigg (1981) showed that crambe as a double crop with soybeans would be profitable for growers at a price of from \$0.29 to \$0.44/kg in Indiana, Illinois, and Kentucky, assuming a 1900 kg/ha yield. At these prices and, assuming a processing cost of about \$0.04/kg of seed, crambe oil could be produced for about \$1.10-\$1.60/kg (1981 dollars). A market for crambe meal would reduce costs of the oil considerably. Although these prices are slightly higher than present rapeseed oil prices, crambe oil has a higher erucic acid content than rapeseed.

Because of risks, such as insects and diseases, involved with producing crambe, several years of commercial-scale trials would be required before crambe could be considered a reliable crop in the United States.

GRAIN AMARANTH

Interest in grain amaranth (*Amaranthus* spp.) as an annual, drought-resistant crop increased in the 1970s after research reports revealed the nutritional value of the tiny grain. The seed contains about 15% high quality protein (5% lysine and 4% sulfur-containing amino acids) and 63% easily digestible carbohydrates. Health food and cereal food industries are interested in grain amaranth for use in breakfast cereals, granola, pablum, or as an ingredient in baked goods or confections.

History

Five centuries ago, grain amaranth was a staple food in Aztec and other Mexican Indian diets. By Cortez's time, amaranth was a crop of equal importance with maize. Amaranth also was used in religious festivals, for which the popped grain was mixed with honey and human blood. The resulting product was shaped into religious idols and eaten. Because of its association with pagan rituals, the Spaniards tried to eliminate the cultivation of grain amaranth. Small plots of amaranth are still cultivated in Mexico, where the grain is mixed with honey and eaten as a candy called alegria (Cole, 1979).

Dr. John Robson, a nutritionist at the University of Michigan, while searching for nutritionally balanced foods, became interested in the potential of grain amaranth. In 1971, Robert Rodale of Rodale Press initiated yield and laboratory studies at the Rodale Research Center (RRC) near Kutztown, Pennsylvania. Since then, RRC plant breeders and scientists have collected amaranth germplasm in Latin America and India and used the material in a comprehensive breeding program (Hemmendinger and Laidig, 1981).

Some public agronomic research on amaranth has been conducted at Iowa State University, Pennsylvania State University, Cornell University, North Dakota State University, the University of California at Davis, and the University of Puerto Rico.

Health Valley Natural Foods of Montebello, California, introduced an amaranth cereal onto the market in 1981, using grain obtained from Mexico and India.

Description

"Grain amaranth" is a name commonly used for certain lines of at least three species of the family Amaranthaceae, viz. *A. hypochondriacus* A., *A. caudatus* L., and *A. cruentus* L. All are annual herbaceous plants with a drought-tolerant taproot and a broad leaf canopy that effectively suppresses weed competition.

In most lines, the grain-like seeds are located in large terminal and lateral inflorescences. The small, ovular seeds range in color from white to black, but the pale seeds are considered best in terms of appearance, flavor, and popping capability (NAS, 1975b). Grain amaranth is a vigorous, fast growing plant that possesses the enhanced photosynthetic efficiency of the C-4 carbon fixation pathway.

Production

Amaranth is a warm-season, temperate zone crop requiring soil temperatures of at least 18 C for optimal seed germination. The small seed requires a favorable seed bed, moist but well aerated. The crop is hardy and drought tolerant and grows in soils that are acid, high in aluminum, or saline (Edwards, 1981). Some amaranth varieties show little day-length sensitivity. Amaranth grows in the northern United States where short growing seasons often constrain production of other crops. Amaranth also has the potential to become an alternative crop for water-short areas of the Southwest.

In 1980 at RRC, infestations of *Lygus lineolaris* caused extensive seed damage. Other insects, birds, and diseases have been identified as potential production problems (Edwards, 1980). Because of amaranth's slow emergence after seeding, weed competition can become critical. No herbicides have been approved for use on amaranth. The crop can be seeded in rows and cultivated after emergence to control weeds.

Combines can be adjusted to harvest the small amaranth seeds. Yields ranging from 500 to 2400 kg/ha have been achieved in the United States.

Processing and Consumption

Grain amaranth can be milled, toasted, or popped using established grain processing techniques and equipment. According to Rodale's Test cereal grain blends to improve the amino acid balance of products such as commercial breakfast cereals, flour, cake mixes, and bakery products. Popped amaranth seeds can be used in snack bars, breadings, and toppings. Amaranth flour can substitute for a portion of wheat flour in home baking, and amaranth seeds can be used in home cooked cereals and pilaf.

Grain Amaranth's PMC System Status

Although there is a great deal of interest in grain amaranth as a high-quality protein grain crop, food companies appear apprehensive about undertaking major product development efforts without some assurances of dependable supplies of amaranth. The major constraints to amaranth's commercialization thus appear to be in the production subsystem (Hemmendinger and Laidig, 1981).

Rodale Press conducts plant breeding and selection research in Pennsylvania. Additional research is required to develop large-seeded cultivars with even grain maturity, consistent height for mechanization, and resistance to pests and diseases.

Research into cultural practices have been concerned mainly with plant density studies (Edwards, 1980). Research is limited or entirely lacking on fertility response, chemical weed and other pest control, response to irrigation, and the effects of grain amaranth on different cropping systems.

Grain amaranth field studies have been conducted under varying climatic regimes in Pennsylvania, California, North Dakota, Iowa, and Puerto Rico. Additional studies are needed to determine the adaptability of grain amaranth cultivars to different environmental conditions.

Applied engineering work is needed to solve planting and harvesting constraints that result from the small seed size and light weight of amaranth grain. Specific harvesting difficulties that need to be overcome include lodging and seed shattering. Research is also needed to determine methods of seed drying, cleaning, and storage. Additionally, although a limited amount of utilization research has been undertaken, more is needed.

OTHER NEW CROP OPPORTUNITIES

In addition to the potential crops discussed above, other underexploited and potential new crops offer exciting opportunities for commercialization (Table 1). Many of these crops will require development of new cultivars that are superior to any currently available. Commercialization will require development of crop production technology, and establishment of new crop marketing infrastructures and product and market development activities. Some of the industrial crops will require new processing technologies.

Table 1. Undeveloped Crop Opportunities

POTENTIAL CROP	USE	STATUS	REFERENCE
Bladderpod (*Lesquerella fendleri* S. Wats.)	Industrial oilseed, hydroxy acids	Screening selection, no cultivars	Theisen et al., 1978
Cold-hardy citrus, *Poncirus trifoliata* L.	Edible fruit	Research, existing cultivars not acceptable	Theisen et al., 1978
Euphorbia (*Euphorbia lathyris* L. and other species)	Terpenoid oil	Research on *E. lathyris*	Calvin, 1980
Goldenrod (*Solidago leavenworthii* L., *S. altissima* L.)	Rubber and resin	Screening selection	Buchanan et al., 1978
Jerusalem artichoke (*Helianthus tuberosus* L.)	Fermentable carbohydrate, biomass	Garden crop, new research interest	Chubey & Dorrell, 1974
Meadowfoam (*Limnanthes alba* R. Br.)	Industrial oilseed, long-chain acids	Research, one cultivar	Theisen et al., 1978
Mesquite (*Prosopis glandulosa* Torr. and other species)	Biomass, feed and forage, arid-land reclamation	Research	Felker, 1981
Mung bean (*Vigna radiata* L.)	Edible seed	6–9 million kg/year U.S. consumption, mainly imported	Theisen et al., 1978
Persimmon (*Diospyros virginiana* L., *D. kaki* L.)	Edible fruit	Oriental persimmon is a specialty crop, several cultivars of each species	Theisen et al., 1978
Pigeon pea (*Cajanus indicus* K. Spreng.)	Food pulse	Important world crop, no U.S. cultivars, imported for U.S. market.	Theisen et al., 1978

POTENTIAL CROP	USE	STATUS	REFERENCE
Quinoa (*Chenopodium quinoa* Willd.)	Food grain	Subsistence crop in South America	Theisen et al., 1978
Rabbitbrush (*Chrysothamus nauseousus* Pall.)	Rubber and resin, biomass	Screening selection, beginning research	Buchanan et al., 1978
Sesame (*Sesamum indicum* L.)	Food oilseed	U.S. production about 8000 tons/year, imports about 26,000 tons/year	Theisen et al., 1978
Stokes aster (*Stokesia laevis* Greene.)	Industrial oilseed, epoxy-acids	Research, genetic selection	Theisen et al., 1978
Sumac (*Rhus glabra* L., *R. typhina* L.)	Tannin, terpenoid oil	Screening selection, beginning research	Buchanan et al., 1978
Chinese tallow tree (*Sapium sebiferum* Roxb.)	Industrial oils, bio-mass	Research interest	Scheld & Cawles, 1981

KEY REFERENCES

Buchanan, R.A., Otey, F.H., and Hamerstand, G.E. 1980b. Multi-use botanochemical crops: An economic analysis and feasibility study. Ind. Eng. Chem. Prod. Res. and Develop. 19:489-496.

Cole, J.N. 1979. Amaranth, From the Past For the Future. Rodale Press, Emmaus, Pennsylvania.

Knox, E.G. and Theisen, A.A., eds. 1981. Feasibility of Introducing New Crops: Production-Marketing-Consumption (PMC) Systems. NSF/PCM-81004. Rodale Press, Emmaus, Pennsylvania and NTIS, Washington, D.C.

National Academy of Science. 1975a. Products from Jojoba: A Promising New Crop for Arid Lands. NAS, Washington, D.C.

______ 1977. Guayule: An Alternative Source of Natural Rubber. NAS, Washington, D.C.

White, G.A. and Higgins, J.J. 1966. Culture of Crambe: A New Industrial Oilseed Crop. Agr. Res. Ser., USDA/ARS Production Research Report 95, Washington, D.C.

______, Cummins, D.G., Whiteley, E.L., Fike, W.T., Greig, J.K., Martin, J.A., Killinger, G.B., Higgins, J.J., and Clark, T.F. 1970. Culture and harvesting methods for kenaf. USDA Production Research Report 113, Washington, D.C.

REFERENCES

Atchinson, J.E. and Collins, T.T. 1976. Worldwide developments in kenaf. In: TAPPI Nonwood Plant Fiber for Pulp Committee Progress Report No. 8, pp. 75-80.

Berkman, L. 1949. Milkweed, a war strategic material and a potential industrial crop for sub-marginal lands in the United States. Econ. Bot. 3:223-230.

Buchanan, R.A. and Duke, J.A. 1981. Botanochemical crops. In: Handbook of Biosolar Resources, Vol. II (T.A. McClure and E.S. Lipinsky, eds.). CRC Press, Boca Raton, Florida.

Buchanan, R.A., Cull, I.M., Otey, F.H., and Russell, C.R. 1978. Hydrocarbon- and rubber-producing crops: Evaluation of U.S. plant species. Econ. Bot. 32:131-153.

Buchanan, R.A., Otey, F.H., and Bagby, M.D. 1980a. Botanochemicals. In: Recent Advances in Phytochemistry, Vol. 14, The Resource Potential in Phytochemistry (T.S. Swain and R. Kleinman, eds.) pp. 1-22. Plenum Press, New York.

Calvin, M. 1980. Hydrocarbons from plants: Analytical methods and observations. Naturwissenschaften 67:525-532.

Chubey, B.B. and Dorrell, D.G. 1974. Jerusalem Artichoke: A potential fructose crop for the prairies. Can. Inst. Food Sci. Technol. J. 7:98-100.

Edwards, A. 1980. Grain amaranth: Characteristics and culture. Rodale Research Report No. 81-86, Emmaus, Pennsylvania.
Felker, F. 1981. Use of tree legumes in semiarid regions. Econ. Bot. 35:174-186.
Foster, K., McGinnes, W., Taylor, J., Mills, J., Wilkinson, R., Lawless, F., Maloney, J., and Wyatt, R. 1980a. A Technology Assessment of Guayule Rubber Commercialization: Final Report for NSF Division of Policy Research and Analysis (PRA78-04295). Office of Arid Lands Studies, Tucson, Arizona and Midwest Research Institute, Kansas City, Missouri.
Foster, K., Anderson, L., Brooks, W., Kassander, H., Mortenson, B., Rawles, R., Sherbrooke, W., Turner, S., Wright, N., Kelso, G., Lawless, E., and Tinberg, C. 1980b. Development of an Indian Reservation-Based Jojoba Industry: Final Report to NSF (ISP77-04295). Office of Arid Lands Studies, Tucson, Arizona and Midwest Research Institute, Kansas City, Missouri.
Gentry, H.S. 1958. The natural history of jojoba (*Simmondsia chinensis*) and its cultural aspects. Econ. Bot. 12:261-295.
Green, J.L., Huffman, E.L., Burks, R.E., and Sheehan, W.C. 1967. Nylon 13/13: Synthesis and polymerization of monomers. J. Polym. Sci. 5:391-394.
Hemmendinger, A. and Laidig, G. 1981. Grain amaranth. In: Feasibility of Introducing New Crops: Production-Marketing-Consumption (PMC) Systems (E.G. Knox and A.A. Theisen, eds.) pp. 135-149, NSF/PCM-81004. Rodale Press, Emmaus, Pennsylvania and NTIS, Washington, D.C.
Hodgson, P., Lawford, W., Perrault, J., Thompson, C., and Hardy, M. 1980. ANPA, International Sales Kenaf Newsprint Project. Proceedings TAPPI 1980 Pulping Conference, Atlanta, pp. 241-247.
Hogan, L. 1979. Jojoba, a new crop for arid regions. In: New Agricultural Crops (G.A. Ritchie, ed.) pp. 177-202. Westview Press, Boulder, Colorado.
Miller, W.P. 1978. Markets, economics, and future growth of a jojoba agroindustry: An overview. In: Proceedings of the Third International Conference on Jojoba (D.M. Yermanos, ed.) pp. 243-247. University of California, Riverside.
Miwa, T.K. and Rothfus, J.A. 1978. In-depth comparison of sulfurized jojoba and sperm whale oils as extreme pressure/extreme temperature lubricants. In: Proceedings of the Third International Conference on Jojoba (D.M. Yermanos, ed.) pp. 243-247. University of California, Riverside.
Murdock, G.P. 1959. Africa: Its People and Their Cultural History. Macmillan Pub., New York.
Mustakes, G., Kopas, K., and Robinson, N. 1965. Prepress solvent extraction of crambe: First commercial trial run of new seed oil. J. Am. Oil Chem. Soc. 45:1-5.
National Academy of Science. 1975b. Underexploited Tropical Plants With Primary Economic Value. NAS, Washington, D.C.
Nieschlag, H.J. and Wolff, I.A. 1971. Industrial uses of high erucic acid oils. J. Am. Oil Chem. Soc. 48:723-727.

Papathanassiou, G.A. and Lessman, K.J. 1966. Crambe. Purdue Univ. Agric. Exp. Sta. Res. Bull. No. 819, Purdue, Indiana.

Scheld, H.W. and Cawles, J.R. 1981. Woody biomass potential of the Chinese tallow tree. Econ. Bot. 35:391-397.

Taylor, C.S. 1981. Kenaf. In: Feasibility of Introducting New Crops: Production-Marketing-Consumption (PMC) Systems (E.G. Knox and A.A. Theisen, eds.) pp. 22-72, NSF/PCM-81004. Rodale Press, Emmaus, Pennsylvania and NTIS, Washington, D.C.

______, Laidig, G.L., and Blase, M.G. 1981. Applying a systems approach to the introduction of new crops. Petroculture 2:22-25.

Theisen, A.A., Knox, E.G., Mann, F.L., and Sprague, H.B. 1978. Feasibility of Introducing Food Crops Better Adapted to Environmental Stress. NSF/RA-780289. Goverment Printing Office, Washington, D.C.

Thornberry, H.H. and Phillippe, M.R. 1965. Crambe: Susceptibility to some plant viruses. Plant Dis. Rep. 49:704-708.

Twigg, J.W. 1981. Crambe. In: Feasibility of Introducting New Crops: Production-Marketing-Consumption (PMC) Systems (E.G. Knox and A.A. Theisen, eds.) pp. 150-172, NSF/PCM-81004. Rodale Press, Emmaus, Pennsylvania and NTIS, Washington, D.C.

Verbiscar, A., Banigan, T., Weber, C., Reid, B., Trei, J., Nelson, E. Raffanf, R., and Kosersky, D. 1980. Detoxification of jojoba meal. J. Agr. Food Chem. 28:181-197.

Weihe, D.L. and Nivert, J.J. 1980. Assessing guayule's potential: An update of Firestone's developmental program and economic studies. Paper presented at the Third International Guayule Conference. Pasadena, California and Firestone Tire and Rubber Co., Akron, Ohio.

Whiteley, E.L. and Rinn, C.A. 1963. *Crambe abyssinica*: A potential new crop for the Blackland. Tex. Agric. Prog. 95:23-24.

Whiting, A.G. 1943. A summary of the literature on milkweeds (*Asclepias* spp.) and their utilization. USDA Bibliographic Bulletin No. 2.

Yermanos, D.M. 1974. Agronomic survey of jojoba in California. Econ. Bot. 28:160-174.

______ 1979. Jojoba: A crop whose time has come. Calif. Agric. 33: 4-7,10,11.

CHAPTER 3

Development of New Varieties via Anther Culture

H. Hu and *J. Z. Zeng*

Since Guha and Maheshwari (1964, 1966) first reported the direct development of haploid embryos from microspores of *Datura innoxia* by anther culture in vitro, many researchers have been interested in the induction of androgenetic haploids. The simplicity and ease with which haploids can be produced in large numbers by this technique has stimulated a great deal of activity in the production and utilization of haploids for fundamental and applied genetics. The anther culture technique has been refined and extended to 171 species belonging to 60 genera and 26 families of angiosperms. Researchers in China began work in this field in 1970. The first pollen-derived plants obtained from cultured anthers in China were rice and wheat. Subsequently, studies on anther culture were also carried out on many species such as maize, rye, Triticale, tobacco, cotton, soybean, rapeseed, rubber tree, Chinese cabbage, and pepper. Induction of pollen-derived plants of many species was first successfully achieved in China. These species are listed in Table 1.

It has been suggested that homozygous dihaploid plants obtained by anther culture could shorten the number of generations needed to develop new varieties. At the present time, there are some problems in the use of haploids, such as: (a) regenerated plants cannot be obtained in some economically important plants, e.g., cotton; (b) the induction frequency of pollen-derived plants in some important crops, e.g., soybean and maize, is low; and (c) some pollen-derived plants possess characteristics with genetic and chromosomal instability. These problems are continually being researched by scientists.

Table 1. Species from which Pollen-Derived Plants Were First Obtained in China

YEAR	SPECIES	REFERENCE
1971	*Capsicum annuum* L.	Wang et al., 1973
	Triticale	Sun et al., 1973
	Triticum aestivum	Ouyang et al., 1973
1973	*Beta vulgaris* L. (2n = 4x = 36)	Institute of Sugarbeet and Sugar Crops, 1978
	Brassica pekinensis	Teng and Kuo, 1978
	Triticum aestivum x *Agropyron glaucum*	Wang et al., 1975b, 1980
1975	*Populus canadenia* x *P. koreana*	North Eastern Forestry Academy, 1977
	P. harbinesis x *P. pyramidalis*	North Eastern Forestry Academy, 1977
	P. nigra	Wang et al., 1975a
	P. simonii x *P. nigra*	Heilungkiang Institute of Forestry, 1975
	P. ussuriensis	Heilungkiang Institute of Forestry, 1975
	Zea mays	Institute of Genetics, 1975
1977	*Brassica chinensis* L.	Chung, 1978
	Hevea brasiliensis	Chen et al., 1978
1978	*P. alba* x *P. simonii*	Chang, 1978
	P. berolinesis	Lu et al., 1979
	P. berolinesis x *P. pyramidalis*	Lu et al., 1979
	P. pekinensis	Lu et al., 1979
	P. pseudo-simonii	Lu et al., 1979
	P. pseudo-simonii x *P. pyramidalis*	Lu et al., 1979
	P. simonii	Lu et al., 1979
	P. simonii x *P. pyramidalis*	Lu et al., 1979
	Rehmannia glutinosa	Chinese Medicine Laboratory, 1978
	Sorghum vulgare	Zhou, 1978
1979	*Beta vulgaris* (2n = 2x = 18)	Shao, 1979
	Citrus microcarpa	Z.G. Chen et al., 1980
	Linum usitatissimum	Sun, 1979
	Medicago denticulata	Xu Su, 1979
	Populus euphratica	Zhu et al., 1980
	P. euramericana	Zhu et al., 1980
	Saccharum sinensis	Z.H. Chen et al., 1979
1980	*Coix lacryma*	Wang et al., 1980
	Glycine max	Yin et al., 1980
1981	*Fragaria orientalis*	Xue et al., 1981
	Lycium barbarum	Gu, 1981
	Vitis vinifera	Zhou and Li, 1981

FACTORS INFLUENCING THE RESPONSE OF CULTURED ANTHERS

The following factors affecting the induction of pollen plantlets have been examined: (a) the appropriate developmental stage of the microspore, (b) the composition of the culture media, and (c) the culture conditions.

Selecting Anthers at the Appropriate Developmental Stage for Inoculation

Although tetrads and mature pollen can develop into pollen plants (Bouharmont, 1977; Kameya and Hinata, 1970; Nakata and Tanaka, 1968), the best yield of pollen plants cannot be obtained at every developmental stage of pollen. For example, in wheat the pollen-derived callus could be induced in cultured anthers inoculated at stages from pollen mother cell to mature pollen. When anthers containing mid- or late-uninucleate microspores were inoculated, the callus induction frequency was the highest (He and Ouyang, 1980). Many other species are also like wheat (Table 2).

Table 2. Appropriate Pollen Phase for Inoculation in Several Cereal Crops

SPECIES	DEVELOPMENT PHASE OF POLLEN	REFERENCE
Coix lacryma	Uninucleate	Wang et al., 1980
Hordeum vulgare	Uninucleate or late uninucleate	Clapham, 1971; Zhou & Yang, 1980
Lolium	Uninucleate	Clapham, 1971
Oryza sativa ssp. *keng*	Mid or late uninucleate	Guha et al., 1970; Institute of Genetics, 1974; Wang et al., 1973
O. sativa ssp. *shien*	Unincleate or late uninucleate	Guha-Mukherjee, 1973; Kwangtung Institute of Botany, 1976
Secale cereale	Late uninucleate	Sun, 1978
Triticale	Uninucleate	Sun, 1973
Triticum aestivum	Mid- or late uninucleate	Ouyang et al., 1973; Pan & Gao, 1980
T. aestivum x *Agropyron glaucum*	Late uninucleate	Wang et al., 1973

Improvement of Culture Media

The culture medium is the main environmental factor for induction and control of the development of intact plantlets from pollen grains.

ELEVATED SUCROSE CONCENTRATION. We noticed in early experiments that the concentration of sucrose was an important factor for induction of pollen plants. When the concentration of sucrose was raised from 0.088 to 0.18 M, pollen plants were obtained in wheat (Ouyang et al., 1973). Sucrose not only regulates osmotic pressure of the medium, but it also may be the most effective carbohydrate source. Based on preliminary observations, researchers took notice of the concentration of sucrose. Up to now, it has been defined that 0.26 M sucrose is optimal for anther culture of wheat and 0.35 M for maize (Institute of Genetics, 1975; Ku et al., 1978; Miao et al., 1978; Department of Biology, 1977). In other species, higher levels of sucrose have also been adopted; for example, in rapeseed, *Triticale*, and barley, sucrose was raised from 0.18 to 0.35 M (Clapham, 1973; Chu et al., 1975; Institute of Genetics, 1977b).

REGULATION OF THE RATIO OF AMMONIUM NITROGEN TO NITRATE NITROGEN. Clapham (1973) and Chu et al. (1975) found that high concentrations of ammonium ions inhibit formation of pollen callus in barley and rice. Based on this result, the N_6 medium with reduced concentration of ammonium ions and with altered ammonium to nitrate nitrogen ratios was developed (Table 3). It has been demonstrated that N_6 medium is more efficient than other synthetic media for anther culture of rice and other cereals (Chu et al., 1978; Chung, 1978; Kwangtung Institute of Botany, 1976; Department of Biology, 1977; Sun et al., 1980; Table 4).

POTATO MEDIUM. Natural extracts derived from potato, sweet potato, yam, tomato, sprouted wheat, rice endosperm, and maize in the grain-filling stage were tested as supplements to synthetic culture media. Natural extracts have two merits—high induction frequency and simplicity of preparation. Of the above natural extracts, potato aqueous extract has attracted significant attention. Potato aqueous extract was first used in anther culture of tobacco (Institute of Tobacco, 1974a). Subsequently, it has been used for anther cultures of rice to replace the principal components of Miller's medium (Kwangtung Institute of Botany, 1975; Institute of Genetics, 1977b). Both of these potato media had approximately the same effects on the induction of pollen embryos or pollen callus as their corresponding synthetic media. But since the contents of the major salts of potato tubers vary widely, the inductive effects of potato medium are not always stable. Recently, potato media was further improved by Chuang et al. (1978). The new potato medium, which is called potato II medium, contains a 10% aqueous potato extract, half-strength macroelements of WH medium, and the Fe salts and thiamine as in MS medium (Table 3). The frequency of callus induction using this medium was much higher than with synthetic media (Table 5).

Improved Culture Conditions

The physical factors of culture such as temperature and light have an influence on the frequency of induction of pollen callus.

Table 3. Components of N_6 and Potato II Media

COMPONENT	N_6 MEDIUM (mM)	POTATO II MEDIUM (mM)[a]
Aqueous potato medium	—	10%
KNO_3	28.0	9.9
$Ca(NO_3)\cdot4H_2O$	—	0.42
$MgSO_4\cdot7H_2O$	1.6	0.51
$(NH_4)_2SO_4$	3.5	0.75
KH_2PO_4	2.0	1.5
KCl	—	0.5
$CaCl_2\cdot2H_2O$	1.13	—
$FeSO_4\cdot7H_2O$	0.1	0.1
$Na_2EDTA\cdot2H_2O$	0.11	0.11
$MnSO_4\cdot H_2O$	0.2	—
$ZnSO_4\cdot7H_2O$	5.2 μM	—
H_3BO_3	0.26	—
KI	4.8 μM	—
Thiamine·HCl	3.0 μM	3.0 μM
Glycine	0.027	—
Pyridoxine·HCl	2.4 μM	—
Nicotinic acid	4.1 μM	—
Sucrose	0.26 M	0.26 M
pH	5.8	5.8

[a]For anther culture of wheat.

TEMPERATURE DURING CULTURE. When culture temperature was elevated, the frequency of callus formation of wheat and rapeseed increased (Genovesi and Magill, 1979; Ho et al., 1978; Keller and Armstrong, 1978). For example, in recent years we found that if wheat anthers were incubated at elevated temperature for a few days prior to transfer to normal temperature, the frequency of callus increased. Yields of green plants have also been observed to increase (Hu et al., 1982; Table 6). When anthers were cultured at 33 C for 8 days and then transferred to 24 C, the frequency of callus increased along with the frequency of differentiating green plantlets (Table 6). From these results it is evident that the physiological condition of the donor plant has an influence on the frequency of callus induction. The frequency of differentiated plantlets from anthers of the main spike is higher than from tillering spikes.

COLD PRETREATMENT. Cold pretreatment before inoculation may increase the frequency of callus induction. The frequency increased about two times when wheat anthers were pretreated at 1-4 C for 48 hr (Institute of Genetics, 1977a). In rice and other cereal crops, the effect of cold pretreatment has also been favorable, so this treatment

Table 4. Application of N_6 Medium to Anther Culture of Cereals (Chu et al., 1975)

AIM OF CULTURE	STAGE OF POLLEN	SUCROSE (M)	SUPPLEMENTS (μM)	
			Essential	Useful
Induction of Pollen Callus				
Oryza sativa	Middle or late uninuclear	0.15	9.0 2,4-D	5.4–11.0 NAA, 300–500 mg/l LH[a]
Triticum, Triticale, Secale	Middle uninuclear	0.23–0.29	9.0 2,4-D	1.4–2.3 KIN, 500 mg/l LH, 0.55 mM myo-inositol, 0.4 vitamin B
Zea mays	Middle uninuclear	0.35–0.44	9.0 2,4-D	2.3 KIN, 500 mg/l CH or LH, 0.5% active carbon
Shoot Differentiation from Callus				
Oryza sativa	Middle or late uninuclear	0.15	0	2.9 IAA or 1.1 NAA, 2.3–4.6 KIN
Triticum, Triticale, Secale	Middle uninuclear	0.23	0	2.9 IAA or 1.1 NAA, 2.3–4.6 KIN
Zea mays	Middle uninuclear	0.35	0	2.9 IAA or 1.1 NAA, 2.3–4.6 KIN
Induction of Pollen Embryo and Plantlet				
Oryza sativa	Middle or late uninuclear	0.088	0	—
Triticum, Triticale, Secale	Middle uninuclear	0.23	0	—
Zea mays	Middle uninuclear	0.35	0	—

[a]LH = Lactalbumin hydrolysate.

Table 5. Effect of Different Media on Induction of Pollen-derived Plants of Wheat Variety Orefen

MEDIUM	NO. ANTHERS INOCULATED	CALLI		PLANTLETS	
		No.	Percent	No.	Percent
N_6	630	48	7.6	8	1.3
Potato II (Sichuan potato)	630	149	23.6	27	4.3
Potato II (Ke-1 potato)	630	162	25.7	32	5.1
W_5	630	46	7.3	10	1.6

Table 6. Effect of Incubation Conditions and Physiological State of Donor Plants on Induction of Pollen-derived Plants of Wheat Variety Orefen

INCUBATION TEMPERATURE[a]	NO. ANTHERS INOCULATED	CALLI		PLANTLETS	
		No.	Percent	No.	Percent
Main Spikes					
33 C	1186	715	60.3	194	16.4
24 C	1186	516	43.5	92	7.8
Tillering Spikes					
33 C	1092	451	41.3	156	14.3
24 C	1134	240	21.2	33	2.9

[a]Incubated for 8 days after inoculation.

has been widely applied to many species (Chu, 1982). The proper temperature and duration of cold pretreatment for several cereal crops is summarized in Table 7.

CHARACTERISTICS OF POLLEN-DERIVED PLANTS

Stability of Homozygous Pollen Plants

Homozygous diploids can be produced by doubling the chromosomes of haploids, thereby producing stable strains. In early investigations of anther culture, a great deal of effort was directed to ascertain if

Table 7. Cold Pretreatment Procedure for Flower Buds

SPECIES	TEMPERATURE	DURATION	REFERENCE
Hordeum vulgare	3 C	48 hr	Bouharmont, 1977
Oryza sativa	10 C	48 hr	Wang et al., 1974
	10 C	4-7 days	Chen et al., 1979
Secale cereale	6 C	3-15 days	Nitzsche & Wenzel, 1977
Triticale	3-5 C	72 hr	Sun et al., 1980
Triticum aestivum	3-5 C	48 hr	Pan, 1975
	4 C	48 hr	Institute of Genetics, 1977

Table 8. Analysis of Characteristics of Pollen-derived Lines (H_3)[a] and their Parents in Kotung 58 x 7479

CHARACTER	Kotung 58 (n = 20)	7479 (n = 20)	Kotung 58 x 7479 (H_3; n = 34)
Plant height (cm)	87.00 ± 3.69	95.15 ± 4.25	96.00 ± 3.00
C.V.[b] (%)	4.24	4.47	4.06
Ear length (cm)	8.42 ± 0.40	8.63 ± 0.39	8.52 ± 0.37
C.V. (%)	4.75	4.52	4.34
Seed weight, 1000 seeds (g)	50.90 ± 3.15	45.02 ± 3.54	55.22 ± 2.69
C.V. (%)	6.19	7.86	4.87

[a]H_3 indicates the third generation of pollen-derived lines from the culture of F_1 hybrid anthers.

[b]C.V. = coefficient of variation.

pollen-derived strains using F_1 hybrids and their parent varieties were genetically stable for tobacco (Institute of Tobacco, 1974b), rice (Hsu et al., 1975; Institute of Genetics, 1976), and wheat (Ouyang et al., 1973). Three characters of pollen-derived lines (H_3) and the parental lines of a wheat cross were analyzed (Table 8). The coefficient of variation of all three characters of pollen-derived lines (H_3) is less than or approximately equal to that of the parents. This is consistent with the prediction that characteristics of pollen-derived plants are uniform and similar to stable inbred varieties. Their haploids do not appear to segregate. In recent years, analysis of genetic characteristics and cytological investigations were further completed with natural populations of pollen-derived strains.

Results obtained from unselected populations of pollen-derived strains of wheat (Hu et al., 1979), rice (Chen et al., 1978), corn (Gu et al., 1980), and tobacco (Xu et al., 1980) during several years indicate that about 90% of the diploid lines had uniform genetic expression. When the calculated data of 444 pollen-derived strains of wheat is compiled (Table 9), it can be seen that about 90% of the strains were completely homozygous and about 10% of the strains segregated for some characteristics.

Table 9. Uniformity in the H_2 Generation of Pollen-Derived Wheat Plants

CHARACTER	NUMBER
Total Strains	444
Uniform (89.2%)	396
Segregating (10.8%)	48
Segregating Characteristics	
Plant height	19
Spike shape	10
Awnness	2
Grain color	5
Fertility	6
Springness or winterness	6

In addition, a cytological investigation of root tip chromosomes from 54 plants (H_1) was completed (Hu et al., 1978). Using the basic chromosome number, pollen-derived plants may be classified into five types (Table 10); approximately 88.9% of the regenerated plants were haploids and homozygous diploids.

In recent years, the cytological investigation of PMC from 72 pollen-derived plants (H_1) was also carried out (Hu et al., 1978). Table 11 indicates that 87.5% of the regenerated plants were 3x and 6x plants and 12.5% were variant plants. These studies verified that a high frequency of haploids and diploids as well as heteroploids could be obtained with anther culture methods.

Among the 44 3x plants observed, several PMCs contained 21 univalents (Fig. 1a). However, in some cells, two kinds of chromosome configurations, 19I + 1II (19 univalents and 1 bivalent) and 17I + 2II were more frequently observed. All 19 pollen-derived 6x plants contained only bivalents (Fig. 1b) indicating the identical, stable chromosome configuration as their parents.

Table 10. Root-Tip Chromosome Numbers of Pollen-Derived Wheat Plants

PLOIDY LEVEL	NUMBER	PERCENT
Haploid (3x)	38	70.4
Diploid (6x)	10	18.5
5x	2	3.7
Mixoploid	3	5.6
Anther Plant	1	1.8
Total	54	100.0

Table 11. PMC Chromosome Counts of Pollen-Derived Wheat Plants

PLOIDY LEVEL	NUMBER	PERCENT
Haploid (3x)	44	61.1
Diploid (6x)	19	26.4
8x	1	1.4
Nullisomics (6x-2)	2	2.8
Mixoploid	6	8.3
Total	72	100.0

Using the corn varieties Ba Tang Bai and Qun Dan 105 as materials, Cao et al. (1981) and Wu et al. (1980) obtained maize clones from pollen using anther culture techniques. Within the last 3 years, Gu et al. (1980) subcultured the clones derived from var. Ba Tang Bai every 4 weeks onto new culture medium. According to the degree of totipotency observed during this period, they can be classified into four clones. Callus of clone No. 1 possessed a great ability to differentiate and regenerate, and green plantlets were easily regenerated. Gu and Zhang (1981) made cytological studies on callus of clone No. 1 and root tip cells of 35 pollen plants regenerated from the clone. About 90% of regenerated plants were haploids (Table 12). The karyotype analysis of No. 1 clone and plants regenerated from this clone were carried out with Giemsa banding technique (Fig. 1c–f). Results indicated that chromosome C-banding patterns of the parent and the regenerated plants were similar. These results further indicated that anther culture methods can be used to recover genetically uniform plants.

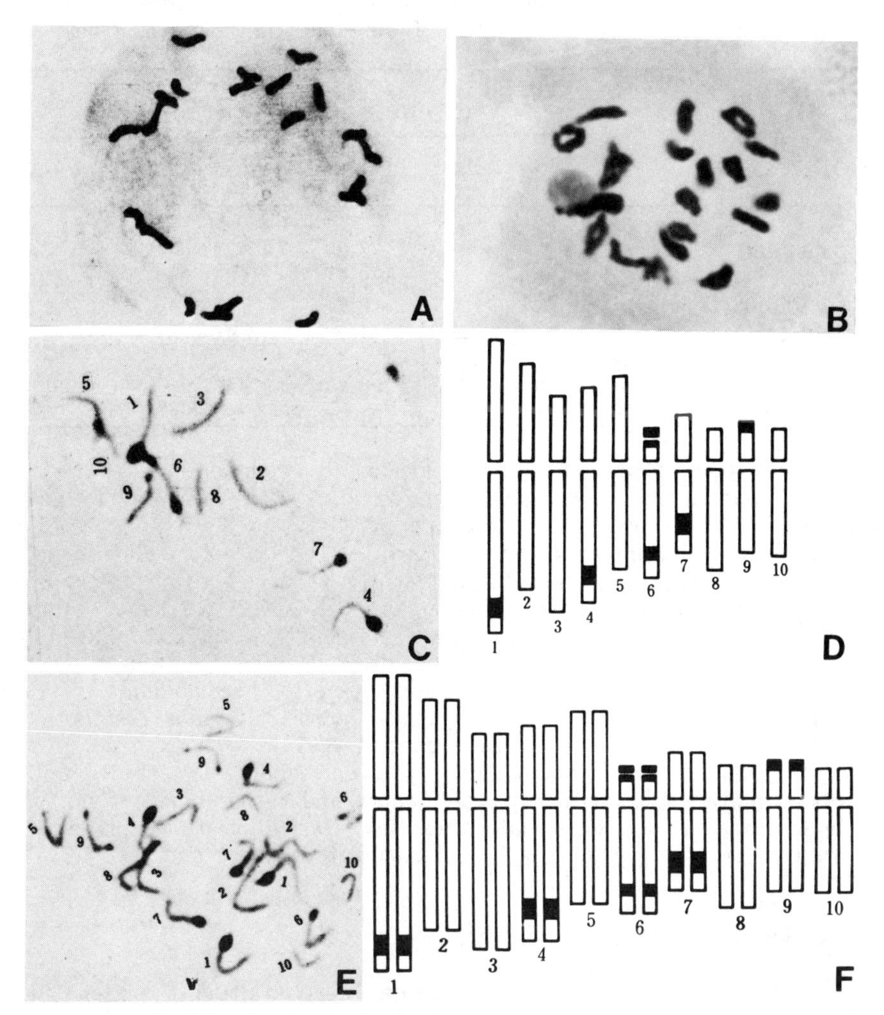

Figure 1. (A) PMC of 3x plant of pollen wheat 21I. (B) PMC of 6x plant of pollen wheat diakinesis, 21II. (C,E) Giemsa C-banding pattern of pollen callus clone No. 1 of maize variety Batangbai; C-haploid, E-diploid. (D,F) Model of Giemsa C-bands in somatic chromosomes of pollen callus No. 1 of maize variety Batangbai; D-haploid, F-diploid.

The Diversity of Phenotypes of Pollen Plants (H_1) from F_1 Hybrids

As F_1 hybrids are often heterozygous for several genes and as recombination occurs during meiosis, plants derived from pollen grains of F_1 hybrids may represent different genetic recombinants. This pheno-

Table 12. Variation in Chromosome Numbers in Maize Callus of Clone No. 1 and Root-tip Cells of 35 Plants Regenerated from Clone No. 1

PLOIDY LEVEL	CALLUS		PLANTS	
	No. Cells	Percent	No. Cells	Percent
Hypohaploid (< 10)	34	7.9	25	3.9
Haploid (= 10)	383	89.7	564	87.4
Hypodiploid (< 20)	2	0.2	10	1.6
Diploid (= 20)	8	1.9	46	7.1
Total	427	99.7	645	100.0

menon has been observed in our experiments with rice and wheat as unique recombinants have been recovered. For example, following culture of F_1 anthers of the wheat cross between Xian nong 5675 (a variety with white glume color, top awn, red grain color, clavate spike shape, and short stalk), and Jili (a variety with red glume color, awn, red grain color, fusiform spike shape, and tall stalk) we have obtained several diploid pollen plants (Institute of Genetics, 1977b). Examining the spike character, plants could be divided into seven forms (Table 13). These seven categories reflect genetic recombinants of the original parents used to produce the F_1 hybrid. Similar results were also obtained in rice (Institute of Genetics, 1974). For example, in the cross Nipponbare x Chienchunpang, the ear characters of H_1 pollen plants exhibited some paternal types (short and rather compact lax) and some maternal types (long and rather lax), while others combined the characters of both parents: long and compact or short and lax.

Expression of Recessive Characters

The frequency of recessive characters derived from a pollen plant population are much higher than within a conventional F_2 population. Since a haploid plant only contains one set of chromosomes, dominant characters cannot mask recessive characters as in diploids, so recessive characters are expressed in high frequency. For example, in a wheat cross between Sonora 62 with red grain color and Hongtu with white grain color, red grain color is a dominant character. We compared the segregation ratio of grain color between F_2 and H_2 populations and found that the frequency of white seeds is much higher in an H_2 population than in an F_2 population (Table 14). In the cross of Sonora 62 x Hongtu, there were 413 plants in the F_2 generation. Plants with red grain numbered 313, and there were 100 plants with white grain. The segregation ratio is 3.1:1, but in H_2 plant lines, the ratio of red to white grains is 0.8:1. The former is not significantly different from a 3:1 ratio ($X = 0.1364$; $p > 0.05$), while the latter is consistent with a

Table 13. Diversity of Characters of H_1 Pollen Plants in Xian nong 5675 x Jili

PLANT MATERIAL	GLUME COLOR	AWN-NESS	GRAIN COLOR	SPIKE TYPE	PLANT HEIGHT (cm)[a]
Xian nong 5675	White	Top	Red	Clavate	41
Jili	Red	Awn	Red	Fusiform	115
H_1 form					
1	White	Top	Red	Clavate	60-69
2	White	Awn	Red	Fusiform	41-49
3	White	Non-awn	Red	Clavate	19-24
4	White	Awn	Red	Oblong	23-32
5	Red	Top	Sterile	Oblong	52
6	Red	Non-awn	Sterile	Clavate	—
7	Red	Top	Sterile	Clavate	—

[a]Form plant heights are rather short. This may be associated with the weak growth of H_1 pollen plants.

Table 14. Comparison of the Segregation Ratio of Grain Color Between F_2 and H_2 Populations

EXPRESSION OF COLOR	Sonora 62 x Hongtu		Orefen x Xiaoyu 759	
	F_2 Plants	H_2 Lines	F_2 Plants	H_2 Lines
Red	313	14	116	27
White	100	18	31	28
Total	413	32	147	55
Red:white ratio	3.1:1	0.8:1	3.7:1	0.9:1

ratio of 1:1 ($X = 0.5$; $p > 0.05$). In the cross Orefen x Xianoyan 759, the result is approximately the same. These experiments illustrate that haploid breeding can raise the selective efficiency of recessive characters.

APPLICATION OF ANTHER CULTURE TO CROP IMPROVEMENT

Recent Varieties Developed from Haploids

The procedures of haploid and conventional breeding are compared in Fig. 2. In some cases haploid breeding may reduce the time for

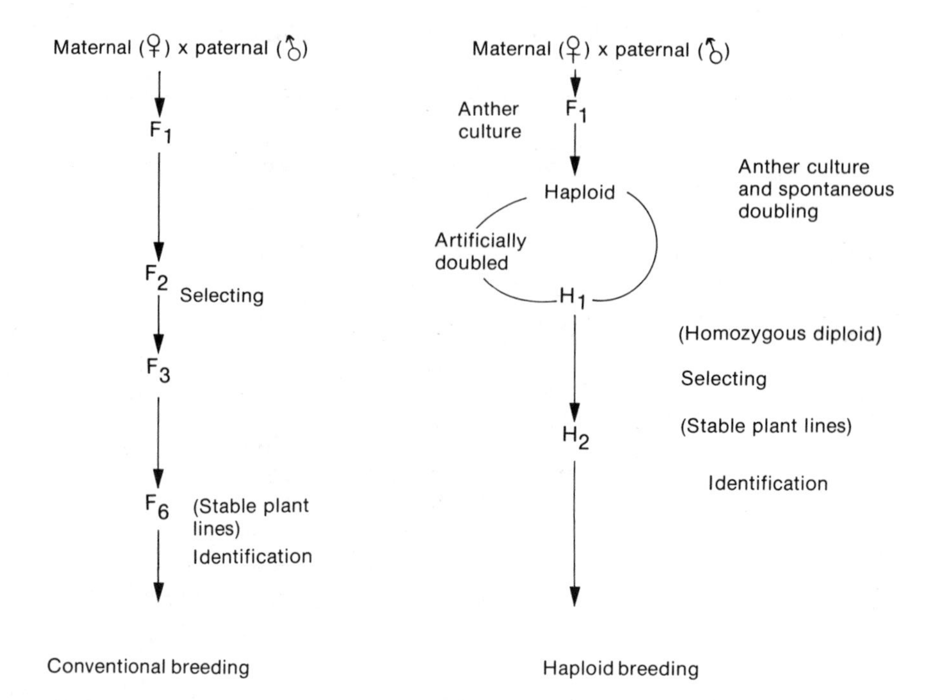

Figure 2. Comparison of procedures between haploid and conventional breeding.

variety development by three to four generations. For example, the varieties of rice Hua Yu No. 1 and No. 2 only took 5 years from the initial crossing to release of the new variety (Institute of Agricultural Academia, 1977). In Fig. 3, the development of Hua Yu No.1, No. 2, and their parents is summarized. Three kinds of breeding methods, anther culture, bulk, and pedigree were used to develop these varieties. Among the three methods, the duration needed for anther culture is the shortest—only four generations.

Many rice varieties and strains have been recently developed through anther culture in China. Hua Yu No. 1 and No. 2 are two new anther-derived varieties. These two varieties were selected by Y. Chen et al. of the Institute of Genetics of the Chinese Academy of Sciences in collaboration with Niu, and coworkers of Tianjin Agricultural Research Institute. These varieties have high yield (7500 kg/ha), resistance to bacterial blight, and wide adaptability. Both have been released in the suburbs of Beijing and Tiangjing. It is expected that the cultivated area of these varieties will reach several thousand hectares in the Tiangjing suburbs in the coming years (Fig. 4A).

Xin Xiu and Tang Huo No. 2 are two other anther-derived varieties of rice. Xin Xiu was developed by Zhang Zhen-hua at Shanghai Academy of Agricultural Science. It has been released and 100,000 ha are cultivated in eastern China. Tang Huo No. 2 was developed by Wu Xiaoyu at Agricultural Research Institute of Tonglin County, An hui

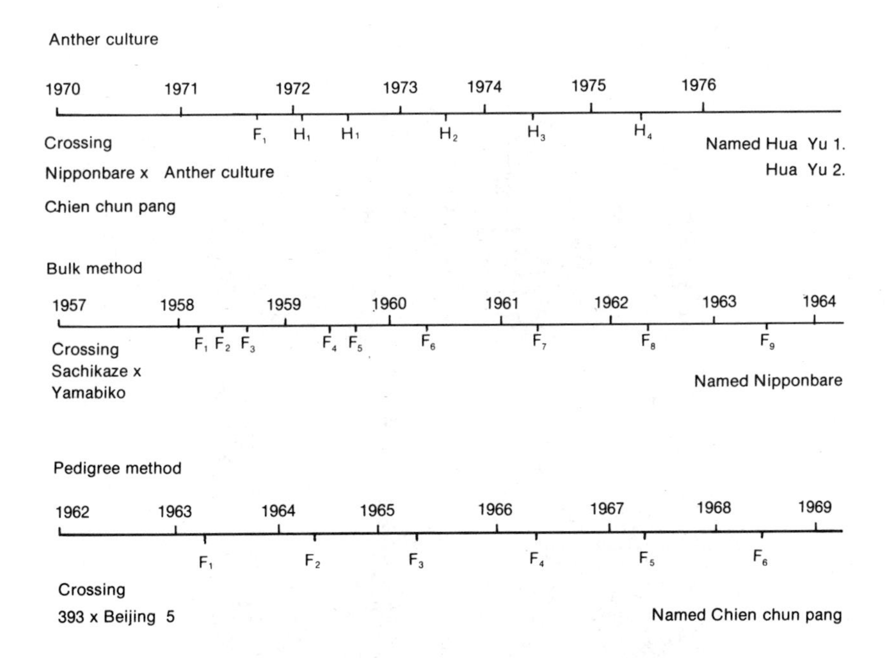

Figure 3. Comparison among three breeding methods.

province. About 6000 ha have been cultivated with this variety. Both varieties were selected for high yield.

F_1 hybrid anthers of winter-type wheat were used as donor tissue for development of the new strain Jingdan 2288 via the anther culture method by Hu Daofun and coworkers. This new variety has large spikes and plenty of grains, vigorous tillering, stripe rust and powdery mildew resistance, short stem, and resistance to lodging. It is a promising new strain of winter wheat for use in the Beijing area (Fig. 4B).

In tobacco, the variety Danyu No. 1 was developed by the Institute of Tobacco in Shantung, China (1974). It combined the merits of both parental lines and possesses characteristics of high yield and disease resistance. It took about 3 years from inoculation of F_1 anthers until varieties were released for production (Institute of Tobacco, 1974a).

Development of Inbred Lines of Maize and Asexual Lines of Sugarcane

Since 1975, many pollen-derived maize plants have been obtained in China. Among them, a line named Qun Hua selected by Wu et al. (1980) is a notable one and has good combining ability. In 62 crosses, about 90% (56 crosses) had increased yield. The cross 411 x Qun Hua was best, as it has both high yield and disease resistance. The selection process of this pure line was reduced by ten generations when compared with conventional methods (Fig. 5).

Figure 4. (A) New rice variety Hua Yu No. 1 growing in the field. (B) New strain of wheat Jingdan 2288 (left) and its parental line (right) growing in the test field.

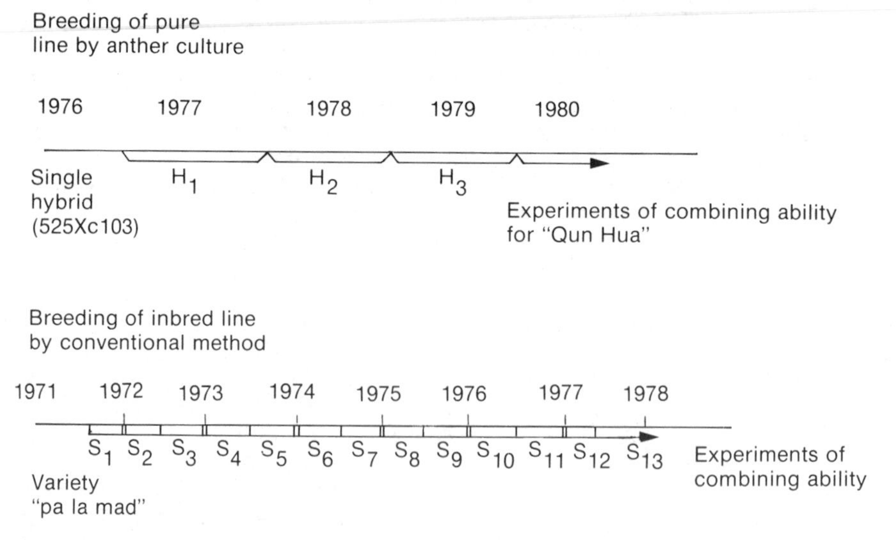

Figure 5. Comparison between anther culture and conventional method in breeding of inbred line.

In addition, the application of pollen-derived sugarcane plants to the development of improved varieties has also been studied. In sugarcane, a pollen plant containing high content of sugar and with tall stem was selected. The asexual line of the pollen plant has been propagated for further study and potential variety release (Z.H. Chen et al., 1980).

Integration of Haploid Plant Production with Conventional Breeding

Six strains of rice have been selected from pollen plants derived from F_1 anther culture of the hybrid rice variety Shan You No. 2. The phenotype and some economic characters of four of these lines are very similar to the donor plants. They are named Shan Hua No. 7706, 78-1, 791, 792. Researchers have attempted to use these strains to replace the F_1 hybrid variety and save the trouble of producing hybrid seed.

Using the conventional crossing method, Z.H. Zhang (1980) at Shanghai Academy of Agricultural Science combined anther culture with multiple crosses and selected the variety Hua Han Zao and other new strains such as Strain 77001. In this way the difficulty of selection using multiple crosses was overcome. Therefore, anther culture has not only increased the efficiency of selection using multiple crosses, but has also significantly reduced the time needed to develop new varieties. The breeding procedure for Hua Han Zao is summarized in Table 15.

Table 15. Breeding Scheme for Hua Han Zao

YEAR	PROCEDURE
1973	Sexual crosses: Jia Nong 485 (Ken) x Laberate F_1 x Tai Nan 13
1974	Anther culture: H_1 production
1975	Sexual cross: $Ke^{C}1669$ x H_1 pollen plant, line 175
	Anther culture: H_1 production
	Reproduction: H_2 production
	Identification and comparison of lines
1976	Reproduction: Line 76057
1977	Comparison test
	Production trial
1978	Regional test
	Multisite production trial
1979	Regional test
	Multisite demonstration and production trial
1980	Regional test
	Multisite demonstration and production trial

Production of Pollen-Derived Aneuploid and Heteroploid Plants

VARIABILITY OF POLLEN-DERIVED PLANTS. Cultured cells of both animal and plant tissue are characterized by instability of chromosome number and structure. The phenomenon is expressed in regenerated pollen-derived plants. Chromosome variations, in addition to gene mutations, have been expressed as changes in ploidy level, fertility, etc., in about 34 species from 24 different genera and 10 families of dicots and monocots. A list of heteroploid plants regenerated from anther culture is compiled in McComb (1978). These heteroploids are not disadvantageous as chromosome variation is often used for genetic investigations and plant breeding.

THE ORIGIN OF POLLEN-DERIVED HETEROPLOID PLANTS. There are various ways in which pollen-derived heteroploid plants might be produced: abnormal meiosis, endomitosis, spindle fusion, nuclear fusion, and endoreduplication. These abnormal phenomena would occur in the following different stages of pollen development.

Early Stages of Pollen Development In Vitro. If the initial uninucleate microspore is not a haploid because of some meiotic abnormality, the derived embryo and pollen-derived plants will express this abnormal chromosome constitution.

In some species, e.g. *N. tabacum*, the initial division of the vegetative nucleus is preceded by degeneration of the cytoplasmic organelles. Thus, endomitosis may lead to a diploid restitution nucleus. This is one possible source of producing doubled dihaploids.

Fusion of nuclei during early nuclear division of the pollen grain is also possible. Zeng and Ouyang (1980) observed various abnormalities in the development in vitro of pollen grains such as asynchronous division at different stages of mitosis, different nuclear fusion types (Fig. 6a,b), and endomitosis. Pollen grains and callus with doubled chromosome number may be formed in this way (Fig. 6c) and nonhaploid or spontaneous doubled dihaploids may be produced.

The generative nucleus sometimes shows endoreduplicated chromosomes on its metaphase plate. The dividing reduplicated generative nucleus might fuse with a dividing vegetative nucleus (Sunderland, 1977). Cell lines derived from such a fusion would be triploid. While polyploids at 2n, 4n, and 8n might be derived from endoreduplication alone, triploidy involves at least one fusion event.

Callus Development In Vitro. The later stage of callus development may be marked by incomplete cytokinensis; therefore, various abnormal mitoses might result. These abnormalities include spindle fusion and endoreduplication leading to euploidy, while multipolar mitosis, bridges, and lagging chromosomes lead to aneuploidy.

THE ORIGIN OF ABNORMAL MITOSES IN VITRO. Abnormal mitoses can result from nuclear endoreduplication, which frequently occurs. It should be noted that variation of chromosome number and structure in vitro occurs more frequently in haploid tissue than in diploid tissue (Sacristan, 1971).

In intact plants, heat shock, x-rays, and hormones can all induce formation of diplochromosomes. Endoreduplicated nuclei also occur during

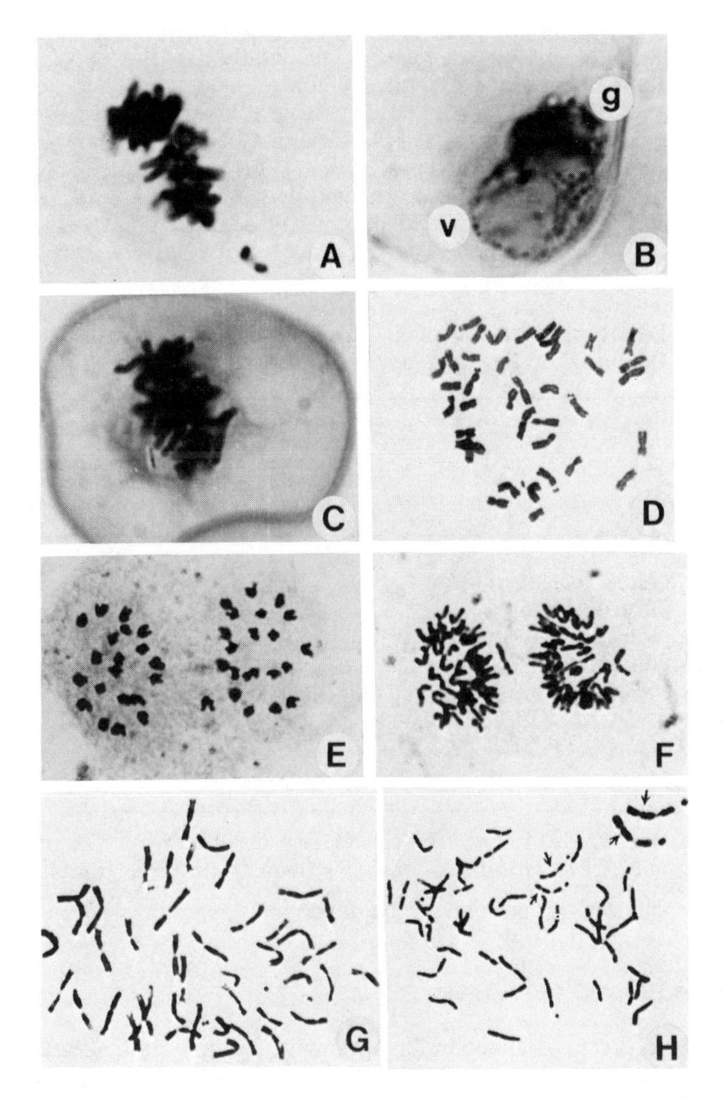

Figure 6. (A) Synchronous division of nuclei in pollen grain of cultured anther. (B) Nuclear fusion in pollen grain of cultured anther. (C) Pollen grain with doubled chromosome number in cultured anther. (D) Spring intervarietal hybrids (Pitic 62 x Hua Pei No. 1) F_1. Root somatic chromosome of nullisomic ($2n = 6x-2 = 40$). (E) Semi-winter hybrids (Orofen x Xiaoyan 759) F_1. Monosomic anaphase I, showing lagging chromosome. (F) Semi-winter intervarietal hybrids (Orofen x Xiaoyen 759) F_1. Monosomic telophase, showing free 6B chromosome with large satellite and lower arm ratio. (G) Trisomic plant derived from variety Orofen ($2n = 6x+1 = 43$). (H) Tetrasomic plant derived from variety Orofen with a pair of dicentric 6B chromosomes (arrow). G = generative nucleus; V = vegetative nucleus.

normal differentiation of somatic tissues. Cells with diplochromosomes can be induced to divide in culture, especially by the supply of cytokinin and auxin (Ghosh and Gadgil, 1979).

Prolonging the length of callus incubation often induces spontaneous chromosome doubling. For example, in wheat, the frequency of chromosome doubling in somatic cells of No. 71-1-11-1 is only 7.25%; meanwhile, the frequency of chromosome doubling of No. 75-1-14, which was derived from the same callus and differentiated into plantlets at a later date, was 23.4% (Table 16). The difference between the two is significant (Hu et al., 1978).

Table 16. Effect of Duration of Callus Incubation on Chromosome Doubling of H_1 Regenerated Plant[a,b]

VARIABLE	H_1 PLANT	
	72-1-11-1	75-1-14
Days from callus formation to plantlet differentiation	31	95
Days from plantlet differentiation to transplantation	64	69
No. of cells observed	69	94
Doubled cells		
Number	5	22
Percent	7.3	23.4

[a]72-1-11-1 and 75-1-14: two H_1 plants regenerated from the same callus induced from anthers of spring wheat (Inia 66 x Hongtu) F_1.

[b]t value = 2.7419; significant at 1% level.

Aneuploids Derived from Wheat Pollen

At present, various aneuploid plants have been obtained using anther culture. This material might be used for genetic investigations, such as gene mapping. Using chromosome engineering methods, alien addition lines, alien substitution lines, and translocation lines could be obtained. Alien addition lines of hybrid wheat are those in which one pair of chromosomes from a given wild parent species has been added to the full chromosome complement of cultivated wheat. An alien substitution line is one in which a pair of alien chromosomes is substituted for a pair of homologous chromosomes of cultivated wheat (Feldman and Sears, 1981). Substitution of one arm of an alien chromosome for an arm of a homologous wheat chromosome (translocation line) can result from the simultaneous misdivision of two univalent chromosomes at meiosis. The misdivision occurs rarely in the cell; therefore, using monosomic 5B on which ph, the major pairing suppressor, is located, genes can be transferred to the chromosome of wheat from its relatives by genetic induction of homologous pairing in hybrids that have one or more alien chromosomes (Feldman and Sears, 1981).

Some valuable alien genes might be transferred to common wheat by using chromosome engineering methods. Possibly wheat varieties that are characterized by high yield and resistant to disease could be obtained. By using Giemsa banding technique, it has been possible to show that both Avrora and Kavkaz varieties possess a single pair of wheat-rye translocation chromosomes that contain at least the major part of the short arm of chromosome 1R (Hu et al., 1978).

In recent years, plants and seed of various aneuploids, such as nullisomic, monosomic, trisomic, tetrasomic, and tetrasomic with dicentric chromosomes (Fig. 6d-h) were obtained in our laboratory. These were obtained via anther culture, by use of intervarietal hybrids and intergeneric hybrids, e.g., hexaploid *Triticale* crosses with common wheat. Changes of chromosome structure have also been observed, so it is expected that alien substitution and translocation lines might be obtained directly by means of anther culture.

CONCLUSIONS

Research in China has resulted in successful development of haploid and doubled-haploid plants from several crop species. These include maize, rye, *Triticale*, tobacco, soybean, rubber tree, pepper, and eggplant. Recent work, however, has been directed toward development of new plant varieties by using anther-derived plants. New varieties of rice, tobacco, and winter wheat and new breeding lines of maize and sugarcane have been developed using doubled-haploids. Anther culture appears to be a particularly powerful tool when integrated with conventional breeding. In this chapter we have also discussed the value of anther culture for development of new variant lines and for recovery of novel recombinants from F_1 hybrid lines.

KEY REFERENCES

Chu, C.C. 1982. Haploids in plant improvement. In: Plant Improvement and Somatic Cell Genetics (I.K. Vasil, W.R. Scowcroft, and K.J. Frey, eds.) pp. 129-158. Academic Press, New York.

Hu Han and Shao, Q.Q. 1981. Advances in plant cell and tissue culture in China. Adv. Agron. 34:1-13.

REFERENCES

Bouharmont, J. 1977. Cytology of microspores and calli after anther culture in *Hordeum vulgare*. Caryologia 30:351-360.

Cao, Z.Y., Guo, C., and Hao, J. 1981. A study of embryogenesis in pollen callus of maize (*Zea mays* L.). Acta Genet. Sin. 8:269-374.

Chang, L.K. 1978. Induction and culture of regenerated plantlets from anthers of genus *Populus*. In: Proceedings of Symposium on Plant Tissue Culture (Hu Han, ed.) pp. 241. Science Press, Beijing.

Chen, Y. and Li, L.T. 1978. Investigation and utilization of pollen-derived haploid plants in rice and wheat. In: Proceedings of Symposium on Plant Tissue Culture (Hu Han, ed.) pp. 199-211. Science Press, Beijing.

Chen, Y., Lu, D.Y., Li., S.Y., Zuo, Q.X., and Zheng, S.W. 1979. Study on effect of cold pretreatment on rice anther and isolated pollen culture in vitro. In: Annual Report of the Institute of Genetics, Academia Sinica, p. 81.

Chen, Z.G., Wang, M.Q., and Liao, H.H. 1980. The induction of Citrus pollen plants in artificial media. Acta Genet. Sin. 7:189-191.

Chen, Z.H., Chen, F., Chien, C., Wang, C., Chang, S., Hsu, H., Ou, H., Ho, Y., and Lu, T. 1978. Induction of pollen plants of *Hevea brasiliensis* Muell-Arg. Acta Genet. Sin. 5:107.

Chen, Z.H., Quian, C.F., Qin, M., Weng, C.H., Suo, C.J., Chen, F.Z., and Deng, Z.T. 1979. The induction of pollen plants of sugarcane. Annual Report of the Institute of Genetics, Academia Sinica. p. 92-93.

Chen, Z.H., Quian, C.F., Qin, M., Xu, X.E., Xiao, Y.L., Lin, M.Y., Deng, Z.T., Wu, S.L., and Huan, N.S. 1980. Recent advances in anther culture of rubber tree and sugarcane. Annual Report of the Institute of Genetics, Academia Sinica 1980, p. 105.

Chinese Medicine Laboratory, He Ze, Shandong, 1978. Plant Magazine 4:8.

Chu, C.C., Wang, C.C., Sun, C.S., Hsu, C., Yin, K.C., Chu, C.Y., and Bi, F.Y. 1975. Establishment of an efficient medium for anther culture of rice through comparative experiments on the nitrogen sources. Sci. Sin. 18:659-668.

Chu, C.C., Sun, C.S., and Wang, C.C. 1978. Cytological investigation on androgenesis of *Triticum aestivum*. Acta Bot. Sin. 20:6-12.

Chuang, C.C., Ouyang, T.W., Chia, H., Chou, S.M., and Ching, C.K. 1978. A set of potato media for wheat anther culture. In: Proceedings of Symposium on Plant Tissue Culture (Hu Han, ed.) pp. 51-56. Science Press, Beijing.

Chung, C.H. 1978. Preliminary study on anther culture of three *Brassica* cultivars. In: Proceedings of Symposium on Plant Tissue Culture (Hu Han, ed.) p. 200. Science Press, Beijing.

Clapham, D. 1971. In vitro development of callus from the pollen of *Lolium* and *Hordeum*. Z. Pflanzenzuecht. 65:385-393.

______ 1973. Haploid *Hordeum* plants from anther in vitro. Z. Pflanzenzuecht. 69:142-155.

D'Amato, F. 1977. Cytogenetics of differentiation in tissue and cell cultures. In: Plant Cell Tissue and Organ Culture (J. Reinert and Y.P.S. Bajaj, eds.) pp. 343-357. Springer-Verlag, Berlin.

Department of Biology, Specialty in Plant Genetics and Breeding, Beijing University, and The Experimental Station of Shuang Qiao Commune, Beijing. 1977. The significance of the embryoid formation in anther cultures of *Zea mays*. Acta Genet. Sin. 4:361-362.

Feldman, M. and Sears, E.R. 1981. The wild gene resources of wheat. Sci. Am. 244:102-113.

Forche, E. et al. 1979. Investigations on ploidy distribution pattern in barley plants (*Hordeum vulgare* L.) derived from anther culture. Z. Pflanzenzuecht. 83:222-235.

Genovesi, A.D. and Magill, C.W. 1979. Improved rate of callus and green plant production from rice anther culture following cold shock. Crop Sci. 19:662-664.

Ghosh, A. and Godgil, V.N. 1979. Shift in ploidy level of callus tissue: A function of growth substances. Indian J. Exp. Biol. 17:562-564.

Gu, M.G. and Zhang, X.Q. 1981. Studies on the karyotype of pollen callus clone of maize in long term subcultures. In: Annual Report of the Institute of Genetics, Academia Sinica, 1980, pp. 97-98.

Gu, M.G., Zhang, X.Q., Cao, Z.Y., and Guo, C.Y. 1980. The totipotency and genetic stability of pollen callus of maize in subcultures. In: Annual Report of the Institute of Genetics, Academia Sinica, 1980, pp. 96-97.

Gu, S.R. 1981. Pollen plants of *Lycium* obtained from anther culture. Acta Bot. Sin. 23:246-248.

Guha, S.R. and Maheshwari, S.C. 1964. Nature London 204:497.

______ 1966. Cell division and differentiation of embryos in the pollen grains of *Datura* in vitro. Nature London 221:97-98.

Guha, S.R., Iyer, R.D., Gupta, N., and Swaminathan, M.S. 1970. Totipotency of gametic cells and the production of haploids in rice. Curr. Sci. 39:174-176.

Guha-MuKherjee, S. 1973. Genotypic difference in the in vitro formation of embryoids from rice pollen. J. Exp. Bot. 24:139-144.

He, D.G. and Ouyang, J.W. 1980. Effect of stages of anther development on anther culture of wheat (*Triticum aestivum* L.). In: Annual Report of the Institute of Genetics, Academia Sinica, 1979, p. 74.

Heilungkiang Institute of Forestry, Tree Improvement Laboratory, 1975. Induction of haploid poplar plants from anther culture in vitro. Sci. Sin. 18:769-777.

Ho, T.F., Pao, J.J., Hu, Y., Wang, Y.F., and Chou, W.Y. 1978. Advances in anther culture of wheat in 1977. In: Proceedings of Symposium on Plant Tissue Culture (H. Hu, ed.) p. 291. Science Press, Beijing.

Hsu, C., Yin, K.C., Chu, C.Y., Bi, F.Y., Wang, S.D., Liu, D.Y., Wang, F.L., Chien, N.F., Chu, C.C., Wang, C.C., and Sun, C.S. 1975. Induction pollen plants of rice and observations of their progenies. Acta Genet. Sin. 2:294-301.

Hu, H., Hsi, T.Y., and Chia, S.E. 1978. Chromosome variation of somatic cell of pollen calli and plants in wheat (*Triticum aestivum* L.). Acta Genet. Sin. 5:23-30.

Hu, H., Xi, Z.Y., Zhuang, J.J., Ouyang, J.W., Zheng, J.Z., Jia, S.E., Tia, X., Jing, J.K., and Zhou, S.M. 1979. Genetic investigation on pollen-derived plants in wheat (*Triticum aestivum* L.). Acta Genet. Sin. 6:322-330.

Hu, H., Xi, Z.Y., Jing, J.K., and Wang, X.Z. 1982. Production of aneuploids and heteroploids of pollen-derived plants. In: Proceedings of the Fifth International Congress Plant Tissue and Cell Culture (A. Fujiyama, ed.) pp. 421-424. Tokyo.

Institute of Agricultural Academia, Rice Group, Tianjing and Institute of Genetics, 2nd Division, 3rd Laboratory, Academica Sinica. 1977.

New success of pollen haploid breeding—the new rice cultivars "Hua yu no. 1" and "Hua yu no. 2". In: Collection of Works of Haploid Breeding, pp. 21-25. Science Press, Beijing.

Institute of Genetics, 2nd Division, 3rd Laboratory, Academia Sinica. 1974. Investigation on the induction and genetic expression of rice pollen plants. Sci. Sin. 2:209-221.

Institute of Genetics, Cell Culture Laboratory and 1st Group of 4th Laboratory, Academia Sinica. 1975. Primary study on induction of pollen plants of *Zea mays*. Acta Genet. Sin. 2:143.

Institute of Genetics, 302 Research Group, Academia Sinica. 1976. Genetic studies on pollen plants in rice (*Oryza sativa* L.). Acta Genet. Sin. 3:277-285.

Institute of Genetics, 301 Group, Academia Sinica. 1977a. Effect of cold pretreatment on anther culture of wheat. Hered. Breed. 4:24-25.

Institute of Genetics, 1st Division, 3rd Laboratory, Academia Sinica. 1977b. Investigation on some induction factors and genetic expression of wheat pollen plants. In: Collection of Works of Haploid Breeding, pp. 65-81. Science Press, Beijing.

Institute of Sugarbeet and Sugar Crops, Breeding Laboratory, Heilungkiang Province. 1978. Induction of pollen plants in sugarbeet. In: Proceedings of Symposium on Plant Tissue Culture (Hu Han, ed.) p. 303. Science Press, Beijing.

Institute of Tobacco, Breeding Group, Shantung Province and Institute of Botany, Academia Sinica. 1974a. Success of breeding the new tobacco cultivar "Tan Yuh No. 1." Acta Bot. Sin. 16:301-303.

Institute of Tobacco, Academy of Agricultural Sciences and Institute of Botany, Academia Sinica. 1974b. The evaluation of the progenies of the pollen plants of tobacco. Acta Genet. Sin. 1:26-39.

Kameya, T. and Hinata, K. 1970. Induction of haploid plants from pollen grains of *Brassica*. Jpn. J. Breed. 20:82-87.

Keller, W.A. and Armstrong, K.C. 1978. High frequency production of microspore-derived plants from *Brassica napus* anther culture. Z. Pflanzenzuecht. 80:100-108.

Ku, M.K., Cheng, W.C., Kuo, L.C., Kuan, Y.L., An, H.P., and Huang, C.H. 1978. Induction factors and morphocytological characteristics of pollen-derived plants in maize (*Zea mays*). In: Proceedings of Symposium on Plant Tissue Culture (Hu Han, ed.) pp. 35-42. Science Press, Beijing.

Kwangtung Institute of Botany, Laboratory of Genetics. 1975. Studies on anther culture in vitro in *Oryza sativa* subsp. shien. I. The role of basic medium and supplemental constituents in callus induced from anther and in differentiation of root and shoot. Acta Genet. Sin. 2: 81-89.

______ 1976. The study on simplified medium in anther culture of *Oryza sativa* subsp. shien. Acta Genet. Sin. 3:169-170.

Lu, C.J., Zhang, J.C., and Liu, Y.X. 1979. Observation of the chromosome of somatic cells of poplar pollen plants. J. North-Eastern For. Inst. 2:14-22.

McComb, J.A. 1978. Variation in ploidy levels of plants derived from anther culture. In: Proceedings of Symposium on Plant Tissue Culture (Hu Han, ed.) pp. 167-180. Science Press, Beijing.

Miao, S.H., Kuo, C.S., Kwei, Y.L., Sun, A.T., Ku, S.Y., Lu, W.L., and Wang, Y.Y. 1978. Induction of pollen plants of maize and observations on their progeny. In: Proceedings of Symposium on Plant Tissue Culture (Hu Han, ed.) pp. 35-42. Science Press, Beijing.

Nakata, M. and Tanaka, M. 1968. Differentiation of embryoids from developing germ cells in anther culture of tobacco. Jpn. J. Genet. 43:65-71.

Nitzsche, W. and Wenzel, G. 1977. Haploids in plant breeding. Z. Pflanzenzuecht. 8:132.

North Eastern Forestry Academy, Group Tree Breeding. 1977. Induction of haploid poplar plantlets from pollen. Acta Genet. Sin. 4:49.

Ouyang, T.W., Hu, J., Chuang, C.C., and Tseng, C.C. 1973. Induction of pollen plants from anthers of *Triticum aestivum* L. cultured in vitro. Sci. Sin. 16:79-95.

Pan, J. and Gao, G.H. 1975. Certain factors affecting the frequency of induction of wheat (*Triticum vulgare*) pollen plants. Acta Bot. Sin. 17:161-166.

______ 1978. The production of wheat pollen embryo and the influence of some factors on its frequency of induction. Acta Bot. Sin. 20:122-128.

Sacristan, M.D. 1971. Karyotypic changes in callus cultures from haploid and diploid plants of *Crepis capillaris* (L) wall. Chromosoma 33: 273-283.

Sun, C.S. 1978. Anther and androgenesis of rye (*Secale cereale* L.). In: Proceedings of Symposium on Plant Tissue Culture, 1977 (Hu Han, ed.) p. 121. Science Press, Beijing.

______, Wang, C.C., and Chu, C.C. 1973. Cytological studies on the androgenesis of *Triticale*. Acta Bot. Sin. 15:165-173.

______, Wang, C.C., and Chu, C.C. 1974. Cell division and differentiation of pollen grains in *Triticale* anther culture in vitro. Sci. Sin. 2: 47.

______, Chu, C.C., Wang, C.C., and Tigerstedt, P.M.A. 1980. Studies on the anther culture of *Triticale*. Acta Bot. Sin. 25-31.

Sun, H.T. 1979. Preliminary report on anther culture of flax. Kexue Tongbao 24:948-950.

Sunderland, N. 1977. Anther and pollen culture. In: Plant Tissue and Cell Culture (H.E. Street, ed.) pp. 223-268. Blackwell Scientific Pub., Oxford.

Teng, L.P. and Kuo, Y.H. 1978. Induction of pollen plants in *Brassica pekinensis* Rupr. In: Proceedings of Symposium on Plant Tissue Culture, 1977 (Hu Han, ed.) p. 198. Science Press, Beijing.

Vunchkova. 1979. Growing haploids from triticale anthers and their cytological characteristic. Dok. Vasknil 10:8-11.

Wang, C.C., Sun, C.S., and Chu, C.C. 1974. On the condition for the induction of rice pollen plantlets and certain factors affecting the frequency of induction. Acta Bot. Sin. 16:43-54.

Wang, C.C., Chu, C.C., and Sun, C.S. 1975a. The induction of populus pollen plants. Acta Bot. Sin. 17:56-62.

Wang, C.C., Chu, C.C., Sun, C.S., Hsu, C., Yin, K.C., and Bi, F.Y. 1975b. Induction of pollen plants from the anther culture of *Triticum vulgare-Agropyron glaucum* hybrid. Acta Genet. Sin. 2:77.

Wang, C.C., Chu, C.C., and Sun, C.S. 1980. Studies on the induction of pollen sporophyte of *Coix lacryma* johic L. Acta Bot. Sin. 22:316-322.

Wang, Y.Y., Sun, C.S., Wang, C.C., and Chien, N.F. 1973. The induction of the pollen plantlets of *Triticale* and *Capsicum annuum* from anther culture. Sci. Sin. 16:147-151.

Wu, J.L., Zhong, Q.L., Nong, F.H., and Chang, T.M. 1980. Embryogenesis in corn culture. Acta Photo. 6(2):221-224.

Xu, H.J., Ai, S., Chen, Z., and Jia, X. 1980. Report on heredity and viability of progenies of pollen-derived tobacco. Zhongguo Yancao 1: 8-10.

Xu, S. 1979. Success of induction of pollen plants in alfalfa. Plant Magazine (6):5.

Xue, G.R., Sei, K.W., and Hu, J. 1981. Obtaining haploid plantlets of strawberry (*Fragaria orientalis* A. Los) by anther culture in vitro. Acta Hortic. Sin. 8:9-12.

Yin, K.C., Li, S.C., Xu, Z., Chen, L., Zhu, Z.Y., and Bi, F.Y. 1980. A study of anther culture of *Glycine max*. Kexue Tongbao 25:976.

Zeng, J.Z. and Ouyang, J.W. 1980. The early androgenesis in vitro of wheat anthers under ordinary and low temperatures. Acta Genet. Sin. 7:165-172.

Zhang, G. 1980. Information of workshop of pollen breeding in corn. Hereditas 2:41.

Zhang, Z.H. 1980. Application of anther culture technique to rice breeding. Symposium Celebrating the 20th Anniversary of the International Rice Research Institute, Los Banos, Philippines.

Zhou, C. and Yang, H.Y. 1980. Anther culture and androgenesis of *Hordeum vulgare* L. Acta Genet. Sin. 7:287.

Zhou, C.J. and Li, P.F. 1981. Induction of pollen plants of grape (*Vitis vinifera* L.). Acta Bot. Sin. 23:79-81.

Zhou, W.B. 1978. A preliminary study on the anther culture of *Sorghum*. Acta Genet. Sin. 5(4):337-338.

Zhu, X.Y., Wang, R.L., and Liang, G.Y. 1980. Sci. Silval Sin. 16:190-197.

SECTION II
Cereals

CHAPTER 4
Forage Grasses

B. S. Ahloowalia

Grasses are the dominant component of the natural pastures and reseeded grasslands. The production of milk, meat, and wool in the world depends largely on grass and legume mixtures, in which grass is the major source of energy supply for animal growth and performance. Therefore, continued improvement of high yielding and better quality forage and fodder grasses by breeding more productive varieties is of crucial importance for increasing the stocking rates (animal number per unit of land) and productivity of cattle. The modern trend in animal management, whereby increasing numbers of cattle are fed conserved grass (hay and silage), increases demand for special-purpose grass cultivars to achieve target milk and meat yields. Reseeded temperate grasslands can produce 10,000 kg/ha per year of dry matter yield with improved varieties and timely agronomic inputs of fertilizer and cutting regimes. Higher yields can be obtained in the tropical and semi-tropical zones than in the temperate climate, but the quality of the dry matter is variable and digestibility is low, resulting in poor animal performance. Hence, the quality, and in particular, the digestibility of the dry matter produced and conserved becomes of crucial importance.

Several key grass species predominate in natural pastures or are grown in reseeded grasslands. These grass species can be grouped as either temperate, cool-season (Table 1) or tropical, warm-season (Table 2), though some grasses, e.g., the fescues and canary grasses, can grow successfully in the transitional zones of subtropical and temperate areas. The most important temperate grasses are wheatgrasses, *Agropyron* spp., smooth bromegrass, *Bromus inermis* Leyss., cocksfoot, *Dactylis glomerata* L., tall fescue, *Festuca arundinacea* Schreb., annual

Table 1. Important Temperate Forage Grasses

GENUS AND SPECIES	COMMON NAME	CHROMOSOME NUMBER (2n)	CENTER OF ORIGIN
Agropyron			
cristatum (L.) Gaertn.	Crested wheatgrass	14, 28, 42	E. Europe, Siberia
elongatum (Host) Beauv.	Tall wheatgrass	14, 56, 70	S. Europe, Asia Minor
repens (L.) Beauv.	Quackgrass	28, 42	Europe, temperate Asia
Agrostis			
alba L.	Redtop	28, 42	Europe, temperate Asia, N. America
palustris Huds.	Creeping bent	28	
stolonifera L.	Creeping bent	28, 35, 42	
tenuis Sibth	Brown top	28	
Alopecurus			
pratensis L.	Meadow foxtail	28	Temperate N. Europe, N. Asia
Arrhenatherum			
elatius (L.) Beauv.	Tall oatgrass	28	Europe & W. Asia
Bromus			
erectus Huds.	Meadow bromegrass	42, 56	N. & Central Europe, temperate Asia
inermis Leyss.	Smooth bromegrass	42, 56, 70	
Cynosurus			
cristatus L.	Crested dogstail	14	S.W. Europe, S.W. Asia
Dactylis			
glomerata L.	Orchard grass (cocksfoot)	28	N. Africa, Europe, Asia
Elymus			
canadensis L.	Canada wild rye	28, 42	N. America
glaucus Buckl.	Blue wild rye	28	N. America
junceus Fisch.	Russian wild rye	14	Asia
Festuca			
pratensis Huds.	Meadow fescue	14, 28	Europe
arundinacea Schreb	Tall fescue	42, 70	N. Africa
ovina L.	Sheeps fescue	14, 21, 28, 42, 70	Temperate Europe

Table 1. Cont.

GENUS AND SPECIES	COMMON NAME	CHROMOSOME NUMBER (2n)	CENTER OF ORIGIN
Festuca			
rubra L.	Red fescue	14, 42, 56, 70	N. Europe, temperate Asia
Holcus			
lanatus L.	Velvet grass	14	Europe, W. Asia, N. Africa
Lolium			
multiflorum Lam.	Italian (annual) ryegrass	14	S. Europe
perenne L.	Perennial ryegrass	14	N. Africa
rigidum Gaud.	Wimmera ryegrass	14	S.W. Asia
Phalaris			
arundinacea L.	Reed canarygrass	14, 28, 35, 42	Temperate Europe, Asia
tuberosa L.	Large canarygrass	28	America, Australia
Phleum			
alpinum L.	Mountain timothy	28	
pratense L.	Timothy	14, 42	Eurasia
Poa			
compressa L.	Canada bluegrass	35, 42, 45, 49, 56	Eurasia
pratensis L.	Kentucky bluegrass[a]	28-124	N. America
trivialis L.	Roughstalk bluegrass	14	

[a]Apomictic.

ryegrass, *Lolium multiflorum* Lam., perennial ryegrass, *Lolium perenne* L., Kentucky bluegrass, *Poa pratensis* L., and reed canarygrass, *Phalaris arundinacea* L. The major tropical grasses are bermudagrass, *Cynodon dactylon* L., guineagrass, *Panicum maximum* L., bahiagrass, *Paspalum notatum* Flugge, kikuyugrass, *Pennisetum clandestinum* Hoschst. ex Chiov., napiergrass, *Pennisetum purpureum* Schumach., sudangrass, *Sorghum bicolor* ex. *sudanense* (Piper) Stapf., Johnsongrass, *Sorghum halepense* (L.) Pers. Other forage grasses are either restricted in usage or grow as companion weeds and do not contribute substantially to dry matter production.

Table 2. Important Tropical Forage Grasses

GENUS AND SPECIES	COMMON NAME	CHROMOSOME NUMBER (2n)	CENTER OF ORIGIN
Andropogon			
gerardii Vitman	Big bluestem	40, 60, 70	N. America
scoparius Michx.	Little bluestem	40	N. America
Bouteloua			
curtipendula (Michx.) Torr.	Side oats grama	28, 35, 40, 42, 45, 56, 70, 98	N. America
gracilis (H.B.K.) Lag. ex. Steud.	Blue gramagrass	21, 28, 35, 42, 61, 77	N. America
Buchloe			
dactyloides (Nutt.)	Buffalograss	56, 60	N. America
Cynodon			
dactylon (L.) Pers.	Bermudagrass	36	Africa, India
Digitaria			
sanguinalis Scop.	Crabgrass	36	Africa
Eragrotis			
curvula (Schrad.) Nees	Weeping lovegrass	40	E. Africa
Panicum			
antidotale Retz.	Blue panicgrass	18	N. India, Arabia
maximum Jacq.	Guineagrass	32	Africa
virgatum L.	Switchgrass	18, 32, 36, 54, 72, 90, 108	N. America
Paspalum			
dilatatum Poir.	Dallisgrass[a]	40	N. Argentina, Uruguay, S. Brazil
notatum Fluegge	Bahiagrass	40	S. America
Pennisetum			
clandestinum Hochst.	Kikuyugrass	36	Africa
americanum (L.) K. Schum.	Pearl millet	14	Africa
purpureum Schumach.	Napiergrass/ elephantgrass	28	Africa

Table 2. Cont.

GENUS AND SPECIES	COMMON NAME	CHROMOSOME NUMBER (2n)	CENTER OF ORIGIN
Sorghastrum			
nutans (L.) Nash	Yellow indiangrass	40	S. United States
avenaceum (Michx.) Nash	Indiangrass	40	S. United States
Sorghum			
bicolor (L.) Moench.	Sweet sorghum	20	Africa
halepense (L.) Pers.	Johnsongrass	40	N. Africa, S. Asia, S. Europe
bicolor (L.) Moench ex. *sudanese* (Piper) Stapf.	Sudangrass	20	Africa
Stipa			
viridula Trin.	Green needlegrass	32	N. America

[a]Apomictic.

A number of cereal and millet crops are also major contributors to forage and fodder production in the world and are extensively used as green fodder, hay, and silage (Table 3). These include oats, *Avena sativa* L., barley, *Hordeum vulgare* L., rye, *Secale cereale* L., sorghum, *Sorghum bicolor* (L.) Moench ex. *vulgare* Pers., pearl millet, *Pennisetum americanum* (L.) K. Schum., and maize, *Zea mays* L.

Sugarcane (*Saccharum officinarum* L. and *S. barberi* Jesw.) is fed to cattle; particularly cane tops are used green or after chopping and often conserved as silage. Chinese cane, *S. sinense* Roxb. is cultivated in China and Japan in the same way as napiergrass. Wild cane, *S. spontaneum* L., which is an obnoxious weed, is also used as a coarse feed for buffaloes in India. To a limited extent, wheat (*Triticum aestivum* L.) cultivars are also used for making hay and silage in Australia.

ORIGIN, DISTRIBUTION AND EVOLUTION OF FORAGE GRASSES

Most cool-season grasses originated in the temperate regions of western Asia and southern Europe. On the other hand, most warm-season tropical grass species originated in Africa, though few species are natives of southern United States and South America (Tables 1 and

Table 3. Important Cereal Grasses Used as Fodder Crops

GENUS AND SPECIES	COMMON NAME	CHROMOSOME NUMBER (2n)	CENTER OF ORIGIN
Avena			
sativa L.	Oat	42	W. & N. Europe
Hordeum			
vulgare L.	Barley	14	Near East
Saccharum			
barberi Jesw.	Indian cane	81-124	N. India
officinarum L.	Noble canes	80	Pacific Islands
robustum Brandes & Jesw.	New Guinea cane	60-80	New Guinea
sinensis Roxb.	Chinese cane	106-120	China
spontaneum L.	Wild cane	40-128	India, S.E. Asia
Secale			
cereale L.	Rye	14	W. Asia
Sorghum			
bicolor (L.) Moench	Sorghum millet	20	E. & Central Africa
Panicum			
miliaceum L.	Broomcorn millet	36	India
Pennisetum			
americanum (L.) K. Schum.	Pearl millet	14	Africa
Zea			
mays L.	Maize	20	S. America
Zea (Euchlaena)			
mexicana Schrad.	Teosinte	20	Central America, Mexico

2) (Whyte et al., 1959; Simmonds, 1976). Since most grasses have a high reproductive potential and their seeds are easily dispersed either by wind or by sticking to animal furs and feathers of birds, grasses have spread all over the land masses. The high reproductive capacity of grass species has resulted in their wide genetic diversity which is reflected in their adaptation and ubiquitous presence in various ecogeographical regions ranging from cold tundra, arid steppes, and dry sahara to the humid savanna (Fig. 1). The migration of grasses probably accompanied the migration of early man, first through the animal herds and later as a mixture in cereal seeds.

Nearly all the temperate grass species have $x = 7$ as their basic chromosome number and most tropical and warm season grass species have genome $x = 8$, 9, and 10. In both the temperate and tropical genera of forage grasses, polyploidy is widespread (Tables 1 and 2) and

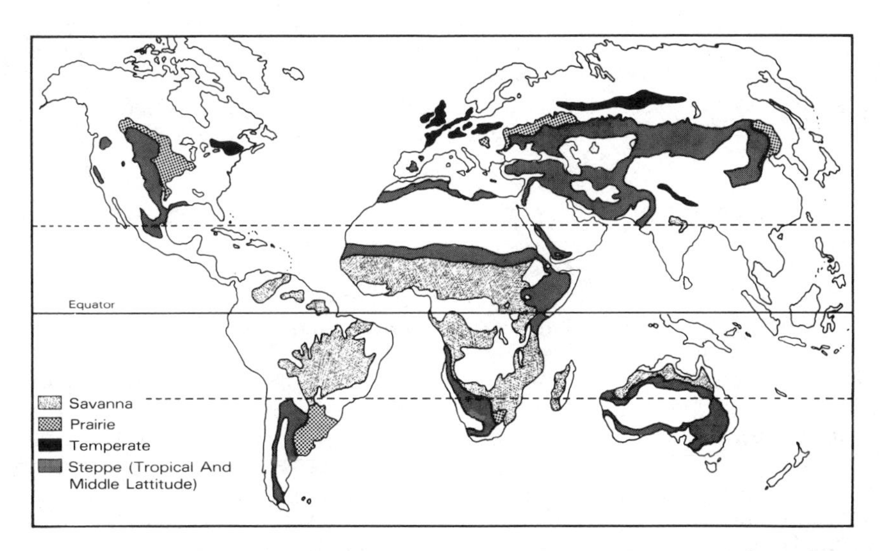

Figure 1. World distribution of natural and reseeded grasslands. (Modified after Trewartha et al., 1967).

has played an unusually large role in their evolution (Stebbins, 1956). Nearly 70% of the species in Gramineae are polyploid as against 30-35% in all the other flowering plants. In general, the chromosome size in the diploid species is relatively larger than that in their related polyploid species and unrelated species with higher chromosome number.

The temperate forage grass species have x = 7 genome in common with that of the temperate cereals such as wheat, oats, barley, and rye species. Some of the forage grasses have been successfully hybridized with the cereal grasses and thus share some of the genetic information with their ancestral species. For example, tall wheatgrass, *Agropyron elongatum* (Host.) Beauv. and *A. intermedium* (Host.) Beauv., have been crossed with species of the genera *Triticum*, *Hordeum*, and *Secale*. Wheatgrasses also cross easily with the other forage grass species such as *Elymus* and *Sitanion* (Stebbins et al., 1946a,b; Dewey, 1964, 1980, 1981); see also Myers (1947) and Carnhan and Hill (1961) for extensive reviews. Most grass species share some genetic information with each other. This is indicated by their promiscuous nature to cross-pollinate between genera. An extensive number of interspecific and intergeneric crosses in forage grasses involving *Festuca*, *Lolium*, *Poa*, *Dactylis*, *Panicum*, *Pennisetum*, *Cenchrus*, etc. have been attempted (Matzk et al., 1980) and have revealed several types of cross-incompatibility barriers between species. These barriers involve pre- and post-zygotic lethality, disturbed embryo and endosperm development, and gametic incompatibility (Matzk et al., 1980). Some of these barriers to crossing can be overcome by special techniques resulting in the production of hybrids (see protocol section).

Most forage grasses are cross-fertilized and wind-pollinated. A few grass species, such as annual ryegrass, perennial ryegrass, meadow fescue, and reed canarygrass are self-incompatible. Only a few forage grasses are apomicts, e.g., *Poa pratensis* L., *Paspalum notatum* Flugge, and *Cenchrus ciliaris* L., the latter being facultatively apomictic. Rarely, forage grasses are self-fertilizing, e.g., *Agropyron trachycaulum* (Link) Malte, *Lolium remotum* Schrank, and *L. temulentum* L.

GRASSLAND PRODUCTIVITY AND OBJECTIVES IN GRASS BREEDING

Grassland productivity depends upon a trio of soil-plant-animal interaction (Heath et al., 1973). Growth of grass is limited in many parts of the world because of the lack of water and fertilizer, overgrazing, and poor animal and grassland management. Stress conditions of drought, soil salinity, alkalinity, dessication of leaves by hot winds in warmer regions, and "leaf burn" and leaf tearing by high-velocity winds in colder climates, decrease yield and impair the quality of the grass. Selection and breeding for tolerance to stress conditions is important for increased grassland productivity. In the temperate climates of northern Europe and North America, low temperature is the major limiting factor to grass growth. Seasonal production of ryegrasses, fescues, orchard grass, etc. is strongly related to the prevailing spring and summer temperatures, and by mid-summer most of the temperate species have produced over 60% of their annual dry matter yield.

Since grasses are ultimately utilized by the animals either as green (in situ grazing) or conserved feed (hay, silage), the primary objective of breeding grass is to increase dry matter yield. Because grass species and varieties vary considerably in quality of the dry matter, particularly at flowering, quality is a relative term. Dry matter digestibility, based on in vitro digestion (Tilley and Terry, 1963) or determined from acid-detergent fiber (Van Soest, 1965) at or around the primary date of flowering (heading date) has been considered as an indicator of the quality of dry matter. Therefore, efficiency of an improved grass cultivar is determined by the amount of digestible dry matter it produces in a unit of time.

Dry matter yield of a grass cultivar is strongly influenced by tillering capacity, growth habit, rate of growth, and its ability to compete with a legume component in mixed stands, as well as its ability to keep out unwanted weeds and other grasses from the canopy. Tolerance to withstand stress such as cold, drought, windburn, freezing injury, animal treading, machinery clipping, mechanical compaction, salinity, and toxic levels of minerals contributes to the making of outstanding grass cultivars under special situations. Seasonal production, such as early growth in spring for early grazing and late autumn production to extend the grazing period in pasture type grasses has resulted in the breeding of special purpose cultivars. Where reseeding is practiced infrequently, persistency and longevity of grass cultivars are of crucial importance.

Grass quality is largely based on its digestibility, which in turn is affected by its carbohydrate, protein, and fiber content. Selection for increased dry matter yield and digestibility, however, is negatively correlated with the protein content. An empirical selection procedure for maintaining high digestibility is to select more leafy plants based mostly on visual score. Many good cultivars of grass have been produced through such empirical selection procedures.

In addition to selection for yield and quality, which are mostly vegetative characters, it is important to breed grass strains that have uniform and rapid germination and are fast in growth. Very little information is available on the correlation of the early seedling characters with the adult-stage competitive ability of grass cultivars. Since most cultivars are propagated by seed, seed yield and nonshattering of seed at maturity are important characters in the breeding and selection of grass cultivars.

Diseases generally do not reduce yield of grasses as they do in cereal crops, since most of the growth is removed by repeated grazing or cutting and epiphytotics do not occur in the same fashion as in cereals. Nonetheless, grass yield and quality can be impaired by fungal, bacterial, viral diseases, and insect pests. Diseases like *Helminthosporium*, powdery mildew, leaf scald, leaf blotch, rusts (black, brown, and crown rust), and bacterial wilt (Table 4; O'Rourke, 1976) can reduce yield and highly susceptible material should be discarded during selection.

GRASS BREEDING: CONVENTIONAL AND NOVEL

Conventional breeding of grass cultivars consists of four basic steps: (a) collection of populations and generation of variation, (b) selection of the desired variants, (c) evaluation of the selected variants to prove their advantage over the existing cultivars, and (d) multiplication of the selected and tested genotypes and their release for cultivation (Fig. 2). Of these basic steps, the first two involve assembling large numbers of genotypes and populations that are grown in a limited area or environment. On the other hand, in the third step a limited number of genotypes are grown over a large area and tested for their performance in several locations and for a number of years. Conventional breeding of cultivars is a rather slow and lengthy process. It usually takes between 18 and 20 years to breed a grass cultivar from the time of initiation to varietal release. Hopefully, tissue culture may positively complement the conventional breeding procedures, and save time, labor and cost, particularly in the first two steps, i.e., generation of variation and selection of desired genotypes.

In the conventional method of grass breeding, new variants are generated from collections of wild and natural land races, ecotypes, and existing cultivars from diverse ecogeographical regions. Large populations of these collections are grown as spaced plants in source nurseries and plants are studied on an individual basis for their growth vigor, leafiness, date of flowering, winter hardiness, drought tolerance, disease resistance, and other component characters that contribute to yield such as tillering capacity, plant height, crown girth, and growth response to repeated cutting and mowing. Most of these characters are visually assessed and scored on a 1-9 scale. Plants selected in this manner are vegetatively multiplied into 5-10 propagules (ramets)

Table 4. Important Diseases of Some Cool-Season Forage Grasses

HOST GRASS	GENUS	DISEASE AND PATHOGEN[a]
Bromegrass	*Bromus*	Leaf spot, *Drechslera festuca* Scharif.
Fescues	*Festuca*	Powdery mildew, *Erysiphe graminis* Dc.; leaf and stem rusts, *Puccinia* spp.; stripe smut, *Ustilago striiformis* (West.) Nielssl.; leaf eye spot, *Helminthosporium vagans* Drechsl.; anthracnose, *Colletotrichum graminicola* (Ces.) G.W. Wils.
Ryegrasses	*Lolium*	Crown rust, *Puccinia coronata* Corda; brown rust, *Puccinia dispersa* Eriks. & Henn.; black stem rust, *Puccinia graminis* Pers.; leaf spot, *Drechslera festuca* Scharif.; ergot, *Cleviceps purpurea* (Fr.) Tul.; blind seed disease, *Gloeotinia temulenta* (Prill. & Delacr.) Wilson, Noble & Gray (syn. *Phialea temulenta* Prill. & Delacr.)
Orchard grass	*Dactylis*	Brown stripe, *Scolecotrichum graminis* Fckl.; leaf scald, *Rhynchosporium orthosporum* Cald.
Timothy	*Phleum*	Stem rust, *Puccinia graminis* var. phleipratensis (Eriks & E. Henn.) Stakman & Piem.

[a]From O'Rourke, 1976.

each and evaluated further for the desired component characters of yield and quality by taking actual measurements of dry matter yield and digestibility. The plants selected on the basis of yield and quality performance are intercrossed to combine the desired traits, and one or more cycles of selection are carried out. Sometimes, clones of selected plants are interpollinated at random to produce a new strain. A very large number of grass cultivars have been produced by this simple procedure of mass selection. In yet another version of grass breeding, selected plants are interpollinated with each other either in known crosses or in a polycross that permits maximum random mating between the clones. The resulting progenies are tested to determine the specific combining ability of the known parents in crosses or general combining ability (mean performance) of the individual plants entering

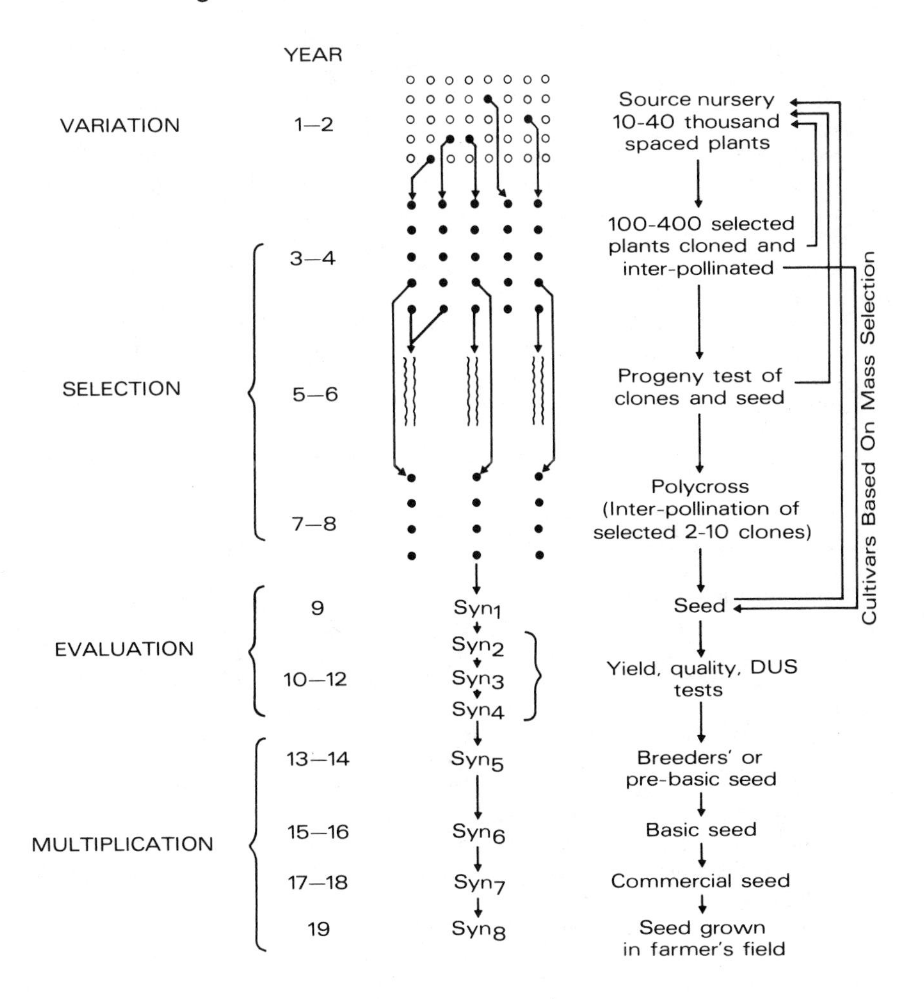

Figure 2. Schematic representation of conventional grass breeding procedure.

the polycross. The plants selected on the basis of progeny test are then intercrossed in a polycross, and equal amount of the resulting seed from each clone is mixed to produce a synthetic variety (syn_1) and tested for its performance in replicated small plot trials. The synthetic strains that show outstanding yield and quality performance are then multiplied for 4-6 generations to build sufficient seed stock to test yield performance in several locations and in larger tests and to examine distinctiveness, uniformity, and stability (DUS). The strains that show above-average performance over the control cultivars are multiplied and released for commercial propagation.

In a number of grasses, variation is often generated through interspecific hybridization and in some cases even by intergeneric hybridization. Such hybrids have been attempted on an extensive scale in the genera *Agropyron* (Stebbins et al., 1946a,b; Dewey, 1964, 1980), *Lolium* (Corkill, 1945; Ahloowalia, 1977), and *Lolium* x *Festuca* crosses (Buckner et al., 1961).

Interspecific and intergeneric hybridization has often been followed up by doubling the chromosome number to produce amphiploids, as in the case of *Lolium* x *Festuca*. Variation may also be generated by producing autopolyploids of grasses that are natural diploids and have a low chromosome number, e.g., *L. perenne* L. and *L. multiflorum* Lam. (Ahloowalia, 1967a,b).

In tissue culture technology, variants can be recovered in several ways, including spontaneous and induced mutagenesis in callus and cell cultures from leaf tissues, buds, embryos, and by cell fusion and cell transformation. The cultures may be initiated from any meristematic tissue of a plant. The variants may arise spontaneously or may be induced by X-rays, gamma rays, neutrons, or by adding chemicals such as ethane methane sulphonate or colchicine to the medium. In floricultural plants a number of variants have been produced by bud culture and irradiation (Broertjes, 1982), techniques as yet to be exploited for graminaceous crop plants.

Tissue culture techniques as yet have not made a significant contribution to the practical breeding of forage grass cultivars. To be of value, tissue culture techniques should permit the achievement of the following sequence of steps. It is essential that (a) aseptic cultures can be established routinely, (b) established cultures can be cloned and maintained by subculture, (c) plants can be regenerated in large numbers, (d) the regenerated plants survive in large numbers on transfer into soil, and (e) the regenerated plants are at least partially fertile, and produce sufficient pollen and seed for propagation of the desired genotypes. The last prerequisite is, of course, not essential in case of vegetatively propagated grasses such as sugarcane. Since most forage and fodder grasses are propagated through seed, fertility of the regenerated plants is of paramount importance.

Although the establishment of aseptic cultures of graminaceous crops was attempted successfully in the 1940s, whole plant regeneration from callus was not achieved until the late 1960s in cereals and early 1970s in forage grasses. As early as 1949, La Rue established maize endosperm cultures (La Rue, 1949; Straus and La Rue, 1954; Straus, 1954). Norstog (1956) reported the culture of endosperm excised from developing seeds of perennial ryegrass. These cultures were maintained for over 2 1/2 years. Norstog (1956) and Norstog et al. (1969) also recorded changes in the chromosome number and structure (anaphase bridges) of the cultured cells. However, no organogenesis was reported in these cultures. Similarly, although Schenk and Hildebrandt (1972) and Atkin and Barton (1973) reported the establishment of callus cultures from caryopsis of several temperate forage grasses (Table 4), no plants were regenerated from these cultures. Earlier, Gamborg et al. (1970) had succeeded in plant regeneration in smooth bromegrass, *Bromus inermis* L; however, regenerated plants were albino. Likewise, the single plantlet regenerated from anther callus of Italian ryegrass, *Lolium multiflorum* Lam. (Clapham, 1971), was also albino. Between

1967 and 1974 several graminaceous crop plants were regenerated from callus cultures. Among these were rice, *Oryza sativa* L. (Nishi et al., 1968; Niizeki and Oono, 1968), wheat, *Triticum* spp. (Shimada et al., 1969), sorghum, *Sorghum bicolor* (L.) Moench (Mastellar and Holden, 1970), broom corn millet, *Panicum miliaceum* L. (Rangan, 1974), and sugarcane, *Saccharum* sp. hybrid (Barba and Nickell, 1969; Heinz and Mee, 1969). In 1975 large-scale regeneration of plants was reported in maize, *Zea mays* L. (Green and Phillips, 1975) and ryegrass, *Lolium* spp. (Ahloowalia, 1975b). Since then a number of forage grass species have been regenerated from callus culture (Table 4). Likewise, the number of cereals regenerated from callus cultures has increased (King et al., 1978; Thomas et al., 1979).

Establishment of Callus Cultures and Plant Regeneration

In most forage grasses (see Table 5), callus can be induced from any meristematic tissue, e.g., apical meristems and young leaves (Dale, 1975; Wu and Antonovics, 1978; Haydu and Vasil, 1981; Lu and Vasil, 1981b), developing embryos and caryopses (Schenk and Hildebrandt, 1972; Atkin and Barton, 1973; Ahloowalia, 1975a,b; Conger and Carabia, 1978; Lowe and Conger, 1979), young inflorescences (Chen et al., 1977; Kasperbauer et al., 1979; Lo et al., 1980; Bajaj and Dhanju, 1981), endosperm (Norstog, 1956), and anthers (Clapham, 1971; Nitzsche, 1972; Pierson and Collins, 1981). In most cases supplementation of MS (Murashige and Skoog, 1962) or LS (Linsmaier and Skoog, 1965) medium with only 2,4-D is sufficient to induce callus formation. The continued proliferation and growth of the induced callus in subsequent subcultures, however, requires either a reduced level of 2,4-D than present in the callus induction medium or additional growth regulators. The amount of 2,4-D used in the medium for callus induction has varied from 2.3 to 67.8 μM depending upon the tissues cultured and species used for callus induction (Table 4). Callus from the endosperm of perennial ryegrass was induced on WH medium supplemented with 5.7 μM IAA and 20% CW or 0.25% yeast extract (Norstog, 1956) without adding 2,4-D. The addition of growth regulators such as IAA, KIN, CW, and yeast extract along with 2,4-D is not essential for callus induction in forage grasses. Addition of KIN as low as 0.47 μM was found to be inhibitory for callus induction in wheat, barley, rice, and bromegrass (Schenk and Hildebrandt, 1972) and a number of forage grasses (Atkin and Barton, 1973). Similarly, Lo et al. (1980) reported that KIN did not enhance callus initiation on medium containing 2,4-D.

Callus growth and proliferation requires a reduced amount of 2,4-D and other nutrients in the medium than that in the callus induction phase. This can be achieved by either extending the duration of callus subcultures or by transfer to half-strength medium with reduced levels of 2,4-D. In most forage grasses, 4.5–9.0 μM of 2,4-D was sufficient for continued growth and proliferation of callus cultures (Schenk and Hildebrandt, 1972; Atkin and Barton, 1973).

Callus cultures, which are usually unorganized masses but are often differentiated into various types of cells, may be cloned to produce

Table 5. Callus Induction in Forage Grasses

SPECIES	MEDIUM	2,4-D (μM)	TISSUE SOURCE	REFERENCE
Agrostis stolonifera L.	LS	18.0	Shoot meristem	Wu & Antonovics, 1978
	LS	2.3	Root meristem	Wu & Antonovics, 1978
A. tenuis Sibtth.	LS, Miller, WH	23.0	Root meristem[a]	Atkin & Barton, 1973
	LS, Miller, WH	4.5	Caryopsis[a]	Atkin & Barton, 1973
Agropyron cristatum (L.) Gaertn.	LS	23.0	Young inflorescence	Lo et al., 1980
A. smithii Rydb.	LS	23.0	Young inflorescence	Lo et al., 1980
Alopecurus arundinaceus Poir.	LS	23.0	Young inflorescence	Lo et al., 1980
Andropogon gerardii Vitman	LS	23.0	Young inflorescence	Chen et al., 1977
Bromus inermis Leyss.	B5	4.5	Mesocotyl, inflorescence	Gamborg et al., 1970
	LS	23.0	Young inflorescence	Lo et al., 1980
	SH	2.3–4.5	Caryopsis[a]	Schenk & Hildebrandt, 1972
Cynosurus cristatus L.	LS	4.5	Caryopsis[a]	Atkin & Barton, 1973
Dactylis glomerata L.	LS	23.0	Caryopsis[a]	Atkin & Barton, 1973
	SH	4.5	Caryopsis[a]	Schenk & Hildebrandt, 1972
	SH	67.8	Caryopsis	Conger & Carabia, 1978
Eleusine coracana Gaertn.	MS	18.0–45.0	Mesocotyl	Rangan, 1976
Festuca arundinacea Schreb.	MS	40.7	Embryos	Low & Conger, 1979
	MS	9.0	Anthers	Kasperbauer et al., 1980

SPECIES	MEDIUM	2,4-D (μM)	TISSUE SOURCE	REFERENCE
Festuca pratensis Huds.	LS	4.5	Caryopsis[a]	Atkin & Barton, 1973
F. rubra L.	LS	4.5	Caryopsis[a]	Atkin & Barton, 1973
Lolium multiflorum x *perenne*	MS	6.8	Caryopsis	Ahloowalia, 1975b
L. multiflorum Lam.	MS	31.6	Embryos	Dale, 1980b
L. multiflorum Lam. x *Festuca arundinacea* Schreb.	MS	9.0–18.0	Internodes, peduncles	Kasperbauer et al., 1979
L. multiflorum Lam.	LS	4.5	Caryopsis[a]	Atkin & Barton, 1973
L. perenne L.	LS	4.5	Caryopsis[a]	Atkin & Barton, 1973
L. temulentum L.	LS	4.5	Caryopsis[a]	Atkin & Barton, 1973
Paspalum scrobiculatum L.	MS	45.0	Mesocotyl	Rangan, 1976
Panicum miliaceum L.	MS	45.0	Mesocotyl	Rangan, 1974
	MS	9.0	Caryopsis	Bajaj et al., 1981b
P. maximum Jacq.	MS	11.3–45.0	Immature embryos, young inflorescences, developing leaves	Lu & Vasil, 1981a,b
P. virgatum L.	LS	23.0	Young inflorescence	Chen et al., 1977
Pennisetum typhoideum Pers.	MS	45.0	Mesocotyl	Rangan, 1976
P. americanum (L.) K. Schum.	LS	11.3	Immature embryos, protoplasts	Vasil & Vasil, 1980
P. purpureum Schum.	MS	23.0	Young inflorescences, seeds	Bajaj & Dhanju, 1981
	MS	2.3–45.0	Leaves	Haydu & Vasil, 1981
P. americanum x *P. purpureum*	MS	11.3	Immature embryos, young inflorescences	Vasil & Vasil, 1981
Phleum pratense L.	SH	4.5	Caryopsis[a]	Schenk & Hildebrandt, 1972
	LS	4.5	Caryopsis[a]	Atkin & Barton, 1973

Table 5. Cont.

SPECIES	MEDIUM	2,4-D (μM)	TISSUE SOURCE	REFERENCE
Poa trivialis L.	LS	4.5	Caryopsis[a]	Atkin & Barton, 1973
Phragmitis communis (Cav.) Trin. & Stend.	MS	4.5	Rhizomes	Sangwan & Gorenflot, 1975
Saccharum sp. hybrid	MS	13.5	Shoot apices, leaves	Heinz & Mee, 1969
Sorghastrum avenaceum (Michx.) Nash.	LS	23.0	Young inflorescence	Chen et al., 1977
Stipa viridula Trin.	LS	23.0	Young inflorescence	Lo et al., 1980

[a]Callus only; no plants were regenerated.

more calli or may be triggered to produce shoots, roots, and somatic embryos (embryoids) that ultimately grow into plants. If the calli produce roots before shoot primordia, it is very unlikely that shoot primordia will develop.

It is becoming increasingly evident that whereas the basal medium for callus induction, callus proliferation, organogenesis, and plantlet formation is the same, the amount and ratio of auxin to KIN and the type of growth regulators required are different and critical for each stage of callus growth and differentiation. For example, whereas addition of 2,4-D is essential for callus induction and proliferation, it is not required at organogenesis and plantlet formation. In general, a relatively high auxin and a low KIN concentration (e.g., 6.8 μM 2,4-D, 37.1 μM IAA, and 10.2 μM KIN in the case of ryegrass) promotes callus induction and proliferation. On the other hand, a relatively high level of KIN (4.6 μM) and low auxin (2.3 μM 2,4-D) in *Lolium* spp. promotes shoot formation. For somatic embryogenesis, a lower level of auxin is required than for shoot organogenesis. Complete removal of 2,4-D often leads to plant formation in callus cultures that have undergone organogenesis.

While callus cultures may be initiated from any meristematic tissue in grasses, not all callus cultures regenerate plants. In general, callus cultures derived from immature embryos of grasses have given the highest rate of plant regeneration; a reduced level of 2,4-D (2.3 μM), reduced strength of the medium (half-strength), and an extended duration of culture period promoted differentiation and organogenesis in forage grass callus cultures (Ahloowalia, 1975b; Chen et al., 1977; Lo et al., 1980). Addition of other growth regulators, e.g., zeatin, also promotes organogenesis in grasses (Ahloowalia, 1975b, 1982; Haydu and Vasil, 1981).

Several graminaceous crop plants have been regenerated from callus cultures, e.g., sugarcane (Heinz and Mee, 1969; Barba and Nickell, 1969; Nickell, 1977; Nadar and Heinz, 1977), rice (Nishi et al., 1968; Oono, 1978), wheat, *Triticum aestivum* L. (Shimada, 1978; Trione et al., 1968; Ahloowalia, 1982), *T. durum* L. (Bennici and D'Amato, 1978), maize (Green and Phillips, 1975; Harms et al., 1976), barley (Cheng and Smith, 1975), oats (Carter et al., 1967; Cummings et al., 1976), sorghum (Mastellar and Holden, 1970; Gamborg et al., 1977; Thomas et al., 1977; Brar et al., 1979; Wernicke and Brettell, 1980), *Panicum miliaceum* L. (Rangan, 1974), *Pennisetum typhoideum* Pers. = *P. americanum* (L.) K. Schum. (Rangan, 1976; Vasil and Vasil, 1980), and Triticale (Sharma et al., 1980), as well as several forage grasses. Some of the graminaceous crops mentioned above are also used as green or dried fodder in many parts of the world.

Perhaps one of the first successes in regeneration of forage grasses was in a group of tropical grass species (Urata and Long, 1968). Although no detailed account of this work was ever published, they reported that callus cultures of several grasses including *Brachiaria ruziziensis* Germain & Evrard, Pangola grass, *Digitaria decumbens* Stent, sugarcane, *Saccharum officinarum* L., wild cane, *S. spontaneum* L. and *Chloris radiata* (L.) Swartz, were obtained on medium containing inorganic salts, sucrose, agar, CW, and 2,4-D. This callus underwent organogenesis into shoot primordia that gave whole plants on transfer to medium lacking 2,4-D. These authors also stated that tissue culture could be useful for the propagation of sterile hybrids and induction of amphiploids by treating callus with colchicine.

Heinz and Mee (1969) reported regeneration of plants from callus cultures established from shoot apices, leaves, and inflorescences of *Saccharum officinarum*, *S. spontaneum*, and *Saccharum* spp. hybrids. The callus cultures were induced on MS medium supplemented with 13.6 μM 2,4-D and 10% (v/v) CW. Transfer of callus to a medium without 2,4-D produced plant differentiation. Some of the regenerated plants were morphological variants of the parental clones, thus establishing the possibility of producing new genotypes in grasses through callus culture.

Following the success with tropical grasses, Gamborg et al. (1970) reported plantlet regeneration from suspension cultures in cool-season grass species of smooth bromegrass; however, all the regenerated plants were albino. It was perhaps also the first time that somatic embryogenesis was obtained in a forage grass species in suspension cultures. The callus cultures had been initiated from immature seeds by Schenk and Hildebrandt (1972) with 2.3 μM 2,4-D, 10.7 μM p-chlorophenoxyacetic acid, and 0.46 μM KIN. These callus cultures were grown in liquid suspensions with 4.5 μM 2,4-D at 28 C in darkness. The cultured cells were washed with B5 medium without 2,4-D and incubated in B5 medium at 28 C and 150 rpm on a gyratory shaker. The suspension cultures yielded somatic embryos within 2 weeks, when these were transferred to B5 solid medium and kept at 18/6 hr light/dark period at 15/18 C. Several dozen plantlets appeared within 30 days, but all plants were albino.

These successes in forage grass regeneration set a trend of callus and cell culture for several other forage grasses (Table 5). During 1975, four authors simultaneously reported plant regeneration from callus culture in Gramineae: ryegrass (Fig. 3; Ahloowalia, 1975b), barley (Cheng and Smith, 1975), indiangrass, *Phragmites communis* L. (Sangwan and Gorenflot, 1975), and maize (Green and Phillips, 1975). Whereas in each of the species callus induction occurred on medium with 2,4-D, reduction or removal of 2,4-D from the medium produced plant regeneration. Ryegrass callus was induced from immature seeds on MS medium supplemented with 6.8 μM 2,4-D, 37.1 μM IAA, and 10.0 μM KIN, and plant regeneration occurred on half-strength MS medium with 3.4 μM 2,4-D, 21.4 μM IAA, and 5.0 μM KIN. Addition of CW (1 ml/l) promoted shoot organogenesis and chlorophyll development in the callus cultures. In barley, callus induction occurred from apical meristems on MS medium with 10 μM IAA, 15 μM 2,4-D, and 1.5 μM 2iP, and plant regeneration was obtained on the same medium without any of these growth regulators (Cheng and Smith, 1975). Addition of 20 μM NAA to the medium containing IAA and 2iP promoted callus proliferation; however, the requirement of 2iP for callus maintenance was not ascertained.

Sangwan and Gorenflot (1975) induced callus cultures from rhizomes of tetraploid (2n = 4x = 48) and octoploid (2n = 8x = 96) *Phragmites communis* L. on MS medium containing 4.5 μM 2,4-D. They also used a

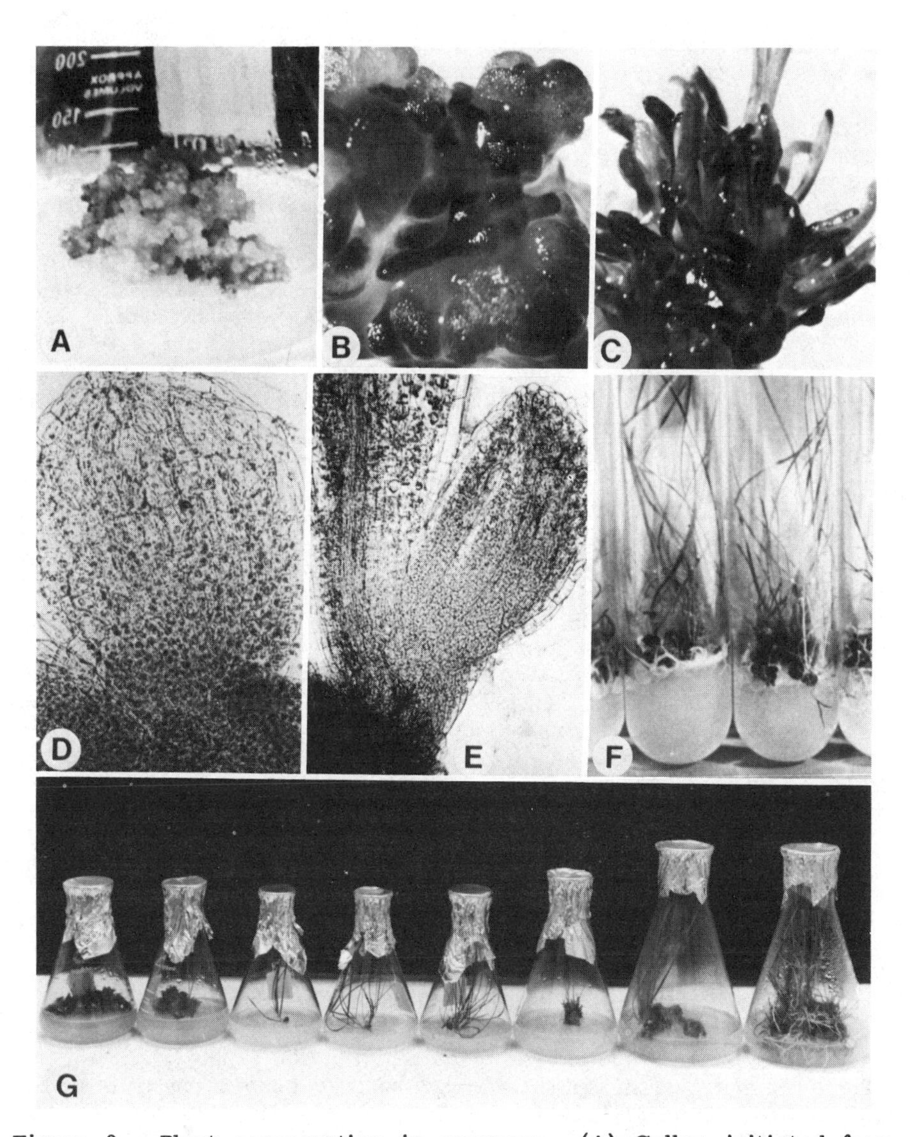

Figure 3. Plant regeneration in ryegrass. (A) Callus--initiated from developing embryo. (B) Organogenesis--formation of multiple shoot-primordia. (C) Shoot formation in large numbers. (D) A plantlet at single-leaf stage. (E) A plantlet at two-leaf stage. (F) Plantlet differentiation in large number in test tubes. (G) Callus proliferation and plant regeneration.

range of 2.3 to 45.2 µM 2,4-D concentrations and obtained callus induction on media with 2.3 to 13.6 µM 2,4-D. Kinetin did not enhance callus formation, and concentrations of 4.6-9.3 µM KIN inhibited callus induction. Higher KIN levels reduced callus production. Organ differentiation occurred on medium with 0.46 µM KIN plus 5% CW (v/v). The highest frequency of organogenesis (shoot formation) was observed on auxin free medium. Root formation was best promoted on medium containing a combination of IAA (22.8 µM) and KIN (0.93 µM). Sucrose requirement of *Phragmites* callus was high being 0.12 M. Cell suspensions were established, and 5 weeks after incubation produced small clumps which later produced only roots. No somatic embryos were observed in these cultures.

Green and Phillips (1975) induced callus cultures from immature embryos of maize on MS medium supplemented with 9.0 µM 2,4-D; shoot organogenesis and leaf development occurred on the same medium with 1.13 µM 2,4-D and plantlets with roots were formed on 2,4-D free medium. In the induced callus cultures, leaf organogenesis and development was promoted when calli were subcultured on MS medium containing 5.4 µM NAA and 0.24 µM 2iP.

Following the publication of these reports, several papers have appeared on plant regeneration in many of the important forage grasses, e.g., cocksfoot, *Dactylis glomerata* L. (Conger and Carabia, 1978), big bluestem, *Andropogon gerardii* Vitman (Chen et al., 1977), creeping bentgrass, *Agrostis stolonifera* L. (Wu and Antonovics, 1978), *A. palustris* Huds. (Krans et al., 1982), indiangrass, *Sorghastrum nutans* (Michx.) Nash (Chen et al., 1979), tall fescue, *Festuca arundinacea* Schreb. (Lowe and Conger, 1979; Conger and McDaniel, 1983), ryegrass x tall fescue hybrids, *Lolium multiflorum* Lam. x *F. arundinacea* Schreb. (Kasperbauer et al., 1979), annual ryegrass, *L. multiflorum* Lam. (Dale, 1980b), creeping foxtail, *Alopecurus arundinaceus* Poir, crested wheatgrass, *Agropyron cristatum* L. (Gaertn.), green needlegrass, *Stipa viridula* Trin., smooth bromegrass, *Bromus inermis* Leyss, and western wheatgrass, *Agropyron smithii* Rydb. (Lo et al., 1980), and napiergrass, *Pennisetum purpureum* Schum. (Bajaj and Dhanju, 1981; Haydu and Vasil, 1981). Likewise, plant regeneration has been reported in several fodder grasses, e.g., oats, *Avena sativa* L. (Cummings et al., 1976), sorghum, *Sorghum bicolor* L. Moench (Gamborg et al., 1977; Brar et al., 1979; Dunstan et al., 1979; Wernicke and Brettel, 1980), and pearl millet, *Pennisetum americanum* (L.) K. Schum. (Rangan, 1976; Vasil and Vasil, 1980). In all these forage and fodder grasses the requirement for callus induction has been addition of 2,4-D with or without IAA and its removal or reduction for plant regeneration (Table 5).

Plant regeneration in grasses through somatic embryogenesis in callus cultures was reported first by Gamborg et al. (1970) in *Bromus*, and since has been confirmed in a number of species, i.e., *Lolium* spp. (Ahloowalia, 1978, 1983), *L. multiflorum* Lam. (Dale, 1980b), *Sorghum bicolor* (L.) Moench (Thomas et al., 1977; Dunstan et al., 1978; Wernicke and Brettall, 1980), *Pennisetum americanum* and *P. americanum* x *P. purpureum* hybrids (Vasil and Vasil, 1981a,b), *Panicum maximum* Jacq. (Lu and Vasil, 1981a,b; Bajaj et al., 1981), maize, *Zea mays* L. (Lu et al., 1982), and *Saccharum officinarum* L. (Ahloowalia and Maretzki, 1983). In contrast to the shoot primordia, which appear as green clusters of meristematic regions dispersed among the callus, somatic embryos appear as bipolar structures and initiate as disc-shaped structures on the surface of the callus cultures. Low levels of auxin and high sucrose concentrations in the medium promote somatic embryogenesis.

Origin of Variation in Callus Cultures

Callus cultures undergo "spontaneous" changes in their genetic makeup. These changes may involve cytoplasmic structures, such as plastids and mitochondria, or chromosomes. The cytoplasmic changes may arise through mutation and asymmetric cell division. The chromosomal changes include chromosomal gain, loss, breakage and reunion (translocations), and increase in chromosome number by either doubling through endomitosis or cell fusion. Cells in a callus subculture may undergo fusion of nuclei, thus increasing chromosome number, and further undergo chromosomal loss, resulting in aneuploids, thus giving rise to an array of variants. This type of variation has been found in ryegrass callus cultures from which over 4000 plants were regenerated over a period of 5 years (Ahloowalia, 1976). An albino mutant sector of the triploid callus (2n = 21) regenerated albino plants, whereas the green sector produced green plants. The regenerated plants had a range in chromosome number, i.e., 2n = 21, 20, 19, 18, 15, 40, 30, 29, 28, 27, and 26, and all originated from the parental callus initially with 2n = 21 chromosomes (Ahloowalia, 1983). Some aneuploid plants showed chromosomal translocations. Similar types of chromosomal changes have been reported in the callus-regenerated plants of durum wheat *Triticum durum* L. 2n = 4x = 28. The regenerated plants had chromosome numbers of 2n = 14-56 (Bennici and D'Amato, 1978). The changed chromosome number in tissue-culture regenerates has also been reported in several other grass species, sugarcane (Heinz and Mee, 1971; Nickell, 1978), rice (Oono, 1978), ryegrass x fescue hybrids (Kasperbauer et al., 1979; cf. Larkin and Scowcroft, 1981 for examples in other species). It is not uncommon to obtain distinctly different phenotypes, perhaps representing genotypes, among the regenerates from the same callus. Such variants have been reported in ryegrass (Ahloowalia, 1976, 1983), *Panicum* spp. (Bajaj et al., 1981), and wheat (Ahloowalia, 1982). Callus can often be triggered to produce somatic embryos, which presumably originate from single cells and may include new variants.

Embryo Culture and Wide Crosses

Novel variation not found in natural gene pools can be generated by producing interspecific and intergeneric hybrids. Such hybrids have been obtained by forage grass breeders in several genera. Embryo culture is one of the earliest techniques used by plant breeders to obtain interspecific and intergeneric hybrids where routine fertilization fails to produce viable offspring. Plant embryos were cultured for the first time by E. Hannig (cf. Rappaport, 1954). As early as 1934 Tukey reported embryo culture of deciduous fruits. By 1945 Randolph had perfected embryo culture to produce plantlets in *Iris*, and Lammerts (1942) had used embryo culture for shortening the breeding cycle of deciduous trees and increasing hybrid seed germination. Several forage legume hybrids were successfully produced through embryo culture by 1953 (Keim, 1953). These studies clearly established that in most situations nutritional, physiological, and genetic incompatibility of the developing embryo with that of the endosperm resulted in early zygotic lethality which could be overcome by embryo culture (Rappaport, 1954). The failure to obtain plants from several interspecific and intergeneric crosses in forage grasses and related cereals can be ascribed to pre- and post-zygotic barriers (Matzk et al., 1980). Even within a species, difference in chromosome number (ploidy) of the parental plants results in early embryonic death; e.g., diploid x tetraploid cross in *Lolium perenne* L. seldom if ever produces triploids unless the developing embryos are cultured (Ahloowalia, 1975a).

Embryo culture as a technique has taken a new dimension in recent years, and hybrids from a series of wide crosses have been produced that 10 years ago were thought to be impossible. Intergeneric crosses between barley x rye *Hordeum vulgare* L. x *Secale cereale* L. (Kruse, 1967; Fedak, 1979), barley x wheat, *H. vulgare* L. x *Triticum aestivum* L. (Kruse, 1973; Islam et al., 1976, 1981; Mujeeb et al., 1978), *H. vulgare* L. x *T. monococcum* L. (Kruse, 1973), *H. vulgare* x *T. turgidum* L. (Mujeeb et al., 1978), and trigeneric hybrids combining barley x wheat x rye, *H. vulgare* L. x *T. aestivum* L. x *S. cereale* (Fedak and Armstrong, 1980) have been obtained by combining in vivo treatment of the pollinated florets with GA_3 and in vitro culture of the excised developing embryos. In some cases, success in crossing wide species has been ascribed to the use of immunosuppressants such as E-amino-caproic acid or chloramphenicol (Bates et al., 1974) and in others to growth regulators such as IAA, GA, and 2,4-D at pre- and post-fertilization stages. In wide crosses of cereals, GA has been used following pollination of barley with wheat or rye. The intergeneric hybrid between sugarcane x maize, *Saccharum officinarum* L. x *Zea mays* L., produced through conventional hybridization, was induced to complete inflorescence emergence by GA treatment (Janaki Ammal et al., 1972), after nearly 30 years of vegetative propagation. Intergeneric hybrids between ryegrass x fescue, *Lolium perenne* L. x *Festuca rubra* L., have been produced through whole ovary culture (Nitzsche and Henning, 1976). More recently hybrids from wide cross between *F. arundinacea* Schreb. and *Dactylis glomerata* L. were obtained through embryo culture technique (Matzk, 1981).

Most of the wide crosses in cereals and grasses are completely sterile. In others, chromosomes are eliminated in the progeny, and often chromosomes of one parent are lost in preference to those of the other parent. In some cases the sterile hybrids can be induced to form callus, and plants can be regenerated (Pickering and Thomas, 1979). In addition, callus cultures of such hybrids produce an array of chromosomal variants either without any treatment, as reported in sugarcane hybrids (Heinz and Mee, 1969, 1971), ryegrass, *Lolium multiflorum* Lam. x *L. perenne* (Ahloowalia, 1976), or colchicine treatment of the callus from interspecific hybrids between *Hordeum vulgare* L. x *H.*

jubatum L. (Orton and Steidl, 1980), and intergeneric hybrids between *Triticum crassum* [(Boiss.) Aitch and Hensl] x *H. vulgare* L. (Nakamura et al., 1981).

Cell and Protoplast Culture

In some grass species, it is now feasible to grow callus in liquid media and obtain single cells or clumps of cells in suspension culture, e.g., in smooth bromegrass, *Bromus inermis* (Gamborg et al., 1970), ryegrass, *Lolium* spp. (Ahloowalia, 1975b), and pearl millet, *Pennisetum americanum* (Vasil and Vasil, 1981b), *Panicum maximum* (Lu and Vasil, 1981b). However, in most grasses single-cell cultures have not responded to plant regeneration in the same manner as in the model dicot plants such as carrot, tobacco, petunia, and potato (Smith, 1974). The best success to date has been in *Pennisetum* (Vasil and Vasil, 1981b) and *Panicum* spp. (Lu and Vasil, 1981a), and in both cases, cell suspensions were obtained from immature embryos (see protocol section), and the plants originated through somatic embryogenesis.

A further refinement of cell culture techniques has been the removal of cell walls with enzymes to produce protoplasts (see protocol section). Protoplast culture has been successfully used in regenerating several interspecific hybrid plants (see Evans, Vol. 1, Chap. 4). In forage grasses and cereals no wide somatic hybrids have so far been produced through protoplast fusion. However, complete plant regeneration from protoplast culture has been reported in pearl millet (Vasil and Vasil, 1980), suggesting that this approach may be useful in the near future.

Since wide crosses produced through conventional sexual crossing have been of little or no value so far, somatic hybrids may not yield any economically important cultivars in the immediate future. However, somatic hybrids that involve cytoplasmic fusion of the donor species, even when most of the nuclear genetic material is eliminated after protoplast fusion, may provide a new series of useful variants that are not produced through sexual crossing. The use of DNA transfer and cell modification in grasses is in its infancy and it will be some time before these techniques have a significant impact on conventional grass breeding.

ANTHER CULTURE

If variation can be produced by tissue culture techniques, the converse of it, homozygosity, can also be greatly speeded up, saving several years of time and labor. Unlike most cereals, forage grasses are cross-pollinated, and homozygosity is not as important as in the breeding of pure line varieties. Nonetheless, for the production of inbred parental lines for selection of specific characters and exploitation of heterosis, homozygous genotypes are necessary.

The regeneration of haploids from anther and microspore culture (Maheshwari et al., 1980) can produce not only rapid homozygosity but also create genetic variation (Burke and Chaplin, 1980; Paepe De et al., 1981; Mix et al., 1978). The microspores from a hybrid or a highly heterozygous genotype of a forage grass species would give rise to an array of homozygous haploid genotypes. Doubling the chromosome number of these haplodis would produce homozygous diploid plants that could be used in conventional breeding programs.

In forage grasses haploid production from in vitro anther culture has been rather limited. The first haploids from anther culture of annual ryegrass, *Lolium multiflorum* Lam., were albino (Clapham, 1971). The anthers were cultured on Linsmaier and Skoog's (1965) medium supplemented with 0.35 M sucrose, 15% CW, and 5.4 µM NAA. About 20% of the anthers developed callus of which 3% formed plants, which were all albino. A single dihaploid plant (2n = 14) was obtained from anther culture of tetraploid (2n = 4x = 28) *L. multiflorum* Lam. (Nitzsche, 1970). Following these early studies, (poly) haploid plants were successfully regenerated from anther culture of hybrids between *L. multiflorum* (4x) x *Festuca arundinacea* (12x), *F. pratensis* (2x) x *L. multiflorum* (4x), and *F. pratensis* (4x) x *L. multiflorum* (4x). However, the success rate was extremely low: only 15 plants, of which 4 were albino, were regenerated from over 55,000 cultured anthers (Nitzsche, 1972; Nitzsche and Wenzel, 1977). Likewise, only a single haploid plant (2n = 3x = 21) was obtained from culturing approximately 65,000 anthers of oats, *Avena sativa* L. (Rines, 1983). More recently Kasperbauer et al. (1980) have regenerated haploids in tall fescue, *F. arundinacea* Schreb. (2n = 6x = 42), by culturing anthers while still attached to panicles. The panicle pieces were cultured on modified MS medium and supplemented with 9.0 µM 2,4-D that was reduced to 1.1 µM once plantlets were formed. Panicle preculture and cold treatment at 5 C for 0-34 days, followed by culture of excised anthers after 2 weeks of preculture on N6 medium, also produced green and albino plants in tall fescue (Pierson and Collins, 1981). In both studies with tall fescue the regenerated plants included aneuploids.

Since some grasses are apomicts (e.g., *Poa*) and in others unreduced egg-cells are known to produce plants, culture of whole ovaries would be valuable to obtain homozygous genotypes. The haploids regenerated from unfertilized ovule-ovary culture would contain maternally inherited characters and therefore may be genetically distinct from the haploids obtained from microspore regenerates. Such studies, however, have not been undertaken in forage grasses. Furthermore, a combination of sexual-crossing and tissue culture techniques for production of haploids as exploited in the barley cross between *H. vulgare* L. x *H. bulbosum* L., in which the set of *H. bulbosum* chromosomes is selectively eliminated (Jensen, 1981), has not been developed in forage grasses.

SELECTION

Once sufficient variation exists in the breeding material, selection of the desired genotypes is necessary. Under field conditions, selection in forage grasses is done on single-spaced plants in the source nursery (Fig. 2). Selection in tissue culture may be carried out at either the cell, callus, or plantlet level. Currently, very few characteristics can be selected at the cellular level. The fact that extremely large numbers of cells and plantlets can be grown in a small space makes tissue culture selection for some characters more efficient than conventional field selection. Moreover, manipulation of the growth medium may provide an efficient and rapid screening of cells of callus. Selection methods have been developed for resistance to toxins produced by pathogens and resistance to stress. Tobacco cells screened on a medium containing methionine sulfoximine—an analog related to the toxin produced by *Pseudomonas tabaci*—regenerated plantlets that showed an increased resistance to the bacterium. Resistance to methionine sulfoximine was determined by a single semidominant gene (Carlson, 1973). Gengenbach et al. (1977) selected maize cell lines resistant to toxin produced by southern corn blight, *Helminthosporium maydis* race T originating from Texas male sterile cytoplasm (cms-T) genotype, which is susceptible to the pathogen. Regenerated plantlets were resistant to the disease.

Earlier, Dix and Street (1975) had shown that it was possible in *Nicotiana sylvestris* Speg. & Comes and *Capsicum annuum* L. to select salt tolerant cell lines which were capable of growing in 0.34 M NaCl. By using elevated levels of copper and zinc in the growth medium, Wu and Antonovics (1978) selected cell lines of creeping bentgrass, *Agrostis stolonifera* L., with high tolerance to these elements. The plants subsequently regenerated from these cells showed similar levels of tolerance to copper and zinc. Likewise, Fitch and Moore (1981) have reported selection for salt tolerance in sugarcane. Plants regenerated from callus were cultured on modified MS medium containing 0.31 or 0.34 M NaCl. The surviving plantlets on subsequent culture on medium with 0.20-0.34 M NaCl were salt tolerant and proved to be salt resistant in field tests. Although in vitro selection for tolerance to stress has not been undertaken, these studies illustrate the potential of tissue culture techniques for selection of specific variants in forage grasses. A number of other characters, such as increased tolerance to herbicides, alkalinity, cold, heat, and increased photosynthesis, are amenable to in vitro selection by including appropriate compounds in the growth medium or by subjecting the cultures to selective growth conditions. Recessive nuclear and cytoplasmic characters may be selected more effectively at the haploid level.

MAINTENANCE OF CLONES AND SEED MULTIPLICATION

The final step, seed multiplication and varietal maintenance, is as important as the breeding procedure itself. In a large number of forage grasses, e.g., ryegrasses, timothy, fescues, and orchard grass, the parental genotypes or clones have to be maintained by vegetative propagation for producing the first generation (Syn 1) seed at regular intervals. It is now possible to multiply and store clones by meristem culture of several grasses, including ryegrass, fescue, timothy, and orchard grass (Dale, 1977, 1980a). The rate of in vitro multiplication is far more rapid than in conventional pots and plots. In addition, propagation of grasses through in vitro techniques allows maintenance of clones free from virus diseases and insects. Even when field-grown clones become infected with viruses, viral eradication can be achieved through thermotherapy and meristem culture. The eradication of viruses by meristem tip culture from parental clones of old, established cultivars of ryegrass has been successful (Dale, 1977; Dale et al., 1980). In some cases tissue culture is the only method to recover healthy clones from the virus-infected plants.

The in vitro propagated clones are more uniform in growth, free from insects and viruses, and have better growth on planting than the conventionally propagated material. The maintenance of grass clones by in vitro culture offers perhaps the simplest and the most useful method in grass breeding procedures, because 3 or more years must elapse before clonal progenies can be tested for their performance, and meanwhile the parental genotypes have to be maintained in the field for recrossing to produce the first generation seed.

PROTOCOLS

Embryo-Callus Induction in Grasses

1. Harvest whole inflorescence 10–12 days after anthesis.
2. Rinse inflorescence in tap water, and wipe floral surface with tissue paper soaked in 95% ethanol.
3. Cut the inflorescence into 2–3 cm long pieces, and dissect out seeds with seed covers (lemma and palea are still attached).
4. Soak in sterile water 4–6 hr or overnight.
5. Remove lemma and palea, and surface sterilize seeds with Clorox (or sodium hypochlorite, 5%) diluted with sterile water (1:4 v/v) for 10–15 min. Surface sterilize individual seeds further by a quick dip (1–3 sec) in 95% alcohol and float in sterile distilled water. To avoid cross infection, float single seeds on separate water droplets in petri dish.
6. Dissect out embryos from large seeded grass species. Culture whole caryopsis in cases of small seeded grasses on medium with 2,4-D. The amount of 2,4-D varies with the species (Table 5). Callus cultures initiate within 6–10 days and may be kept for the next 30 days before subculture.

Cell Suspension Culture

1. Place callus pieces induced from immature embryos (about 1 cm^3) in 150–200 ml flasks with 50 ml liquid medium supplemented with 2,4-D (low level, 0.45–0.9 μM) and CW.
2. Place flasks on gyratory shaker at 150 rpm for 6–10 days in dark at 15 C.
3. Remove flasks from shaker and let stand for half an hour.
4. Decant approximately top half of the supernatant and reject.

5. Filter the remaining culture suspension through presterilized steel mesh. Large callus pieces stay up; fine cell suspension settles.
6. Centrifuge the filtrate for 3 min at 100 x g and remove the supernatant with a Pasteur pipette.
7. Add fresh medium and bring to a known volume. Determine cell number/ml of medium. Adjust density to 10,000-15,000 cells/ml. Plate cells in droplets on solid medium with 1% CW + 9.1 μM ZEA, 2.3 μM 2,4-D, seal and incubate upside down in diffuse light. Observe colony formation after 14 days. For selection to salt tolerance, use medium with 0.03-0.05 M NaCl.
8. Some cell colonies may undergo somatic embryogenesis and produce profuse shoot primordia. Replant selected plantlets on selective medium (i.e., high salt concentration). Surviving plantlets may be cultured for another 30 days or more before transplanting into soil.

Protoplast Isolation and Culture in Grasses

1. Take 10 ml of 4-6-day-old cell suspension cultures (as grown in previous protocol for embryo-callus induction).
2. Mix with 50 ml filter-sterilized nondesalted enzyme mixture containing 2% cellulase, 1% macerozyme, 0.5% driselase, 0.5% Rhozyme, 0.25 M sorbitol, 0.25 M mannitol, 1.38 mM glucose, 3 mM MES buffer in LS medium without growth regulators. Adjust to pH 5.5-5.6.
3. Incubate for 15-20 hr at 15 C in dark.
4. Filter the mixture through cheesecloth and then through 100 and 50 μm stainless steel mesh.
5. Pour into centrifugation tubes, spin at 100 x g for 3 min. Remove supernatant. Add 5 ml MS medium, 0.4 M glucose, 3.5 mM sucrose, and 4.5 μM 2,4-D; recentrifuge. Repeat the process three times to remove the last traces of enzymes.
6. Bring pellet to a known volume of about 4-5 ml. Break the pellet by a gentle shaking. Determine protoplast density.
7. Plate protoplasts onto solid media (see Vasil and Vasil, 1980, for additional details).

CONCLUSIONS AND FUTURE PROSPECTS

This brief review on grass improvement suggests that tissue culture techniques may provide novel ways to generate variation, carry out in vitro selection for some characters, and multiply disease-free clones. These techniques are only complementary to and not a replacement of the established conventional breeding methods. Compared to breeders of major graminaceous crops such as rice, wheat, maize, oats, barley, and sugarcane, grass breeders have not utilized the genetic diversity available to them in natural gene pools. Interspecific and intergeneric hybrids also have not been extensively used in developing new grass cultivars. Tissue culture methods are likely to play a key role in obtaining wide variation by removing barriers to interspecific and intergeneric crosses, whether by sexual crossing and embryo culture or through somatic hybridization. The culture of developing embryos for producing plantlets either directly or through callus, cell and protoplast culture may shorten the breeding and selection process, as would in vitro selection of plants tolerant to stress. The technology of somatic cell fusion and plant regeneration in grasses is not sufficiently advanced to make significant contributions in breeding new and novel cultivars. Grass and cereal cultivars based on in vitro culture techniques have yet to become a commercial success. The ultimate measure of success in plant breeding is the release of new cultivars and the acreage these cultivars cover and the time these cultivars last before being replaced by better types. So far no specific cultivar in cereals and grasses can be credited to any tissue culture work. However, time, labor, and space-saving through tissue culture may cut down on the number of years required to develop new cultivars. The adoption of tissue culture technology by breeders could thus immediately enhance the rate of release of conventionally bred grass cultivars and in due course provide novel recombinants not obtained by conventional breeding.

ACKNOWLEDGMENTS

Part of this review was compiled during sabbatic leave at the Department of Genetics and Pathology, Hawaiian Sugar Planters' Association. I wish to thank Dr. D.J. Heinz, Vice President and Director of Research, and Dr. Andrew Maretzki for their valuable suggestions. My sincere thanks are also due to Mrs. Betsy Sakamoto, who undertook the onus of typing the script.

KEY REFERENCES

Carnhan, H.L. and Hill, H.D. 1961. Cytology and genetics of forage grasses. Bot. Rev. 27:1-162.

Heath, M.E., Metcalfe, D.S., and Barnes, R.E. 1973. Forages: The Science of Grassland Agriculture, 3rd ed. Iowa State Univ. Press, Ames.

Larkin, P.J. and Scowcroft, W.R. 1981. Somaclonal variation—A novel source of variability from cell cultures for plant improvement. Theor. Appl. Genet. 60:197-214.

Schenk, R.U. and Hildebrandt, A.C. 1972. Medium and techniques for induction and growth of monocotyledonous and dicotyledonous plant cell cultures. Can. J. Bot. 50:199-204.

Thomas, E., King, P.J., and Potrykus, I. 1977. Shoot and embryo-like structure formation from cultured tissues of *Sorghum bicolor*. Naturwissenschaften 64:587.

REFERENCES

Ahloowalia, B.S. 1967a. Chromosome association and fertility in tetraploid ryegrass. Genetica 38:471-484.

______ 1967b. Colchicine induced polyploids in ryegrass. Euphytica 16:49-60.

______ 1975a. Triploid hybrids between Italian and perennial ryegrass. Euphytica 24:413-419.

______ 1975b. Regeneration of ryegrass plants in tissue culture. Crop Sci. 15:449-452.

______ 1976. Chromosomal changes in parasexually produced ryegrass. In: Current Chromosome Research (K. Jones and P.E. Brandham, eds.) pp. 115-122. Elsevier/North Holland Biomedical Press, Amsterdam.

______ 1977. Hybrids between tetraploid Italian and perennial ryegrass. Theor. Appl. Genet. 49:229-235.

______ 1978. Novel ryegrass genotypes regenerated from embryo callus culture. In: Fourth International Congress Plant Tissue Cell Culture (abst.) p. 162. Calgary.

______ 1982. Plant regeneration from callus culture in wheat. Crop Sci. 22:405-410.

______ 1983. Spectrum of variation in somaclones of triploid ryegrass. Crop Sci. 23:1141-1147.

______ and Maretzki, A. 1983. Plant regeneration via somatic embryogenesis in sugarcane. Plant Cell Rep. 2:21-25.

Atkin, R.K. and Barton, G.E. 1973. The establishment of tissue culture of temperate grasses. J. Exp. Bot. 24:689-699.

Bajaj, Y.P.S. and Dhanju, M.S. 1981. Regeneration of plants from callus cultures of napier grass (*Pennisetum purpureum*). Plant Sci. Lett. 20: 343-345.

Bajaj, Y.P.S., Sidhu, B.S., and Dubey, V.K. 1981. Regeneration of genetically diverse plants from tissue cultures of forage grass--*Panicum* spp. Euphytica 30:135-140.

Barba, R. and Nickell, L.G. 1969. Nutrition and organ differentiation in tissue cultures of sugarcane, a monocotyledon. Planta 89:299-302.

Bates, L.S., Campos, A.V., Rodriguez, R.R., and Anderson, R.G. 1974. Progress towards novel cereal grains. Cereal Sci. Today 19:283-286.

Bennici, A. and D'Amato, F. 1978. In vitro regeneration of *Durum* wheat plants. I. Chromosome numbers of regenerated plantlets. Z. Pflanzenzuecht. 81:305-311.

Blakeslee, A., Belling, J., Farnham, M.E., and Berger, A.D. 1922. A haploid mutant in *Datura stramonium*. Science 55:646-647.

Blakeslee, A.F., Morrison, G., and Avery, A.G. 1927. Mutations in a haploid *Datura*, and their bearing on the hybrid-origin theory of mutants. J. Hered. 18:193-199.

Brar, D.S., Rambold, S., Gamborg, O., and Constabel, F. 1979. Tissue culture of corn and sorghum. Z. Pflanzenphysiol. 95:377-388.

Broertjes, C. 1982. Significance of in vitro adventitious bud technique for mutation breeding of vegetatively propagated crops. In: Induced Mutations in Vegetatively Propagated Plants II, pp. 1-10. International Atomic Energy Agency, Vienna.

Buckner, R.C., Hill, H.D., and Burrus, P.B., Jr. 1961. Some characteristics of perennial and annual ryegrass x tall fescue hybrids and of the amphidiploid progenies of annual ryegrass x tall fescue. Crop Sci. 1:75-80.

Burk, L.G. and Chaplin, J.F. 1980. Variation among anther-derived haploids from a multiple disease-resistant tobacco hybrid. Crop Sci. 20:334-338.

Carlson, P.S. 1973. Methionine sulfoximine-resistant mutants of tobacco. Science 180:1366-1368.

Carter, O., Yamada, Y., and Takahashi, E. 1967. Tissue culture of oats. Nature 214:1029-1030.

Chen, C.H., Stenberg, N.E., and Ross, J.G. 1977. Clonal propagation of big bluestem by tissue culture. Crop Sci. 17:847-850.

Chen, C.H., Lo, P.F., and Ross, J.G. 1979. Regeneration of plantlets from callus cultures of Indiangrass. Crop Sci. 19:117-118.

Cheng, T.Y. and Smith, H.H. 1975. Organogenesis from callus culture of *Hordeum vulgare*. Planta 123:307-310.

Clapham, D. 1971. In vitro development of callus from the pollen of *Lolium* and *Hordeum*. Z. Pflanzenzuecht. 65:285-292.

Conger, B.V. and Carabia, J.V. 1978. Callus induction and plantlet regeneration in orchardgrass. Crop Sci. 18:157-159.

______, and McDaniel, J.K. 1983. Use of callus cultures to screen tall fescue seed samples for *Acremonium coenophialum*. Crop Sci. 23:172-174.

Corkhill, L. 1945. Short rotation ryegrass, its breeding and characteristics. N. Z. J. Agric. 71:465-470.

Cummings, D.P., Green, C.E., and Stuthman, D.D. 1976. Callus induction and plant regeneration in oats. Crop Sci. 16:465-470.

Dale, P.J. 1975. Meristem tip culture in *Lolium multiflorum*. J. Exp. Bot. 26:731-736.

______ 1977. Meristem tip culture in *Lolium*, *Festuca*, *Phleum* and *Dactylis*. Plant Sci. Lett. 9:333-338.

______ 1980a. A method for in vitro storage of *Lolium multiflorum* Lam. Ann. Bot. 45:497-502.

______ 1980b. Embryoids from cultured immature embryos of *Lolium multiflorum*. Z. Pflanzenphysiol. 100:73-77.

______, Cheyne, V.A., and Dalton, S.J. 1980. Pathogen elimination and in vitro plant storage in forage grasses and legumes. In: Tissue Culture Methods for Plant Pathologists (D.S. Ingram and J.P. Helgeson, eds.) pp. 119-124. Blackwell Scientific, Oxford.

Dewey, D.R. 1964. Genome analysis of *Agropyron repens* x *Agropyron cristatum* synthetic hybrids. Am. J. Bot. 51:1052-1068.

______ 1980. Hybrids and induced amphiploids involving *Agropyron curvifolium*, *A. repens*, and *A. desertorum*. Crop Sci. 20:473-478.

______ 1981. Cytogenetics of *Agropyron ferganense* and its hybrids with six species of *Agropyron*, *Elymus*, and *Sitanion*. Am. J. Bot. 68:216-225.

Dix, P.J. and Street, H.E. 1975. Sodium chloride-resistant cultured cell lines from *Nicotiana sylvestris* and *Capsicum annuum*. Plant Sci. Lett. 5:231-237.

Dunstan, D.I., Short, K.C., and Thomas, E. 1978. The anatomy of secondary morphogenesis in cultured scutellum tissues of *Sorghum bicolor*. Protoplasma 97:251-260.

Dunstan, D.I., Short, K.C., Dhaliwal, H., and Thomas, E. 1979. Further studies on plantlet production from cultured tissues of *Sorghum bicolor*. Protoplasma 101:355-361.

Fedak, G. 1979. A viable hybrid between *Hordeum vulgare* and *Secale cereale*. Cereal Res. Commun. 6:353-358.

______ and Armstrong, K.C. 1980. Production of trigeneric (barley x wheat) x rye hybrids. Theor. Appl. Genet. 56:221-224.

Fitch, M.M. and Moore, P.H. 1981. The selection for salt resistance in sugarcane (*Saccharum* spp. hybrids) tissue cultures (abstr.). Plant Physiol. 67:26.

Gamborg, O.L., Constable, F., and Miller, R.A. 1970. Embryogenesis and production of albino plants from cell cultures of *Bromus inermis*. Planta 95:353-358.

Gamborg, O.L., Shyluck, J.P., Brar, D.S., and Constable, F. 1977. Morphogenesis and plant regeneration from callus of immature embryos of sorghum. Plant Sci. Lett. 10:67-74.

Gengenbach, B.G., Green, C.E., and Donovan, C.M. 1977. Inheritance of selected pathotoxin resistance in maize plants regenerated from cell cultures. Proc. Natl. Acad. Sci. USA 74:5113-5117.

Green, C.E. and Phillips, R.L. 1975. Plant regeneration from tissue cultures of maize. Crop Sci. 15:417-421.

Gosch-Wackerle, G., Avivi, L., and Galun, E. 1979. Induction, culture, and differentiation of callus from immature rachises, seeds, and embryos of *Triticum*. Z. Pflanzenphysiol. 91:267-278.

Guha, S. and Maheshwari, S.C. 1964. In vitro production of embryos from anthers of *Datura*. Nature 204:497.

Harms, C.T., Lorz, H., and Potrykus, I. 1976. Regeneration of plantlets from callus cultures of *Zea mays* L. Z. Pflanzenzuecht. 77:347-351.

Haydu, Z. and Vasil, I.K. 1981. Somatic embryogenesis and plant regeneration from leaf tissues and anthers of *Pennisetum purpureum* Schum. Theor. Appl. Genet. 59:269-273.

Heinz, D.J. and Mee, G.W.P. 1969. Plant differentiation from callus tissue of *Saccharum* species. Crop Sci. 9:346-348.

______ 1971. Morphologic, cytogenetic and enzymatic variation in *Saccharum* species hybrid clones derived from callus tissues. Am. J. Bot. 58:257-262.

Islam, A.K., Shepherd, K.W., and Sparrow, D.H.B. 1976. Addition of individual barley chromosomes to wheat. In: Barley Genetics III (Proceedings 3rd International Barley Genetic Symposium) (H. Gaul, ed.) pp. 260-270. Karl Thiemig, Munich.

______ 1981. Isolation and characterization of euplasmic wheat-barley chromosome addition lines. Heredity 46:161-174.

Janaki Ammal, E.K., Jagathesan, D., and Sreenivasan, T.V. 1972. Further studies in *Saccharum-Zea* hybrid. I. Mitotic studies. Heredity 28:141-142.

Jensen, C.J. 1981. Use of cell and tissue culture techniques in plant breeding and genetics. In: Genetic Engineering for Crop Improvement (K.O. Rachie and J.M. Lyman, eds.) pp. 87-104. Rockefeller Foundation, New York.

Kasperbauer, M.J., Buckner, R.C., and Bush, L.P. 1979. Tissue culture of annual ryegrass x tall fescue F_1 hybrids: Callus establishment and plant regeneration. Crop Sci. 19:457-460.

Kasperbauer, M.J., Buckner, R.C., and Springer, W.D. 1980. Haploid plants by anther-panicle culture of tall fescue. Crop Sci. 20:103-107.

Keim, W.F. 1953. An embryo culture technique for forage legumes. Agron. J. 45:509-510.

Kimber, G. and Riley, R. 1963. Haploid angiosperms. Bot. Rev. 29: 490-531.

King, P.J., Potrykus, I., and Thomas, E. 1978. In vitro genetics of cereals: Problems and perspectives. Physiol. Veg. 16:381-399.

Krans, J.V., Henning, V.T., and Torres, K.C. 1982. Callus induction, maintenance and plantlet regeneration in creeping bentgrass. Crop Sci. 22:1193-1197.

Kruse, A. 1967. Intergeneric hybrids between *Hordeum vulgare* L. spp. *distichum* (v. Pallas, 2n = 14) and *Secale cereale* L. (v. Petkus, 2n = 14). Royal Veterinary and Agriculture College Yearbook, pp. 82-92. Copenhagen.

______ 1973. *Hordeum* x *Triticum* hybrids. Hereditas 73:157-161.

Lammerts, W.E. 1942. Embryo culture—an effective technique for shortening the breeding cycle of deciduous trees and increasing germination of hybrid seed. Am. J. Bot. 29:166-171.

La Rue, C.D. 1949. Cultures of the endosperm of maize. Am. J. Bot. 34:585-586.

Linsmaier, E.M. and Skoog, F. 1965. Organic growth factor requirement for tobacco tissue cultures. Physiol. Plant. 18:100-127.

Lo, P.F., Chen, C.H., and Ross, J.G. 1980. Vegetative propagation of temperate forage grasses through callus culture. Crop Sci. 20:363-367.

Lowe, K.W. and Conger, B.V. 1979. Root and shoot formation from callus cultures of tall fescue. Crop Sci. 19:397-400.

Lu, C. and Vasil, I.K. 1981a. Somatic embryogenesis and plant regeneration from freely suspended cells and cell groups of *Panicum maximum* Jacq. Ann. Bot. 48:543-548.

______ 1981b. Somatic embryogenesis and plant regeneration from leaf tissues of *Panicum maximum* Jacq. Theor. Appl. Genet. 59:275-280.

______, and Ozias-Akins, P. 1982. Somatic embryogenesis in *Zea mays* L. Theor. Appl. Genet. 62:109-112.

Maheshwari, S.C., Tyagi, A.K., Malhotra, K., and Sopory, S.K. 1980. Induction of haploidy from pollen grains in angiosperms—The current status. Theor. Appl. Genet. 58:193-206.

Mastellar, V.J. and Holden, D.J. 1970. The growth of and organ formation from callus tissues of sorghum. Plant Physiol. 45:362-364.

Matzk, F. 1981. Successful crosses between *Festuca arundinacea* Schreb. and *Dactylis glomerata* L. Theor. Appl. Genet. 60:119-122.

______, Grober, K., and Zacharias, M. 1980. Ergebnisse von Art- und Gattungskreuzungen mit Gramineen im Zusammenhang mit den naturlichen Isolationsmechanismen zwischen den Arten. Kulturpflanze 28: 257-284.

Mix, G., Wilson, H.M., and Foroughi-Wehr, B. 1978. The cytological status of plants of *Hordeum vulgare* L. regenerated from microspore callus. Z. Pflanzenzuecht. 80:89-99.

Mujeeb, K.A., Thomas, J.B., Rodriguez, R., Waters, R.F., and Bates, L.S. 1978. Chromosome instability in hybrids of *Hordeum vulgare* L. with *Triticum turgidum* and *T. aestivum*. J. Hered. 69:179-182.

Murashige, T. and Skoog, F. 1962. A revised medium for rapid growth and bioassay with tobacco tissue culture. Physiol. Plant. 15:473-495.

Myers, W.M. 1947. Cytology and genetics of forage grasses. Bot. Rev. 13:319-421.

Nadar, H.M. and Heinz, D.J. 1977. Root and shoot development from sugarcane callus tissue. Crop Sci. 17:814-816.

Nakamura, C., Keller, W.A., and Fedak, G. 1981. In vitro propagation and chromosome doubling of a *Triticum crassum* x *Hordeum vulgare* intergeneric hybrid. Theor. Appl. Genet. 60:89-96.

Nickell, L.G. 1977. Crop improvement in sugarcane: Studies using in vitro methods. Crop Sci. 17:717-719.

Niizeki, H. and Oono, K. 1968. Induction of haploid rice plant from anther culture. Jpn. Acad. Proc. 44:554-557.

Nishi, T., Yamada, Y., and Takahashi, E. 1968. Organ redifferentiation and plant restoration in rice callus. Nature 219:508-509.

Nitzsche, W. 1970. Herstellung haploider Pflanzen aus Festuca-Lolium Bastarden. Naturwissenschaften 57:199-200.

______ 1972. Deutsches Weidelgras (*Lolium perenne* L.) aus Kreuzungen zwischen Wiesenschwingel (*Festuca pratensis* Huds.) und Welschem Weidelgras (*Lolium multiflorum* Lam.). Habilitationsschrift Friedrich-Wilhelms-Univ., Bonn.

______ and Hennig, L. 1976. Fruchtknotenkultur bei Grasern. Z. Pflanzenzuecht. 77:80-82.

______ and Wenzel, G. 1977. Haploids in Plant Breeding. Verlag Paul Parey, Berlin and Hamburg.

Norstog, K.J. 1956. The growth of ryegrass endosperm in vitro. Bot. Gaz. 117:253-259.

______, Wall, W.E., and Howland, G.P. 1969. Cytological characteristics of ten-year-old rye-grass endosperm tissue cultures. Bot. Gaz. 130:83-86.

Oono, K. 1978. Test tube breeding of rice by tissue culture. Trop. Agric. Res. Ser. 11:109-124.

O'Rourke, C.J. 1976. Diseases of Grasses and Forage Legumes in Ireland. Agric. Inst. Dublin.

Orton, T.J. and Steidl, R.P. 1980. Cytogenetic analysis of plants regenerated from colchicine-treated callus cultures of an interspecific *Hordeum* hybrid. Theor. Appl. Genet. 57:89-95.

Paepe De, R., Bleton, D., and Gnangbe, F. 1981. Basis and extent of genetic variability among doubled haploid plants obtained by pollen culture in *Nicotiana sylvestris*. Theor. Appl. Genet. 59:177-184.

Pickering, R.A. and Thomas, H.M. 1979. Crosses between tetraploid barley and diploid rye. Plant Sci. Lett. 16:291-296.

Pierson, P.E. and Collins, G.B. 1981. Anther culture of tall fescue *Festuca arundinacea* Schreb. following panicle pre-culture. Agronomy Abstracts, p. 70, Crop Sci. Soc. Am. 26th Annual Meeting. Agronomy Society of America, Madison, Wisconsin.

Randolph, L.F. 1945. Embryo culture of Iris seed. Bull. Am. Iris Soc. 97:33-45.

Rangan, T.S. 1974. Morphogenic investigations on tissue cultures of *Panicum miliaceum*. Z. Pflanzenphysiol. 72:456-459.

_____ 1976. Growth and plantlet regeneration in tissue cultures of some Indian millets: *Paspalum scrobiculatum* L., *Eleusine coracana* Gaertn. and *Pennisetum typhoideum* Pers. Z. Pflanzenphysiol. 78:208-216.

Rappaport, J. 1954. In vitro culture of plant embryos and factors controlling their growth. Bot. Rev. 20:201-225.

Rines, H.W. 1983. Oat anther culture: Genotype effects on callus initiation and the production of a haploid plant. Crop Sci. 23:268-272.

Sangwan, R.S. and Gorenflot, R. 1975. In vitro culture of *Phragmites* tissues. Callus formation, organ differentiation and cell suspension culture. Z. Pflanzenphysiol. 75:256-269.

Schaeffer, G.W., Baenziger, P.S., and Worley, J. 1979. Haploid plant development from anthers and in vitro embryo culture of wheat. Crop Sci. 19:697-702.

Sears, R.G. and Deckard, E.L. 1982. Tissue culture variability in wheat: Callus induction and plant regeneration. Crop Sci. 22:546-550.

Sharma, G.C., Bello, L.L., and Sapra, V.T. 1980. Genotypic differences in organogenesis from callus of ten triticale lines. Euphytica 29:751-754.

Shimada, T. 1978. Plant regeneration from the callus induced from wheat embryo. Jpn. J. Genet. 53:371-374.

_____, Sasakuma, T., and Tsunewaki, K. 1969. In vitro culture of wheat tissues. I. Callus formation, organ redifferentiation and single cell culture. Can. J. Genet. Cytol. 11:294-304.

Simmonds, N.W. 1976. Evolution of Crop Plants. Longman, London and New York.

Smith, H.H. 1974. Model systems for somatic cell plant genetics. BioScience 24:269-275.

Stebbins, G.L. 1956. Cytogenetics and evolution of grass family. Am. J. Bot. 43:890-905.

_____, Valencia, J.I., and Valencia, R.M. 1946a. Artificial and natural hybrids in the Gramineae, tribe Hordeae. I. *Elymus*, *Sitanion* and *Agropyron*. Am. J. Bot. 33:338-351.

_____, Valencia, J.I., and Valencia, R.M. 1946b. Artificial and natural hybrids in the Gramineae, tribe Hordeae. II. *Agropyron*, *Elymus* and *Hordeum*. Am. J. Bot. 33:579-586.

Straus, J. 1954. Maize endosperm tissue grown in vitro. II. Morphology and cytology. Am. J. Bot. 41:833-839.

_____ and La Rue, C.D. 1954. Maize endosperm grown in vitro. I. Culture requirements. Am. J. Bot. 41:687-694.

_____ 1979. Improvement of crop plants via single cells in vitro: An assessment. Z. Pflanzenzuecht. 82:1-30.

Tilley, J.M.A. and Terry, R.A. 1963. A two stage technique for in vitro digestion of forage crops. J. Br. Grassl. Soc. 18:104-111.

Trewartha, G.T., Robinson, A.H., and Hammond, E.H. 1967. Elements of Geography, 5th ed. McGraw-Hill, New York.

Trione, E.J., Jones, L.E., and Metzger, R.J. 1968. In vitro culture of somatic wheat callus tissue. Am. J. Bot. 55:529-531.

Tukey, H.B. 1934. Artificial culture methods for isolated embryos of deciduous fruits. Proc. Am. Soc. Hortic. Sci. 32:313-322.

Urata, U. and Long, P.P. 1968. A method of propagating grasses by tissue culture. In: Agronomy Abstracts, p. 24, Crop Sci. Soc. Am. Annual Meeting. Agronomy Society America, Madison, Wisconsin.

Van Soest, P.J. 1965. Use of detergents in analysis of fibrous feeds. III. Study of effects of heating and drying on yield of fiber and lignin in forages. J. Assoc. Off. Agric. Chem. 48:785-790.

Vasil, V. and Vasil, I.K. 1980. Isolation and culture of cereal protoplasts. II. Embryogenesis and plantlet formation from protoplasts of *Pennisetum americanum*. Theor. Appl. Genet. 56:97-99.

______ 1981a. Somatic embryogenesis and plant regeneration from tissue cultures of *Pennisetum americanum* and *P. americanum* x *P. purpureum* hybrid. Am. J. Bot. 68:864-872.

______ 1981b. Somatic embryogenesis and plant regeneration from suspension cultures of pearl millet (*Pennisetum americanum*). Ann. Bot. 47:669-678.

Wernicke, W. and Brettell, R. 1980. Somatic embryogenesis from *Sorghum bicolor* leaves. Nature 287:138-139.

Whyte, R.O., Moir, T.R.G., and Cooper, J.P. 1959. Grasses in Agriculture. Food and Agriculture Organization, Rome.

Wu, L. and Antonovics, J. 1978. Zinc and copper tolerance of *Agrostis stolonifera* L. in tissue culture. Am. J. Bot. 65:268-271.

CHAPTER 5
Millets

T. S. Rangan and *I. K. Vasil*

The millets constitute a major source of energy and protein for millions of people in Africa and Asia. Probably the most important character of millets is their ability to survive and tolerate continuous or intermittent drought conditions as a result of low and uncertain rainfall. Consequently, they form the most important food crops of the arid and semiarid tropics (SAT), which in general include the Sahara and large regions of Botswana, Ethiopia, Kenya, Lesotho, Mozambique, Somalia, Swaziland, Tanzania, Uganda, Zambia, and Zimbabwe in Africa, and in India the states of Andhra Pradesh, Bihar, Karnataka, Madhya Pradesh, Maharashtra, Orissa, Rajasthan, and Tamil Nadu. Although in the Western hemisphere the millets are grown mainly for feed grain production, the food, fodder, feed, and other uses of these crops play an important role in the economy of the developing nations of Africa and Asia.

The term millet is likely derived from the Latin "millesimum," meaning a thousandth part. Millet is, therefore, a name applied generally to a number of cereals characterized by their small seeds. Table 1 lists some of the important millets along with their common names. Teff (*Eragrostis teff*) is not strictly considered a millet but is an extremely important crop in Ethiopia, the only country in which it is known to be widely cultivated.

HISTORY OF THE CROP

Sorghum and the other millets have received scant attention compared with cereals such as rice, wheat, and maize. It would appear

Table 1. Some Important Millets

BOTANICAL NAME	COMMON NAME
Sorghum bicolor (L.) Moench	Sorghum
Pennisetum americanum (L.) K. Schum.	Pearl millet
Pennisetum purpureum Schum.	Napier or elephantgrass
Panicum miliaceum L.	Common millet
Panicum maximum Jacq.	Guineagrass
Eleusine coracana (L.) Gaertn.	Finger millet
Setaria italica (L.) Beanv.	Foxtail millet
Paspalum scrobiculatum L.	Kodo millet

that sorghum had its origin in Africa, more specifically in the region of present-day Ethiopia and Sudan, where pearl millet, finger millet, and teff have been cultivated for many centuries. The first known record of cultivated sorghum appears in a 2700-year-old carving discovered in an Assyrian palace. The ancestors of present-day sorghum may have been carried across North Africa to West Africa about 3000 B.C. Historical records suggest that sorghum reached India from East Africa about 2000 B.C. It may also have been carried from East Africa to Arabia from where with the dhow traffic it probably spread to the Persian Gulf and the Near East. The spread along the coast of Southeast Asia and China occurred around the beginning of the Christian era. It is also likely that sorghum reached China via the silk trade routes (Purseglove, 1972). Sorghum arrived in North America as "guinea corn" from West Africa along with the slave trade, from where it spread to Central and South America (Hulse et al., 1980).

Possibly the earliest mention of pearl millet is by Al-Idrisi, an Arab scholar. In an account of his travels in Abyssinia (the upper Nile) he mentions two types of millets, durra and dokhn, that were commonly cultivated. The first name refers to sorghum and the second to pearl millet (Brunken et al., 1977). The available archeological evidence is rather vague. Pottery impressions of pearl millet discovered in Mauritania have been dated to 1100-1000 B.C. Archeological remains as carbonized grains indicate that pearl millet reached the northwest coast of India by 1000 B.C. Faced with the wide diversity found in pearl millet today, one could conclude that the crop is probably a product of multiple domestication (Purseglove, 1972).

Common millet (*Panicum miliaceum*), unknown in a wild state, is the "milium" of the Romans and the true millet of history. It was cultivated in Europe by the Lake Dwellers and also domesticated in central and eastern Asia. Because of its ability to mature quickly, it was often grown by nomads.

Finger millet (*Eleusine coracana*) is either of Indian or African origin and is believed to have evolved from the wild species *E. indica* (Mehra, 1963). Uganda and the neighboring regions are thought to be center of origin, not only because of its ancient cultivation in the region, but also because of the part it plays in religious and tribal ceremonies. *E. coracana* might have been taken to India at a very early date, probably around 3000 B.C., and reached Europe about the beginning of the Christian era (Purseglove, 1972).

Foxtail millet (*Setaria italica*) is domesticated in eastern Asia, where it has been cultivated since ancient times. It was one of the five sacred plants in China as early as 2700 B.C. (Hill, 1952).

Though the cereals provide excellent food, they demand a truly sedentary population. Generally, they can be sown and reaped only at definite times of the year and will not last if they are left beyond the time of harvesting. In dry regions of the world with land unsuitable for intensive cultivation, shifting cultivation with substitute plants became necessary. Sorghum and millets are grown in a shifting cultivation, especially in Africa, and as much use as possible is made of these plants. Not only do they serve as producers of grain, but their straw is used for making baskets as well as for the construction of the roofs of houses. Thus in these regions even the poorer cereals are extremely important crop plants.

GEOGRAPHICAL DISTRIBUTION

Sorghum and the millets form the most important food crops in the SAT regions of Asia and Africa (Table 2). The sorghum and millet production zone almost encircles the earth, beginning in China, passing through India and extending to Africa, southern United States, and Latin America. Sorghum is a staple food in Africa, especially the savannah regions in Niger, Chad, Upper Volta, Mali, Sudan, and Yemen. Pearl millet, although widely introduced throughout the tropics and subtropics, has achieved little importance outside Africa and India.

Common millet is suited to a dry climate and is grown north of most of other millets in Mongolia, Manchuria, Japan, India, and eastern and central Russia. It is also found in parts of the Middle East, Syria, Iran, Iraq, and Afghanistan. According to Matz (1969), it is considered to be the largest millet crop grown in the United States. It is one of the most popular cereals in northern China, commanding a price equal to that of wheat.

Finger milet is an important staple food in East and Central Africa and India. It is also grown in many other tropcial countries, but it is not considered an important crop outside Africa and India.

Foxtail millet is used as food grain in Asia, Southeast Europe, North Africa, and India and is the most important millet in Japan.

Echinocloa frumentacea, also described as Japanese barnyard millet, is the fastest growing of all millets and can be grown to maturity in 45 days. Though not of major importance, it is grown in the Orient, India, and, as a forage crop, in the United States. It is frequently grown in Egypt as a reclamation crop on land too saline for rice cultivation.

The nine most important millet-producing countries are China including Manchuria, India, Nigeria, the USSR, Ethiopia, Niger, Chad, Mali, and Tanzania. Together these countries produce 42.34 million metric

Table 2. Area, Yield, and Production of Sorghum and Millets[a]

REGIONS	SORGHUM[b]			MILLETS[c]		
	Area (10^3 ha)	Yield (kg/ha)	Production (10^3 t)	Area (10^3 ha)	Yield (kg/ha)	Production (10^3 t)
World	43,929	1,179	51,812	72,808	707	51,461
Africa	13,939	704	9,813	16,325	650	10,615
North and Central America	7,760	2,884	22,382	—	—	—
South America	2,421	2,696	6,525	241	1,275	307
Asia	18,956	591	11,202	53,191	676	35,966
Europe	151	3,079	465	27	1,384	38
USSR	196	1,531	300	2,999	1,501	4,500
Oceania	1,590	2,224	1,124	25	1,400	35
Developed countries	6,937	2,914	20,214	61	1,176	72
Developing countries	36,683	850	31,173	38,428	558	21,452

[a] After Hulse et al., 1980.

[b] Sorghum includes various types of *Sorghum vulgare*, e.g., milo, durra, jowar, kafir corn.

[c] Millets include *Pennisetum americanum*, *Eleusine coracana*, *Panicum miliaceum*, *Setaria italica*, and *Echinocloa frumentacea*.

tons of grain on 63.08 million ha annually or 23% of the total production and 93% of the area devoted to millets in the world. Secondary producers, all in Asia and Africa, include Pakistan, North Korea, Uganda, Upper Volta, Senegal, and Sudan with annual production of 253,000 metric tons on 378,000 ha for a total of about 5.6 and 5.5%, respectively, of world production and area (Rachie and Majumdar, 1980).

ECONOMIC IMPORTANCE

The term sorghum includes at least four groups of cultivated annual plants: (a) grain sorghum, (b) sweet sorghum as forage and for animal feed, (c) sudan grass for pasture, hay, and silage, and (d) broom corn, for making of brooms.

Sorghum and the millets provide about 16% of total world cereal consumption for humans (Arnould and Miche, 1971). In the developing countries they provide at least 25% of the total cereal intake (Table 3). Sorghum is important as food for people in developing countries and as feed in more affluent nations. About 35% of the world acreage of sorghum and millets is in India, and almost all the grain except that used for seed is for human consumption.

The world production of sorghum has increased more than 50% since 1960. In the United States, sorghum (750 million bushels per year) is the third most important cereal crop, following corn and wheat. Sorghum production in terms of percentage of total production for the 1975 crop was North America 37%, Central America 7%, South America 11%, Africa 20%, West Africa less than 1%, South Asia 21%, East Asia 1%, Oceania 2%, and Europe less than 1% (Rooney et al., 1980).

Sorghum is the fourth most important world cereal, following wheat, rice, and maize. Because of its drought resistance it is the crop par excellence for dry regions and areas with unreliable rainfall. In the developed countries sorghum is used as a stock feed, but in developing countries it is too valuable as human food to be used for this purpose (Table 4). The most attractive feature of sorghum and other millets is their ability to survive and yield grain during continuous or intermittent drought stress. It can remain dormant during periods of stress and renew growth when conditions become favorable. It is more tolerant to flooding than maize but does not grow at its best under prolonged wet conditions. Sorghum grows successfully on any soil types but best on medium-textured, light-textured, or sandy soil. It also tolerates medium-high pH conditions in soil. It feeds heavily on soil, and a crop yielding 6000 kg/ha removes about 105 kg of nitrogen, and 15 kg each of potassium and phosphorus (Hulse et al., 1980).

The technology of sorghum production in the developed countries has progressed far beyond those in Africa and Asia, and the average yields are about five times more than those of the African, Asian, or Latin American regions. Average yield in the United States is about 3500 kg/ha, in contrast to reported average yields of about 700 kg/ha in Asia and Africa. It is difficult to obtain accurate statistics of world acreage and production of sorghum, as it is often included with millets. It seems reasonable to suppose that the total area under sorghum

Table 3. Consumption and Utilization of Sorghum and Millets[a]

REGION	CONSUMPTION (10^3 t)	HUMAN NUTRITION (%)[b]	ANIMAL FEED (%)	FERMENTED BEVERAGES AND INDUSTRIAL USES (%)
Europe	3,683	1.0	95.7	0.2
North America	16,662	0.5	97.6	1.1
Central America	1,883	13.3	83.9	—
South America	1,115	—	96.2	—
Africa	20,214	85.2	1.0	5.7
Asia	23,258	78.1	15.0	—
Oceania	312	—	99.0	—
World total	67,130	53.3	39.4	2.0

[a]After Hulse et al., 1980.

[b]All percentages by weight of total consumption.

Table 4. Percentage Distribution of Sorghum Usage[a]

USE	DEVELOPED COUNTRIES (%)	DEVELOPING COUNTRIES (%)	TOTAL (%)
Human nutrition	1.1	82.5	53.3
Animal feed	93.8	7.7	39.4
Fermented beverages and industrial uses	1.6	2.3	2.0

[a]After Hulse et al., 1980.

cultivation in the world is over 50 million ha. China has probably the largest acreage (about 20 million ha), India being the next largest producer with about 15 million ha. In parts of Africa, although maize has tended to replace sorghum as the staple cereal, some 15 million ha are still under sorghum cultivation, especially in the drier regions of West Africa and Sudan. About 7 million ha are grown annually in the United States, of which about 75% is grown for grain, mainly stock feed.

Sorghum is divided into several races distributed mainly in Africa and India (Harlan, 1972; Harland and De Wet, 1972). The basic races are bicolor, guinea, caudatum, kafir, and durra, while the hybrid races are obtained as a result of crossing between the various basic races.

There is a major difference in the types of plants that are grown and the uses to which the grain is put in the developing and developed countries. In the developed countries the emphasis has moved toward development of hybrids with high uniformity, early maturity, and short stature suitable for mechanized harvesting. The grain is used almost exclusively for livestock and poultry feed. Approximately 2.5 million bushels of sorghum grain are used for production of alcohol annually; alcohol yield is about 6 gal/bu. In the developing nations, Africa and Asia in particular, the grain is still used for human food and for processing of beverages (Hulse et al., 1980; Rooney et al., 1980). The plant in these areas tends to be quite tall, and the preferred grain type is light in color with hard or corneous endosperm. In some areas red or brownish types might be preferred because of their apparent superior quality for making beer. They are also less palatable to grain-eating birds, which can be serious pests in certain parts of Africa.

Pearl millet, also called the bulrush millet, is the most widely grown of all the millets. According to Purseglove (1972), pearl millet includes a number of cultivated races, and the cultivars vary in several characters such as height, thickness and the degree of branching of the stem, size and shape of inflorescence, and shape and color of the grain. *Pennisetum americanum* is not known to occur in a wild state, and the wild progenitors are not known with certainty. There are divergent views regarding the origin of pearl millet. Some consider the various races or species to be polyphyletic in origin, some derived from one or several species and the other races by hybridization. Others consider the pearl millet to have been derived from two sources, the tall, widely spread *P. purpureum* and the xerophytic group of small species of the section of *Penicillaria*. Cytogenetical studies of *P. americanum* and *P. purpureum* hybrids do not support the view that the latter is the ancestor of the former (Purseglove, 1972; Muldoon and Pearson, 1979).

Pearl millet is one of the most drought tolerant of all cereals. While it can survive under conditions of lower total rainfall than sorghum, it does not appear to have the ability of temporary dormancy during periods of drought. It grows well on light soils, and semiarid conditions, although it is grown with irrigation in India.

Reliable figures for the production of pearl millet are difficult to obtain, as they are often included with those of sorghum and other millets. The crop also does not appear in international trade. The average annual area under pearl millet cultivation in India is about 11 million ha, with a production of 3-4 million tons. The grain may be fed to poultry and other livestock. The green plant provides a useful fodder, and it is sometimes grown for this purpose. The yield under dry cultivation is 770-1100 kg/ha and with irrigation 1100-2200 kg/ha. Partial or complete replacement of rice by pearl millet in a vegetarian diet is said to appreciably increase the nutritive value of the diet.

Hybrid pennisetum is an interspecific hybrid between pearl millet (*Pennisetum americanum*) and napier or elephant grass (*P. purpureum*). Hybridization between these two species probably occurs naturally, since both species are protogynous and cross pollinate. Hybrid pennisetum is an erect grass that grows to a height of 4-5 m. In gross morphology it resembles napier grass but also possesses some of the leafiness of pearl millet. It is sterile, but induction of polyploidy restores fertility (Muldoon and Pearson, 1979).

Hybrid pennisetum is used for direct animal consumption and as conserved fodder. It has also been suggested that a particular rhizomatous strain might be useful in soil conservation. Because of its high dry matter yields, it is a potentially important crop for energy (biomass) production.

Panicum miliaceum displays perhaps the lowest water requirement of any cereal. It generally matures between 60 and 90 days and may be grown where the climate is too hot, the soil too poor, and the rainy season short. It can grow on any kind of soil except coarse sand. Many common millets have different plant coloration, hairiness, and seed color. Some high-yielding selections are reported from India, and some Russian varieties are said to mature earlier.

The husked grain is considered nutritious and is eaten whole, boiled, or cooked like rice. It is fermented into a kind of beverage in Ethiopia. In the United States it is particularly valued as hog feed. The grain contains 10-18% of proteins, with a biological value of 56%. Feeding trials with rats indicate that it has a lower supplementary value than the protein of wheat or rice. The grain also contains starch as the major carbohydrate, with minor amounts of sugars and dextrins (Deshaprabhu, 1966; Purseglove, 1972).

Guinea grass (*P. maximum*) is grown mainly as green fodder but also makes excellent hay and silage. The crop can give continuous supply of green fodder for several years. It is more nutritious than napier grass (*Pennisetum purpureum*) and compares favorably with sorghum and maize fodder.

In India the yields of common millet are reported to be between 450 and 650 kg/ha when rain-fed and up to 2 tons/ha when irrigated. As with pearl millet, separate production figures for *Panicum* and other minor millets are not available. In India 4-5 million ha of minor millets are grown annually, with a yield of about 3 million tons which include *P. miliaceum, Echinocloa frumentacea, Paspalum scrobiculatum,* and *Setaria italica.*

SORGHUM

Sorghum is an annual plant with a single stem. A single radicle is produced by the seedling, followed by adventitious fibrous roots from the lowest nodes of the stem. The stem has a number of root primordia above the attachment of the leaf sheath and those near the base of the plant grow out through the sheath to produce prop roots. The stem is solid, usually erect, dry or juicy, with leaves developing alternatively in two ranks. The inflorescence is a panicle, and the central rachis of the panicle has primary, secondary, and sometimes tertiary branches bearing racemes of spikelets.

The spikelets are borne in pairs, one of which is sessile and hermaphrodite and the other pedicelled and male or sterile. Each spikelet contains two florets, the lower of which is sterile and the upper a perfect floret. The caryopsis is rounded and bluntly pointed, about 4-8 mm in diameter, varying in size, shape, and color with cultivar. The grain is usually partially covered by glumes. The seed coat consisting of a fused pericarp and testa comprises 6% of the total weight of the grain and varies in color from pale yellow through various shades of red and brown to a deep shade of purple brown. The outer endosperm may be hard and corneous but is white and floury inside. One kilogram of sorghum grain contains 25,000-70,000 seeds. Some cultivars show seed dormancy for the first month after harvesting. The seed will remain viable for considerable periods provided it is properly stored (Purseglove, 1972).

The proteins in sorghum can be divided into two main classes: (a) the structural proteins present in the embryo that are of relatively constant composition and (b) the storage proteins in the endosperm and the aleurone layer that provide nutrition for the developing seedling and vary in composition according to genetic and agronomic history. As the seed matures and as the proportion of protein nitrogen in the endosperm increases, the composition of the endosperm storage protein changes. Glutamic acid and proline increase, while lysine and other basic amino acids and methionine decrease in proportion to total nitrogen. A comparative value of nutrients in sorghum, cereals, and other millets is given in Table 5. Protein content and composition within the endosperm are influenced by both genetic background and the conditions of growth. Grain protein content is directly controlled by the plant's ability to take up and transfer nitrogen from roots and leaves to the seed. It is also believed that many hybrids manifest their vigor in high grain yield rather than in a higher protein content. A relatively consistent finding with sorghum and pearl millet is that nitrogen applied as foliar spray results in a greater increase in grain protein content than nitrogen applied to the soil.

Lysine is the first limiting amino acid in all cereals, followed by threonine and tryptophan. Lysine is negatively correlated to protein content. Like other cereals, the percentage of lysine in the protein decreases as protein content increases, even though the total lysine as a percentage of dry weight increases. Sorghum proteins contain slightly more tryptophan than corn proteins (Rooney et al., 1980). Sorghum, foxtail, little, and Japanese barnyard millets have the lowest lysine content among the known cereal grains. In recent years sorghum with higher than normal lysine in endosperm proteins have been found to occur naturally in Ethiopia and have been induced by chemical mutation in the United States. The high lysine character is believed to be controlled by a single recessive allele which appears to suppress the formation of prolamine in the endosperm storage proteins. Though in normal sorghum lysine may be affected by the environment, it is reported that both the naturally occurring and chemically induced high lysine character is comparatively independent of environment (Hulse et al., 1980).

Sorghum is an extremely variable crop. Many cultivars are recognized in Africa, India, the United States, and elsewhere, and new cultivars are being produced. A world collection of sorghum germplasm at ICRISAT (International Crops Research Institute for the Semi-Arid Tropics, Pantancheru, India) contains 17,986 accessions from 68 countries (ICRISAT, 1981). Both grain and wild sorghums cross readily, and thus there is considerable scope for improvement by selection and breeding. Local cultivars in many areas have become adapted to their environment, particularly with regard to climatic conditions and resistance to local pests and diseases. In the tropics cultivars were developed that could survive and yield even under infertile conditions. In many countries, particularly in tropical Africa, small farmers usually grow a mixture of cultivars, as these are better adapted and more flexible to environmental fluctuations.

Sorghum has possibilities for extension into areas where it is not presently grown. It may find a place in the rice-growing areas in south and east Asia, as a crop that can be grown in the latter part of the rainy season or in the dry season, utilizing the accumulated subsoil moisture.

The development of varieties that are somewhat shorter in stature, less sensitive to photoperiod, and maturing within a given period of time would seem to have much promise for the African region. Such varieties must have the capacity to resist damage from grain molds. Likewise, these cultivars must also have the capacity to withstand drought periods and still produce a reasonably good crop. There are a

Table 5. Nutrition of Various Cereals[a,b]

Cereals	Moisture	Calories	Protein (g)	Fat (g)	Carbo-hydrate (g)	Fiber (g)	Ca (mg)	Fe (mg)	Thia-mine (mg)	Ribo-flavin (mg)	Niacin-amide (mg)
Sorghum	12	355	10.4	3.4	71	2.0	32	4.5	0.5	0.1	3.5
Pearl millet	12	363	11.0	5.0	69	2.0	25	3.0	0.3	0.1	2.0
Finger millet	12	336	6.0	1.5	75	3.0	350	5.0	0.3	0.1	1.4
Wheat	13	344	11.5	2.0	70	2.0	30	3.5	0.4	0.1	5.0
Maize (whole and parboiled)	12	363	10.0	4.5	71	2.0	12	2.5	0.3	0.1	2.0
Rice (parboiled)	12	354	8.0	1.5	77	0.5	10	2.0	0.2	0.05	2.0

[a]After Hulse et al., 1980.

[b]Representative values per 100 g of edible portion.

number of pests and diseases that need to be overcome. Since pesticides are expensive and are difficult to obtain, importance should be placed on obtaining plants able to resist pests such as shootfly, stemborer, and grain midge and diseases such as rust and downy mildew.

There are promising possibilities for improving the nutritional qualities of sorghum. Certain lines derived from collections in Ethiopia have been found to have an inherently high protein content (16-17%), a relatively high percentage of lysine (3.25%) and a biological value approximating or exceeding that of opaque-2 maize (Hulse et al., 1980). The characters for high protein and high lysine are being incorporated into crosses and into populations out of which selection will be made for future varieties. It is important to realize that high lysine in itself has no readily apparent advantage to the consumer, and this must be incorporated into grain types having the same grain quality as existing cultivars if it is to be accepted.

Recent advances in tissue and cell culture have prompted speculation about the methods by which cereal species could be improved by in vitro technology. A necessary prerequisite for such an application is the availability of relatively homogeneous cell populations and their ability to regenerate into plants under defined conditions. However, in crop plants especially the cereals, the tissue and cell culture technology is limited (Vasil and Vasil, 1980).

Strogonov et al. (1968) reported the formation of callus from tillering nodes of *Sorghum vulgare*, which subsequently differentiated shoot buds. Mastellar and Holden (1970) obtained a compact yellow callus from germinated seedlings. Upon transfer to a medium with low concentration of IAA, 2,4-D, or NAA root formation occurred. However, if transferred to a medium supplemented with 27.0 μM NAA (instead of 2,4-D), shoot buds differentiated and subsequently developed into plants. Rogers et al. (1974) obtained callus and roots but no shoot buds.

Mascarenhas et al. (1975a,b) induced the formation of callus from seedling segments. The cultures initially remained unorganized but subsequently developed roots. When callus growing on agar medium was transferred to agitated liquid medium, of the same composition, there was a morphological change. The callus differentiated a large number of roots that continued to grow as roots with a number of lateral roots. However, if retransferred to agar, they reverted to unorganized callus and grew as callus. Such differentiation and its reversal could be repeated several times.

As with other millets, the most successful explant for in vitro studies of *Sorghum* appears to be the immature embryo. Thomas et al. (1977) reported the formation of large numbers of shoots and embryo-like structures on cultured tissue derived from immature embryos of *S. bicolor*. Immature embryos (excised from seeds 10-30 days after pollination) cultured on a 2,4-D containing medium produced an intense white callus tissue. A large number of structures, some clearly resembling sexually produced embryos, were formed in the cultures. These structures gave rise to hundreds of shoots bearing leaves. Subsequently, similar structures were also induced on scutellar callus from embryos excised from 2-year-old seeds.

Gamborg et al. (1977) observed the formation of leafy shoots from immature embryos excised 12-18 days after pollination. Although cytokinins were not obligatory for callus initiation and growth, shoot formation occurred only when ZEA or 2iP was present. Small pieces of callus with leafy shoots when placed on medium supplemented with IAA, NAA, or 2,4-D resulted in plant regeneration. IAA proved superior to both NAA and 2,4-D in ensuring complete plant development. The plantlets produced were grown to maturity, and although some of the plants were sterile, the others produced normal seeds. Meiotic metaphase from each of the 25 plants regenerated from callus revealed the chromosome number 2n = 20, corresponding to that of normal plants.

The observations of Dunstan et al. (1978) were somewhat similar to those of Thomas et al. (1977). They observed that the cells of the scutellum of the immature embryo formed compact structures that either directly gave rise to leaves or closely resembled sexual embryos. It was suggested that the shoots and embryos arise directly from the scutellum and not from the callus and that to a limited extent these were capable of further proliferation (see also King et al., 1978). In a subsequent study Dunstan et al. (1979) postulated two distinct pathways leading to plantlet production in immature embryo cultures of *Sorghum*. Generally, the plantlets arose from scutellar cells and in some instances from single cells. During repeated subculture in the presence of 2,4-D or NAA, the growth of the shoot was suppressed but not the formation of axillary buds. With any change in the level of growth regulators these organized primordia developed into shoots. In this respect these cultures are similar to shoot tip cultures, and this phenomenon has been termed microtillering (Thomas et al., 1977).

Wernicke and Brettell (1980) reported somatic embryogenesis in leaf cultures of *Sorghum*. Segments of young leaves in the presence of 2,4-D proliferated to form a callus that after transfer to an auxin-free medium formed shoots and plantlets. After prolonged subculture the callus formed somatic embryos. They suggest, albeit in the absence of histological evidence, that the callus from the leaves consists of a mass of proliferating primordia. The presence of 2,4-D in the medium totally inhibits the morphogenetic expression of the primordia, and this inhibition is released upon transfer to a low or auxin-free medium.

Subsequently, Brettell et al. (1980) also observed the formation of embryo-like structures and shoot primordia in immature inflorescence segments of *S. bicolor*. On a medium supplemented with 2,4-D, organized cell divisions in the superficial cell layers resulted in a compact tissue that gave rise to embryo-like structures (some with well-defined shoot-root axis). These structures germinated to give rise to plantlets that were grown to maturity. Similar responses were also observed in cultures of individual florets.

PENNISETUM

Pennisetum is a tall, erect annual, 0.5-4.0 m in height. A single seminal root is produced, followed by fibrous adventitious and prop roots from lowest nodes of the stem. The stem is solid, and slender or stout, with prominent nodes with rings of silky hair. The leaves are found in two vertical rows on either side of the stem. The inflorescence which resembles a bulrush is a contracted panicle or a false spike bearing densely packed clusters of spikelets. The spikelets are borne in pairs, and the lower floret is usually male, while the upper floret is perfect. The spike emerges about 10 weeks after sowing, and grain begins to develop about 8 days after the emergence of the spike. It takes about 40 days from fertilization to "seed" ripening. The caryopsis is obovoid or elliptic, small, and variable in size and color. The caryopses of pearl millet are larger (305 g/1000 seeds) than other millets (Rachie and Majumdar, 1980).

Pearl millet grain almost never enters international trade and is largely consumed within the regions of production. The world production exceeds 10 million metric tons from about 21 million ha. Since it is almost entirely a rain-fed crop, grown under uncertain moisture conditions, the regional production tends to fluctuate widely. The average annual world production of millets was estimated at about 44 million metric tons on 68.8 million ha for the period 1967-1971. Approximately 85% of this production was for human food, 6% for seed, and 9% for livestock feed (Rachie and Majumdar, 1980).

Though most pearl millet grain is consumed as human food, it also has excellent fodder qualities if used before grain formation, and these features make pearl millet a most versatile crop. *Pennisetum* grain is among the most nutritious of the major cereal grains and highly palatable. Its protein content is not only high but is of exceptionally good quality, being seriously deficient only in lysine. It also has good amounts of phosphorous and iron and reasonable quantities of thiamine, riboflavin, and nicotinic acid. Pearl millet contains 67-72% carbohydrates, of which starch is the major constituent. The amino acid profile compares most favorably with those of sorghum, wheat, maize, and rice. According to Aykroyd et al. (1963), *Pennisetum* is higher than wheat in all essential amino acids except phenylalanine and exceeds wheat in lysine by 64%. It is equivalent or superior to parboiled milled rice in 5 out of 7 essential amino acids, histidine, lysine, methionine, threonine, and valine. It is superior to sorghum in 7 out of 10 essential amino acids, being slightly lower in phenylalanine, methionine, and leucine. The grain is exceptionally high in fat content, a characteristic believed to contribute to its high acceptability, palatability, and energy level.

The protein efficiency ratio (PER) for unsupplemented pearl millet is higher than that of wheat or sorghum. The addition of lysine to a pearl millet diet results in a significant increase in the PER, thereby indicating that in many millets lysine is the limiting amino acid. The amino acid profiles in millet may be complemented by other plant proteins, and inclusion of legumes in the diet dramatically improves the PER (Hoseney and Varriano-Marston, 1980).

Pennisetum americanum already has a high degree of variability within the germplasm collection, and it is difficult to maintain the stability of these collections because of natural cross pollination. A world collection of pearl millet germplasm at ICRISAT has 14,074 accessions from 25 countries (ICRISAT, 1981). Some of the same pests are attracted to millets as sorghum, although millets appear to be more resistant to shootfly and stem borer. In recent years, in India, there have been serious attacks of downy mildew and ergot. The development of varieties that are resistant to these diseases under a range of conditions will be very useful.

Cytoplasmic male sterile lines have been identified and could play an important role in hybridization and population improvement programs. A major difficulty that needs to be overcome is to find an acceptable substitute for the presently known male sterile lines that are susceptible to ergot. The major breeding objectives in research programs in different regions of the world are: increasing grain yield and quality; increasing fodder yield and quality; incorporating resistance to diseases and pests; increasing stress tolerance, especially to heat, drought, and low temperature; and raising the response to better management.

Rangan (1976) reported that tissue cultures obtained from mesocotyl explants of *Pennisetum typhoideum* (=*americanum*) developed shoot buds when transferred to 2,4-D free or low auxin supplemented medium. Complete plants were obtained that were grown to maturity as pot plants. Bajaj and Dhanju (1981) observed regeneration of plants in callus cultures obtained from immature seeds or segments of young inflorescences of *P. purpureum*.

Vasil and Vasil (1981a) observed somatic embryogenesis and subsequent plant regeneration in callus cultures established from immature embryos and young inflorescence segments of *P. americanum* and *P. americanum* x *P. purpureum* hybrid. Immature embryos excised from the inflorescence 7-15 days after pollination were found to be ideal for culture. The behavior of the embryos was closely related to the manner in which they were placed on the medium. In the face-up position (when the scutellum is in contact with the medium and the embryo axis is away from the medium), the embryos germinated precociously and gave rise to a weak seedling. In the face-down position (when the scutellum was away from the medium and the embryo axis in contact with the medium), the peripheral region of the scutellum became swollen and showed a number of infoldings subsequently giving rise to a compact pale-yellow callus. In 3-4-week-old cultures cup-shaped structures developed at the periphery of the scutellum. When the scutellar callus or small segments of the callus were transferred to a 2,4-D-free medium, tube-like structures, reminiscent of coleoptiles, emerged from the center of the cup-shaped scutellar structures. Many of the scutellar cups with the coleoptiles were associated with roots and the entire structure resembled the very early stages of germination of a zygotic embryo. On culture the scutellum of each embryo gave rise to an embryogenic callus tissue that formed well-organized embryos, each with its own scutellum, coleoptile, and shoot-root axis (Fig. 1). The somatic embryos arise from single cells (Vasil and Vasil, 1982b). The in vitro regenerates were eventually grown in the greenhouse, and root tip squashes revealed a normal diploid number of 2n = 14.

Inflorescence segments of *P. americanum* and that of hybrid *Pennisetum* were cultured on MS medium supplemented with 2,4-D with or without CW (Vasil and Vasil, 1981a). The compact embryogenic callus obtained was very similar to the scutellar callus and gave rise to many

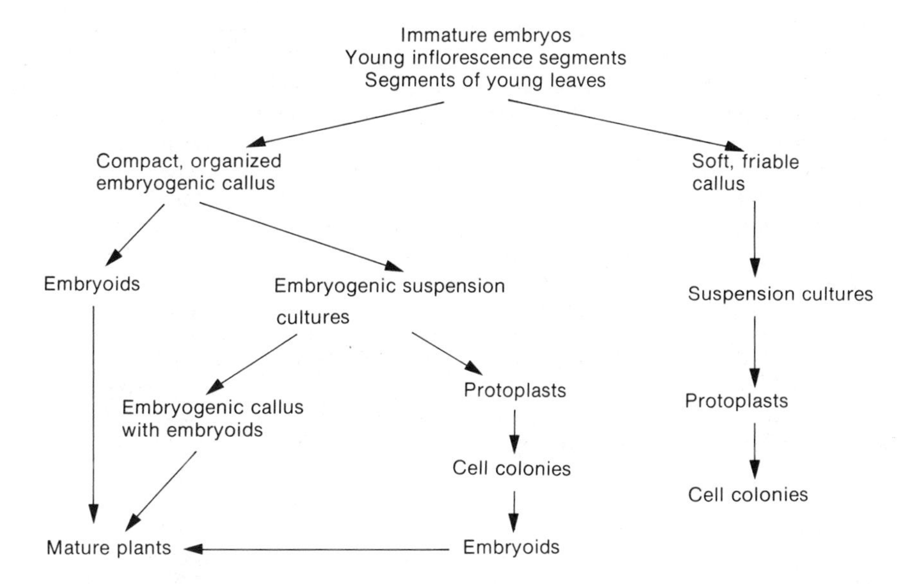

Figure 1. Scheme for somatic embryogenesis in *Pennisetum americanum*, *P. purpureum*, and *Panicum maximum*.

cup-shaped structures. When transferred to a medium containing 0.04 µM ABA + 2.65 mM of ammonium malate, and 0.18 mM sucrose, fully organized somatic embryos developed which subsequently germinated and grew into mature plants.

Vasil and Vasil (1981b, 1982a) also reported somatic embryogenesis and plantlet regeneration from embryogenic cell suspension cultures of *P. americanum*. Suspension cultures were established from embryogenic callus obtained from the scutellum of immature embryos and were composed of large, highly vacuolated cells as well as richly cytoplasmic embryogenic cells. The embryogenic cells contained plastids and starch and occurred in groups of four or more cells and occasionally as single cells. Structures resembling various stages of embryogenic development were found in suspension cultures. When plated on a 2,4-D free medium containing abscisic acid, embryos with typical organization of cereal embryos were produced and germinated to produce plantlets that could be successfully transferred to soil. These embryos apparently developed from single embryogenic cells either directly or after the formation of proembryonal masses.

Haydu and Vasil (1981) indicated the formation of embryogenic callus from leaf explants of *P. purpureum*. The callus formed embryos that later developed into plants. Histological studies revealed that the callus originated from the cells of the mesophyll as well as the lower epidermis. A similar embryogenic callus was also obtained from the somatic tissues of anthers and induced to form plants through somatic embryogenesis. The plants obtained from leaf and anther culture were shown to have the normal chromosome number of $2n = 4x = 28$.

Although considerable progress has been made in the use of plant protoplast research, such success has been confined to plants such as carrot, tobacco, *Datura*, and *Petunia*, and the isolation and culture of protoplasts of cereals has proved to be very difficult. Vasil and Vasil (1979, 1980) reported the successful culture and plantlet regeneration from protoplasts isolated from embryogenic suspension cultures of *Pennisetum americanum*. The cultures comprised large and vacuolated cells and small densely cytoplasmic embryogenic cells. Protoplasts from the large vacuolated cells never divided and eventually degenerated. However, protoplasts from the embryogenic cells regenerated cell walls and formed 15-20 celled masses within 2 weeks, after isolation and culture. Further cell divisions resulted in the formation of organized structures resembling the early stages of embryogenesis. In 4-5 weeks large cell masses consisting of tightly packed embryogenic cells were obtained. These cell masses when transferred to a hormone-free medium formed embryoids that further developed into plantlets. The protocols for establishment of cell cultures and plant regeneration from protoplasts are described in a later section.

PANICUM

The genus *Panicum* consists of about 500 species of annual and perennial grasses in the tropics and warm temperature regions. *Panicum miliaceum* and *P. sumatrense* are small millets grown as cereals, and a number of species such as *P. maximum* are important fodder plants.

Panicum miliaceum is a shallow-rooted annual, usually free tillering, with cylindrical, hollow internodes. The inflorescence is a panicle bearing spikelets, each bearing two flowers. The lower floret is sterile, and the upper floret is fertile. The caryopsis is somewhat oval, whitish, and smooth.

Common millet is usually grown as a rain-fed crop and does not appear subject to any serious pests and diseases. It is cultivated mainly in eastern Asia, including Mongolia, Japan, India, eastern and central Russia, and the Middle East. Reliable production figures for the crop are not available. In India yields are reportedly between 450 and 650 kg/ha when rain-fed and 1000-2000 kg/ha when irrigated. The crop appears to have received little attention from plant breeders.

The analysis of the husked grain is as follows: water 11.9%, protein 12.5%, fat 1.1%, carbohydrate 68.9%, fiber 2.2%, and ash 34%. The protein content varies from 10 to 18% (Deshaprabhu, 1966).

Rangan (1974) observed that callus cultures established from mesocotyl explants of *Panicum miliaceum* developed shoot buds when transferred to 2,4-D free or low auxin (NAA, IAA) medium. Plantlets were obtained and grown to maturity. Bajaj et al. (1981) also reported plantlet regeneration through shoot bud formation in callus cultures established from embryos, shoot tips, and inflorescence segments of *P. maximum*, *P. miliaceum*, and *P. antidotle*. The regenerants showed variation in morphological characters, such as size, leaf shape, and tillering.

Lu and Vasil (1981a,b, 1982; Lu, 1981) induced somatic embryogenesis in cultures obtained from immature embryos, inflorescence segments, and leaf tissue of *P. maximum*. In cultures of immature embryos the scutellum produced a compact callus that formed several cup-shaped structures which subsequently differentiated into embryoids. A reduction in the auxin concentration or its complete omission from the medium resulted in the germination of these embryoids. The plantlets formed in vitro were successfully transplanted to soil. The plants showed a normal tetraploid chromosome number of 2n = 4x = 32. Interestingly, in tissues obtained from mature embryos, plant regeneration was achieved through the organization of shoot buds, followed by the differentiation of adventitious roots.

The developmental stage of the inflorescence was found to be critical for induction of embryogenic callus (Lu and Vasil, 1982). The best response was obtained with inflorescences in which the floral primordia were just being formed and had not completely differentiated. As with the scutellar callus from immature embryos, with a reduction in the amount of auxin in the medium embryoids with typical organization of grass embryos were formed which germinated to form plantlets and were successfully transferred to soil.

Transverse segments 2-mm thick from 1-4 of the youngest whorled leaves of *Panicum maximum* were cultured on MS medium supplemented with different combinations of 2,4-D and CW. A 3-cm-long region of the leaf starting from about the level of the shoot meristem was found to be most responsive for the induction of callus formation. After 1 week, the area of the veins became swollen, followed by the appearance of callus which developed only on the abaxial surface (lower) and not on the adaxial (upper) surface. The initial callus tissue was soft and transluscent and after about two weeks localized areas of white and embryogenic callus and many embryoids appeared on the surface. When the embryoids forming on the 2,4-D + CW medium were transferred to MS medium supplemented with 1 mg/l gibberellic acid, they matured and germinated to form plantlets. The plantlets were transplanted to potting soil and finally moved to the greenhouse and grown to maturity. The regenerated plants were found to have the normal chromosome number of 2n = 4x = 32 (Lu and Vasil, 1981b).

The compact white embryogenic callus was produced by cell division in the lower epidermis and in mesophyll cells. Periclinal divisions occurred in cells of the lower epidermis resulting in the formation of radially oriented files of cells. Divisions in mesophyll tissue were restricted to cells located in the lower half of the leaf, especially those on either side of the vascular bundles.

Lu and Vasil (1981b) also obtained the formation of somatic embryos and plantlet regeneration from suspension cultures established from embryogenic calli derived from immature embryos and young inflorescences of *Panicum maximum*. The suspension cultures consisted of two types of cells: small, richly cytoplasmic, and often starch-containing embryogenic cells, and large, vacuolated nonembryogenic cells. Single embryogenic cells and various stages of embryoid formation up to the differentiation of a lateral notch were seen in suspension cultures. When the suspensions were plated in an agar medium without any hormone or with only 0.9 μM 2,4-D, fully differentiated embryoids were formed which germinated to form plantlets. Embryogenic suspension cultures from immature embryos as well as inflorescence segments gave rise to plants that showed the normal somatic chromosome number. Protoplasts isolated from embryogenic suspension cultures were also successfully cultured to produce an embryogenic callus tissue, embryoids, and plantlets (Lu et al., 1981).

ELEUSINE CORACANA

Also known as African millet or finger millet, *E. coracana* is an important staple food in parts of East and Central Africa and India and is the principal cereal grain in Uganda and Zambia. Finger millet is grown mainly in the tropics and requires a well-distributed rainfall during the growing season, without any drought, but it cannot tolerate heavy rainfall. The crop is grown on a variety of soils, but reasonably fertile, free-draining sandy loam ones are preferred. Mehra (1963) recognizes two groups of cultivars: (a) the African highland types with grains enclosed within the florets and (b) the Afro-Asiatic types with mature grains exposed out of the florets. The cultivars vary in many characters such as height, ranging from dwarf to tall types, color of the vegetative organs varying from green to purple, degree of inflorescence and color of grain varying from almost white through orange-red, brown, and purple to almost black. In India a number of improved cultivars have been produced by selection and breeding. The time from sowing to maturity varies from 3.5 to 6 months, depending upon genotype and environment. The yields are variable; in India finger millet cultivation covers about 2-2.5 million ha with a yield of about 600-800 kg/ha.

The grains are consumed as porridge and are also used for making beer. The great advantage of finger millet is that it can be stored for periods of up to 10 years without deterioration or insect damage. The crop also suffers little from bird damage, and the straw can be used as fodder.

The protein content shows considerable variation, ranging from 6 to 11%. The grain is rich in cysteine, tyrosine, tryptophan, and methionine but low in lysine. It is also a rich source of calcium, phosphorous, and iron (Sastri, 1952).

Rangan (1976) reported that cultured mesocotyl explants from 5-day-old seedlings of *E. coracana* developed roots on a low auxin medium but at higher concentrations produced yellowish, nodular callus. On transfer to an auxin-free or low-auxin medium the callus developed shoot buds that on isoaltion and transfer produced plantlets. The developing plantlets showed precocious tillering and flowering and could be transplanted to soil and grown to maturity.

PASPALUM

The genus *Paspalum* consists of about 250 species of grasses, mostly perennial, that are found mainly in tropical America but also in the Old World. While some species such as *P. conjugatum* (buffalo grass), *P. dilatatum* (Dallis grass), and *P. notatum* (Bahia grass) are used for pasture or forage and anti-erosion control, *P. scrobiculatum* (kodo millet) is a minor grain crop in India.

Paspalum scrobiculatum is an annual tufted grass suited to dry conditions. Several types of kodo millet are recognized in India differing in crop duration, number of rows of grain, and grain color. The grains are consumed by the very poor. Kodo millet is hardy and drought resistant and is often grown on poor soils, the crop receiving very little attention and maturing in 4-6 months. The yield of grain varies from 250 to 1000 kg/ha (Deshaprabhu, 1966).

The whole grain contains water (11.6%), protein (10.6%), fat (4.2%), carbohydrate (59.2%), fiber (10%), and ash (4.4%).

As in finger millet, somatic tissue cultures of *P. scrobiculatum* established from mesocotyl explants developed shoot buds on transfer to a hormone-free or low-hormone-containing medium (Rangan, 1976). Addition of cytokinins such as KIN or SD8339 (6-benzyl-9-tetrahydro-pyrane-adenine) significantly improved shoot bud induction. The plantlets obtained were successfully transplanted to soil.

PROTOCOLS

Establishment of Pearl Millet Cell Suspension Cultures

1. Aseptically germinate pearl millet seeds on 1% agar.
2. Transfer hypocotyls from 3-day-old seelings to LS agar medium containing 1.1 μM 2,4-D, 5.4 μM NAA, and 0.47 μM KIN. Culture in the dark at 27 C.
3. Friable callus will develop from hypocotyl sections within 2 weeks. This callus should be subcultured every 4 weeks to fresh medium.
4. Suspension cultures are initiated by placing callus in 35 ml of liquid medium described above. Flasks are shaken at 150 rpm in darkness.
5. Once established, the suspensions are subcultured every 4 days.
6. Cell suspension cultures can also be established from immature embryo- or inflorescence-derived callus. Immature embryos are cultured 10-20 days after pollination. Segments of inflorescences are cultured when the individual floral primordia are just beginning to be formed. The tissues are cultured on LS medium containing 11.3 μM 2,4-D and 5% CW. Callus derived from the scutellum of immature embryos or from the young inflorescence segments is used to start cell suspensions which are subcultured every 5-6 days and are maintained in the dark at 27 C with shaking (150 rpm) (Vasil and Vasil, 1980, 1982).
7. Alternatively, cell suspensions may also be initiated from Guinea grass using callus obtained from immature embryos or young inflorescences (Lu and Vasil, 1980; Lu et al., 1981).

Pearl Millet and Guinea Grass Protoplast Isolation, Culture, and Regeneration

1. Subculture cell suspensions every other day 1 week before protoplast isolation. This increases the density of embryogenic cells

and reduces the number of enlarged, thick-walled cells in the suspension.

2. Mix 0.75–1.5 ml of cell suspension with 8–10 ml of enzyme solution (2% Onozuka R-10 Cellulase, 0.15% Pectolyase, 0.25% Driselase, 0.5% Rhozyme, 0.15 M Sorbitol, 0.15 M Mannitol, 7 mM $CaCl_2 \cdot 2H_2O$, 0.7 mM $NaH_2PO_4 \cdot H_2O$, and 3 mM MES buffer, pH 5.6).
3. Incubate at 25 C, on a gyrotary shaker at 60 rpm for 4–6 hr.
4. Filter protoplast preparation through Miracloth, then, 100 µm and 50 µm stainless steel filters to remove undigested cells and debris.
5. Wash three times by centrifugation (100 x g for 3 min) in Kao and Michayluk's (1975) modified nutrient medium (Vasil and Vasil, 1980) containing 0.3 M glucose, 0.9 µM 2,4-D and 2.3 µM ZEA.
6. Culture protoplasts in the modified Kao and Michayluk medium in liquid droplets or in a thin layer. Seal petri plates with Parafilm and incubate at 27 C in darkness for 3 days. First divisions occur after 4 days.
7. Add fresh medium to protoplast cultures after 7 days, then every 3–4 days.
8. Transfer regenerated callus masses and embryos to MS agar medium containing 1.1–4.5 µM 2,4-D after 3 weeks. Embryoids develop into plantlets on hormonefree MS agar medium. For more details, see Vasil and Vasil (1980) and Lu et al. (1981).

Initiation of Somatic Embryogenesis in *Pennisetum americanum*, *P. purpureum*, and *Panicum maximum*

1. Immature embryos, young inflorescence segments, and pieces of young leaves on a 2,4-D-supplemented medium develop soft, friable, nonembryogenic callus as well as compact, embryogenic callus.
2. Suspension cultures established from soft, friable callus and protoplasts isolated from these suspension cultures develop into nonmorphogenic cell colonies.
3. The compact, embryogenic callus forms embryoids, which differentiate on a low-auxin supplemented medium, germinate and eventually grow into mature plants.
4. Suspension cultures established from embryogenic callus also remain embryogenic and when plated form embryogenic calli with embryoids.
5. Protoplasts isolated from embryogenic suspension cultures form cell colonies with embryoids, which germinate and grow to form plantlets (see Fig. 1).

CONCLUSIONS

The recent upsurge of interest in plant cell culture techniques has been prompted by their presumed usefulness in the improvement of crop plants. However, to attain most of the objectives often discussed, it is essential to have the ability to regenerate plants from single cells.

Plant regeneration has been achieved from cultured tissues of several important species of cereals and grasses. Generally, regeneration takes place by the organization of shoot meristems that are multicellular in origin. However, this type of regeneration is not only sporadic and ephemeral but may also result in the formation of chimeral and mixoploid plants not suitable for genetic studies. On the other hand, regeneration through somatic embryogenesis would be advantageous, as the embryos arise from single cells. While somatic embryogenesis is fairly common in dicotyledons, a few examples are known among the monocotyledons. Recent reports of somatic embryogenesis in the Gramineae are, therefore, very encouraging.

Vasil and Vasil (1981a,b) suggest that somatic embryogenesis may not be as rare and as difficult to obtain in cereals and grasses as previously believed. This view is supported by the fact that embryo formation and plant regeneration has been achieved from callus, cell suspension, and protoplast cultures obtained from immature embryos, young inflorescences, and young leaves of a number of species.

KEY REFERENCES

Dunstan, D.I., Short, K.C., Dhaliwal, H., and Thomas, E. 1979. Further studies of plantlet production from cultured tissue of *Sorghum bicolor*. Protoplasma 101:355-362.

Green, C.E. and Phillips, R. 1975. In vitro plant regeneration from tissue cultures of maize. Crop Sci. 15:417-421.

Hulse, J.H., Laing, E.M., and Pearson, O.E. 1980. Sorghum and the Millets: Their Composition and Nutritive Value. Academic Press, New York.

Lu, C.-Y., Vasil, V., and Vasil, I.K. 1981. Isolation and culture of protoplasts of *Panicum maximum* Jacq. (Guinea grass): Somatic embryogenesis and plantlet formation. Z. Pflanzenphysiol. 104:311-318.

Vasil, V. and Vasil, I.K. 1980. Isolation and culture of cereal protoplasts. Part 2. Embryogenesis and plantlet formation from protoplasts of *Pennisetum americanum*. Theor. Appl. Genet. 56:97-99.

______ 1982. The ontogeny of somatic embryos of *Pennisetum americanum* (L.) K. Schum. I. In cultured immature embryos. Bot. Gaz. 143:454-465.

REFERENCES

Arnould, J.P. and Miche, J.C. 1971. Review of the economy and utilization of millets and sorghums in the world. Agron. Trop. 26:865-887.

Aykroyd, W.R., Gopalan, C., and Balasubramaniam, C. 1963. The Nutritive Value of Indian Foods and the Planning of Satisfactory Diets.

Special Report Series No. 42. Nutritional Research Laboratories, Hyderabad, ICMR, New Delhi.

Bajaj, Y.P.S. and Dhunju, M.S. 1981. Regeneration of plants from callus culture of napier grass (*Pennisetum purpureum*). Plant Sci. Lett. 20: 343-345.

Bajaj, Y.P.S., Sidhu, B.S., and Dubey, V.K. 1981. Regeneration of diverse plants from tissue cultures of forage grass—*Panicum* spp. Euphytica 30:135-140.

Brettell, R.I.S., Wernicke, W., and Thomas, E. 1980. Embryogenesis from cultured immature inflorescence of *Sorghum bicolor*. Protoplasma 104:141-148.

Brunken, J., De Wet, J.M., and Harlan, J.R. 1977. The morphology and domestication of pearl millet. Econ. Bot. 31:163-174.

Deshaprabhu, S.B.D. (ed.) 1966. The Wealth of India: A Dictionary of Indian Raw Materials and Industrial Products, Vol. 8. Publication and Information Directorate, CSIR, New Delhi.

Dunstan, D.I., Short, K.C., and Thomas, E. 1978. Anatomy of secondary morphogenesis in cultured scutellum tissue of *Sorghum bicolor*. Protoplasma 97:251-260.

Dunstan, D.I., Short, K.C., Dhaliwal, H., and Thomas, E. 1979. Further studies on plantlet production from cultured tissue of *Sorghum bicolor*. Protoplasma 101:355-362.

Gamborg, O.L., Shyluk, J.P., Brar, D.S., and Constabel, F. 1977. Morphogenesis and plant regeneration from callus of immature embryos of *Sorghum*. Plant Sci. Lett. 10:67-74.

Green, C.E. and Phillips, R. 1975. In vitro plant regeneration from tissue cultures of maize. Crop Sci. 15:417-421.

Harlan, J.R. 1972. A new classification of cultivated sorghums. In: Sorghum in Seventies (N.G.P. Rao and L.R. House, eds.) pp. 512-516. Oxford and IBH Publishing, New Delhi.

______ and De Wet, J.M.J. 1972. A simplified classification of cultivated sorghums. Crop Sci. 12:172-176.

Haydu, Z. and Vasil, I.K. 1981. Somatic embryogenesis and plant regeneration from leaf tissue and anthers of *Pennisetum purpureum* Schum. Theor. Appl. Genet. 59:269-273.

Hill, A.F. 1952. Economic Botany. McGraw Hill, New York.

Hoseney, R.C. and Varriano-Marston, E. 1980. Pearl millet: Its chemistry and utilization. In: Cereals for Food and Beverages. Recent Progress in Cereal Chemistry and Technology (G.E. Inglett and L. Munck, eds.) pp. 461-494. Academic Press, New York.

Hulse, J.H., Laing, E.M., and Pearson, O.E. 1980. Sorghum and the millets: Their composition and nutritive value. Academic Press, New York.

ICRISAT (International Crops Research Institute for the Semi-Arid Tropics) 1981. Annual Report 1979/80. Patanacheru, A.P. India.

King, P.J., Potrykus, I., and Thomas, E. 1978. In vitro genetics of cereals—Problems and perspectives. Physiol. Veg. 16:381-400.

Lu, C.-Y. 1981. Somatic Embryogenesis in *Panicum maximum* Jacq. (Guinea grass). Ph.D. Thesis, Univ. of Florida, Gainesville.

_____ and Vasil, I.K. 1981a. Somatic embryogenesis and plant regeneration from leaf tissue of *Panicum maximum* Jacq. Theor. Appl. Genet. 59:275-280.

_____ and Vasil, I.K. 1981b. Somatic embryogenesis and plant regeneration from freely-suspended cells and cell groups of *Panicum maximum* Jacq. Ann. Bot. 48:543-548.

_____ and Vasil, I.K. 1982. Somatic embryogenesis and plant regeneration in tissue cultures of *Panicum maximum* Jacq. Am. J. Bot. 69: 77-81.

_____, Vasil, V., and Vasil, I.K. 1981. Isolation and culture of protoplasts of *Panicum maximum* Jacq. (Guinea grass): Somatic embryogenesis and plantlet formation. Z. Pflanzenphysiol. 104:311-318.

Mascarenhas, A.F., Pathak, M., Hendre, R.R., and Jagannathan, V. 1975a. Tissue culture of maize, wheat, rice and sorghum. I. Initiation of viable callus and root culture. Indian J. Exp. Biol. 13:103-107.

Mascarenhas, A.F., Hendre, R.R., Pathak, M., Nadgir, A.L., and Jagannathan, V. 1975b. Tissue culture of maize, wheat, rice and sorghum. III. Growth and nutrition of root culture of maize, wheat and sorghum in agitated liquid medium. Indian J. Exp. Biol. 13:112-115.

Mastellar, V.J. and Holden, D.J. 1970. The growth of and organ formation from callus tissue of sorghum. Plant Physiol. 45:362-364.

Matz, S.A. 1969. Millet, wild rice, adlay and rice grass. In: Cereal Science (S.A. Matz, ed.) pp. 224-236. AVI Pub., Westport, Connecticut.

Mehra, K.L. 1963. Differentiation of the cultivated and wild *Eleusine* species. Phyton (Buenos Aires) 20: 189-198.

Muldoon, D.K. and Pearson, C.J. 1979. The hybrid between *Pennisetum americanum* and *Pennisetum purpureum*. Herbage Abstr. 49: 189-199.

Purseglove, J.W. 1972. Tropical Crops. Monocotyledons 1. John Wiley and Sons, New York.

Rachie, K.O. and Majumdar, J.V. 1980. Pearl Millet. Pennsylvania State Univ. Press, University Park.

Rangan, T.S. 1974. Morphogenic investigations on tissue cultures of *Panicum miliaceum*. Z. Pflanzenphysiol. 72:456-459.

_____ 1976. Growth and plantlet regeneration in tissue cultures of some Indian millets: *Paspalum scrobiculatum* L., *Eleusine coracana* Gaertn. and *Pennisetum typhoides* Pers. Z. Pflanzenphysiol. 78:208-216.

Rogers, M.S., Gal, H.L., and Horner, H.T., Jr. 1974. Callus formation and differentiation in tissue cultures of normal and cytoplasmic male sterile sorghum, pepper, sunflower and tobacco. In Vitro 9:463-467.

Rooney, L.W., Khan, M.N., and Earp, C.F. 1980. The technology of sorghum products. In: Cereals for Food and Beverages: Recent Progress in Cereal Chemistry and Technology (G.E. Inglett and L. Munck, eds.) pp. 513-554. Academic Press, New York.

Sastri, B.N. ed. 1952. The Wealth of India: A Dictionary of Indian Raw Materials and Industrial Products, Vol. 3. Publ. and Information Directorate, CSIR, New Delhi.

Strogonov, B.P., Komizerko, E.I., and Butenko, R.G. 1968. Culturing of isolated glasswort, sorghum, sweet clover and cabbage tissues for comparative study of their salt resistance. Sov. Plant Physiol. 15: 173-177.

Thomas, E., King, P.J., and Potrykus, I. 1977. Shoot and embryo-like structure formation from cultured tissue of *Sorghum bicolor*. Naturwissenschaften 64:587.

Vasil, V. and Vasil, I.K. 1979. Isolation and culture of cereal protoplasts. I. Callus formation from pearl millet (*Pennisetum americanum*) protoplasts. Z. Pflanzenphysiol. 92:379-383.

______ 1980. Isolation and culture of cereal protoplasts. II. Embryogenesis and plantlet formation from protoplasts of *Pennisetum americanum*. Theor. Appl. Genet. 56:97-99.

______ 1981a. Somatic embryogenesis and plant regeneration from tissue cultures of *Pennisetum americanum* and *P. americanum* x *P. purpureum* hybrid. Am. J. Bot. 68:864-872.

______ 1981b. Somatic embryogenesis and plant regeneration from suspension cultures of pearl millet (*Pennisetum americanum*). Ann. Bot. 47:669-678.

______ 1982a. Characterization of embryogenic cell suspension cultures derived from cultured inflorescences of *Pennisetum americanum* (Pearl millet). Am. J. Bot. 69:1441-1449.

______ 1982b. The ontogeny of somatic embryos of *Pennisetum americanum* (L.) K. Schum. 1. In cultured immature embryos. Bot. Gaz. 143:454-465.

Wernicke, W. and Brettel, R.I.S. 1980. Somatic embryogenesis from *Sorghum bicolor* leaves. Nature 287:138-139.

CHAPTER 6
Rice

Y. Yamada and *W. H. Loh*

Rice, which has sustained mankind since prehistory, is the primary source of calories for 40% of the world's population. The importance of rice to the survival of earlier civilizations is such that the grain has attained a semisacred status among various Asian cultures, particularly those of Japan, Indonesia, and Sri Lanka.

DISTRIBUTION AND ECONOMIC IMPORTANCE

The total annual world production of rice between 1976 and 1978 was 363,940,000 tons. As seen in Table 1, over 90% of the world's rice crop is harvested in Asia, more than half of it in India and China. Only wheat ranks higher than rice in total acreage harvested. Nutritionally, rice generates more calories per unit area than either maize or wheat.

Rice is unique among the world's major cereal crops in that it grows submerged under water. The world's rice lands can be classified into four categories according to the water regimes and primary types of rice planted in each of these geotypes. The new technology in rice production that will be discussed in this chapter is restricted to the shallow water region under irrigation, which comprises roughly 25% of the world's rice acreage.

The major Asian rice-producing nations can be divided into three basic categories, as shown in Table 2. China, Taiwan, Iran, North and South Korea, and Japan are examples of countries with high average yields and a demonstrated potential for moderate increases in annual

Table 1. World Rice Production, 1976-1978[a]

	Acreage (10^6 ha)	Percent Total	Harvest (10^6 t)	Percent Total	Average (t/ha)
Asia	129.1	89.9	335.8	91.5	2.6
East Asia	41.4	28.8	160.5	43.7	3.9
Japan	2.7	1.9	15.6	4.3	5.8
China (PRC)	36.0	25.1	129.7	35.3	3.6
Southeast Asia	33.7	23.5	71.6	19.5	2.1
Philippines	3.5	2.5	6.9	1.9	2.0
Vietnam	5.2	3.6	11.0	3.0	2.1
Thailand	8.6	6.0	15.9	4.3	1.9
Burma	5.1	3.5	9.5	2.6	1.9
Indonesia	8.5	5.9	24.0	6.5	2.8
South Asia	53.4	37.2	101.6	27.7	1.9
Bangladesh	10.0	7.0	18.5	5.0	1.9
India	39.6	27.6	74.7	20.4	1.9
Africa	4.9	3.4	7.8	2.1	1.6
North Africa	0.5	0.3	2.5	0.7	5.0
Sub-Sahara Africa	4.5	3.1	5.3	1.4	1.2
West Africa	2.4	1.7	2.9	0.8	1.2
Central and East Africa	2.0	1.0	2.5	0.7	1.3
North and South America	16.0	11.0	32.7	8.8	2.0
United States	1.0	0.7	5.3	1.4	5.3
Brazil	5.5	3.8	7.7	2.1	1.4
Latin America	7.5	5.2	13.7	3.7	1.8

[a]Adapted from DeDatta, 1981.

rice production. These fall into category I. Category II includes India, Indonesia, Pakistan, Malaysia, Philippines, and Afghanistan, all of which have low to medium rice yields, but show consistent yield increases in the last decade. This increase is defined as a 15% gain in yield between 1961-65 and 1971-75. Nations with very low average yields and which can expect only small increases in rice production over the next decade fall into category III, and include Vietnam, Bangladesh, Burma, Thailand, Kampuchea, Sri Lanka, Laos, and Nepal.

While these nations differ considerably as to their level of technology, the key demographic factor is the amount of land they have under irrigation. Asian farmers who grow rice on rain-fed regions must make substantial sacrifices to satisfy the moisture requirement of rice. In order to coordinate panicle formation with the annual wet season, consecutive plantings are not possible in rain-fed fields. In addition,

Table 2. Classification of Rice Cultivation of Major Asian Producers[a]

CATEGORY	LAND UNDER IRRIGA-TION (%)	FERTILI-ZATION RATE (kg/ha)	AVERAGE YIELD (t/ha)	POTENTIAL INCREASE
I	95	275	3.0	Moderate
II	40[b]	31	1.5-2.5	Constant
III	25[b]	20	2.0	Low

[a]Adapted from Chandler, 1979.

[b]These results are exaggerated since they include lands that receive irrigation only as a rainy season supplement.

the rainy season is characterized by overcast skies, which substantially lower the incident solar radiation and increase the potential for disease. With irrigation, these areas could support 2-3 crops a year.

In the twelve-year span between 1960-62 and 1974-76, the world's average population increased 30%. During the same period, the world's rice consumption increased 42%. The increase in per capita consumption can be traced to a rise in the real income of much of the world's population during the past 15 years. In every continent except Asia, rice consumption had increased dramatically.

Since the end of World War II, three nations have emerged as the world's leading rice exporters: China, Thailand, and the U.S. China's rice exports peaked at 2.14 million metric tons (MMT) in 1972 but annual harvests in that country fluctuate dramatically. The Chinese government has a policy of selling rice in exchange for lower-priced wheat and other foodstuffs. Thailand has been a major rice exporter since the days of colonial rule. However, since it topped the world in 1964 with a record of 1.9 MMT exported, annual export tonnage has declined steadily.

Although it produces only 1.7% of the world total, the U.S. exports 60% of its annual harvest. Not only is the U.S. the world's leading exporter, averaging 1.8 MMT annually, it has become the world's most reliable supplier of rice. Unlike many countries where rice is part of the staple diet, with exports of secondary concern, the American rice grower regards rice as a cash crop.

BREEDING AND CULTURE

Rice technology is one of the most refined in the world. Breeding programs began before the turn of the century, with the establishment of research centers in India, Indonesia, and West Bengal, present-day Bangladesh. Through conventional breeding, the earlier plant breeders were able to select for disease resistance, tolerance to drought, flood, salinity, and soil alkalinity.

Only two species of *Oryzae* are cultivated; O. *sativa* L. and O. *glaberrima* L. The latter species is grown only in West Africa and has been rapidly displaced by O. *sativa*. *Oryza sativa* consists of three distinct land races: Indica, Japonica, and Javanica. Most geneticists agree that the Indica race, grown primarily in the humid tropics, is the progenitor of the other two races. Japonica has been adapted for growth in the cooler climates of Japan and northern China. The tall and vigorous Javanica race is grown exclusively in Indonesia.

The protocol for breeding a self-pollinating species like rice is relatively straightforward. Lines with desirable traits are allowed to self-pollinate for several generations in order to select against deleterious traits. Prior to making a cross, the rice breeder opens the florets of the recipient lines and removes the anthers. Cross pollination takes place when ripe anthers from the donor line are broken over the mature stigmas of the emasculated recipient florets. The inflorescences are then covered with paper bags to prevent chance cross-pollination. Since the parental lines are homozygous, all F_1 individuals are identical. The desirable gene combinations are selected from the F_2 and allowed to self until pure lines are once more established. Desirable traits are considered to be stable by the F_5 or F_6 generation.

Rice yield can be calculated by the following formula, adapted from Metsushima (1980):

Grain yield/hectare = (No. of plants/square meter) x
(No. of panicles/plant) x
(No. of grains/panicle) x
(% of ripened grains) x
(weight of 1000 grains)/1000

It would seem that one quick way to increase yield is to increase the number of plants per unit area. Close planting of traditional varieties, however, usually fails to increase yields. The native rice plant is a very vigorous grass with an extensive root system, so that close planting quickly exhausts the available soil nutrients. Potential yield increases are offset by lower productivity as adjacent plants shade each other. Application of fertilizers produces an excess of vegetative growth with very little effect on grain yield. The increased canopy of fertilized plants increases lodging prior to harvest.

Major increases in rice yields were not recorded until the release of Taichung Native 1 (TN1) in 1960. TN1 was a dwarf strain by O. *sativa* that demonstrated fertilizer responsiveness through high grain/straw ratios. Although it was widely planted in Taiwan and India, TN1 did not gain wide recognition as the prototype for modern rice.

In 1962 workers at the International Rice Research Institute (IRRI) in the Philippines crossed Peta, a tall Indonesian heavy yielder, with a dwarf Taiwanese variety called Dee-Geo-Woo-Gen (DGWG) to produce IR8-288-3. The name was later shortened to IR-8 when released. This was a semidwarf strain of Indica rice that responded to fertilization with even heavier yields than TN1. Its short stature and thick stem enabled it to bear a heavy panicle of grain above the water. A smaller root system and short, medium-width, upright leaves permitted close planting with minimal shading of adjacent plants. To date, no new variety of rice has ever surpassed IR-8 in yield potential.

Release of IR-8 was soon followed by other IRRI varieties with similar growth habits, accompanied by improvements in grain quality, disease resistance, photoperiod insensitivity, and shorter maturation dates. The greater adaptability of these strains enabled farmers to adopt the varieties throughout Asia. The simultaneous release of semi-dwarf rice plants and high-yielding wheat cultivars from Mexico registered dramatic increases in world food harvest, an event popularly dubbed as the Green Revolution.

Since 1975, IRRI has altered its policy of producing named varieties for distribution. Instead, it has concentrated its efforts on providing local research efforts with germplasm for production of varieties compatible with local climates. Through centuries of unconscious selection, traditional varieties of rice are uniquely adapted to the local soils, pests, insect, and climatic peculiarities. Currently, most modern rice varieties are the result of the introduction of the semi-dwarf habit of IR-8 and TN1 into local lines.

CURRENT RESEARCH PRIORITIES

Conventional Breeding

Building on the achievements of the Green Revolution, rice breeders have targeted specific traits for future breeding programs. Many new rice cultivars can reach seed maturation within 90 to 100 days after germination. Combined with photoperiod insensitivity, farmers may be able to harvest two or more crops per year, providing that irrigation is available year around. Yet, most of the earlier varieties seem to lack both insect and disease resistance. One of the goals of the current breeding effort is to produce early-maturing strains (90-95 days) with good plant habit and disease resistance.

Up to now, breeding for disease and pest resistance in rice has involved the use of vertical and multiline approaches. In the vertical strategy, one gene is responsible for resistance. The aim of the breeder is to introduce that gene, in as many copies as possible, into the host crop. The alternative has been to produce lines with similar agronomic features, but different genes for resistance. When planted together, the crop represents a land race with polyallelic resistance to pathogens. Thus, some yields are guaranteed even if pathogens were able to evolve and overcome a resistant phenotype. The ideal approach will be to develop horizontal resistance, which incorporates many minor genes—each providing a moderate level of resistance—into the new rice varieties.

While most new rice varieties possess short, thick straw for supporting heavy panicles, there is a tight linkage between genes coding for weak stem and disease resistance. Disease resistant lines usually have a weak stem, resulting in a high incidence of lodging.

Since rice farms already occupy the choicest lands and utilize the best water resources, future increases in rice acreage will have to come in marginal areas where irrigation and other technological inputs would be far more expensive. Adapting the rice plant to the environment will, in the long run, be far cheaper than the reverse process.

Mutational Breeding

In addition to conventional crosses, rice breeders have induced mutations through the use of ionizing radiation and chemical mutagens to generate more desirable plant types. Mutational breeding has several advantages over conventional breeding techniques (IAEA, 1971).

The genetic changes induced by physical and chemical mutagens are often point mutations. Thus it is often possible to confer specific improvements without affecting other agronomic traits. Mutational breeding is most applicable for the improvement of specific traits in well-adapted strains. In addition, the technique may be used to generate novel genotypes quickly. Mutational breeding may also be valuable in breaking linkages between beneficial and deleterious traits, such as the linkage between weak stem and disease resistance in new semidwarf rice.

Mutational breeding has already met with some success in rice. Calrose 76, a short variant induced through mutagenesis of Calrose, has enjoyed wide acceptance by California growers (Rutger and Brandon, 1981). Li et al. (1968) have also produced short-stature mutants with increased panicle weight in Taiwan. In Japan, blast-resistant mutants have been produced by both gamma-irradiation and mutagenesis with ethyl methane sulphonate (Yamasaki and Kawai, 1968). In Korea, X-irradiation has produced variants that mature 4-8 days earlier than the parent variety, but with identical yields (Ree, 1968).

REVIEW OF THE LITERATURE

Diploid Culture

The differentiation of rice shoots and roots and the regeneration of whole plants from rice callus were among the first successful regeneration experiments with tissue cultures of cereals. Until 15 years ago, there was very little success in inducing cereal callus and in maintaining good growth (Straus and LaRue, 1954; Norstog, 1956; Carew and Schwarting, 1958; Tamaoki and Ullstrup, 1958). The problem of inducing callus growth was resolved by using higher concentrations of auxins in the tissue culture media (Yamada et al., 1967b; Carter et al., 1967) than were used for dicotyledonous tissue. Cereal-derived callus can be routinely induced when there is a high enough concentration of auxin in the culture medium.

Studies of rice tissue culture began in Japan during the 1960s. Furuhashi and Yatazawa (1964) first succeeded in inducing callus from rice stem nodes. They emphasized that yeast extract was an essential factor in long-term callus proliferation (Yatazawa et al., 1967). A completely synthetic medium capable of inducing and maintaining callus growth was reported by Yamada et al. (1967).

Kawata and Ishihara (1968) induced callus from the root tips of rice plants with 2,4-D plus IAA at several concentrations. Approximately one year after the callus was induced, it was transferred onto a differentiation medium containing 0.15 M sucrose, 1% casein hydrolysate, 5.7 μM IAA, but no 2,4-D. One plantlet was induced to differentiate and grew to about 10 cm in height. Sekiya et al. (1977) induced callus and achieved rapid cell proliferation with both IAA and indole-3-valeric acid. Hendre et al. (1975) were able to induce rice callus with 2,4-D, IBA, and NAA. Callus proliferation could be maintained with either maltose, glucose, or sucrose as the carbon source. No stimulation was noted by adding amino acid supplements to the medium.

Nishi et al. (1968) obtained callus from the roots of rice on Linsmaier-Skoog (LS) (1965) medium containing 9.0 μM 2,4-D. When this callus was transferred to a similar medium without auxin and incubated in the light, it differentiated into shoots and roots, and subsequently formed whole plants. The regenerated plants were diploid, although a few showed peculiar phenotypes (dwarfed, twisted, flag-leaf shortened, and albino).

Histological study of shoot organogenesis from embryo-derived callus revealed that leaf primordia originated within minute indentations on the surface of the callus. Normal leaf primordia subsequently formed around the abnormal apices (Tamura, 1968). When transferred to shoot-induction medium, organogenesis can be detected within 3 days by radial divisions in the callus (Nakano and Maeda, 1979). Procambial and provascular tissue can be seen in the radial tissue within 9 days. During this time, the outermost cell layer of the callus mass becomes stratified and meristematic. Leaf primordia initiate from the surface layer of cells, which has also developed trichomes and chloroplasts.

Nishi and Mitsuoka (1969) reported regeneration of plants of various ploidy levels from callus induced from different parts of a rice plant. The medium used was LS with 9.0 μM 2,4-D and no cytokinins. Diploid plants were recovered from callus induced from both embryo and shoot node explants. Anther callus could be induced to differentiate haploid, triploid, and pentaploid, as well as diploid, plants. Ovary callus regenerated diploids and tetraploids. Plants of the same ploidy exhibited morphogenic properties unique to the ploidy level, regardless of the source of the callus.

Nishi et al. (1973) recovered plants grown from callus induced on LS medium with 4.5 μM 2,4-D without other auxins or cytokinins. The callus was induced from roots of germinating seeds in 10 days, and subcultured after 2 months on a medium containing no auxins. After another 2 months, differentiation of shoots and roots ocurred in every test tube. Plantlet formation from callus decreased in the second and third subcultures to 95% and 64%, respectively.

Plants have also been regenerated from callus of diverse somatic origin. Mascarenhas et al. (1975a,b) were able to regenerate plants from seedling mesocotyl segments by transferring the proliferating callus onto 0.54 μM NAA. Diploid plants have been regenerated from scutellar tissue and cotyledonary nodes in culture (Henke et al., 1978; Wu and Li, 1971). Bhattacharya and Sen (1980) were able to regenerate plants from leaf sheath cells in culture. Bajaj et al. (1980) have recently regenerated triploid plants from both immature and mature endosperm tissue. The resulting triploid plants exhibit broader leaves, higher growth rates and increased tillering when compared to the diploid parents.

Reviewing the previous experiments on morphogenesis in rice tissue culture, Nishi et al. (1973) proposed that the degree of organ induction depended solely on the level of auxin used, suggesting that differentiation in rice depended on only one exogenous factor—auxin. In their experiments, a higher percentage of organ differentiation was always observed at lower concentrations of auxin. This occurred regardless of which auxin was used to induce callus formation. Callus originating from every type of tissue differentiated when the callus-inducing auxin was removed from the culture medium.

Interestingly, Cornejo-Martin et al. (1979) reported that ethylene (0.36 μM) alone will promote shoot regeneration in rice tissue culture. Optimal rates of organogenesis require 0.18 μM ethylene and 2% CO_2. In general, shoot regeneration could be induced with an ethylene/CO_2 ratio of 2.5 x 10^{-4} to 1. Carbon dioxide alone had no effect on organogenesis.

Haploid Culture

Niizeki and Oono (1968) were the first to culture *Oryza* anthers successfully to produce haploid plants. Approximately 1-2 days prior to heading, anthers containing mature pollen grains were placed onto a medium containing 9.4 μM NAA, 10.0 μM KIN, and 4.9 μM 2,4-D, supplemented with 3 ppm yeast extract (YE). The anther walls blackened, but pale yellow callus emerged after 4-8 weeks of culture. Haploid plants were regenerated after transferring the callus into a similar medium lacking 2,4-D.

Callus induction from anther explants proved to be the major barrier in haploid rice culture. Induction rates vary from 0.21 to 2.2% (Myint and deFossard, 1974). Once callus was induced, plants regenerated at high frequencies without auxins, cytokinins, or organic supplements.

Woo and Hsu (1974) attempted to regenerate haploid plants from Indica x Japonica hybrids as a means of preventing segregation in subsequent generations. Only one viable haploid plant was regenerated from the callus which originated from the anthers of the F_1 progeny. The plant was placed in 0.05% colchicine in order to double the chromosomal complement and then was selfed. The F_2 generation is identical to the F_1, suggesting that the colchicine-doubled parent was completely homozygous. However, sterility continues to be a problem in subsequent generations.

Zhou and Yang (1981) cultured ovaries with attached receptacles aseptically removed from florets at the late uninucleate pollen stage. The ovaries were cultured in 0.6 μM 2-methyl-4-chlorophenoxyacetic acid. Callus induction occurred at a frequency of 4.4%, and plants were regenerated after the auxin concentration was lowered to 0.16 μM. Plants originating from the female gametophyte exhibited varied ploidy levels.

Yin et al. (1976) described a haploid rice breeding program in China which has released several varieties, including Tanfeng 1, for cultivation. Anthers at the uninucleate stage were cultured in medium containing 9.0 μM 2,4-D. According to their protocol, callus induction occurred at rates as high as 37.7%, averaging 16.3%. Plant regeneration occurs in the absence of 2,4-D. Despite the high incidence of albinism (50% of the regenerate), some promising haploid varieties were chosen for testing after chromosome doubling. Field tests indicate yields could run as high as 8.34 t/ha for several varieties developed from haploid culture. Other cultivated rice varieties that have been produced from haploid cultures include Huayu 1 and Huayu 2. A summary of haploid studies in China is presented by Hu et al. (1978).

Deka and Sen (1976) were the first to report isolation of protoplasts from leaf sheath and stem callus of rice. The callus was treated with 2% pectinase and 3% cellulase in pH 5.4 buffer, with 0.45 M mannitol as the osmoticum. When transferred to SH medium containing 2.3 μM 2,4-D, 2.9 μM IAA, and 0.47 μM KIN, cell wall synthesis took place within 24 hr. No plants, however, were regenerated despite callus proliferation.

Rice protoplasts have also been isolated from pollen callus, growing on 18.0 μM 2,4-D and 2% CW (Peking Institute of Botany, 1975). Cell wall regeneration and subsequent cell division were reported in medium containing 0.23 μM 2,4-D and 1.3 μM BA. No rice plants have been regenerated from cultured protoplasts.

Agricultural Applications

Cognizant of the major advances in the development of tissue culture, IRRI established a tissue culture laboratory in 1979. This group has screened 85 lines of the Indica varieties for callus growth and regeneration from anther and pollen explants. Haploid callus exhibits better growth and higher regeneration efficiency than diploid tissue. While the frequency of regeneration varies, certain strains possess regeneration efficiences ranging up to 53%, as compared to a typical rate of 3-10% (Anonymous, 1981).

Using haploids, F_2 variations can be discerned immediately after regeneration. Each plant becomes true breeding through duplication of the genome. Scientists at IRRI estimate that pure lines can be developed within 15 to 21 months in the tropics, and within 3 years in temperate zones. This contrasts with a 7-year minimum through conventional breeding.

One of the first efforts of the IRRI tissue culture breeding program was to screen for salt-resistant varieties (Anonymous, 1981). Callus has been induced from seeds and placed on high salt medium. By March, 1981, 2000 plants have been regenerated from salt-resistant callus. In June, seedling regenerates were transferred to salt-water culture. The survivors have produced seed, and their progeny are currently being tested in the field.

See Chapter 3 for an additional and extensive discussion of the utilization of anther culture for the development of new varieties in rice.

PROTOCOLS

Plant Material

Almost any part of a rice plant can be used to initiate callus cultures (see Fig. 1); however, seeds are a particularly common and easy material to use. To induce callus from specific tissue, such as nodes, shoots, or root tips, one must first grow plants from sterilized seeds in a flask containing a sterile agar salt medium. From this plant, a segment of the nodal or root segment is removed aseptically and placed on the appropriate agar medium for callus induction.

It is very difficult to sterilize any part of a plant grown under field conditions. Growing a plant from sterilized seeds is preferable in terms of the efficiency and ease of callus induction. The preparation of rice seeds is the same for both callus induction and seed germination under sterile conditions.

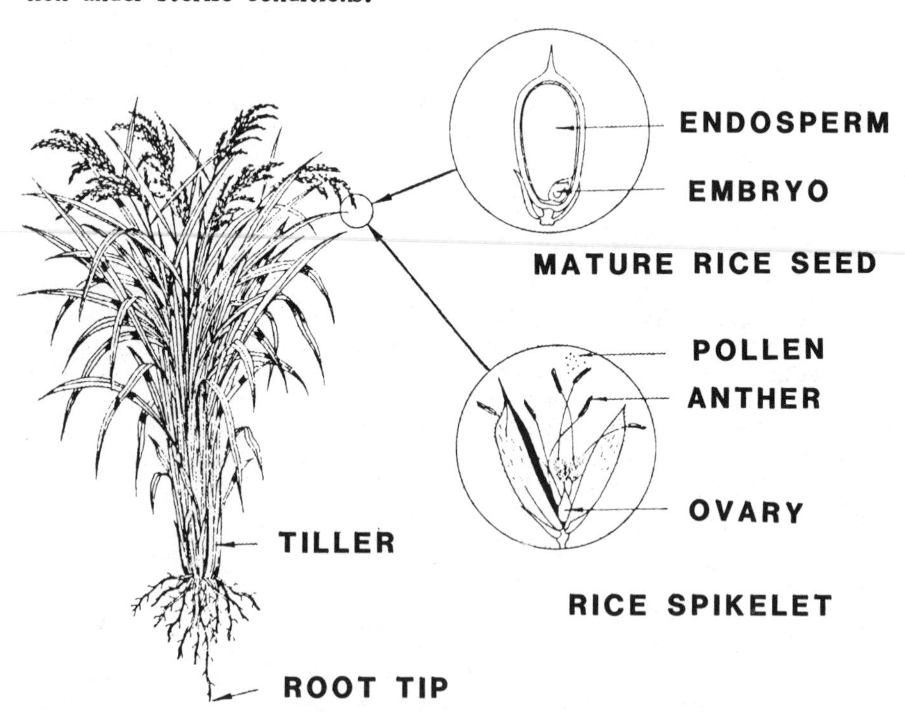

Figure 1. Somatic and sexual tissue of *Oryza sativa* suitable for callus induction and subsequent regeneration.

1. Collect fully mature seeds and remove the hull.
2. Rinse the brown rice grains in 95% ethanol to remove the wax from the surface of the grains. If the seeds are rinsed longer than 10 sec, they will die.
3. Decant the alcohol and cover the top of the beaker with a double layer of gauze. Secure this covering with a rubber band.
4. Rinse the beaker under running water for 2-10 hr to wash the grains.
5. Submerge the grains in sterile petri dishes in 70% alcohol for 5 min.
6. Discard the alcohol and add a 3% sodium hypochlorite solution. Commercial bleach can be used at half strength as a substitute. Leave the petri dish covered for 50 min, occasionally shaking the chlorine solution. Do not leave seeds in bleach for longer than an hour.
7. Rinse the grains three times in sterile, distilled water.
8. Place the seeds on agar medium to obtain young plants or on tissue culture medium containing auxins to obtain callus.

Media

The optimal concentrations of auxin for both haploid and diploid cultures are much higher than those used to culture dicotyledonous plants. The concentration of 2,4-D should be about 22.6 µM, that of NAA 537 µM, and that of IAA 570 µM. These auxin concentrations are optimal for both callus induction and proliferation. Chu (1978) and others have reported that the N_6 medium is superior to MS medium for anther and pollen culture. The formulation of N_6 medium is provided in Table 3.

Culture Conditions

The most suitable temperature for the induction of rice callus is between 25 and 30 C. Growth in the dark is superior to that in light. The callus may form as a friable or compact mass. The degree of friability of the cells affects the ease of preparing a liquid suspension culture. To generate cell suspensions, transfer a soft, friable callus to an aliquot of liquid medium at the rate of 10^{5-6} cells per ml. The culture is usually shaken at 120 rpm.

Plant Regeneration

Plant regeneration or shoot organogenesis can be induced by decreasing the concentration of 2,4-D or substituting a weaker auxin, such as IAA or NAA. Exogenously supplied cytokinins do not appear to be essential to obtain organogenesis, but they may accelerate organogenesis.

Table 3. N_6 Medium for Anther Culture of Cereals[a]

COMPONENT	CONCENTRATION
$(NH_4)_2SO_4$	3.5 mM
KNO_3	0.03 M
KH_2PO_4	3.0 mM
$MgSO_4 \cdot 7H_2O$	0.75 mM
$CaCl_2 \cdot 2H_2O$	1.13 mM
$MnSO_4 \cdot 4H_2O$	0.02 mM
$ZnSO_4 \cdot 7H_2O$	5.2 μM
H_3BO_3	0.025 mM
KI	4.8 μM
$FeSO_4 \cdot 7H_2O$	5 ml solution[b]
Glycine	0.027 mM
Thiamine·HCl	3.0 μM
Pyridoxine·HCl	2.4 μM
Nicotinic acid	4.1 μM
Sucrose	0.15 M
Agar	0.1-1%
pH	5.6

[a] Adapted from Chu, 1978.

[b] Five ml of solution obtained by dissolving 0.02 M $FeSO_4 \cdot 7H_2O$ and 0.022 M Na_2-EDTA in 1 liter distilled water.

The first observable signs of regeneration are green spots on calli cultured under lights. The green spots later develop into shoots. When one shoot dominates, other shoots will not grow vigorously. The relationship of the plant hormones in the culture medium and organogenesis in cereal callus differ from that in dicotyledonous plants. The number of regenerating shoots from a cereal callus is much lower than from a dicotyledonous callus. Care should be taken that differentiation of only one shoot per callus does not represent elongation of explant apex that was previous inhibited by high auxin concentrations (see discussion by King et al., 1978).

FUTURE PROSPECTS

Generating Variation

With the development of a reproducible system for plant regeneration in rice from anther culture, tissue culture gives the plant geneticist greater flexibility in selecting his breeding strategy, and shorter generation times in producing pure lines. Plants can be routinely regenerated from embryo-derived callus or from callus of cultured anthers of rice (Fig. 2).

Figure 2. (A) Callus induction from rice plant in 45.0 μM 2,4-D and 0.05 μM KIN (from Yamada et al., 1967). (B) Callus growth in subculture. (C) Regeneration of plantlet from rice callus. 1-Organogenesis on the callus surface. 2-Adventive shoot formation in rice callus. 3-Abnormal leaf structure associated with adventive shoot apex (from Tamura, 1968). (D) Regeneration of plants of various ploidy from cultured anthers, embryos, and ovaries of rice (from Nishi and Mitsuoka, 1969).

One of the most useful features of tissue culture in plant breeding projects is its ability to generate variation. Point mutations and chromosome recombination have been detected in plants regenerated from callus and suspension cultures (Shepherd, 1982). Oono (1978) has shown that rice plants regenerated from tissue culture will also exhibit a great deal of variation.

Breeding programs owe their major expense to the maintenance of large crop populations prior to evaluation. Using tissue culture techniques, genetic selection occurs within the laboratory and permits selection of rare mutations at frequencies of 10^{-5} to 10^{-6}. Both recessive and co-dominant genes can be expressed in haploid culture systems, and cell densities approaching 10^5 cells per ml are not uncommon. Thus, the routine use of haploid cell culture techniques would make it possible to screen individual variants on a scale heretofore impossible.

Once haploid cell cultures are established, resistant individuals may be selected on the basis of salinity, herbicide resistance, pH imbalance, etc. Plant tissue culturists have already developed a large background of expertise for selection of specific biochemical mutants. For example, Heyser and Nabors (1982) has reported the development of salt and drought tolerant cell lines of rice through long-term embryogenic culture, but no plants have been regenerated.

Schaeffer and Sharpe (1981) have been able to increase total protein content in rice by treating anther cultures with S-aminoethyl-L-cysteine (S-ACE), a cysteine analog. The variant individuals retained resistance to S-ACE after 30 subcultures without the analog. After selfing, the F_2 generation contained individuals with a 10% increase in free lysine and a 48% increase in protein content. The experimental scheme is similar to that of Chaleff and Carlson (1974) who failed to regenerate plants from their mutant lines.

Selection of useful biochemical mutants through plant tissue culture is a technique that is just beginning to show results. Many types of mutants can be selected for and are awaiting imaginative screening systems. Rice tissue culture has been used to regenerate plants of various ploidy from callus originating from different sexual tissue (Nishi and Mitsuoka, 1969). This development is worth further research since polyploidy has been used elsewhere to improve plant vigor.

Somatic Mutagenesis

The ability for rapid clonal propagation through tissue culture makes it an ideal method for mutational research. In conventional mutational breeding, ionizing radiation and chemical mutagens cause point mutations at random in the seed. Since the apical meristem is multicellular, a chimeric plant with mutated sectors will grow from the mutagenized seed (Broertjes and Keen, 1980). Desirable traits are lost unless the change occurs in a germ line cell. Somatic mutations will not be transmitted by sexual hybridization. With tissue culture techniques, such mutations may be saved through clonal propagation of the variant sector.

Mutational breeding of rice seeds dropped in popularity as a result of the paucity of beneficial variants. Mutagenesis alone is not enough for selection of rare traits. Tissue culture offers the mutational breeder the luxury of numbers, since mutagenized cell populations may be evaluated in the laboratory, rather than in the field.

Embryo Culture and Protoplast Culture

In addition to point mutations, tissue culture offers the potential for making major changes in the plant genome. Out of 20 *Oryzae* species, only 2 have been cultivated. Many of the wild species may contain valuable genes controlling disease resistance, tolerance to salinity, metal toxicity, etc. For example, prior to 1976, resistance to grassy stunt virus was only found in *Oryzae nivira*. That resistance locus has subsequently been transferred to several of the IRRI varieties. However, conventional crossing between disparate genomes is not possible, and tissue culture offers two potential solutions to this problem.

Wide crosses often fail as a result of endosperm incompatibility. Through embryo culture, it may be possible to rescue the hybrid embryo and culture it in vitro. Development of the embryo occurs entirely within the culture flask and the resulting hybrid is transferred into the greenhouse only when it is capable of autotrophic growth.

The other approach involves protoplast fusion. Once the cell walls of suspension cells have been stripped enzymatically, plant protoplasts can be fused mechanically. In some cases, the resulting hybrid protoplast can be induced to regenerate a cell wall. If these cells can be induced to divide, often callus will form. Regeneration of the hybrid callus will produce plants containing genomes from two different species. Many interspecific and even intergeneric hybrids have been produced through the fusion of protoplasts, however, no hybrids with graminaceous species have been reported.

Germplasm Preservation

The most important function of tissue culture in crop breeding, however, may lie in the preservation of exotic germplasm. Sala et al. (1979) have developed a technique for freeze-preservation of rice using cell suspension cultures. Through the use of DMSO and a quick-freezing technique, they have been able to preserve rice tissue with minimal damage. Upon thawing, log phase growth resumed within one cell generation, suggesting that no genetic changes, including selection for cold resistance, could have occurred in the preservation process.

Upwards of 60,000 germlines of *O. sativa* have been preserved at IRRI, as well as an expanding collection of *O. glaberrima* genotypes. Yet very few genes have been identified for resistance to insects such as rice gall midge (Rao et al., 1980). None of the cultivars tested by IRRI have exhibited a high level of resistance to sheath blight (IRRI, 1978). Given the enormous expense associated with maintenance and periodic rejuvenation of seed stock, tissue culture offers a quick and inexpensive means of preserving rare and valuable genotypes for future analysis and use.

Breeding Applications

Despite the tremendous advances in rice breeding, yield increases generated by the introduction of semi-dwarf varieties have been restricted to the one-quarter of the world's rice lands which are under irrigation. One-third of the world's rice acreage is rainfed, and another third is classified as deepwater zones (Barker and Herdt, 1979). Rainfed regions are subject to periodic droughts while deepwater zones represent lowlands prone to annual monsoonal floods.

RAINFED RICE CULTURE. The modern semidwarf varieties of rice can tolerate a maximum of 15 cm of water. Some of the taller semidwarfs can survive a maximum of 50 cm of water. Traditional varieties of rice can grow in 1 m of water, while many of the floating types will survive in water over 1.5 m deep.

In general, rainfed rice fields are planted only with traditional varieties. Without irrigation facilities, the average rice farmer cannot plant high-yielding rice. Rather than risk a total loss due to climatic peculiarities, growers plant traditional rice varieties and settle for yields averaging 1 t/ha. The only exceptions are the Philippines, where modern varieties are planted on 70% of rainfed fields, and Vietnam, where 30% of the Mekong Delta is planted with the semi-dwarf rice varieties.

Several research priorities have been suggested for improving rice productivity. Since the lack of water is the major reason for decreased yields, development of drought-tolerant varieties could be important. The ideal varieties must have intermediate stature and quickly achieve tall seedling height to resist early floods. They should also exhibit tolerance to short periods of submergence. Other factors such as photoperiod sensitivity, early maturity, and tolerance to adverse soil nutrients are also important, but these will probably await further research to pinpoint the causal factors in lowered yields.

DEEPWATER RICE CULTURE. In rainfed regions that are subject to monsoonal floods, commonly referred to as deepwater fields, farmers have traditionally grown floating rice varieties. Initially, rice breeders were able to augment low yields by introducing elongation genes into semi-dwarf high-yielding varieties. As the breeding effort intensified, traditional deepwater varieties were crossed with semi-dwarf breeds. The deep-rooted progeny from these crosses tend to retain their semi-dwarf habit under ideal conditions. When the water level rises, however, they are able to effect rapid elongation of the internodal regions and keep the panicles above the water. Yet, high yields could be produced with these varieties only if the water level did not drop subsequently. Future breeding efforts are aimed at producing deepwater varieties that will resist lodging even if the water level drops before harvest.

Breeders are developing methods for evaluating nodal tillering ability in traditional rice types. These genes will permit the elongated rice plant to root at the stem nodes as the water level drops, and regain dwarf stature at harvest time. Other desirable traits include development of small erect leaves to increase photosynthetic efficiency; resistance to stem borers, a major pest in deepwater farming; and fertilizer responsiveness, since flooded soils are usually high in nitrogen.

GENETIC DIVERSITY. Despite the intensive cross breeding efforts of the last 25 years, the genetic base of modern cultivated rice has narrowed over recent years. An analysis of Asian rice-breeding programs (Hargrove, 1979) revealed that in 1965, TN1 was the most popular donor variety, participating in 22% of all crosses. Almost as popular, IR-8 was involved in 20% of the crosses, while other IRRI varieties contributed to 14% of the total gene pool. In 1975, these ratios have been reduced to 1%, 3%, and 44%, respectively, as IRRI began to encourage local breeding efforts to introduce modern rice genomes into traditional varieties (Hargrove and Cabanilla, 1979).

However, since all IRRI genomes owe their semi-dwarf stature to IR-8, over 80% of existing varieties of modern rice contain the DGWG dwarf genes. Thus, a substantial proportion of inbreeding has occurred. In addition, most of the new semi-dwarf varieties belong to the Indica land race. Using tissue culture techniques to preserve rare and potentially valuable genes and to generate variation in existing genomes will be essential to the vitality of future rice breeding efforts.

KEY REFERENCES

DeDatta, S.K. 1981. Principles and Practices of Rice Production. John Wiley, New York.

Deka, P.C. and Sen, S.K. 1976. Differentiation in calli originated from isolated protoplasts of rice (*Oryza sativa* L.) through plating technique. Molec. Gen. Genet. 145:239-244.

Nishi, T. and Mitsuoka, S. 1969. Occurrence of various ploidy plants from anther and ovary culture of rice plants. Jpn. J. Genet. 44:341-346.

Nishi, T., Yamada, Y., and Takahashi, E. 1973. The role of auxins in differentiation of rice tissue cultured in vitro. Bot. Mag. Tokyo 86: 183-184.

Sala, F., Cella, R., and Rollo, F. 1979. Freeze-preservation of rice cells grown in suspension culture. Physiol. Plant. 45:170-176.

Schaeffer, G.W. and Sharpe, F.T., Jr. 1981. Lysine in seed protein from S-aminoethyl-L-cysteine resistant anther-derived tissue cultures of rice. In Vitro 17:345-352.

Yamada, Y., Nishi, T., Yasuda, T., and Takahashi, E. 1967a. The sterile culture of rice cells, *Oryza sativa* L., and its application. In: Advances in Germfree Research and Gnotobiology (M. Miyakawa and T.D. Luckey, eds.) pp. 377-386. CRC Press, Cleveland.

Yin, K.C., Hsu, C., Chu, C.Y., Pi, F.Y., Wang, S.T., Liu, T.Y., Chu, C.C., Wang, C.C., and Sun, C.W. 1976. A study of the new cultivar of rice raised by haploid breeding method. Sci. Sinica 19:227-242.

REFERENCES

Anonymous. 1981. Tissue culture work promises faster, lower-cost breeding. IRRI Reporter (June issue).

Bajaj, Y.P.S., Saini, S.S., and Bidani, M. 1980. Production of triploid plants from the immature and mature endosperm culture of rice. Theor. Appl. Genet. 58:17-18.

Barker, R. and Herdt, R.W. 1979. Rainfed Lowland Rice as a Research Priority: An Economist's View. IRRI Research Paper Series 26. International Rice Research Institute, Los Banos, Philippines.

Bhattacharya, S. and Sen, S.K. 1980. Potentiality of leaf sheath cells for regeneration of rice (*Oryza sativa* L.) plants. Theor. Appl. Genet. 58:87-90.

Broertjes, C. and Keen, A. 1980. Adventitious shoots: Do they develop from one cell? Euphytica 29:73-87.

Carew, D.P. and Schwarting, A.E. 1958. Production of rye embryo callus. Bot. Gaz. 119:237-239.

Carter, O., Yamada, Y., and Takahashi, E. 1967. Tissue culture of oats. Nature 214:1029-1030.

Chaleff, R.S. and Carlson, P.S. 1974. In vitro selection for mutants of higher plants. In: Genetic Manipulation with Plant Material (L. Ledoux, ed.) pp. 351-363. Plenum Press, New York.

Chandler, R.F., Jr. 1979. Rice in the Tropics: A Guide to the Development of National Programs. Westview Press, Boulder.

Chu, C.C. 1978. The N_6 medium and its applications to anther culture of cereal crops. In: Proceedings of Symposium on Plant Tissue Culture (1978 May 25-30, Peking) pp. 43-50. Science Press, Peking.

Cornejo-Martin, M.J., Mingo-Castel, A.M., and Primo-Millo, E. 1979. Organ redifferentiation in rice callus: Effects of C_2H_4, CO_2 and cytokinins. Z. Pflanzenphysiol. 94:117-123.

Furuhashi, K. and Yatazawa, M. 1964. Indefinite culture of rice stem node callus. Kagaku 34:623 (in Japanese).

Hargrove, T.R. 1979. Diffusion and adoption of semidwarf rice cultivars as parents in Asian rice breeding programs. Crop Sci. 19:571-574.

______ and Cabanilla, V.L. 1979. The impact of semidwarf varieties on Asian rice-breeding programs. BioScience 29:731-735.

Hendre, R.R., Mascarenhas, A.F., Pathak, M., and Jagannathan, V. 1975. Tissue culture of maize, wheat, rice and sorghum. II. Growth and nutrition of callus cultures. Indian J. Exp. Biol. 13:108-111.

Henke, R.R., Mansar, M.A., and Constantin, M.J. 1978. Organogenesis and plantlet formation from organ- and seedling-derived calli of rice (*Oryza sativa*). Physiol. Plant. 44:11-13.

Heyser, J. and Nabors, M.W. 1982. Responses of cereal tissue cultures to salt and drought. Plant Physiol. 69(Suppl.):32.

Hu, H., Hsi, T.Y., Tseng, C.C., Ouyang, T.W., and Ching, C.K. 1978. Applications of anther culture to crop plants. In: Frontiers of Plant Tissue Culture 1978 (T.A. Thorpe, ed.) pp. 123-130. Univ. Calgary Press, Alberta.

International Atomic Energy Agency. 1971. Rice Breeding with Induced Mutations. IAEA, Vienna.

International Rice Research Institute. 1978. Annual Report for 1977. IRRI, Los Banos, Philippines.

Kawata, S. and Ishihara, A. 1968. Regeneration of rice plant, *Oryza sativa* L., in the callus derived from the seminal root. Proc. Jpn. Acad. 44:549-553.

King, P.J., Potrykus, I., and Thomas, E. 1978. In vitro genetics of cereals: Problems and perspectives. Physiol. Veg. 16:381-399.

Li, H.W., Hu, C.H., and Wu, H.P. 1968. Induced mutation breeding of rice in the Republic of China. Report of an FAO/IAEA research coordination meeting on the use of induced mutations in rice breeding (1967 June 5-9, Taipei) pp. 17-23. International Atomic Energy Agency, Taipei.

Linsmaier, E.M. and Skoog, F. 1965. Organic growth factor requirements of tobacco tissue cultures. Physiol. Plant. 18:100-127.

Mascarenhas, A.F., Pathak, M., Hendre, R.R., and Jagannathan, V. 1975a. Tissue culture of maize, wheat, rice, and sorghum. I. Initiation of viable callus and root cultures. Indian J. Exp. Biol. 13:103-107.

Mascarenhas, A.F., Pathak, M., Hendre, R.R., Ghugale, D.D., and Jagannathan, V. 1975b. Tissue culture of maize, wheat, rice and sorghum. IV. Studies of organ differentiation in tissue cultures of maize, wheat and rice. Indian J. Exp. Biol. 13:116-119.

Matsushima, S. 1980. Rice Cultivation for the Millions: Diagnosis of Rice Cultivation and Techniques of Yield Increase. Japan Science Society Press, Tokyo.

Myint, A. and deFossard, R.A. 1974. Induction of haploid callus from rice anthers and regeneration of plants. In: Haploids in Higher Plants: Advances and Potential (K.J. Kasha, ed.) p. 139. Univ. Guelph Press, Guelph, Ontario.

Nakano, H. and Maeda, E. 1979. Shoot differentiation in callus of *Oryza sativa* L. Z. Pflanzenphysiol. 93:449-458.

Niizeki, H. and Oono, K. 1968. Induction of haploid rice plants from anther culture. Proc. Jpn. Acad. 44:554-557.

Nishi, T., Yamada, Y., and Takahashi, E. 1968. Organ redifferentiation and plant restoration in rice callus. Nature 219:508-509.

______ 1973. The role of auxins in differentiation of rice tissue cultured in vitro. Bot. Mag. Tokyo 86:183-184.

Norstog, K. 1956. Growth of rye-grass endosperm in vitro. Bot. Gaz. 117:253-259.

Oono, K. 1978. Test tube breeding of rice by tissue culture. Trop. Agric. Res. Ser. 11:109-124.

Peking Institute of Botany. 1975. Isolation and culture of rice protoplasts. Somatic Hybridization Research Group and Cytobiochemistry Research Group, Academia Sinica. Sci. Sinica 18:779-789.

Rao, U.P., Kalode, M.B., Sastry, M.V.S., and Srinivasan, T.E. 1980. Breeding for gall midge resistance in rice. Indian J. Genet. Plant Breed. 40:581-584.

Ree, J.H. 1968. Rice breeding problems in Korea. Report of an FAO/IAEA research co-ordination meeting on the use of induced mutations in rice breeding (1967 June 5-9, Taipei) pp. 119-126. International Atomic Energy Agency, Taipei.

Rutger, J.N. and Brandon, D.M. 1981. California rice culture. Sci. Am. 244:42-51.

Sekiya, J., Yasuda, R., and Yamada, Y. 1977. Callus induction in tobacco, pea, rice and barley plants by auxins and their analogues. Plant Cell Physiol. 18:1155-1157.

Shepherd, J.F. 1982. The regeneration of potato plants from leaf-cell protoplasts. Sci. Am. 246:154-166.

Straus, J. and LaRue, C.D. 1954. Maize endosperm tissue grown in vitro. I. Culture requirements. Am. J. Bot. 41:687-694.

Tamaoki, T. and Ullstrup, A.J. 1958. Cultivation in vitro of excised endosperm and meristem tissues of corn. Bull. Torrey Bot. Club 85: 260-272.

Tamura, S. 1968. Shoot formation in callus originated from rice embryo. Proc. Jpn. Acad. 44:544-547.

Yamada, Y., Tanaka, K., and Takahashi, E. 1967b. Callus induction in rice, *Oryza sativa* L. Proc. Jpn. Acad. 43:156-160.

Yamasaki, Y. and Kawai, T. 1968. Artificial induction of blast-resistant mutations in rice. Report of an FAO/IAEA research co-ordinating meeting on the use of induced mutations in rice breeding (1967 June 5-9, Taipei) pp. 65-72. International Atomic Energy Agency, Taipei.

Yatazawa, M., Furuhashi, K., and Suzuki, T. 1967. Growth of callus tissue from rice roots in vitro. Plant Cell Physiol. 8:363-373.

Woo, S.C. and Hsu, H.Y. 1974. Doubled haploid rice from Indica and Japonica hybrids through anther culture. In: Haploids in Higher Plants: Advances and Potential (K.J. Kasha, ed.) p. 141. Univ. of Guelph Press, Guelph, Ontario.

Wu, L. and Li, H.W. 1971. Induction of callus tissue initiation from different somatic organs of rice plants by various concentrations of 2,4-dichlorophenoxyacetic acid. Cytologia 36:411-416.

Zhou, C. and Yang, H.Y. 1981. Induction of haploid rice plantlets by ovary culture. Plant Sci. Lett. 20:231-237.

SECTION III
Legumes

CHAPTER 7
Alfalfa

T. McCoy and *K. Walker*

HISTORY OF THE CROP AND ECONOMIC IMPORTANCE

The forage-ruminant livestock complex represents the world's major system of food production. This is because 75% of all foods of animal origin are produced from ruminant animals. Forages of all types provide about 60-80% of the consumption of feed units by ruminant animals. In other terms, forages provide about 18% of the food bulk in the human diet and 36% of the total protein. As a result forages are the largest single source of human nutritional components.

Alfalfa (*Medicago sativa* L.) is one of the world's most productive forage species. Worldwide, alfalfa is grown on about 35 million hectares (Bolton et al., 1972). Historically, alfalfa has been an important forage crop for centuries. An excellent description of alfalfa's historical development is available (Bolton et al., 1972), precluding the necessity for an in-depth description here. Alfalfa's center of origin is Iran, and the species was not introduced into North America until the eighteenth century. Most of the genetic diversity currently present in alfalfa varieties traces to nine distinct germplasm sources that were introduced into North America between 1850 and 1947 (Barnes et al., 1977).

Current production practices with alfalfa range from intensely managed, high yielding farms to minimally maintained areas in pasture and rangeland. In the United States there are approximately 12 million hectares of solid stand alfalfa with a total production of approximately 80 million tons of forage. The value of this production to the farmer is approximately $4.4 billion. The majority of alfalfa grown for forage is grown in the upper midwest (Minnesota and Wisconsin) and the Northeast. An interesting problem with alfalfa is the need to produce seed outside of the area where alfalfa is grown for forage. The majority of alfalfa seed is grown in the western United States in areas of dry weather during the summer months. The U.S. production of seed approximates 102 million pounds which in 1981 had a total wholesale value of $115 million. About 40% of the alfalfa seed is produced in California, making the state the largest production area.

Since the major geographical limitations to alfalfa production are a result of climatic factors the majority of alfalfa is grown in the temperate areas of the world. In the Northern Hemisphere, the northern limitation of production is determined by the severity of winter. Winter-hardy varieties are essential in northerly climates. The fall dormancy of a cultivar is a valuable indicator of winter-hardiness (Barnes et al., 1978). The more dormant a variety is in the fall the less likely it is to suffer winter injury. Rainfall is another major climatic factor determining the success of alfalfa. In high rainfall areas such as some areas of the southeastern United States a plethora of foliar diseases makes economical production of alfalfa difficult.

The majority of alfalfa grown worldwide is used to feed livestock, and is either grazed or harvested and preserved as hay or silage. Alternatively, alfalfa may be dehydrated to produce a protein supplement to be fed to animals. About 1.7 million tons of dehydrated alfalfa are produced annually (Kohler et al., 1972). More recently, alfalfa sprouts have been utilized for human consumption. In California alone, approximately 1.5 million pounds of alfalfa seed are used to produce alfalfa sprouts annually, with an estimated market value of $8.5 million (Hesterman and Teuber, 1981). In the United States it is likely that greater than 6 million pounds of alfalfa seed are used each year to produce alfalfa sprouts.

Major quality considerations of alfalfa include quality and quantity of protein and the need to reduce the components that cause bloat. Overall, alfalfa protein is of fairly high quality although noticeably deficient in methionine. Cell culture technology has recently been utilized in an attempt to isolate amino acid overproducers (Maliga, 1980). In alfalfa, selection for resistance to ethionine (an analog of methionine) has resulted in a cell line with a tenfold increase in soluble methionine and a 40% increase in free methionine and protein amino acids (Reisch et al., 1981). Ethionine resistance was maintained in some plants following plant regeneration, but the inheritance of resistance has not been determined (Reisch, 1980).

Important events in the induction of pasture bloat are plant cell rupture by chewing and the initial rate of microbial digestion. Howarth et al. (1981) claim that selection for slower rates of digestion and/or greater resistance to mechanical rupture are good approaches to the development of a bloat-safe alfalfa cultivar. Results of their research indicate that saponins, once considered a major factor in the incidence of bloat, are not likely to be involved as causal agents.

Until the late 1960s the major efforts in alfalfa improvement were by public (University and USDA) alfalfa breeders. Recently industry has placed a greater emphasis on alfalfa breeding as indicated by the great increase in privately developed alfalfa varieties during the 1970s (Barnes et al., 1977). A wide diversity of alfalfa varieties are grown nationally and internationally. Exact acreages of specific varieties are difficult to obtain because of the nature of alfalfa production.

KEY BREEDING AND CROP IMPROVEMENT PROBLEMS

Alfalfa is a very complex species from a crop improvement perspective. Complexities of alfalfa breeding include autotetraploid genetics, a myriad of insect and disease problems, perennial growth habit, the necessity of insect pollination, seed production outside the area of forage production and the complexities of nitrogen fixation involving the symbiotic relationship with *Rhizobium*. We will first discuss genetic concepts of autotetraploidy. Second, some of the key insect and disease problems will be discussed including strategies for their possible solution. Finally, the potential for improved nitrogen fixation will be evaluated. Alternative breeding methods will be discussed in the protocols section.

Genetics and Breeding of Alfalfa

The breeding of alfalfa has been very successful, given the complexities of autotetraploid genetics. Conventional plant breeding programs have been responsible for developing a number of high-yielding, pest-resistant cultivars. An excellent review of classical alfalfa breeding concepts is available (Busbice et al., 1972) and no attempt will be made to discuss details of alfalfa breeding here. The objective of this section is to concentrate on some of the genetic concepts that must be considered in alfalfa improvement due to its autotetraploid genetic structure.

When compared to a diploid organism that can have only 3 allelic states—homozygous recessive (aa), heterozygous (Aa), or homozygous dominant (AA)—autotetraploid alfalfa can have up to five allelic states: nulliplex (aaaa), simplex (Aaaa), duplex (AAaa), triplex (AAAa), or quadriplex (AAAA). In addition to the different genotypic structures, when considering two alleles, an additional confounding factor is the possible presence of multiple alleles, with the potential of up to four different alleles in a tetraploid for a given locus. The terminology for describing multiple allelic states is as follows: mono-allelic ($a_i a_i a_i a_i$), di-allelic simplex or triplex ($a_i a_i a_i a_j$), di-allelic duplex ($a_i a_i a_j a_j$), tri-allelic ($a_i a_i a_j a_k$) and tetra-allelic ($a_i a_j a_k a_l$).

The importance of multiple allelic states in alfalfa has only recently (since the 1960s) been realized. Demarly (1963) produced theoretical models and some practical results indicating an association between tetra-allelic states and higher yields. Following this report the hypothesis was developed that higher yields could be realized in plants where all polymorphic loci are tetra-allelic.

There are various lines of evidence indicating the importance of maximum heterozygosity in alfalfa. First, inbreeding depression is much greater than expected in alfalfa based on the coefficient of inbreeding for a two-allele model (Busbice and Wilsie, 1966). The rapid inbreeding depression observed by Busbice and Wilsie (1966) was attributed to the rapid rate at which first order interactions are lost from tri- and tetra-allelic loci. More direct evidence for the importance of maximizing heterozygosity to maximize yield was given by Dunbier and Bingham (1975). Utilizing haploid derived parental material Dunbier and Bingham (1975) demonstrated that populations with a greater frequency of tetra-allelic loci outyielded populations where the frequency of tetra-allelic loci was less. The differences were greatest when comparing seed yield differences, but forage yields were also greater (although not significant) with a higher frequency of tetra-allelic loci. Double cross populations in an autotetraploid are thus considered to outyield single cross populations due to the greater frequency of tetra-allelic loci (Dunbier and Bingham, 1975).

Bingham (1980) summarized other experiments which consistently indicated that double cross populations outyield single cross populations. This fact provides strong evidence for the hypothesis of maximum heterozygosity, because under a two allele model the single cross populations should be superior. Although the maximum heterozygosity hypothesis is dependent upon the existence of multiple alleles, the definitive proof of the existence of multiple alleles has only recently been documented. Quiros and Morgan (1981) demonstrated the presence of multiple alleles in determining isozymes in *Medicago*. Establishment of biochemical markers at various chromosomal sites in the alfalfa genome may allow a more definitive test of the relationship between yield and maximum heterozygosity.

Given the hypothetical importance of maximum heterozygosity a breeding methodology problem becomes how to best maximize heterotic combinations. Within a conventional breeding scheme one methodological shift could be to select for pest resistance and agronomic traits in at least four different populations instead of one (Dunbier and Bingham, 1975). The four populations are then synthesized by first producing two separate single cross populations and then allowing the two single crosses to intermate to produce a double cross population. Although some sibbing would occur in the synthesis of the double cross it is hoped that these populations would have a greater frequency of tetra-allelic loci than would equilibrium populations produced by the standard development of synthetic varieties. Bingham (1980) claimed that as much as a 15% gain in yield has been realized by the double cross modification. Alternative methods of maximizing heterozygosity in alfalfa populations will be discussed in the protocol section.

There are several other complex aspects of autotetraploid genetics including: (a) a requirement for large populations to detect segregates in autotetraploids versus diploids based on a two allele model (Busbice et al., 1972), (b) the requirement for double backcrosses in linkage studies to identify gametic genotypes (Bingham et al., 1968), (c) the possiblity of double reduction, and (d) the possiblity of numerical nondisjunction.

Breeding for Disease Resistance

A number of plant diseases and insects are responsible for extensive economic losses in alfalfa production. Barnes et al. (1977) indicate that 21 diseases, 10 insects, and 3 nematodes are major problems on alfalfa. The diversity of insect and disease problems when coupled with the perennial growing habit has resulted in alfalfa breeders concentrating on these problems. On a worldwide basis, the most serious alfalfa diseases and their causal agents are: bacterial wilt (*Corynebacterium insidiosum*), Phytophthora root rot (*Phytophthora megasperma*), Fusarium wilt (*Fusarium oxysporum*), Verticillium wilt (*Verticillium albo-atrum*), anthracnose (*Colletotrichum trifolii*), and common leafspot (*Pseudopeziza medicaginis*). A recent publication describes the symptoms, causal organism, disease cycle, and control methods of these diseases and other alfalfa diseases (Graham et al., 1979).

BACTERIAL WILT. Lack of bacterial wilt resistance was a major obstacle to alfalfa expansion in North America. Once identified in 1925, the widespread occurrence of the disease was realized and attempts to incorporate bacterial wilt resistance were initiated. Historically, breeding for bacterial wilt resistance represents the beginning of breeding for alfalfa disease resistance (Kehr et al., 1972). Currently almost all newly released alfalfa cultivars have some level of bacterial wilt resistance.

Inheritance of bacterial wilt resistance in two alfalfa gene pools has been analyzed in a quantitative and qualitative manner (Viands et al., 1979; Viands and Barnes, 1980). The quantitative analysis suggested different inheritance modes in the two gene pools. The different inheritance modes are likely a function of different resistance mechanisms in the two gene pools (Viands et al., 1979). In the qualitative analysis Viands and Barnes (1980) identified three genes conditioning resistance to bacterial wilt. One gene was dominant and in the absence of the other genes resistance was expressed in the simplex condition. The two other genes were additive and of smaller gene effects than the dominant gene (Viands and Barnes, 1980).

The most effective control measure for bacterial wilt as well as for other diseases is the use of resistant varieties. Several breeding methods including modified backcross, recombining selected clones, and recurrent phenotypic selection have been used to develop bacterial wilt resistant cultivars (Kehr et al., 1972).

PHYTOPHTHORA ROOT ROT. Phytophthora root rot caused by *P. megasperma* f. sp. *medicaginis* is a serious alfalfa disease worldwide. It occurs in almost any area where soil water is excessive. The first alfalfa germplasm with high levels of resistance to *Phytophthora* was developed in the early 1970s by recurrent selection within existing cultivars (Frosheiser and Barnes, 1973). Lu et al. (1973) concluded that susceptibility was conditioned by a single tetrasomic gene (Pm) with incomplete dominance such that nulliplex (pm pm pm pm) plants were highly resistant, simplex (Pm pm pm pm) plants were moderately resistant and duplex (Pm Pm pm pm), triplex (Pm Pm Pm pm), and quadruplex (Pm Pm Pm Pm) plants were increasingly more susceptible.

Irwin et al. (1981a,b) conducted an inheritance study of *Phytophthora* resistance in diploid and in tetraploid alfalfa. The results of the study at the tetraploid level indicated resistance was conditioned by two incompletely dominant complementary genes, such that one locus must be at least duplex and one locus must be simplex to have a resistant phenotype (Irwin et al., 1981b). At the diploid level resistance was conditioned by two independent complementary dominant genes in a diploid population containing cultivated germplasm (Irwin et al., 1981a). However, in a wild diploid relative of alfalfa, *M. coerulea*, inheritance of resistance was more complex and epistatic relationships were important.

FUSARIUM WILT. Fusarium wilt caused by *F. oxysporum* Schlecht f. sp. *medicaginis* (Weimer) Snyd. & Hans. occurs worldwide. It is a warm climate disease that is most severe in the south and southwestern United States. As with most alfalfa diseases the only practical control method is to use resistant varieties. Most varieties grown in the southwestern United States have a high level of Fusarium resistance. Hijano (1981) concluded that resistance to Fusarium wilt was controlled by at least two genes. One is a completely dominant gene and the other has an additive type of gene action. Once again the presence of more than one gene involved in resistance is indicative of more than one mechanism for resistance.

Because *Fusarium* produces toxic components that may be involved in disease, it may be possible to select resistant plants in vitro. Preliminary results indicate that alfalfa cell lines resistant to a toxic component(s) of culture filtrates of *Fusarium* can be selected in callus culture (McCoy, unpublished results). The disease response at the whole plant level remains to be tested. However, based on successful selection of resistant plants by selection in vitro for cell lines resistant to culture filtrates or toxins produced by a pathogen (Gengenbach et al., 1977, 1981; Behnke, 1979, 1980; Sacristan, 1982) this may be a useful approach in selection of *Fusarium*-resistant plants.

VERTICILLIUM WILT. Verticillium wilt caused by *Verticillium albo-atrum* Reinke & Berthe. is a serious disease problem in northern Europe. Currently Verticillium wilt is a major concern in the United States since its discovery in the northwest United States and its more recent discovery in major production areas in the North Central states. Resistant cultivars have been developed in Europe, and development of Verticillium resistant varieties has become a major effort of alfalfa breeders. Although the inheritance of resistance has not yet been determined, resistance to this disease has high heritability and phenotypic recurrent selection seems to be an effective means of developing resistant germplasm (Kehr et al., 1972).

ANTHRACNOSE. Anthracnose caused by *Colletotrichum trifolii* Bain is a highly destructive disease of alfalfa particularly in humid high rainfall areas of the world. As with previously mentioned diseases the best control method is development of resistant varieties. Devine et al. (1971) indicated the mode of inheritance of anthracnose resistance differed among alfalfa populations. They found that high levels of resistance can be developed in most populations by recurrent phenotypic selection (Devine et al., 1971). A later study of inheritance of resistance to anthracnose indicated resistance was conditioned by a single completely dominant gene (Campbell et al., 1974).

COMMON LEAFSPOT. Common leafspot caused by *Pseudopeziza medicaginis* (Lib.) Sacc. is one of the most devastating foliar diseases of alfalfa. Precise inheritance data on resistance is not available, although recurrent selection is an effective means of incorporating resistance to this disease (Kehr et al., 1972). Resistance to foliar pathogens is essential not only because of yield reduction due to defoliation, but also because of the reduction in nutritional quality (Willis et al., 1969).

It is obvious from the previous reviews that the most common procedure for developing resistance to alfalfa diseases is phenotypic recurrent selection. Recurrent selection has also been effective in developing insect resistance to such pests as the spotted alfalfa aphid, potato leafhopper and the pea aphid. However, the most challenging problem facing alfalfa breeders is the development of high-yielding, multiple pest resistant varieties. Hanson et al. (1972) were successful in developing populations of alfalfa resistant to rust, common leafspot, bacterial wilt, and anthracnose, and the insect pests; spotted alfalfa aphid and potato leaf hopper by recurrent phenotypic selection.

The concern with regard to breeding for pest resistance in alfalfa has prevented breeders from concentrating on genes involved in increasing yield per se (Barnes et al., 1977). The incorporation of resistance to a single disease is a fairly straightforward protocol, but the concurrent development of resistance to several key diseases and insects and development of a higher-yielding capacity is a major challenge. Utilization of emerging technology (discussed later) may enable breeders to maintain a high level of pest resistance, but yet develop methodology for capitalizing on heterotic combinations responsible for greater yielding capacity.

Breeding for Increased Nitrogen Fixation

Another major interest in alfalfa improvement is nitrogen (N_2) fixation. With the high cost of energy and subsequent high cost of nitrogen fertilizer, interest in alfalfa as a rotation crop to supply nitrogen for subsequent crops has been revitalized. Research aimed at improving N_2 fixation can be divided into two major areas: increased nitrogen fixation via selecting improved strains of *Rhizobium* species, or via improving the plants capacity to fix nitrogen.

Before 1976 almost all research was aimed at developing improved strains of *Rhizobium*. However, since the mid 1970s research conducted by USDA researchers at the University of Minnesota has indicated various methods for increasing N_2 fixation by selecting plants with specific morphological traits. Viands et al. (1981) concluded that first selecting large plants with a large nodule mass, and then selecting among these plants for nitrogenase activity is an effective method for breeding plants with increased N_2 fixation.

Future research aimed at discovering the basic mechanisms involved in N_2 fixation should further improve the ability to develop more efficient N_2 fixation in alfalfa plants. Manipulation and selection of bacterial strains has resulted in mutant strains with an increased capacity for N_2 fixation. However, techniques have to be developed that will enable new improved bacterial strains to outcompete indigenous strains already present in the soil. Isolating plants that can only form symbiotic relationship with the improved strains and not with native strains may be a successful approach (Viands et al., 1979).

REVIEW OF THE LITERATURE

Nature of the Regeneration System

Since the first report by Saunders and Bingham (1972) the regeneration of alfalfa from in vitro cultured cells and tissues has advanced rapidly. Reports include regeneration from long-term callus culture (Stavarek et al., 1980), suspension culture (McCoy and Bingham, 1977), and protoplast-derived cell aggregates (Johnson et al., 1981; Kao and Michayluk, 1980; dos Santos et al., 1980).

Present technology is not adequate for the regeneration of the full range of alfalfa germplasm. However, selection for regeneration potential has been pursued successfully (Bingham et al., 1975) and a number of strains with improved regeneration capacity have been released. Two of these strains, Regen S and Regen Y were developed by E.T. Bingham at the University of Wisconsin. These strains are tetraploid with purple and yellow flowers, respectively. A diploid clonal strain, HG-2, was also developed in Bingham's laboratory (McCoy and Bingham, 1977). Mitten et al. (1981) have recently reported on the potential for in vitro regeneration of 35 land race cultivars and early introductions. These 35 cultivars represent the range of germplasm sources available to the breeder. A high proportion of regenerator genotypes were found among the cultivars Lakak, Norseman, Turkestan, and Nomad. These cultivars originated in the region bounded by the Turko-Iranian border to the west and Kashmir on the east. It is of interest to note that this geographic region is considered to be the site of origin of the species. Most of the remaining cultivars possessed regeneration capacity at a low genotypic frequency (1-10%) and at a low quantitative level.

The lack of regeneration potential among target germplasm sources may be an early limitation to exploiting somatic cell genetic improvements in alfalfa. However, Reisch and Bingham (1980) have reported that regeneration potential is adequately heritable and may be improved in a population by recurrent selection. These subjects will be more fully discussed in a subsequent section.

Factors Controlling Alfalfa Regeneration

Three factors appear to control alfalfa regeneration (via somatic embryogenesis) in vitro. They are the structural quality and quantity of the exogenous growth regulator used as auxin source, the concentration of exogenous cytokinin and the level of reduced nitrogen in the regeneration medium (Walker et al., 1978, 1979; Walker and Sato, 1981).

Briefly, a two-step regeneration procedure has been developed using a population of cells grown on Schenk and Hildebrandt salts containing 25 µM NAA and 10 µM KIN. When grown in this fashion, the tissue has a high embryo-regeneration capacity after a short exposure (3-4 days) to Schenk and Hildebrandt salts containing 50 µM 2,4-D and 5 µM KIN. Optimal regeneration occurs on a medium without any growth regulators and a level of NH_4^+ between 25 mM and 100 mM. When the induction medium contains 50 µM KIN and 5 µM 2,4-D, roots will form when the regeneration medium ammonium concentration is less than 12.5 mM. Somatic embryo formation occurs when the NH_4^+ concentration is above 25 mM. Optimal conditions for regeneration with respect to these factors varies from genotype to genotype. Qualitative differences in response among different genotypes will also remain even when comparing respective optima.

Regeneration from Protoplasts

Three reports of regeneration of somatic embryos from cell cultures derived from protoplasts have recently appeared (Johnson et al., 1981; Kao and Michayluk, 1980; dos Santos et al., 1980). In each of these cases, young leaf material was used as the protoplast source. All three reports utilized comparable conditions for protoplast preparation. Efficiencies based on starting cell number were not determined. Culture techniques were also similar among the reports. The growth regulator conditions of the various protoplast culture media were similar as all contained 2,4-D, NAA, and KIN (Table 1).

When taken as a whole, these reports provide an adequate basis for protoplast manipulation in alfalfa. The conditions and requirements for protoplast isolation and culture are summarized in Table 2. It is clear from the published reports that a relatively wide range of isolation and culture conditions will support protoplast survival, growth, and subsequent plant regeneration. The lack of quantitative data on callus and plant regeneration frequencies for the base genotypes used in the protoplast studies has made it impossible to interpolate an optimal culture and regeneration sequence. Johnson et al. (1981) selected optimal regeneration genotypes from Regen S for their work. Kao and Michay-

Table 1. A Medium for Culturing Protoplasts[a]

Mineral Salts (mM)			
NH_4NO_3	7.5	KI	4.5 μM
KNO_3	18.8	H_3BO_3	.05
$CaCl_2 \cdot 2H_2O$	4.1	$MnSO_4 \cdot H_2O$	.06
$MgSO_4 \cdot 7H_2O$	1.2	$ZnSO_4 \cdot 7H_2O$	7.0 μM
KH_2PO_4	1.25	$Na_2MoO_4 \cdot 2H_2O$	1.0 μM
KCl	4.0	$CuSO_4 \cdot 5H_2O$	0.1 μM
Sequestrene* 330 Fe	28 mg/l	$CoCl_2 \cdot 6H_2O$	0.1 μM
Sugars (mM)			
Glucose	0.38	Mannose	0.7
Sucrose	0.37	Rhamnose	0.76
Fructose	0.7	Cellobiose	0.37
Ribose	0.83	Sorbitol	0.69
Xylose	0.83	Mannitol	0.69
Organic Acids[b] (mM)			
Sodium pyruvate	0.045	Malic acid	0.075
Citric acid	0.05	Fumaric acid	0.086
Vitamins (μM)			
Inositol	0.55 mM	Biotin	0.02
Nicotinamide	8.1	Choline chloride	3.6
Pyridoxine·HCl	4.9	Riboflavin	0.26
Thiamine·HCl	29.6	Ascorbic	5.7
D-Calcium pantothenate	1.0	Vitamin A	0.017
		Vitamin D_3	0.013
Folic acid	0.45	Vitmin B_{12}	0.007
p-Aminobenzoic acid	0.07		

[a]After Kao and Wetter, 1977.

[b]Adjusted to pH 5.5 with NH_4OH.

luk (1980) and dos Santos et al. (1980) acknowledged the effect of genotype but did not provide any insight into the effect of genotype on protoplast preparation, plating efficiency, or regeneration frequency. It may be concluded that optimal conditions for regeneration of plants from protoplasts of diverse genotypes have not, as yet, been identified.

Regeneration of Long-Term Cultures

Despite the development of optimal sequences by Walker et al. (1979) it must be stated that a wide range of culture conditions have been identified which will enable the successful regeneration of alfalfa in vitro. Stavarek et al. (1980) have reported an alternative treatment

Table 2. Enzyme Solution for Isolation of Alfalfa Protoplasts[a]

COMPONENT	FINAL CONCENTRATION[b]
ENZYMES	
Cellulases (Onozuka R10, Rhozyme)	0.5–5%
Pectinase (Macerozyme)	0.25–1.2%
OSMOTICUM	
Sorbitol or Mannitol	0.6 M
BUFFER	
NaH_2PO_4	0.7 mM
MES	3.0 mM

[a]After Kao and Michayluk, 1980; dos Santos et al., 1980; Johnson et al., 1981.

[b]Two-fold concentrated stocks are usually made.

sequence used to obtain plants from long-term cultures of alfalfa cells. The cell line used as the object of these investigations had been in continuous culture for 32 months. In addition, the cells had been exposed to 1% NaCl for a program of resistance selection. Stavarek et al. (1980) discovered that to obtain shoots from this material, a medium with high cytokinin (KIN at 27.9–100 µM) and low auxin was required for the initial stages of regeneration. Further stages of leaf and root formation necessitated the transfer of the tissues to growth regulator-free media. The significance of these observations to general problems of regeneration from long-term culture is not clear. The authors do not report when, during the 32-month culture cycle, high regeneration potential was lost. It is not possible to determine whether or not the long exposure to NaCl containing medium had an adverse effect on subsequent regenerability.

Walker has observed the loss or significant reduction in regenerative potential among some long-term cell lines of the genotype RA3 (unpublished observations). Normally, these lines do retain a residual ability to regenerate plants and when cell cultures are reestablished from these plants high frequency regeneration is often, but not always, restored. This procedure could be particularly useful for germplasm targeted for additional in vitro research. The ability to restore the capacity to regenerate by reestablishing callus from a regenerated plantlet may suggest that lost regenerative potential is a result of non-genic shifts in the competence of the cell population. Alternatively, the long-term cultured line of cells may be composed of different types of cells with inherently different capacities. These possibilities must be given due consideration when the problem of lost regeneration potential is encountered.

Regeneration Breeding and Genetics

Bingham at Wisconsin not only pioneered regeneration work with alfalfa, but has contributed the only detailed report on regeneration breeding for this species (Bingham et al., 1975). Only a small number of genotypes capable of regeneration were found among the cultivars screened for that property in vitro. It has been demonstrated since, that the bulk of alfalfa germplasm regenerates poorly. The three exceptional cultivars, which contain from 50 to 80% regenerator genotypes are Ladak, Turkestan, and Norseman (Mitten et al., 1981). In improving regeneration by breeding, Bingham and his colleagues identified five fertile regenerable clones (genotypes), four from cv. Saranac and one from cv. DuPuits. These were randomly intercrossed. The progeny were rescreened for regeneration. Two cycles of recurrent selection increased the frequency of regenerator genotypes from 12 to 67%. Subsequently, Walker et al. (1978) found that with modified regeneration techniques the percentage of regenerator genotypes in Bingham's material exceeded 67%, and may reach as high as 100%. This work by Bingham et al. (1975) is particularly important. First, it demonstrated the high heritability of alfalfa regeneration in vitro. Second, it established that the bulk of alfalfa germplasm is poorly regenerable in vitro. Finally, a yellow flowered regenerator stock produced from the Canadian cultivars Rambler, Roamer, and Drylander, when crossed to the Flemish derivative (Regen S) indicated that the genetic control of regeneration was similar in both.

In a subsequent paper, Reisch and Bingham (1980) examined the qualitative inheritance of in vitro regeneration in a diploid line of alfalfa. The data generally supported a two gene model for qualitative control of regeneration. However, the analysis is founded on the assumption that the progeny genotypes may be grouped on the percentage basis of responding replicate tests performed. This manner of grouping results does not consider that the quantitative frequency of embryos formed/plate may be quite low or quite high. A more acceptable presentation of these data might have been on a quantitative basis of the absolute number of embryos produced per gram fresh weight of cells examined.

Genetic Selection In Vitro

Very little information on the genetic improvement of alfalfa using tissue culture can be found in the published literature. Reisch et al. (1981) selected alfalfa cell lines that were resistant to ethionine, an analog of methionine. One cell line had a tenfold increase in soluble methionine, presumably due to altered feedback control of an enzyme involved in methionine production. This cell line also had a 43% increase in total free amino acids and a 40% increase in protein amino acids. Reisch (1980) tested 25 cell lines from 25 plants regenerated from ethionine resistant cultures for their response on ethionine containing medium. Only seven of these plants produced cell lines that were still resistant to ethionine. The lack of resistance could be due to the original selected cell lines being a mixture of resistant and susceptible cells. Alternatively, resistance in vitro could be due to epigenetic changes in gene expression which were lost upon plant regeneration. No inheritance studies of the ethionine resistance response have been reported. An interesting sidelight of the ethionine selection work was the regeneration of a wide array of variant morphological types from some of these cultures (Reisch and Bingham, 1981).

Croughan et al. (1978) were successful in selecting a salt tolerant cell line of alfalfa. This cell line was resistant to 1% NaCl and in addition had optimal growth at 0.5% NaCl. The selected cell line grew poorly in the absence of salt. A recent report (Croughan et al., 1981) indicated that plants have now been regenerated from salt tolerant cell lines using the procedure of Staverek et al. (1980) for regenerating plants from long-term cultures. However, the response of regenerated plants to salt has not been tested yet, nor have any genetic studies been conducted.

PROTOCOLS FOR ALFALFA BREEDING AND TISSUE CULTURE

General Comments on the Utilization of Diploids in Breeding Highly Heterozygous Tetraploids

The concept of breeding a polyploid crop at the diploid level as a means of ultimately producing an improved polyploid product is not new. The methodology was given the name "analytic breeding" by Chase in 1963. Analytic breeding is a method of plant improvement proposed for polyploid species (Chase, 1963) in general, and for potatoes in particular. Alfalfa, a natural autotetraploid like potato, is adaptable to this breeding system. The analytic breeding method involves three major steps: (a) transferring germplasm from the tetraploid to the diploid level, (b) breeding and selection are then conducted at the diploid level, and (c) the improved diploid germplasm is transferred back to the tetraploid level. The major advantage of an analytic breeding approach is that the method circumvents complexities of tetrasomic inheritance such as the requirement for large populations to detect rare segregates, the need for double backcrosses to identify heterozygotes, the possibility of preferential pairing or double reduction, and the possibility of numerical nondisjunction.

The two major technological necessities for utilization of an analytic breeding scheme are a method of efficiently transferring germplasm to the diploid level, and then a method of transferring the improved diploid germplasm relatively intact to the tetraploid level. First, we'll consider two methods currently available for transferring tetraploid germplasm to the diploid level. One method is to extract maternal haploids (2n = 2x = 16) following 4x-2x crosses (Bingham, 1971); the other method is to utilize a ploidy level bridge cross by first producing a triploid (2n = 3x = 24) from a 4x-2x or 2x-4x cross and then utilize 3x-2x crosses to produce diploids (Bingham and Saunders, 1974). A complication of the latter approach is that aneuploids (generally trisomics) as well as diploids are frequently produced as a result of 3x-2x crosses (Binek and Bingham, 1970). The bridge cross method is also undesirable because the diploid produced by one cycle of 4x-2x followed by 3x-2x crosses contains only about one third of the tetraploid germplasm, thus requiring several bridge crosses to effectively incorporate the tetraploid germplasm.

The 4x-2x cross method of producing haploids is a desirable method of transferring genes to the diploid level, because only genes from the tetraploid parent are present in the raw haploid, eliminating the need for multiple crosses. The raw haploids are generally male-sterile, and therefore haploid x haploid matings are impossible. However, 2x haploids can be crossed with 2x *M. falcata* to produce a population of 50% cultivated genes from cultivated tetraploid germplasm and 50% "wild" genes from *M. falcata*. Subsequent backcrossing of the hybrids to additional raw haploids was successful in establishing diploid populations with levels of cultivated germplasm ranging from 50 to 98.5% (Bingham and McCoy, 1979). These populations of cultivated alfalfa at the diploid level (CADL) can be crossed with additional raw haploids of other tetraploid populations to incorporate desirable germplasm from the tetraploid level.

A major disadvantage of the 4x-2x cross method of haploid production is the low efficiency of haploid production. On the average only one haploid is produced per 1000 pollinations in 4x-2x crosses (Bingham, 1971). In terms of time and effort this requires approximately 1.5 hr of pollination when the 4x parent is male-sterile. However, recent attempts to improve the efficiency of the system have shown that over 30 haploids per 1000 pollinations can be obtained when the best diploid pollinator (one genotype of *M. glomerata*) is used with a tetraploid female homozygous for a gene (*jp*) that results in no post meiotic cytokinesis (McCoy and Smith, 1983). Future technological advances such as the ability to produce haploids of any tetraploid genotype via anther or microspore culture would greatly increase the efficiency of germplasm transfer across ploidy levels.

Given the ability to transfer germplasm from the tetraploid to the diploid level, breeding and selection can then be carried out at the diploid level. Once desirable improved diploid germplasm is obtained, several methods currently exist for transferring this germplasm to the tetraploid level. One method is to double the chromosome number in somatic cells by colchicine treatment. Another approach is to utilize 2n (unreduced) gametes from the diploid parent. Interploidy bridge crosses whereby the triploid from the 4x-2x cross is crossed with the 4x parent could also be used. Finally, somatic hybrids produced via protoplast fusion should be feasible given the potential to regenerate alfalfa plants from protoplasts (Kao and Michayluk, 1980; dos Santos et al., 1980; Johnson et al., 1981). The methods of transfer vary considerably in regard to the degree of heterozygosity retained. Colchicine doubling is the least favorable method of transfer because this procedure is an inbreeding process whereby the tetraploid product can only be diallelic at any locus. The bridge cross method is also undesirable in terms of transferring the diploid genotype relatively intact.

The most ideal method of germplasm transfer would be the somatic fusion of two unrelated highly heterozygous diploids. Although regeneration of plants from protoplasts is possible, somatic fusion in *Medicago* has not been demonstrated. Regeneration of plants from protoplasts is also likely to be genotype dependent similar to plant regeneration from callus (Reisch and Bingham, 1980); therefore, further technological advances are required before the ability to fuse and regenerate protoplasts of any diploid is feasible.

The most desirable technique currently available for germplasm transfer across ploidy levels is 2n gamete formation. 2n gametes are gametes with the sporophytic chromosome number. A low frequency of 2n gametes occurs in many plant species, and 2n gametes are recognized as the principal polyploidizing mechanism in plant evolution (Harlan and deWet, 1975). Single gene control of 2n gamete formation has been described in *Datura* (Satina and Blakeslee, 1935), *Zea mays* (Rhoades and Dempsey, 1966), *Solanum* (Mok and Peloquin, 1975a), and *Medicago* (McCoy, 1982).

A mechanism that results in 2n gamete formation by restitution of the first meiotic division is the most desirable for transferring the maximal amount of heterozygosity from the selected diploid to the tetraploid. The amount of heterozygosity that is transmitted via first division restitution (FDR) 2n gametes is 100% from the centromere to the first crossover, 50% between the first and second crossovers, and 75% beyond the second crossover. In alfalfa where there is generally one chiasma per bivalent more than 80% of the heterozygosity in the diploid is likely transmitted by FDR 2n gametes (Bingham, 1980). The significant aspect of this mechanism of germplasm transfer across ploidy levels is that not only are intralocus heterotic combinations retained but interlocus heterotic combinations are retained as well.

A recent study of 2n pollen formation in diploids of alfalfa has shown 2n pollen formation to be controlled by a single recessive gene, designated rp (McCoy, 1982). This gene conditions 2n pollen formation by a first division restitution mechanism (Vorza and Bingham, 1979). A preliminary yield study was conducted comparing families from 4x x 2x crosses where the 2x parent formed FDR 2n pollen, versus families from 4x x 4x crosses where the 4x parent was the colchicine doubled counterpart of the 2n pollen forming diploid. In all cases the families produced via 2n pollen outyielded the families from the colchicine doubled counterpart (unpublished data), indicating the superiority of FDR 2n pollen for transferring heterotic combinations in the diploid to the tetraploid progeny.

At the current technological level 2n gametes can be utilized to transfer germplasm from the diploid to the tetraploid level relatively intact. However, an ultimate breeding scheme where 2n gametes could be used to synthesize a maximally heterozygous tetraploid variety may be feasible at a future date. This breeding scheme (Fig. 1) would utilize 2n eggs and 2n pollen to form the tetraploid hybrid. Interestingly, plants that form 2n gametes also form n gametes. Diploid plants homozygous for rp produce 2n pollen at a frequency of 4-56% depending on the genotype and the environment (McCoy, 1982). Therefore 2n pollen producing plants can be maintained at the diploid level and breeding and selection can be conducted in populations containing the rp gene. The breeding scheme would involve crossing two selected and unrelated

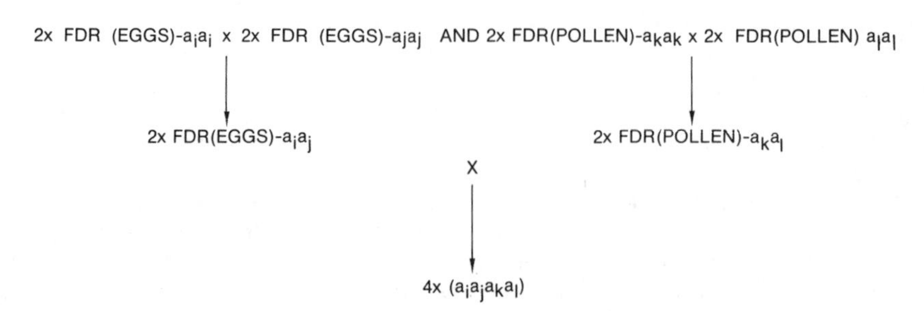

Figure 1. An ultimate breeding scheme utilizing 2n gametes.

2n pollen producers together and two selected and unrelated 2n egg producing diploids. The resultant 2n pollen producing progeny would be crossed with progeny of the 2n egg producers to form a highly heterozygous tetraploid population.

PROTOCOLS

In this section we shall outline a number of protocols useful for the manipulation of alfalfa cells in culture. The protocols should not be considered the only, and in some cases, even the best means to manipulate alfalfa cells in vitro. These reflect the collective biases and experience of the authors.

Establishing a Callus In Vitro

1. Basic nutritional mixture—Schenk and Hildebrandt (1972) supplemented with 25 µM NAA and 10 µM KIN (SH-NAA/KIN medium). This medium has been found to be superior to B5 or Blaydes medium for callus growth. The growth regulator selected for use will not provide the most rapid growth. However, regeneration of cells on this medium has been found to remain at a high level for more than 2 years. More rapid rates of growth may be obtained with 2,4-D at approximately 10 µM.
2. Culture conditions—temperatures between 25 and 30 C are satisfactory. Light is not required.
3. Explant source—Virtually any portion of the plant will form callus on SH NAA/KIN medium. The authors have tested root, stem, leaf lamina, petiole, cotyledon, ovary, and anther tissue. Callus from any explant source will regenerate in vitro.
4. Obtain explants. Young tissue in a vigorous phase of growth is superior.
5. Sterilize for 10 min in commercial bleach diluted 50–70% with water.
6. Rinse explant tissues thoroughly with sterile water.
7. Place on agar solidified (0.8%) SH-NAA/KIN medium in either petri dishes or culture tubes.

8. Place cultures in light or dark.
9. Subculture every 14-21 days after initiation.

Plant Regeneration

Regardless of the growth conditions used for cells, a short exposure to high concentrations of 2,4-D followed by its removal has considerably enhanced the quantitative frequency of regeneration.

1. Culture media—SH supplemented with 25-50 μM 2,4-D and 1-10 μM KIN for embryo induction, followed by SH supplemented with 25 mM NH_4^+ is desirable. Other sources of reduced N have been used successfully, e.g., yeast extract and N-Z amine ^{R}(Hamkalo-Scheffield).
2. Obtain cells from appropriate source. Transfer to solid or liquid induction medium as desired at 75-300 mg FW/plate or 30 mg FW/ml, respectively.
3. After 3 to 4 days, collect cells. (Whether on solid or in liquid medium, cells may be dispersed in a liquid medium for convenient inoculation of culture plates.)
4. Inoculate regeneration plates at a density of 75 mg FW/10 ml of medium.
5. Place in the light. Regeneration will be optimal between 18 and 24 days post-transfer.
6. To obtain plantlets, separate and transfer embryos after 3 weeks of regeneration to fresh regeneration medium.
7. As embryos precociously germinate in vitro, transfer to a half-strength SH medium with 0.27 μM NAA and 29 μM GA_3 to encourage continued growth and rooting.
8. Plantlets may be transferred to the greenhouse when 3-5 cm in height.

Protoplast Isolation and Culture

1. Culture medium—a variety of culture media have been used to isolate protoplasts. The basal salt mixture does not appear to be that critical. The nature and concentration of components added to the isolation mixture are outlined in Table 2.
2. Leaves have routinely been used in the preparation of protoplasts. They are washed, sterilized, and preincubated in plasmolyzing solution prior to the addition of enzymes. The lower epidermis is sometimes peeled to aid in digestion. The leaf material is placed on the enzyme mixture at about 1 g/10-25 ml medium.
3. The protoplasts are released by gentle agitation of the digestion solution after 4-16 hr.
4. The protoplasts are collected by prefiltration on a nylon mesh screen, resuspended, and floated after centrifugation in a solution of sucrose (21% w/v) or other dense material.

5. Culture of protoplasts has usually been on a modified Kao and Wetter (1977) medium (see Table 1 for details of medium) supplemented with a variety of growth regulators including 2,4-D and a cytokinin. The hormones were added in the range from 1-5 μM.

The growth of protoplasts is influenced by plating density. It should be noted that the reports in the literature refer to obtaining protoplasts from leaf mesophyll tissue only. Whether or not this is the best or only tissue source of viable protoplasts has not been addressed.

FUTURE PROSPECTS

The prospects are bright for the use of cell culture techniques in fundamental as well as exploitive research. A full spectrum of techniques is available for the manipulation of cells in vitro. However, to date, the techniques have been applied to only a narrow spectrum of alfalfa germplasm. Extension of these techniques to include the broadest range of germplasm may occur through more fundamental research on the mechanism of in vitro regeneration and culture, or through the genetic approach of manipulating the principles controlling regeneration and culture.

The control of somatic embryogenesis in alfalfa and carrot would appear to share a common mechanism (Walker and Sato, 1981), thus raising the possibility of extending these techniques to other crops. There are a number of reports on the in vitro regeneration of other forage legumes. These include various members of the clover family (Beach and Smith, 1979; Phillips and Collins, 1980; Gresshoff, 1980; Oswald et al., 1977; Mokhtarzadek and Constantin, 1978), *Lotus corniculatus* (Tomes, 1979), *Stylosanthes hamata* (Scowcroft and Adamson, 1976), as well as others. However no concerted effort has been made to determine whether or not all of these forage legumes share any common mechanism of in vitro regeneration.

In summary, in view of the work on alfalfa, carrot, and a number of other species, the possibility of developing a more general model for somatic embryogenesis would seem more tangible than ever before.

KEY REFERENCES

Bingham, E.T. 1980. Maximizing heterozygosity in autotetraploids. In: Polyploidy: Biological Relevance (W.H. Lewis, ed.) pp. 471-489. Plenum Press, New York.

______, Hurley, L.V., Kaatz, D.M., and Saunders, J.W. 1975. Breeding alfalfa which regenerates from callus tissue culture. Crop Sci. 15: 719-721.

Kao, K.N. and Michayluk, M.R. 1980. Plant regeneration from mesophyll protoplasts of alfalfa. Z. Pflanzenphysiol. 96:135-141.

McCoy, T.J. and Bingham, E.T. 1977. Regeneration of diploid alfalfa plants from cells grown in suspension culture. Plant Sci. Lett. 10:59-66.

Walker, K.A. and Sato, S.J. 1981. Morphogenesis in callus tissue of *Medicago sativa*. The role of ammonium ion in somatic embryogenesis. Plant Cell Tissue and Organ Culture 1:109-121.

REFERENCES

Barnes, D.K., Bingham, E.T., Murphy, R.P., Hunt, O.J., Beard, D.F., Skrdla, W.H., and Teuber, L.R. 1977. Alfalfa Germplasm in the United States: Genetic Vulnerability, Use, Improvement and Maintenance. USDA/ARS Technical Bulletin 1571.

Barnes, D.K., Smith, D.M., Stucker, R.E., and Elling, L.J. 1978. Fall dormancy in alfalfa: A valuable predictive tool. Report of the 26th Alfalfa Improvement Conference, p. 34. USDA Manual ARM-NC-7, Brookings, South Dakota.

Beach, K.H. and Smith, R.R. 1979. Plant regeneration from callus of red and crimson clover. Plant Sci. Lett. 16:231-237.

Behnke, M. 1979. Selection of potato callus for resistance to culture filtrates of *Phytophthora infestans* and regeneration of resistant plants. Theor. Appl. Genet. 55:69-71.

______ 1980. General resistance to late blight of *Solanum tuberosum* plants regenerated from callus resistant to culture filtrates of *Phytophthora infestans*. Theor. Appl. Genet. 56:151-152.

Binek, A. and Bingham, E.T. 1970. Cytology and crossing behavior of triploid alfalfa. Crop Sci. 10:303-306.

Bingham, E.T. 1971. Isolation of haploids of tetraploid alfalfa. Crop Sci. 11:433-435.

______ and McCoy, T.J. 1979. Cultivated alfalfa at the diploid level: Origin, reproductive stability, and yield of seed and forage. Crop Sci. 19:97-100.

______ and Saunders, J.W. 1974. Chromosome manipulations in alfalfa: Scaling the cultivated tetraploid to seven ploidy levels. Crop Sci. 14:474-477.

______, Burnham, C.R., and Gates, C.E. 1968. Double and single backcross linkage estimates in autotetraploid maize. Genetics 59:399-410.

Bolton, J.L., Goplen, B.P., and Baenziger, H. 1972. World distribution and historical developments. In: Alfalfa Science and Technology (C.H. Hanson, ed.) pp. 1-34. American Society of Agronomy, Madison.

Busbice, J.H. and Wilsie, C.P. 1966. Inbreeding depression and heterosis in autotetraploids with application to *Medicago sativa* L. Euphytica 15:52-67.

Campbell, T.A., Schillinger, J.A., and Hanson, C.H. 1974. Inheritance of resistance to anthracnose in alfalfa. Crop Sci. 14:667-668.

Chase, S.S. 1963. Analytic breeding in *Solanum tuberosum* L. A scheme utilizing parthenotes and other diploid stocks. Can. J. Genet. Cytol. 5:359-363.

Croughan, T.P., Stavarek, S.J., and Rains, D.W. 1978. Selection of a NaCl tolerant line of cultured alfalfa cells. Crop Sci. 18:959-963.

Demarly, Y. 1963. Genetique des tetraploids et amelioration des plants. Ann. Amelior. Plant. 13:307-400.

Devine, T.E., Hanson, C.H., Ostazeski, S.A., and Campbell, T.A. 1971. Selection for resistance to Anthracnose (*Colletotrichum trifolii*) in four alfalfa populations. Crop Sci. 11:854-855.

dos Santos, A.V.P., Outka, D.E., Cocking, E.C., and Davey, M.R. 1980. Organogenesis and somatic embryogenesis in tissues derived from leaf protoplasts and leaf explants of *Medicago sativa*. Z. Pflanzenphysiol. 99:261-270.

Dunbier, M.W. and Bingham, E.T. 1975. Maximum heterozygosity in alfalfa. Results using haploid-derived autotetraploids. Crop Sci. 15: 527-531.

Frosheiser, F.I. and Barnes, D.K. 1973. Field and greenhouse selection for *Phytophthora* root rot resistance in alfalfa. Crop Sci. 13:735-738.

Gengenbach, B.G., Green, C.E., and Donovan, C.M. 1977. Inheritance of selected pathotoxin resistance in maize plants regenerated from cell cultures. Proc. Natl. Acad. Sci. USA 74:5113-5117.

Gengenbach, B.G., Connelly, J.A., Pring, D.R., and Conde, M.F. 1981. Mitochondrial DNA variation in maize plants regenerated during tissue culture selection. Theor. Appl. Genet. 59:161-167.

Graham, J.H., Frosheiser, F.I., Stuteville, D.L., and Erwin, D.C. 1979. A Compendium of Alfalfa Diseases. American Phytopathological Society, St. Paul, Minnesota.

Gresshoff, P.M. 1980. In vitro culture of white clover: Callus, suspension, protoplast culture, and plant regeneration. Bot. Gaz. 141:157-164.

Hanson, C.H., Busbice, T.H., Hill, R.R., Jr., Hunt, O.J., and Oakes, A.J. 1972. Directed mass selection for developing multiple pest resistance and conserving germplasm in alfalfa. J. Environ. Qual. 1: 106-111.

Harlan, J.R. and deWet, J.M.J. 1975. On O. Winge and a prayer: The origins of polyploidy. Bot. Rev. 41:361-390.

Hesterman, O.B. and Teuber, L.R. 1981. Factors affecting yield and quality of alfalfa sprouts. In: Report of the 27th Alfalfa Improvement Conference, p. 54. USDA manual ARM-NC-19.

Hijano, E.H. 1981. Inheritance of *Fusarium* Resistance in Alfalfa. PhD Thesis, Univ. of Minnesota, St. Paul.

Howarth, R.E., Lees, G.L., and Goplen, B.P. 1981. An overview of legume, pasture bloat research at Saskatoon. In: Report of the 26th Alfalfa Improvement Conference (D.K. Barnes, ed.) p. 55. USDA manual ARM-NC-19.

Irwin, J.A.G., Maxwell, D.P., and Bingham, E.T. 1981a. Inheritance of resistance to *Phytophthora megasperma* in diploid alfalfa. Crop Sci. 21:271-276.

______ 1981b. Inheritance of resistance to *Phytophthora megasperma* in tetraploid alfalfa. Crop Sci. 21:277-283.

Johnson, L.B., Stuteville, D.L., Higgins, R.K., and Skinner, D.Z. 1981. Regeneration of alfalfa plants from protoplasts of selected Regen S clones. Plant Sci. Lett. 20:297-304.

Kao, K.N. and Wetter, L.R. 1977. Advances in techniques of plant protoplast fusion and culture of heterokaryocytes. International Cell Biology (B.R. Brinkley and K.R. Porter, eds.). Rockefeller Univ. Press, Boston.

Kehr, W.R., Frosheiser, F.I., Wilcoxson, R.D., and Barnes, D.K. 1972. Breeding for disease resistance. In: Alfalfa Science and Technology (C.H. Hanson, ed.) pp. 335-354. American Society of Agronomy, Madison.

Kohler, G.O., Bickoff, E.M., and Beeson, W.M. 1972. Processed products for feed and food industries. In: Alfalfa Science and Technology (C.H. Hanson, ed.) pp. 659-676. American Society of Agronomy, Madison.

Lu, N.S.J., Barnes, D.K., and Frosheiser, F.I. 1973. Inheritance of Phytophthora root rot in alfalfa. Crop Sci. 13:714-717.

Maliga, P. 1980. Isolation, characterization, and utilization of mutant cell lines in higher plants. Int. Rev. Cytol. Suppl. 11A:225-250.

McCoy, T.J. 1982. The inheritance of 2n pollen formation in diploid alfalfa, *Medicago sativa*. Can. J. Genet. Cytol. 24:315-323.

______ and Smith, L.Y. 1983. Genetics, cytology, and crossing behavior of an alfalfa (*Medicago sativa*) mutant resulting in failure of the post-meiotic cytokinesis. Can. J. Genet. Cytol. 25:390-397.

Mitten, D.H., Sato, S.J., and Skokut, T.A. 1981. Survey of in vitro regenerative capacity of alfalfa germplasm sources. Report of the 17th Central Alfalfa Improvement Conference, p. 13.

Mok, D.W.S. and Peloquin, S.J. 1975. The inheritance of three mechanisms of diplandroid (2n pollen) formation in diploid potatoes. Heredity 35:295-302.

Mokhtarzadeh, A. and Constantin, M.J. 1978. Plant regeneration from hypocotyl- and anther-derived callus of berseem clover. Crop Sci. 18:567-572.

Oswald, T.H., Smith, A.E., and Phillips, D.V. 1977. Callus and plantlet regeneration from cell cultures of ladino clover and soybean. Physiol. Plant. 39:129-134.

Phillips, G.C. and Collins, G.B. 1980. Somatic embryogenesis from cell suspension cultures of red clover. Crop Sci. 20:323-326.

Quiros, C.F. and Morgan, K. 1981. Peroxidase and leucine-aminopeptidase in diploid *Medicago* species closely related to alfalfa: Multiple gene loci, multiple allelism, and linkage. Theor. Appl. Genet. 60: 221-228.

Reisch, B. 1980. The Selection and Regeneration of Ethionine Resistant Alfalfa Cell Lines and the Genetic Control of Regeneration from Alfalfa Callus Cultures. Ph.D. Thesis, Univ. of Wisconsin, Madison. Diss. Abstr. Int. 41(9B):3266.

______ and Bingham, E.T. 1980. The genetic control of bud formation from callus cultures of diploid alfalfa. Plant Sci. Lett. 20:71-77.

______ and Bingham, E.T. 1981. Plants from ethionine-resistant alfalfa tissue cultures: Variation in growth and morphological characters. Crop Sci. 27:783-788.

______, Duke, S.H., and Bingham, E.T. 1981. Selection and characterization of ethionine-resistant alfalfa (*Medicago sativa* L.) cell lines. Theor. Appl. Genet. 59:89-94.

Rhoades, M.M. and Dempsey, E. 1966. Induction of chromosome doubling at meiosis by the elongate gene in maize. Genetics 54:505-522.

Sacristan, M.D. 1982. Resistance responses to *Phoma lingam* of plants regenerated from selected cell and embryogenic cultures of haploid *Brassica napus*. Theor. Appl. Genet. 61:193-200.

Satina, S. and Blakeslee, A.F. 1935. Cytological effects of a gene in *Datura* which causes dyad formation in sporogenesis. Bot. Gaz. 96: 521-532.

Saunders, J.W. and Bingham, E.T. 1972. Production of alfalfa plants from callus tissue. Crop Sci. 12:804-808.

Scowcroft, W.R. and Adamson, J.A. 1976. Organogenesis from callus cultures of the legume, *Stylosanthes hamata*. Plant Sci. Lett. 7:39-42.

Schenk, R.U. and Hildebrandt, A.C. 1972. Medium and techniques for induction and growth of monocotyledonous and dicotyledonous cell cultures. Can. J. Bot. 50:199-204.

Stavarek, S.J., Croughan, T.P., and Rains, D.W. 1980. Regeneration of plants from long-term cultures of alfalfa cells. Plant Sci. Lett. 253-261.

Tomes, D.T. 1979. A tissue culture procedure for propagation and maintenance of *Lotus corniculatus* genotypes. Can. J. Bot. 57:137-140.

Viands, D.R. and Barnes, D.K. 1980. Inheritance of resistance to bacterial wilt in two alfalfa gene pools: Qualitative analysis. Crop Sci. 20:48-54.

Viands, D.R., Barnes, D.K., Stucker, R.E., and Frosheiser, F.I. 1979a. Inheritance of resistance to bacterial wilt in two alfalfa gene pools: Response to selection and quantitative analysis. Crop Sci. 19:711-714.

Viands, D.R., Vance, C.P., Heichel, G.H., and Barnes, D.K. 1979b. A plant-mediated ineffective nitrogen fixation trait in alfalfa (*Medicago sativa* L.). Crop Sci. 19:905-908.

Viands, D.R., Barnes, D.K., and Heichel, G.H. 1981. Nitrogen Fixation in Alfalfa: Responses to Bidirectional Selection for Associated Characteristics. USDA Technical Bulletin 1643.

Vorza, N. and Bingham, E.T. 1979. Cytology of 2n pollen formation in diploid alfalfa, *Medicago sativa*. Can. J. Genet. Cytol. 21:525-530.

Walker, K.A., Yu, P.C., Sato, S.J., and Jaworski, E.G. 1978. The hormonal control of organ formation in callus of *Medicago sativa* L. cultured in vitro. Am. J. Bot. 65:654-659.

Walker, K.A., Wendeln, M.L., and Jaworski, E.G. 1979. Organogenesis in callus tissue of *Medicago sativa*. The temporal separation of induction processes from differentiation processes. Plant Sci. Lett. 16:23-30.

Willis, W.G., Stuteville, D.L., and Sorensen, E.L. 1969. Effects of leaf and stem diseases on yield and quality of alfalfa forage. Crop Sci. 9:637-640.

CHAPTER 8
Peanut

Y. P. S. Bajaj

The cultivated peanut or groundnut (*Arachis hypogaea* L.) is a protein-rich oilseed crop. It is a native of Brazil and is now being grown in tropical and subtropical as well as warm temperate zones. However, commercial production is found only between 40° N and 40° S latitude. Archaeological evidence has shown the pre-Columbian cultivation of peanut in Peru as early as 600 A.D. Around 1550 A.D., evidence of peanut culture was found in eastern Brazil and the West Indies; thereafter it was introduced to other parts of the world. In 1980, an area of about 19,775,000 ha was under groundnut cultivation, and 18,901 thousand metric tons were harvested with an average yield of 956 kg/ha (FAO, 1980). Asia is the largest producer, followed by Africa, North and Central America, and South America (Table 1). Among the individual countries, India is the largest producer in the world, followed by China, the United States, Sudan, and Nigeria.

Peanut is relatively photoinsensitive, and possesses higher oil and protein content than other legumes. It is also an excellent source of some essential nutrients such as carbohydrates, trace elements, and vitamins. Its cultivation improves soil fertility, as peanut fixes atmospheric nitrogen with the help of bacteria in the root nodules. In addition, it is grown for forage purposes and to check soil erosion in hilly areas. The cultivated peanut, however, is susceptible to several diseases and insect pests.

The systematic position of the genus was not clearly defined until 1894, when Taubert (see Raman, 1976) divided the tribe Hedysaneae of *Papilionaceae* into six subtribes, thus including the genera *Zornia*,

Table 1. World Peanut Production[a]

LOCATION	AREA UNDER CULTIVATION (1000 ha)	YIELD (kg/ha)	PRODUCTION (1000 MT)[b]
World	19,775	956	18,901
Asia	11,950	997	11,910
Africa	6,360	755	4,801
North Central America	728	1,687	1,228
South America	681	1,296	883
Australia	32	1,201	38
Europe	13	2,058	27
India	7,500	853	6,400
China	2,955	1,249	3,692
USA	569	1,830	1,042
Sudan	960	844	810
Indonesia	500	1,500	750
Nigeria	600	950	570
Senegal	1,000	500	500
Burma	525	941	494
USSR	1	1,182	1

[a]Adapted from FAO, 1980.

[b]Includes shell.

Chapmannia, *Stylosanthes*, and *Arachis* in the last subtribe *Stylosanthineae*. The genus *Arachis* has both diploid (2n = 20) and tetraploid (2n = 40) species. The important diploid species include *A. benthamii*, *A. diogoi*, *A. duranensis*, *A. helodes*, *A. lutescens*, *A. repens*, *A. rigoni*, *A. villosa*, whereas the important tetraploid species are *A. glabrata*, *A. hagenbeckii*, *A. hypogaea*, *A. monticola*, *A. nambyquarae*, *A. rasteiro*, *A. salvagem*, *A. marginata*, and *A. prostrata*. In addition, *A. pusilla* has been reported to be an octaploid form (Krapovickas and Rigoni, 1952).

Peanut is attacked by a large number of diseases and a variety of pests (Table 2) which cause tremendous losses in yield. The most damaging diseases are the foliar types such as leaf spots (*Cercospora arachidicola; C. personata*) and rust (*Puccinia arachidis*), which are now becoming a worldwide problem (Subrahmanyam et al., 1978). Other fungi causing serious disorders in groundnut are *Aspergillus niger*, *Fusarium* spp., *Pythium* spp., and *Rhizoctonia* spp. Apart from these, the attack of *Aspergillus flavus* produces a toxic metabolite (aflatoxin) that affects human health. Major viral diseases are bud necrosis caused by tomato spotted wilt virus (TSWV) and peanut mottle virus (PMV). Among various kinds of insect pests, the aphids, jassids, thrips, and

Table 2. Some Common Diseases and Pests of Peanuts

COMMON NAME	ORGANISM
Disease	
Afla root disease	*Aspergillus flavus*
Collar rot	*A. niger*
Seed rot	*Aspergillus* sp.
Rust	*Puccinia arachidis*
Leaf spot (Tikka disease)	*Cerospora arachidicola*
	C. personata
Bud blight	Tomato spotted wilt virus
Root rot (wilt rot)	*Rhizoctonia destrens*
Insect pests	
Leaf weber	*Anarsia epippias* (Meyrick)
White grub	*Holotrichia consanguinea*
Aphid	*Aphis craccivora*
Red-hair caterpillar	*Amsacta* sp.
Leaf miner	*Stomopteryx subsceviella*
Termite	*Odontotermes obesus*
Thrips	*Caliothrips indicus*

termites cause the greatest losses. The wild species possess desirable characters, which have not been reported in commercial varieties. For instance, *A. monticola* is resistant to *Cercospora* and *A. glabrata* is resistant to many leaf-spot diseases. *Arachis hagenbeckii* and *A. prostrata* are drought-resistant. Other species such as *A. diogoi*, *A. glabrata*, *A. hagenbeckii*, and *A. marginata* possess better mineral, protein, and fat content. *Arachis villosa* possesses higher oil content, resistance to drought, and resistance to pests and foliar diseases. Genetic manipulation is required to incorporate these desirable characters into the commercially grown varieties because of the differences in the ploidy levels between species and the barriers to interspecific hybridization (Gregory et al., 1965; Wilson, 1973; Moss, 1980).

BREEDING AND CROP IMPROVEMENT

The major breeding objectives in peanut improvement programs are to develop varieties with higher and more stable yields, resistance to various diseases and pests, and drought-tolerance. Higher protein and oil contents with medium-sized seeds would be highly desirable for industrial uses. Early maturing cultivars are also desirable.

The conventional methods of crop improvement include introduction, selection (pure line, mass selections, or modified selection methods), hybridization, and recombination followed by pedigree, bulk, or backcross techniques. These methods have been inadequate in developing varieties that perform as well as those that are now available in crops such as wheat and rice. Introductions must above all overcome the problem of acclimatization. In addition, selection procedures do not create new variability. It is difficult in peanut, a self-pollinated crop, to accomplish most interspecific crosses. The limited success in peanut sexual hybridization may relate to the basic procedures in which undesirable plants are discarded in early backcross generations. Using this method it is difficult to recover desirable plants, as in most instances undesirable characters are often linked with desirable ones. While some improved cultivars have been produced, the strong selection procedures were responsible for eroding genetic diversity. This practice has been used extensively in India and in some African countries, resulting in elimination of certain old races. Hence, there is an urgent need to generate additional genetic variability and also to utilize the already existing variability by adopting unconventional methods.

Progress in plant cell, tissue, and organ culture (see Reinert and Bajaj, 1977; Evans et al., 1983) has opened up several new possibilities for the induction of genetic variability and the selection of desirable mutants. Some of these in vitro methods that can be applied to peanut include: (a) cloning and production of disease-free plants, (b) removal of sexual incompatibility, (c) somatic hybridization and genetic engineering, (d) production of haploids, triploids, and polyploids, (e) selection of mutants resistant to salt, drought, herbicides, diseases, and pests, and (f) freeze preservation of germplasm. Since peanut is a self-pollinated crop, the in vitro production of haploids and their subsequent chromosome doubling can reduce the time needed for pedigree or backcross methods to obtain true breeding homozygous lines. Moreover, haploids could also be used in mutagenic experiments. Meristem culture could be used for obtaining virus-free plants, which is especially important in the effort to develop quality seeds in higher quantity. Wide hybridization, for transferring desirable characters from the wild species into cultivated crops, could be achieved using embryo culture or protoplast fusion. Cell cultures can be treated with various kinds of mutagens and desirable mutants/variants could also be recovered.

Not much work has been reported on protoplast, cell, tissue, and organ culture of *Arachis* spp., but the accumulated literature, especially during the last decade, demonstrates the potential and promise of the use of in vitro techniques for plant regeneration via tissue culture, and ultimately for peanut crop improvement (Table 3).

REVIEW OF THE LITERATURE

Regeneration of Plants from Explants

Clonal propagation of F_1 hybrids, haploids, and other rare plants through the culture of segments and tissues would facilitate peanut breeding and crop improvement programs. In this respect various segments excised from in vitro-grown seedlings have been cultured (Bajaj et al., 1981a) on different media, and their growth response is presented in Table 4. The excised segments of epicotyl and mesocotyl

Table 3. In Vitro Studies on *Arachis*

SPECIES	EXPLANT	MEDIUM	RESPONSE	REFERENCE
A. glabrata	Anthers	MS + IAA (22.0 μM) + KIN (9.3 μM)	Pollen embryos and callus	Bajaj et al., 1980a
A. hypogaea	Anthers	MS + NAA (11.0 μM) + KIN (9.3 μM) + 2,4-D (9.0 μM)	Callus with haploid and varying levels of ploidy	Martin & Rabechault, 1976
	Anthers	MS + BA (0.4 μM)	Diploid callus and plantlets	Mroginski & Fernandez, 1980
	Anthers	MS containing IAA, NAA, and BA (see protocol)	Plantlets and callus, various ploidy level	Bajaj et al., 1981b
	Gynophores	MS + NAA (0.5–11.0 μM) + KIN (0.5–9.3 μM)	Pod formation	Ziv & Zamski, 1975
	Gynophores	MS + NAA (3.7 μM) + KIN (2.3 μM)	Pod, roots, callus	Bajaj et al., 1981a
	Ovules	MS + KIN + GA	Seedlings	Martin, 1970
	Immature embryos	MS	Plantlets	Singh et al., 1980
	Embryos, various stages	MS + IAA + KIN	Callus, plantlets	Bajaj et al., 1981a
	Pollen embryos (frozen at -196 C)	MS + NAA + BA	Plants, callus	Bajaj, 1982a
	Seedling	MS containing IAA, KIN, 2,4-D, and CH (see protocol)	Complete plants from seedling explants; root or shoot from callus	Bajaj et al., 1981a
	Meristems (frozen at -196 C)	MS + IAA (11.0 μM) + KIN (2.3 μM)	Plantlets	Bajaj, 1979c

Table 3. Cont.

SPECIES	EXPLANT	MEDIUM	RESPONSE	REFERENCE
A. hypogaea	Immature leaves	Mod. MS containing NAA (0.05-22.0 μM) and BA (4.4-13.2 μM)	Plantlet regeneration	Mroginski et al., 1981
	Root segments	NM + 2,4-D + KIN	Root callus inoculated with *Rhizobium* showed nitrogenase activity	Ranga Rao & Subba Rao, 1976
	Pericarp	WH + CW (15%) + CH (400 mg/l) + 2,4-D (9.0 μM)	Rhizogenesis in callus	Rangaswamy et al., 1965
	Phloem tissue	WH + CW (7%)	Growth of cells in suspension	Steward et al., 1958
	Cotyledon callus	LS + NAA (11.0 μM) + KIN (2.3 μM) + amino acids	Callus for biochemical studies	Verma & Van Huystee, 1970
	Cotyledon callus	WH + CW (7%)	Homogeneous cell line was derived	Verma & Van Huystee, 1971a
	Cotyledon callus	MS + NaFe-EDTA (5 mg/l) + $MgSO_4$ (1.5 mM)	Chlorophyll in callus	Kumar, 1974a
	Cotyledon callus	MS + 2,4-D (9.0 μM) + NAA (11.0 μM) + KIN (9.3 μM)	Vigorous callus growth	Guy et al., 1978
	Cotyledon callus	MS + 2,4-D (9.0 μM) + NAA (11.0 μM) + KIN (9.3 μM)	Rhizogenesis	Guy et al., 1980

SPECIES	EXPLANT	MEDIUM	RESPONSE	REFERENCE
A. hypogaea	Hypocotyl callus	MS + Thiamine HCl (3 μM) + Nicotinic acid (8 μM) + Pyridoxine HCL (4.9 μM) + Inositol (0.55 mM) + Calcium pantothenate (5.2 μM)	Growth studies	Kumar, 1974b
	Cell suspension	WH + CW (7%)	Protein synthesis and respiration were increased with 500 rad irradiation	Verma & Van Huystee, 1971b,c
	Cell suspension	WH + CW (7%)	Rapid synthesis and release of peroxidase by cell cultures	Van Huystee & Turcon, 1973
	Cell suspension	WH + CW (7%)	Synthesis of porphyrin and peroxidase by cultured cells	Van Huystee, 1977
	Palisade cells	Heller's macroelements + special microelements + CH + 2,4-D + KIN + 0.1 μM NH_4Cl	Division of cells	Ball & Joshi, 1965
	Palisade cells	Heller's macroelements + special microelements + CH + 2,4-D + KIN + 0.1 μM NH_4Cl	Cell growth	Joshi & Ball, 1968

Table 3. Cont.

SPECIES	EXPLANT	MEDIUM	RESPONSE	REFERENCE
A. hypogaea	Mesophyll cells	Minerals + sucrose (.06 M) + CH (400 mg/l) + 2,4-D (4.5 μM) + Thiamine HCl (1.0 μM) + Myoinositol (0.55 mM)	Cell growth	Joshi & Noggle, 1967
	Mesophyll cells from cotyledon callus	Modified MS + sucrose (0.12 M), pH 6.0, temp. 25 ± 1 C, and light intensity 4000 lux	Good callus growth	Kumar, 1974c
	Protoplasts from callus cells	Not given	Protoplast fusion	Hildebrandt & Schenk, 1970
A. hypogaea x *A. villosa*	Protoplasts from seedlings	Cellulase + macerozyme + mannitol + PEG	Fusion, hybrid protoplasts	Bajaj & Gosal, 1983
A. monticola	Ovules	MS + CH (400 mg/l)	Swelling, occasional callusing	Sastri et al., 1980
A. villosa	Anthers	MS + NAA (11.0 μM)	Callus	Mroginski & Fernandez, 1980
	Anthers	MS containing IAA, NAA, and BA (see protocol)	Plantlets and callus, various ploidy level	Bajaj et al., 1981b
	Pollen embryos (frozen at -196 C)	MS + NAA + BA	Plants, callus	Bajaj, 1982a

Table 4. Growth Response of Excised Segments of In Vitro Grown Seedlings of *A. hypogaea* on Various Media[a]

	MS MEDIUM			CH MEDIUM			MS_1 MEDIUM		
EX-PLANTS	Segments cultured	Segments respond-ing	Percent response	Segments cultured	Segments respond-ing	Percent response	Segments cultured	Segments respond-ing	Percent response
Mesocotyl	9	6	66.6	13	10	77.0	9	7	77.7
Epicotyl	7	4	57.0	9	6	66.6	8	3	37.5
Hypocotyl	8	3	37.5	9	5	55.5	7	0	0
Leaf	13	10	77.0	9	6	66.6	6	0	0
Petiole	8	3	37.5	10	6	60.0	7	0	0

[a]Bajaj et al., 1981a.

started to elongate within a week. In the epicotyl, the original meristem unfolded, whereas in the mesocotyl a bud developed which eventually formed the plant (Fig. 1A-C). The excised cotyledons, petiole, and root (Fig. 1D-F) proliferated, sometimes profusely, to form a mass of callus.

Of the various segments cultured, the mesocotyl and epicotyl yield the best growth response in terms of callus, root and shoot formation on CH medium. This was followed by MS_3 and MS_1 media.

In Vitro Culture of Gynophores

The in vitro behavior of the gynophore (Fig. 1G-I; Table 5) is dependent on its stage of development. Very young explants at 7 days after pollination (DAP) did not elongate but formed callus, with no enlargement of the ovary. Medium-aged (10 DAP) and older (15 DAP) explants proliferated continuously. On NAA (2.7 μM) + KIN (2.3 μM), the gynophore remained green and proliferated to form callus (Fig. 1H). When the concentrations of NAA and KIN were raised to 11 and 9.3 μM, respectively, there was an enlargement of the ovary and the formation of roots (Fig. 1I). Pod formation appears to be parthenocarpic as no seeds were observed. The gynophores display a positive geotropic response. These studies, under controlled conditions, should aid in understanding the factors influencing the formation of pods (Ziv and Zamski, 1975).

Establishment of Callus Culture

Actively growing callus and cell cultures have been obtained from a variety of tissues by several researchers (Table 3) and used for physiological and morphogenetic studies. All the segments, embryos, and ovules produced rapidly growing, loose and friable callus (Fig. 2A). However, of the segments cultured, the best growth of the callus was obtained from the mesocotyl on MS + CH (400 mg/l) + 2,4-D (4.5 μM). The addition of KIN (0.5-9.3 μM) resulted in callus that was compact and slow growing, characteristics that increased with the increase of KIN concentration. The callus has been maintained on medium with CH (200 mg/l) and 2,4-D (2.3 μM), and an actively growing cell suspension can be obtained.

Differentiation of Plants from Callus

Thus far, there has not been much success in the differentiation of complete plants from callus cultures. In one early study callus was observed to undergo partial morphogenesis, but only root initials were observed (Rangaswamy et al., 1965). Later Bajaj et al. (1981a) observed that when the callus was transferred to MS + IAA + KIN medium, occasionally roots and shoots were observed (Fig. 2A-C). These studies on organ differentiation are comparable with the results obtained with anther-derived callus (Fig. 3A-E; Bajaj et al., 1981b).

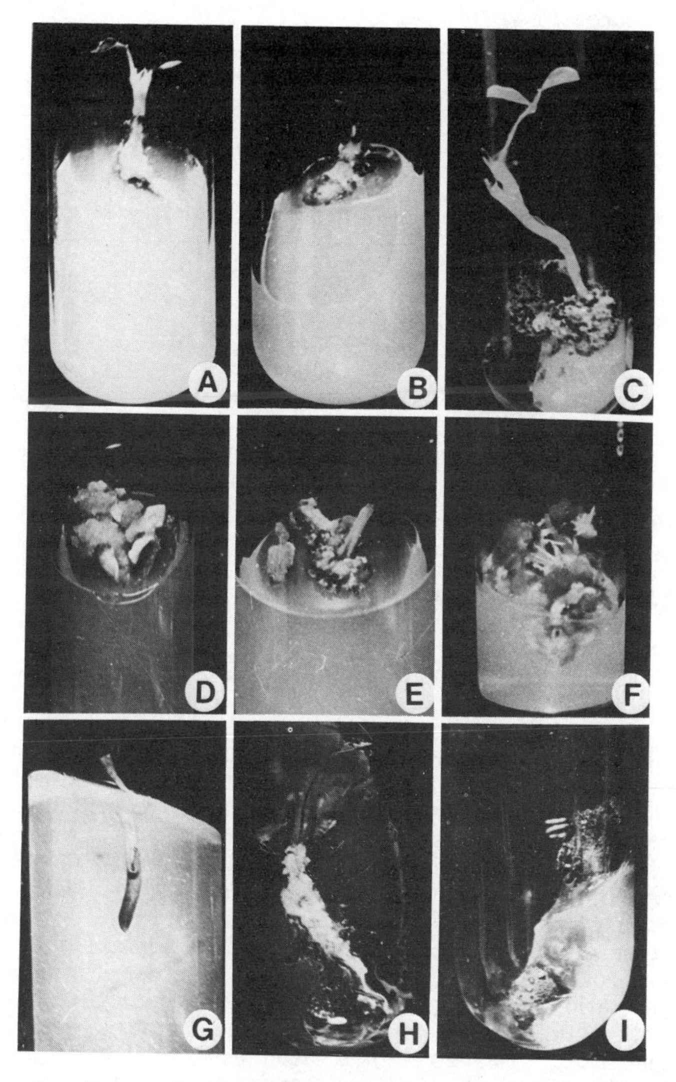

Figure 1. In vitro growth response of various explants of *A. hypogaea* on different media. (A-C) Regeneration of plants from mesocotyl (A) and epicotyl (B,C) segments cultured on MS + IAA (11.0 µM) + KIN (2.3 µM). (D-F) induction of callusing from excised cotyledons, petiole, and root segments, respectively. (G) Young gynophore (10 days after pollination) at the time of culturing on MS + NAA (2.7 µM), and (H) after 4 weeks, showing elongation and formation of callus. (I) Relatively old gynophore (15 days after pollination) 8 weeks after culture on MS + NAA (11.0 µM) + KIN (9.3 µM) showing the development of pod and roots (Bajaj et al., 1981a).

Table 5. Growth Response of Gynophore Explants of *A. hypogaea* Cultured In Vitro[a]

	MEDIA		
EXPLANT	MS	MS + KIN (2.3 µM) + NAA (2.7 µM)	MS + KIN (9.3 µM) + NAA (11.0 µM)
Very young gynophores (7 DAP)[b]	Callus	Callus	Callus
Young gynophores (10 DAP)	Callus and rhizogenesis	Callus and rhizogenesis	Elongation of gynophore
Old gynophores (15 DAP)	Slight enlargement	Elongation of gynophore	Ovary enlargement

[a]Bajaj et al., 1981a.

[b]DAP = days after pollination.

Anther Culture and the Production of Haploids

Haploids are of great significance and their importance in basic research and in crop improvement programs has been emphasized (Bajaj, 1983a). In peanuts, efforts are being made to transfer desirable characters from wild (normally diploid) species to commercially grown *A. hypogaea* (tetraploid). However, it has been difficult or nearly impossible in most cases. To achieve this goal, current methods are via the triploid, autotetraploid, or amphidiploid routes. However, the possibility also exists that dihaploids can be obtained in *A. hypogaea* (4n) through anther or isolated pollen culture, which then could be used for crossing with other wild species with the same chromosome number. Moreover, mutation breeding has given some encouraging results. A strain NC 4X with high yield and good seed quality was developed after X-irradiation. This strain has been used to develop many commercial varieties (Gregory, 1968). In breeding programs, the culturing of anthers from F_1 hybrid plants, with subsequent doubling of chromosome number, can reduce the time necessary to achieve homozygosity. It would also be desirable to develop several lines of *A. hypogaea* with 2n = 20 chromosome number for wide hybridization.

Haploids may occur spontaneously as a result of parthenogenesis, or they may be induced experimentally by a number of methods such as distant hybridization, thermal shock, irradiation, use of abortive pollen, delayed pollination, or spraying with various chemicals. However, such methods are laborious, cumbersome, and time-consuming, due to the low frequency of haploid production. In this regard, in vitro culture of

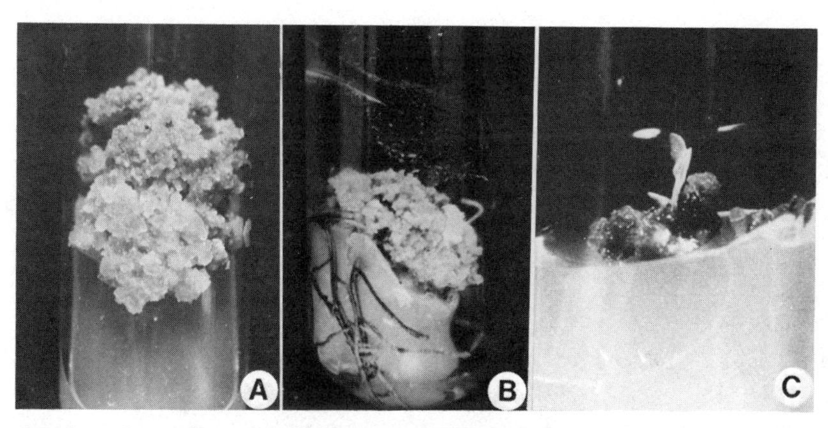

Figure 2. Growth and differentiation in embryo-derived callus culture of *A. hypogaea*. (A) Mass of callus obtained from an immature embryo cultured on MS + CH (500 mg/l) + 2,4-D (4.5 μM) + KIN (9.3 μM), 3 weeks after culture. (B) Rhizogenesis in callus subcultured on MS + IAA (22.0 μM) + KIN (9.3 μM), 3 weeks after culture. (C) Differentiation of shoots from 6-week-old callus subcultured on MS + IAA (22.0 μM) + KIN (9.3 μM) (Bajaj et al., 1981a).

excised anthers and pollen has attracted considerable attention. Excised anther culture is relatively simple, quick, and efficient, and haploid tissues and plants have been obtained in a number of crops (see Bajaj, 1983a). This has led to early release of high yielding cultivars of rice, wheat, and tobacco. This outcome has encouraged plant breeders to explore the use of in vitro methods for peanuts (Bajaj et al., 1980a, 1981b).

The excised anthers of *A. glabrata*, *A. hypogaea* (Bajaj et al., 1980a), and *A. villosa* (Bajaj et al., 1981b) cultured at the first pollen mitosis stage on MS + IAA (11 μM) + KIN (9.3 μM) or BA (8.8 μM) showed initial callus development within 3 weeks in about 13 and 11% of the cases (Table 6). A mass of callus was, however, formed in 5 weeks (Fig. 4A). The callus was white to creamy.

Microscopic studies showed pollen at various stages of embryogenesis (Fig. 4A-I). The pollen underwent segmentation, and repeated nuclear and cell divisions to form pollen embryos. A similar mode of embryogenesis and pollen-callus formation was observed in another grain legume, *Cajanus cajan* (Bajaj et al., 1980b). The pollen embryos mostly underwent proliferation to form callus, which was more compact than that obtained from the anther tissues. However, with the passage of time the two types of callus became mixed, and could not be distinguished.

The callus, when subcultured on MS + NAA (5.4 μM) + BA (8.8 μM), started to differentiate within a week to form shoots (Fig. 3C,D). In some cases, gynophore-like structures and multiple shoots developed during the next 3 weeks. If the differentiating shoots were left on the same medium (BA, 8.8 μM), they started turning yellow and ultimately died. However, after transfer to a basal medium containing NAA (2.7-5.4 μM) rhizogenesis was initiated in some cases. The rooted plantlets

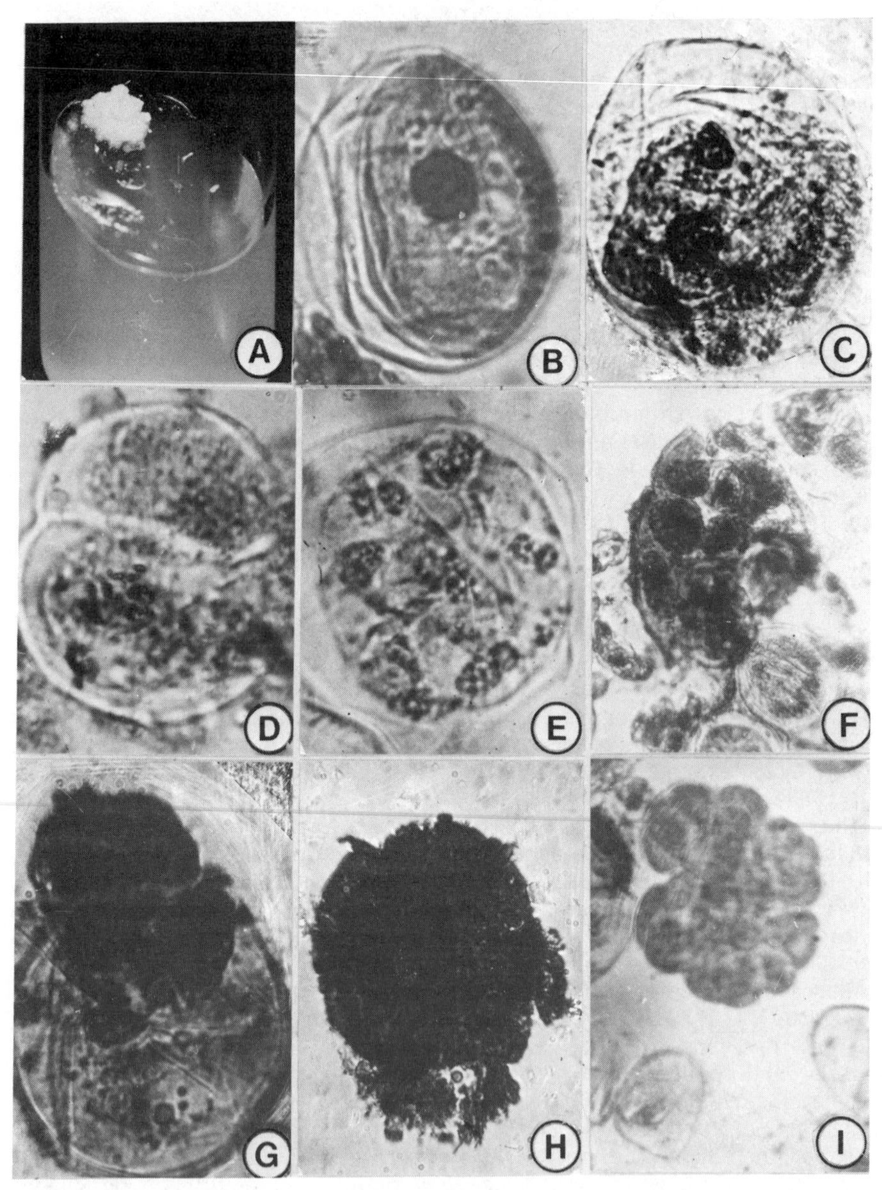

Figure 3. In vitro culture of anthers of *A. hypogaea* and *A. villosa.* (A) Excised anthers of *A. villosa* 3 weeks after culture on MS + IAA (22.0 µM) + KIN (9.3 µM) showing proliferation. (B) Anther-derived callus of *A. hypogaea.* (C,D) Differentiation of shoots and root from callus subcultured on MS + NAA (5.4 µM) + BA (8.8 µM). (E) A plant of *A. villosa* obtained from a callus-differentiated shoot transferred to MS + NAA (5.4 µM), transferred to pot. Total time taken, from the culture of anther to this stage is 23 weeks (Bajaj et al., 1981b).

Table 6. In Vitro Growth Responses of the Excised Anthers of *Arachis*[a]

SPECIES	NUMBER CULTURED[b]	NUMBER CALLUSED	PERCENT CALLUSED	BEHAVIOR OF TRANSFERRED CALLUS[c]
A. hypogaea	750	100	13	About 18% of the calli underwent rhizogenesis; occasional shoot formation
A. villosa	853	95	11	40–70% of the calli differentiated shoots; no root formation

[a]Bajaj et al., 1981b.

[b]MS + IAA (22.0 μM) + KIN (9.3 μM).

[c]MS + NAA (11.0 μM) + BA (8.8 μM).

Figure 4. Induction of pollen embryogenesis in the excised anthers of *A. hypogaea* and *A. glabrata*. (A) Initiation and formation of a mass of callus from excised anthers 4 weeks after culture on MS + IAA (22.0 μM) + KIN (9.3 μM). (B) A uninucleate pollen in an anther at the time of culture. (C-I) Various stages in the early androgenesis with pollen undergoing repeated divisions to form multinucleate and multicellular bodies, leading to the formation of pollen embryos (Bajaj et al., 1980a).

could then be transferred to the pot (Fig. 3E). Thus the elapsed time from the start of anther culture to the regeneration of plants was about 8-12 weeks.

There were discernible differences in the growth in culture and morphogenetc response of *A. villosa* and *A. hypogaea* (Table 6). The anthers of *A. villosa* underwent little proliferation, but regenerated shoots and plantlets frequently, whereas the anthers of *A. hypogaea* proliferated profusely to form a mass of callus, which underwent rhizogenesis and occasionally formed shoots.

The chromosome number, both in the callus cells and the root tips of the regenerated plants, showed a wide range of variation (Mroginski and Fernandez, 1980; Bajaj et al., 1981b), with abundant polyploids and aneuploids. In *A. hypogaea*, chromosome number varied from 20 to 80 (x = 10) (Fig. 5).

Embryo Culture and Interspecific Hybridization

Naturally occurring wild germplasm has not been successfully incorporated into *A. hypogaea*. This is mainly due to cross-incompatibility. Some of the diploid species (*A. duranensis* x *A. villosa*), tetraploid species (*A. nambyquarae* or *A. rasteiro* x *A. hypogaea*) (Stokes and Hull, 1930; Srinivasmurthy and Iyengar, 1950) or diploids with tetraploids (Raman, 1976) hybridize freely. However, most of the species are incompatible or difficult to hybridize with *A. hypogaea*. Thus the wild germplasm cannot be successfully incorporated into the cultivated peanut to the desired extent.

In peanut improvement programs, increased yield, high oil and protein, resistance to various pathogens, and early maturity are the main objectives. Germplasm with these characters is available among

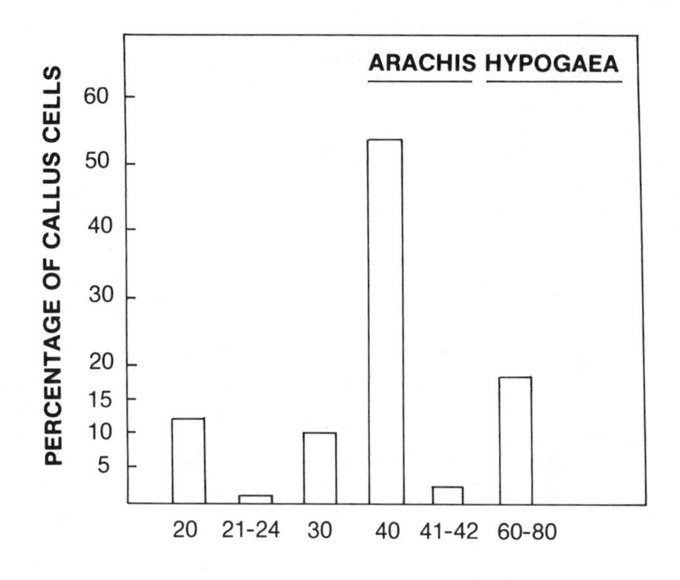

Figure 5. Histogram showing the extent of the range of genetic variability in anther-derived callus of *A. hypogaea*. Data based on 886 dividing cells (Bajaj et al., 1981b).

the wild species (Smartt and Gregory, 1967; Spielman et al., 1979). However, in most cases, the crosses are either incompatible, or the frequency of hybridization is low. Interspecific hybridization in the genus *Arachis* (*A. hypogaea* x *A. correntina*) was first reported in 1952 by Krapovickas and Rigoni and was later demonstrated by other workers (Kumar et al., 1957; Smartt and Gregory, 1967). This subject has been reviewed recently (Smartt, 1979). The frequency of hybridization and the extent of compatibility among various species has not been discussed in these reports. As the types of interspecific hybrids are limited, there is a need to explore nonconventional methods. It is expected that these problems can be overcome by the culture of hybrid embryos (Raghavan, 1977) and also by utilizing various other in vitro techniques (Reinert and Bajaj, 1977). At present, there are no published reports on the culture of hybrid embryos in *Arachis*. Work on these lines has been initiated in our laboratory (Bajaj et al., 1982). Observations on the culture of embryos of the cultivated tetraploid *A. hypogaea* (2n = 40), the wild diploid (2n = 20) species *A. villosa* (resistant to drought, pests, and the tikka disease, and containing high oil content; Cherry, 1977), and their interspecific hybrids are summarized in Table 7 and illustrated in Figs. 6, 7, and 8.

Table 7. Growth Response of Hybrid Embryos (*A. hypogaea* x *A. villosa*) Cultured on Various Media[a,b]

MEDIA	EMBRYOS CULTURED	EMBRYOS SHOWING GROWTH	PERCENT RESPONSE	REMARKS
MS + 2,4-D (4.5 μM) + KIN (2.3 μM) + CH (500 mg/l)	156	12	7.6	Few embryos started to proliferate in the third week, and a mass of callus was formed in 6 weeks
MS + IAA (22 μM) + KIN (9.3 μM)	306	19	6.2	Embryos developed into plants in 5–6 weeks; 10-week-old plants were 8 cm long with 3–5 pairs of leaves
MS + IAA (11.0 μM) + KIN (2.3 μM)	207	5	2.4	Ten-week-old plants were 5 cm long with 2–3 pairs of leaves

[a]30–35 days after pollination.

[b]Bajaj et al., 1982.

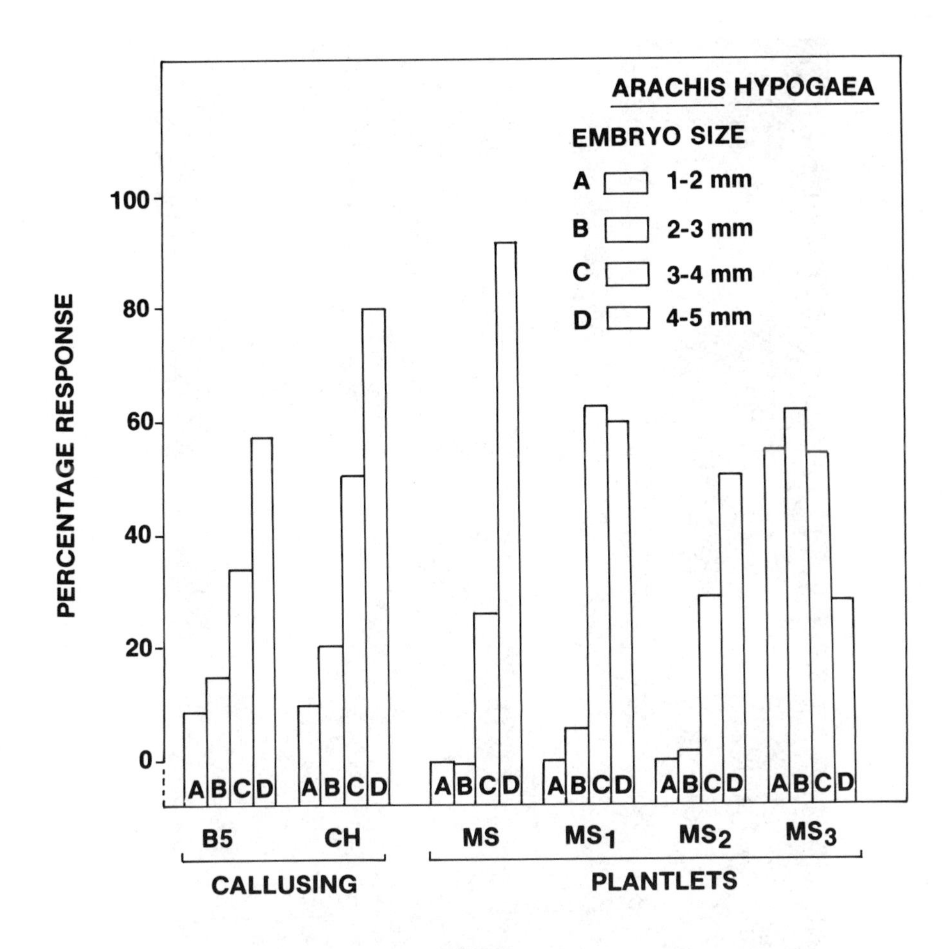

Figure 6. Growth response of the immature and maturing embryos of *A. hypogaea* cultured at various stages of development on the following media: B5—Gamborg et al., 1958; CH—MS + casein hydrolysate (400 mg/l) + 2,4-D (4.5 μM) + KIN (0.5 μM); MS_1—MS + IAA (11.0 μM) + KIN (2.3 μM); MS_3—MS + IAA (22.0 μM) + KIN (9.3 μM) (Bajaj et al., 1981a).

CULTURE OF EMBRYOS OF *ARACHIS HYPOGAEA* AND *ARACHIS VILLOSA*. The embryos that were cultured at various stages of development showed differential growth response (Fig. 6). On MS + 2,4-D medium the embryos proliferated, but growth was considerably enhanced, and callus developed profusely, when the medium was supplemented with CH. The embryos of *A. hypogaea* produced a mass of friable callus in 4 weeks, and grew faster than *A. villosa*. On medium

Figure 7. Interspecific hybridization between *A. hypogaea* x *A. villosa* through the culture of excised embryos. (A) An immature hybrid embryo (30 days after pollination) at the time of culture. (B) Excessive proliferation of a hybrid embryo cultured on MS + 2,4-D (4.5 µM) + casein hydrolysate (400 mg/l) + KIN (4.6 µM). (C-E) Various stages in the development of a hybrid plant from an embryo cultured on MS + IAA (22.0 µM) + KIN (9.3 µM) after 2, 4, and 6 weeks, respectively. (F) Hybrid plant established in the soil. (G-I) The embryo-derived plants of *A. hypogaea* (2n = 40), *A. hypogaea* x *A. villosa* cross (2n = 30), and *A. villosa* (2n = 20), respectively (Bajaj et al., 1982).

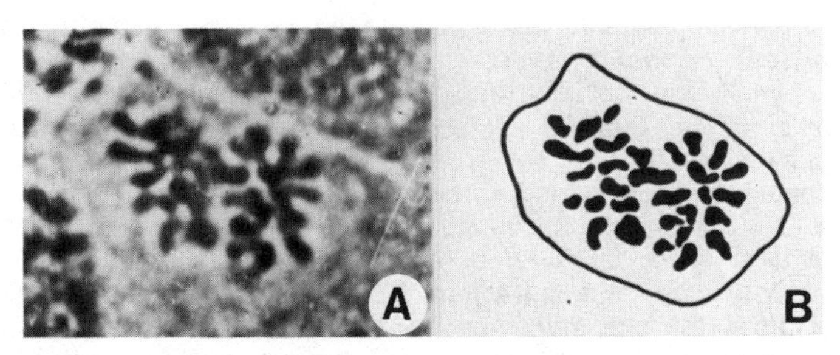

Figure 8. Root tip squashes of the hybrid plant (*A. hypogaea* x *A. villosa*) with triploid number of chromosomes (2n = 30), (A) photographed under high power, and (B) camera lucida drawing (Bajaj et al., 1982).

supplemented with IAA (22 µM) + KIN (9.3 µM), complete plants were obtained from the embryos. As in the case of pollen embryos (Bajaj et al., 1980a), differential growth response was observed for zygotic embryos of the two species.

CULTURE OF HYBRID EMBRYOS. The response of the hybrid embryos on various media is summarized in Table 7. The embryos cultured on MS + IAA (22 µM) + KIN (9.3 µM) (MS_3 medium) produced plantlets (Fig. 7), whereas the addition of 2,4-D and CH caused excessive proliferation (Fig. 7B). On MS_3, the embryos swelled and started to grow within 2 weeks (Fig. 7C-E). The epicotyl turned green and elongated to form a shoot (Fig. 7C), and growth of the radicle was slow. The tap root that formed was thick, short, and very slow growing. However, when such a plantlet was transferred to a basal medium + NAA (5.4 µM), the roots started to elongate. The hybrid plants were triploids with 2n = 30 (Fig. 8), i.e., an intermediate chromosome number between the two parents. The plants were erect, and resembled *A. hypogaea* with respect to the size of their leaves (Fig. 7H). The leaves were dark green and somewhat smaller than those of *A. hypogaea*. The plants obtained by Krapovickas and Rigoni (1952) and Kumar et al. (1957) were also phenotypically similar to *A. hypogaea*.

These investigations on the culture of the embryos of the cross between *A. hypogaea* and *A. villosa*, and the regeneration of hybrid plants, demonstrate the possibility of incorporating alien germplasm into the cultivated peanut by wide hybridization. The extension of this work to other species would be desirable.

Induction of Mutations and Other Desirable Variability

The success of any crop improvement program depends not only on the availability of genetic variation, but also on the extent of genetic variability existing among cultivated varieties. With the continual depletion of naturally occurring reservoirs of germplasm, efforts are being made to supplement the basic gene pools by inducing variation through nonconventional methods. In this connection various in vitro techniques are being optimistically explored. Plant cells and protoplasts offer a unique tool for isolating induced as well as spontaneously occurring variations.

The regeneration of phenotypically different plants from old cell cultures is a common phenomenon. After prolonged culture, the cells undergo various types of nuclear and chromosomal changes, such as endomitosis, polysomy, and mutations (D'Amato, 1977). Such a situation can be exploited for the selection of desirable variants (Skirvin, 1978). Moreover, through the manipulation of various growth regulators and drugs, additional changes can be induced and useful variants can be recovered, as has been done in the case of sugarcane (Liu and Chen, 1976; Heinz et al., 1977), potato (Secor and Shepard, 1981), and maize (Gengenbach et al., 1977).

As most mutations are recessive and are not expressed in the diploid cells in the presence of an unmutated dominant gene, haploid cultures are of special significance in the induction and easy detection of mutants. The general scheme for the isolation of variant cells is to expose the cultures to the conditions which favor growth of only mutant cells. Cell populations are then subjected to various types of stress factors such as salts, heavy metals, toxins, herbicides, extremes of temperature, or amino acid analogs. Haploid cell cultures have been employed by various workers to study the effects of both irradiation and chemical mutagens. Single cells, protoplasts, and isolated pollen have the advantage over the entire plant that they can be plated and screened in large numbers, using techniques similar to those used for microorganisms. Work has recently been conducted on cell lines to obtain mutants in several plant species that are resistant to salts (Nabors et al., 1980), pathotoxins (Bajaj, 1981a), herbicides (Gressel et al., 1978), chilling (Dix and Street, 1976), viruses (Heinz et al., 1977), various drugs (Widholm, 1977), and nematodes (Wenzel and Uhrig, 1981). However, no work has been reported on peanut.

Another possibility for widening the base of gene pools is through protoplast fusion (Bajaj, 1977a), whereby somatic hybrids and cybrids can be obtained between sexually incompatible species. In this regard the fusion of protoplasts from wild species that are resistant to diseases with those of susceptible, cultivated peanut would be rewarding. Such studies have been conducted on tobacco (Evans et al., 1981) and potato (Butenko and Kuchko, 1979). Work on these lines has been initiated on peanut, and some interesting results have been obtained with the fusion of protoplasts of a wild species, *A. villosa* (resistant to leaf spot caused by *Cercospora arachidicola*) with those of highly susceptible *A. hypogaea* (Bajaj and Gosal, 1983). Fairly high quantities of protoplasts could be obtained from the young leaves and segments of in vitro-grown plantlets by overnight treatment with a mixture of Onozuka R10 cellulase and macerozyme, followed by fusion with polyethylene glycol.

Other characters, especially high protein content, could hopefully be incorporated through protoplast fusion or by the uptake and incorporation of desired genetic information.

Cryopreservation of Germplasm

Since the first successful attempt to preserve isolated plant cells in liquid nitrogen (Quatrano, 1968), the technology of cryopreservation (Bajaj, 1979a,b; Bajaj and Reinert, 1977) has been refined to the point where entire plants can be obtained from cell suspensions (Bajaj, 1976a), pollen embryos (Bajaj, 1976b, 1977b), and the excised meristems that have been freeze-preserved in liquid nitrogen for various lengths of time (Seibert, 1976). The freeze preservation of plant cell and tissue cultures has the potential to become a practical method for the conservation and the international exchange of germplasm.

FREEZE PRESERVATION OF MERISTEMS. Although callus and cell cultures, embryos, and other tissues are able to withstand superlow temperatures and retain their regeneration potential, excised meristems have the following advantages over cell cultures in the long-term conservation of germplasm (Bajaj, 1979c): (a) in some cases, it is difficult to regenerate entire plants from the callus, whereas meristematic tips are relatively easy to regenerate; (b) meristems are genetically more stable than callus cultures--when subcultured over an extended period the latter show chromosomal aberration and changes in nuclear and ploidy levels; (c) it is a quicker method for vegetative propagation; (d) excised meristems may yield pathogen-free plants; (e) meristems taken especially from haploid plants would ensure the maintenance of the haploid level; (f) the meristematic cells are small, densely cytoplasmic, and thin-walled—thus a higher survival would be expected; and (g) whereas cell cultures require a slow and controlled rate of cooling, meristems can withstand sudden freezing—thus elaborate and expensive cryostats are not needed for the preservation of meristems.

Excised meristems of *A. hypogaea* subjected to quick freezing in liquid nitrogen showed up to 23-31% survival (Bajaj, 1979c). Later, by manipulation of various factors affecting the freezing method, survival was raised to 42% and normal seeds were obtained from plants regenerated from meristems frozen for up to 20 months (Fig. 9; Bajaj, 1982b, 1983b).

The ability to freeze meristems, to revive them, and to regenerate plants may depend on a number of factors, some of which are described in the following sections.

Preculture. Excised meristems precultured for 4 days on MS + IAA (11.0 μM) + BA (0.88 μM) containing 3% DMSO showed enhanced survival. However, DMSO at 10% proved inhibitory. Thus for all further studies meristems were precultured on 3% DMSO and those showing signs of growth were used for subsequent freezing experiments.

Cryoprotectants. Results on the effect of various cryoprotectants are summarized in Table 8. A combination of 5% each of sucrose, glycerol, and DMSO resulted in the highest survival of 42%; this is a considerable improvement over sucrose (5%) + glycerol (5%), which gave 29% survival.

Figure 9. Various stages in the regeneration of plants of *A. hypogaea* from frozen (-196 C) meristems. (A-C) Frozen-thawed meristems after 6, 9, and 11 weeks of culture, respectively; note the initial general callusing (perhaps due to cryoinjury) of the meristem in A. (D,E) Normal growth of the retrieved meristem (preserved for 20 months), and the regeneration of a complete plant 4 and 12 weeks after culture on MS + IAA + BA, respectively. (F,G) Test-tube plants 3 and 7 weeks after transferring to soil, respectively. (H) Normal seeds obtained from such plants (total time 24 weeks from the culture of excised meristems to seed setting) (Bajaj, 1983b).

Table 8. Effect of Various Cryoprotectants on the Survival of Meristems of *A. hypogaea*[a]

CRYOPROTECTANT[b]	MERISTEMS FROZEN[c]	MERISTEMS SHOWING GROWTH	PERCENT SURVIVAL[d]
Sucrose + glycerol	31	9	29
Sucrose + glycerol + DMSO	33	14	42

[a]Bajaj, unpublished.

[b]All ingredients added at 5% level.

[c]Freeze-preserved in liquid nitrogen for 7 months.

[d]Survival was judged by the capacity of the meristems to green, increase in size, and form callus or shoots.

Thawing. Among frozen cultures thawed at different temperatures, best results were obtained with cultures thawed at 35 C, followed by 40 C and 25 C.

The success of this procedure emphasizes the beneficial influence of preculture and the combination of cryoprotectants on the ability to freeze-preserve meristems. The regeneration of complete plants from such cultures and the formation of normal seeds demonstrate the potential of cryopreservation of meristems for the conservation of plant species.

FREEZE-PRESERVATION OF POLLEN EMBRYOS. Pollen embryos and segments of androgenic anthers of *A. hypogaea* and *A. villosa* (Table 9) freeze-preserved for one year have been revived and entire plants regenerated (Bajaj, 1982a). Anthers cultured for 4-5 weeks on MS + IAA (22.0 μM) + KIN (9.3 μM) were cut into 2-4 segments each, and used for freezing studies (Bajaj et al., 1980a). Segments from 50 anthers were either wrapped in aluminum foil (Withers, 1979), or were put in 1 ml of the cryoprotectant solution with 5% each of sucrose, glycerol, and dimethyl sulfoxide. A suspension of pollen embryos was prepared from some segments. The material was frozen, and the survival was judged by (a) increase in size of the embryos, (b) callus formation, (c) appearance of green nodules, and (d) formation of roots and shoots.

Two species were studied in detail. The ability to freeze-preserve the two species differed (see Table 9), *A. hypogaea* being more sensitive (14% survival) than *A. villosa* (31%). The retrieved anthers underwent proliferation and the callus occasionally differentiated shoots. The anther segments wrapped in the aluminum foil (dry method) yielded

Table 9. Survival of Frozen Pollen Embryos and Segments of the Androgenic Anthers of *Arachis*[a]

	A. hypogaea	*A. villosa*
Pollen embryos frozen[b]	73	121
Pollen embryos survived	21	46
Percent survival	29	38
Anther segments frozen[b]	107	120
Anther segments resumed growth	15	37
Percent survival	14	31

[a]Bajaj, 1982a.

[b]Frozen and stored at -196 C for 2 months.

better results than the anthers frozen in liquid. In the latter case anthers had a tendency to become spongy.

Thus the survival of pollen embryos, and the regeneration of plants from cultures freeze-preserved for one year, suggest the application of cryogenic methods for the conservation of haploid cultures.

GERMPLASM BANKS. Current technological progress and practices have threatened the extinction of rare plant species. This diminishing of genetic resources has caused international concern for the conservation of rare and important plant species, and efforts have been made to develop methods for the long-term conservation of germplasm. Traditionally seeds have been stored for this purpose. However, in some cases seeds are "recalcitrant;" they undergo deterioration in a couple of years, and thus the germplasm cannot be preserved. In such plants the genetic stocks could be maintained through cryostorage of seeds, cells, tissues, and organs (Fig. 10).

For the long-term conservation of germplasm and the establishment of gene banks, the use of meristems is a much better system. They are genetically more stable than callus cultures, which undergo genetic erosion and are then unfit for the maintenance of clonal stocks. The work done so far on the revival of meristems of diverse genera has established the feasibility of employing cryostorage for the maintenance of germplasm (Seibert, 1976; Bajaj, 1981b; Sakai et al., 1978). However, before cryopreservation is recommended as a general tool for the long-term conservation of germplasm, two points require critical study. First, the viability of the retrieved cultures should be high, so that there is no selection pressure. This can be achieved by optimizing various factors, such as the nature, age, and physiological condition of the culture, and the effects of cryoprotectants, freezing, storage, and thawing. By manipulation of such factors in strawberry culture, 80% meristem survival has been obtained (Sakai et al., 1978), and this

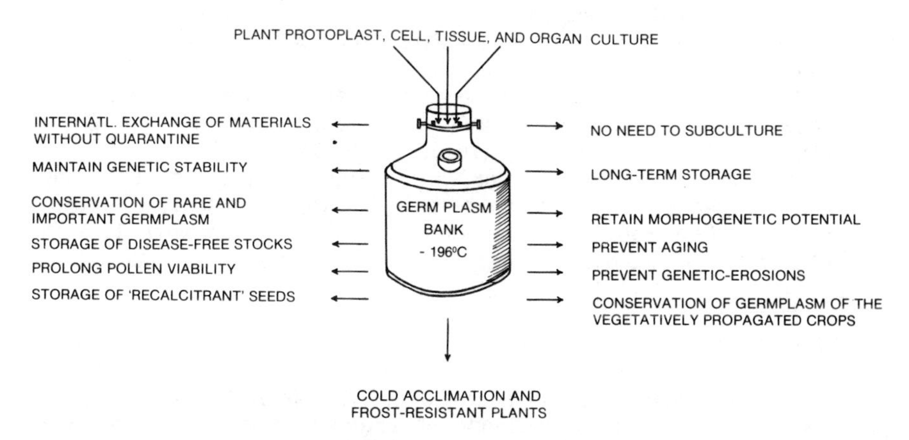

Figure 10. Prospects of the freeze-preservation of protoplasts, cells, tissues, and organs and the establishment of "germplasm banks" (modified after Bajaj, 1979a).

points to the possibility of achieving high viability in other plants. Second, it is necessary to ensure that during long-term cryopreservation there is no genetic deterioration in the germplasm. Although to date no detailed work has been undertaken to study various cytological aspects of the retrieved cultures, it is satisfying to note that in preliminary studies on plants regenerated from frozen meristems of potato (Bajaj, 1981b), and *Arachis* and *Cicer* (Bajaj, 1983b) (cryostored for 24 and 20 months, respectively), no changes in chromosome numbers were observed.

PROTOCOLS

Regeneration of Plants from Anther-Derived Callus of *A. villosa*

1. Culture anthers at first pollen mitosis on MS medium with IAA (22.0 µM), KIN (9.3 µM), or BA (8.8 µM). Incubate at 23–27 C in diffuse light. White compact callus growth emerges within 5 weeks.
2. Subculture callus to MS medium with NAA (5.4 µM) and BA (8.8 µM).
3. Shoot differentiation occurs within 1 week.
4. Transfer regenerated shoots to basal medium with NAA (2.7–5.4 µM) for further elongation and root development.

Regeneration of Plants from Seedling Explants and Callus Culture of *A. hypogaea*

1. Surface sterilize seed in chlorine for 10 min. Rinse 3 times in sterile distilled water.
2. Excise embryo from seed and culture on hormoneless MS medium for 3-5 weeks.
3. Cut juvenile seedlings into segments and culture mesocotyl, epicotyl, or leaf onto MS medium with IAA (11.0 μM) and KIN (2.3 μM) for plant regeneration. Incubate at 25 ± 2 C.
4. Callus formation will occur on seedling segments and excised embryos placed on MS medium with CH (500 mg/l), 2,4-D (4.5 μM), and KIN (9.3 μM). Transfer callus to MS medium with IAA (22.0 μM) and KIN (9.3 μM) for occasional shoot and root formation.

FUTURE PROSPECTS

The yield of peanut has been static during the last few years. It is realized that the routine methods of breeding and crop improvement may not be sufficient to produce significant yield increases. Thus an urgent need exists to generate genetic variability by adopting innovative approaches like in vitro techniques. The progress made in the area of protoplast, pollen, cell, and tissue cultures, and the production of haploids, somatic hybrids, cybrids, and disease- and salt-resistant mutants in various crops emphasizes the need to extend these methods to peanut improvement programs.

In peanut, the use of simple techniques like the culture of hybrid embryo and ovule should be fully exploited to incorporate the germplasm from incompatible wild species into *A. hypogaea.* In addition, studies on the fusion of protoplasts between disease-resistant wild species and *A. hypogaea* would help to induce genetic diversity.

The production of haploids and homozygous plants through anther culture would enable early release of varieties. At present, many diploid wild species are brought to the level of tetraploidy before crossing with the cultivated peanut. Perhaps a better approach would be to evolve diploid *A. hypogaea* plants (2n = 20) and then cross them with the wild diploids; thus tetraploidy in *A. hypogaea* will be restored. This is within the purview of excised anthers, or by selecting cell lines with 20 chromosomes and regenerating plants from them. The incorporation of in vitro methods into research programs thus would be highly rewarding for the improvement of peanuts.

KEY REFERENCES

Bajaj, Y.P.S., Ram, A.K., Labana, K.S., and Singh, H. 1981b. Regeneration of genetically variable plants from anther-derived callus of *Arachis hypogaea* and *A. villosa.* Plant Sci. Lett. 23:35-39.

Bajaj, Y.P.S., Kumar, P., Singh, M.M., and Labana, K.S. 1982. Interspecific hybridization in the genus *Arachis* through embryo culture. Euphytica 31:365-370.

Smartt, J. and Gregory, W.C. 1967. Interspecific cross compatibility between the cultivated peanut, *Arachis hypogaea* L. and other members of the genus *Arachis*. Oleagineux 22:455-459.

Spielman, I.V., Burge, A.P., and Moss, J.P. 1979. Chromosome loss and meiotic behaviour in interspecific hybrids in the genus *Arachis* L. and their implications in breeding for disease resistance. Z. Pflanzenzuecht. 83:236-250.

Wilson, C.T. 1973. Peanut: Culture and Uses. American Peanut Research Education Association. Stillwater, Oklahoma.

REFERENCES

Bajaj, Y.P.S. 1976a. Regeneration of plants from cell suspensions frozen at -20, -70 and -196 C. Physiol. Plant. 37:263-268.

______ 1976b. Gene preservation through freeze-storage of plant cell, tissue and organ culture. Acta Hortic. 63:75-84.

______ 1977a. Protoplast isolation, culture and somatic hybridization. In: Applied and Fundamental Aspects of Plant Cell, Tissue, and Organ Culture (J. Reinert and Y.P.S. Bajaj, eds.) pp. 467-496. Springer-Verlag, Berlin, Heidelberg, New York.

______ 1977b. Survival of *Atropa* and *Nicotiana* pollen-embryos at -196 C. Curr. Sci. 46:75-84.

______ 1979a. Technology and prospects of cryopreservation of germplasm. Euphytica 28:267-285.

______ 1979b. Establishment of germplasm banks through freeze-storage of plant tissue culture and their implication in agriculture. In: Plant Cell and Tissue Culture: Principles and Applications (W.R. Sharp, P.O. Larsen, E.F. Paddock, and V. Raghavan, eds.) pp. 745-774. Ohio State Univ. Press, Columbus.

______ 1979c. Freeze preservation of meristems of *Arachis hypogaea* and *Cicer arietinum*. Indian J. Exp. Biol. 17:1405-1407.

______ 1981a. Production of disease-resistant plants through cell culture: A novel approach. J. Nucl. Agric. Biol. 10:1-5.

______ 1981b. Regeneration of plants from potato meristems freeze-preserved for 24 months. Euphytica 30:141-145.

______ 1982a. Regeneration of plants from pollen-embryos of *Arachis*, *Brassica* and *Triticum* spp. cryopreserved for one year. Curr. Sci. 52:484-486.

______ 1982b. Cryopreservation of germplasm of potato, cassava, peanut and chickpea. In: Proceedings International Congress Plant Tissue and Cell Culture (A. Fujiwara, ed.) pp. 799-800. Jap. Assn. Plant Tissue Culture, Tokyo.

______ 1983a. In vitro production of haploids. In: Handbook of Plant Cell Culture, Volume 1, Techniques and Applications (D. Evans, W.R.

Sharp, P. Ammirato, and Y. Yamada, eds.) pp. 228-287. Macmillan Pub., New York.

______ 1983b. Production of normal seeds from plants regenerated from the meristems of *Arachis hypogaea* and *Cicer arietinum* cryopreserved for 20 months. Euphytica 32:425-430.

______ and Gosal, S.S. 1983. Somatic hybridization and embryo culture studies on *Arachis hypogaea* x *Arachis villosa*. Am. J. Bot. 70:83-84.

______ and Reinert, J. 1977. Cryobiology of plant cell cultures and establishment of gene-banks. In: Applied and Fundamental Aspects of Plant Cell, Tissue, and Organ Culture (J. Reinert and Y.P.S. Bajaj, eds.) pp. 757-777. Springer-Verlag, Berlin, Heidelberg, New York.

______, Labana, K.S., and Dhanju, M.S. 1980a. Induction of pollen-embryos and pollen-callus in anther cultures of *Arachis hypogaea* and *A. glabrata*. Protoplasma 103:397-399.

______, Singh, H., and Gosal, S.S. 1980b. Haploid embryogenesis in anther cultures of pigeon-pea (*Cajanus cajan*). Theor. Appl. Genet. 58:157-159.

______, Kumar, P., Labana, K.S., and Singh, M.M. 1981a. Regeneration of plants from seedling-explants and callus cultures of *Arachis hypogaea* L. Indian J. Exp. Biol. 19:1026-1029.

Ball, E. and Joshi, P.C. 1965. Divisions in isolated cells of palisade parenchyma of *Arachis hypogaea*. Nature 207:213-214.

Butenko, R.G. and Kuchko, A.A. 1979. Somatic hybridization of *Solanum tuberosum* L. and *S. chacoense* by protoplast fusion. In: Advances Protoplast Research, pp. 293-299. Akad. Kiado, Budapest.

Cherry, J.P. 1977. Potential sources of peanut seed proteins and oil in the genus *Arachis*. J. Agric. Food Chem. 25:186-193.

D'Amato, F. 1977. Cytogenetics of differentiation in tissue and cell cultures. In: Applied and Fundamental Aspects of Plant Cell, Tissue, and Organ Culture (J. Reinert and Y.P.S. Bajaj, eds.) pp. 343-357. Springer-Verlag, Berlin, Heidelberg, New York.

Dix, P.J. and Street, H.E. 1976. Selection of plant cell lines with enhanced chilling resistance. Ann. Bot. 40:903-910.

Evans, D.A., Flick, C.E., and Jensen, R.A. 1981. Disease resistance: Incorporation into sexually incompatible somatic hybrids of the genus *Nicotiana*. Science 213:907-909.

Evans, D.A., Sharp, W.R., Ammirato, P.V., and Yamada, Y. (eds.) 1983. Handbook of Plant Cell Culture, Volume 1, Techniques and Applications. Macmillan Pub., New York.

FAO, 1980. Production Year Book, Rome.

Gengenbach, B.G., Green, C.E., and Donovan, C.M. 1977. Inheritance of selected pathotoxin resistance in maize plants regenerated from cell cultures. Proc. Nat. Acad. Sci. 74:5113-5117.

Gregory, M.P., Gregory, W.C., and Smartt, J. 1965. The interspecific hybridization problems in *Arachis hypogaea*. Am. J. Bot. 52:625.

Gregory, W.C. 1968. A radiation breeding experiment with peanuts. Rad. Bot. 8:81-147.

Gressel, J., Zilkah, S., and Ezra, G. 1978. Herbicide action, resistance and screening cultures vs. plants. In: Frontiers of Plant Tissue Culture (T.A. Thorpe, ed.) pp. 427-436. Univ. Calgary Press, Calgary.

Guy, A.L., Heinis, J.L., and Pancholy, S.K. 1978. Induction and biochemical parameters of callus growth from three peanut cultivars. Peanut Sci. 5:78-82.

Guy, A.L., Bleen, T.C., and Pancholy, S.K. 1980. Partial morphogenesis in peanut (*Arachis hypogaea* L.) callus cultures. Proc. Am. Peanut Res. Educ. Soc. 12:23-29.

Heinz, D.J., Krishnamurthy, M., Nickell, L.G., and Maretzki, A. 1977. Cell, tissue and organ culture in sugarcane improvement. In: Applied and Fundamental Aspects of Plant Cell, Tissue, and Organ Culture (J. Reinert and Y.P.S. Bajaj, eds.) pp. 3-17. Springer-Verlag, Berlin, Heidelberg, New York.

Hildebrandt, A.C. and Schenk, R.U. 1970. Hybrid plants by vegetative reproduction. World Crops 22:63.

Joshi, P.C. and Ball, E. 1968. Growth of isolated palisade cells of *Arachis hypogaea* L. on simple defined medium in vitro. Science 158: 1575-1578.

Joshi, P.C. and Noggle, G.R. 1967. Growth of isolated mesophyll cells of *Arachis hypogaea* in simple defined medium in vitro. Science 158: 1575-1577.

Krapovickas, A. and Rigoni, V.A. 1952. Estudies citologicos en al Genero *Arachis*. Rev. Invest. Agric. (Buenos Aires) 5:289-294.

Kumar, A. 1974a. Effect of some iron salts and magnesium sulphate on the growth and chlorophyll development of tissue cultures. Indian J. Exp. Biol. 12:595-596.

______ 1974b. Vitamin requirements of callus tissue of *Arachis hypogaea* L. Indian J. Exp. Biol. 12:465-466.

______ 1974c. In vitro growth and chlorophyll formation in mesophyll callus tissues on sugar-free medium. Phytomorphology 24:96-101.

Kumar, L.S.S., D'Cruz, R., and Oke, J.G. 1957. A synthetic allohexaploid in *Arachis*. Curr. Sci. 26:121-122.

Liu, M.C. and Chen, W.H. 1976. Tissue and cell culture as aids to sugarcane breeding. I. Creation of genetic variation through callus culture. Euphytica 25:393-403.

Martin, J.P. 1970. Culture in vitro d'ovules Arachide. Oleagineux 25: 155-156.

______ and Rabechault, H. 1976. The culture in vitro of groundnut stamens (*Arachis hypogaea* L.). II. Establishment of tissue culture and organogenesis. Oleagineux 31:19-25.

Moss, J.P. 1980. Wild species in the improvement of groundnuts. In: Advances in Legume Science, Vol. I (Summerfield and Bunting, eds.) pp. 525-535. International Legume Conference, Kew, England.

Mroginski, L.A. and Fernandez, A. 1980. Obtainment of plantlets by in vitro culture of anthers of wild species of *Arachis*. (Leguminosae). Oleagineux 35:89-92.

Mroginski, L.A., Kartha, K.K., and Shyluk, J.P. 1981. Regeneration of peanut (*Arachis hypogaea*) plantlets by in vitro culture of immature leaves. Can. J. Bot. 59:826-830.

Nabors, M.W., Gibbs, S.E., Bernstein, C.S., and Meis, M.E. 1980 NaCl-tolerant tobacco plants from cultured cells. Z. Pflanzenphysiol. 97: 13-17.

Quatrano, R.S. 1968. Freeze-preservation of cultured flax cells utilizing dimethyl sulfoxide. Plant Physiol. 43:2057-2061.

Raghavan, V. 1977. Applied aspects of embryo culture. In: Applied and Fundamental Aspects of Plant Cell, Tissue, and Organ Culture (J. Reinert and Y.P.S. Bajaj, eds.) pp. 375-397. Springer-Verlag, Berlin, Heidelberg, New York.

Raman, V.S. 1976. Cytogenetics and Breeding in *Arachis*. Today and Tomorrow's Printers and Publishers, New Delhi.

Ranga Rao, V. and Subba Rao, N.S. 1976. Studies on the infection of legume root callus with *Rhizobium*. Z. Pflanzenphysiol. 80:14-20.

Rangaswamy, N.S., Rangan, T.S., and Rao, P.S. 1965. Morphogenesis of peanut (*Arachis hypogaea* L.) pericarp in tissue cultures. Naturwissenschaften 52:624-625.

Reinert, J. and Bajaj, Y.P.S. (eds.) 1977. Applied and Fundamental Aspects of Plant Cell, Tissue, and Organ Culture. Springer-Verlag, Berlin, Heidelberg, New York.

Sakai, A., Yamakawa, M., Sakata, D., Harada, T, and Yakuwa, T. 1978. Development of a whole plant from an excised strawberry runner apex frozen to -196 C. Low Temp. Sci. Ser. B 36:31-38.

Sastri, D.C., Nalini, M.S., and Moss, J.P. 1980. In vitro culture of *Arachis* ovaries and ovules. In: Plant Tissue Culture, Genetic Manipulation and Somatic Hybridization of Plant Cells (P.S. Rao, M.R. Heble, and M.S. Chadha, eds.) pp. 366-373. Proceedings National Symposium BARC, Bombay.

Secor, G.A. and Shepard, J.F. 1981. Variability of protoplast-derived potato clones. Crop Sci. 21:102-105.

Seibert, M. 1976. Shoot initiation from carnation shoot apices frozen to -196 C. Science 191:1178-1179.

Singh, A.K., Sastri, D.G., and Moss, J.P. 1980. Utilization of wild *Arachis* species at ICRISAT. In: Proceedings International Workshop ICRISAT, pp. 82-90. Hyderabad, India.

Skirvin, R.M. 1978. Natural and induced variation in tissue culture. Euphytica 27:241-266.

Smartt, J. 1979. Interspecific hybridization in the grain legumes: A review. Econ. Bot. 33:329-337.

Srinivasmurthy, T. and Gopala Iyengar, K. 1950. Natural hybrids in *A. nambyquarae*. Curr. Sci. 19:62.

Steward, F.C., Marion, O.M., and Smith, J. 1958. Growth and organized development of cultured cells. I. Growth and division of freely suspended cells. Am. J. Bot. 45:693-903.

Stokes, W.E. and Hull, F.H. 1930. Peanut breeding. J. Am. Soc. Agron. 22:1004-1019.

Subrahmanyam, P., Gibbons, R.W., Nigam, S.N., and Rao, V.R. 1978. Screening methods and further sources of resistance to peanut rust. Proc. Am. Peanut Res. Educ. Assoc. (Abst.) 10:64.

Van Huystee, R.B. 1977. Porphyrin and peroxidase synthesis in cultured peanut cells. Can. J. Bot. 55:1340-1344.

______ and Turcon, G. 1973. Rapid release of peroxidase by peanut cells in suspension culture. Can. J. Bot. 51:1169-1175.

Verma, D.P.S. and Van Huystee, R.B. 1970. Relationship between peroxidase, catalase, and protein synthesis during cellular development in cell cultures of peanut. Can. J. Biochem. 48:444-449.

_____ 1971a. Derivation, characteristics and large scale culture of cell line from *Arachis hypogaea* L. cotyledons. Exp. Cell Res. 69: 402-408.

_____ 1971b. Induction of giant cells in suspension cultures of *Arachis hypogaea* L. by massive irradiation. Rad. Res. 48:518-530.

_____ 1971c. Abberrant recovery of protein synthesis after massive irradiation of *Arachis hypogaea* L. cells in vitro. Rad. Res. 48:531-541.

Wenzel, G. and Uhrig, H. 1981. Breeding for nematode and virus resistance in potato via anther culture. Theor. Appl. Genet. 59:333-340.

Widholm, J.M. 1977. Selection and characterization of biochemical mutants. In: Plant Tissue Culture and its Biotechnological Applications (W. Barz, E. Reinhard, and M.H. Zenk, eds.) pp. 112-122. Springer-Verlag, Berlin, Heidelberg, New York.

Withers, L.A. 1979. Freeze preservation of somatic embryos and clonal plantlets of carrot (*Daucus carota*). Plant Physiol. 63:460-467.

Ziv, M. and Zamski, E. 1975. Geotropic responses and pod development in gynophore explants of peanut (*Arachis hypogaea* L.) cultured in vitro. Ann. Bot. 39:579-585.

SECTION IV
Vegetables

CHAPTER 9
Cole Crops

S. Y. Zee and *B. B. Johnson*

Traditionally, cole crops refer to plants within the genus *Brassica*. They include cabbage, mustard, broccoli, cauliflower, brussels sprouts, turnips, and rape. The cole crops are cultivated widely and are adaptable to most climates. Optimum performance occurs for these crops when grown under moderately cool weather conditions.

Breeding objectives vary from species to species. Tissue culture technology coupled with conventional breeding will make some of these objectives easier to attain. For example, a large percentage of the cole crops now planted are hybrid (F_1) varieties, produced by combining two self-incompatible parent lines via manual pollination. This approach requires extensive hand labor and furthermore, the inbred parent lines are difficult to maintain. Tissue culture has proven useful in the maintenance of the self-incompatible parent lines (Anderson and Carstens, 1977; Clare and Collin, 1973, 1974; Dunwell and Davies, 1975; Kartha et al., 1974a; Kuo and Tsay, 1977; Miszke and Skucinska, 1976; Pow, 1969; Walkey and Woolfitt, 1970; Watts and George, 1963).

Rape (*B. napus*) poses the problem of having undesirable metabolites (euric acid and glucosinolates as well as steroids) that limit its use as either an animal food or a source of vegetable oil (Thomas and Wenzel, 1975). Tissue culture techniques may contribute to solving these problems in a number of ways. Among these are the production of haploids with desirable characteristics (Thomas and Wenzel, 1975), the production of plants from cells that have undergone interspecific or intergeneric fusion, or transformed cells (Kartha et al., 1974c). Another possibility is to form an artificial hybrid between fodder kale (*B. oleracea*) and fodder radish (*Raphanus sativus*). Such an artificial

Raphanobrassica might have resistance to Plasmodiphora disease (Gatenby and Cocking, 1977).

Some of the cole crops are adversely affected by high or low temperature extremes; e.g., the breeding of autumn cauliflower in Britain is difficult because plants selected to be seed parent plants generally succumb to cold in the field or after transplanting to the greenhouse (Watts and George, 1963). Tissue culture techniques can overcome this limitation.

In contrast, in the Far East, Chinese cabbage has seasonal production limitations because of high temperatures. A breeding goal is to develop cultivars that are heat tolerant, disease resistant, and high yielding. Seed parent plants need to be selected from plants produced during the hot, humid season. Most varieties of cabbage die after ripening their seed. It would be advantageous if selected plants could be propagated vegetatively and maintained from year to year (North, 1953).

Embryo culture has been used to save plants from crosses between *B. chinensis* and *B. pekinensis* (Inomata, 1975), *B. oleracea* and *B. campestris* (Feng and Chen, 1981; Harberd, 1969; Inomata, 1968, 1977a,b, 1978a,b; Matsuzawa, 1978; Nishi et al., 1959), and *B. oleracea* and *B. pekinensis* (Inomata, 1975) that would otherwise have failed. Ovules have also been fertilized and cultured in vitro (Kameya and Hinata, 1970a).

Protoplasts are of interest because of their usefulness in genetic engineering and somatic hybridization. Callus has been produced from protoplasts of marrow stem kale (Gatenby and Cocking, 1977). Fused protoplasts of rape and soybean undergo mitotic division (Kartha, 1974b). Plants have been regenerated from rape protoplasts (Kartha et al., 1974c).

Somatic embryogenesis, which can be used to propagate plants or to obtain virus-free plants, has been reported for cauliflower (Pareek and Chandra, 1978).

Haploids for use in producing homozygous lines have been obtained by anther or pollen culture for rape (Hansson, 1978; Keller and Armstrong, 1977, 1978; Thomas and Wenzel, 1975; Wenzel et al., 1977), wild yellow mustard (Keller et al., 1975), and Chinese cabbage (Deng and Guo, 1977). In addition, haploids have been obtained by propagating spontaneously occurring haploid plants (Stringham, 1977; Thomas et al., 1976).

Tissue culture methods for plant regeneration in cole crops are summarized in Table 1, while the various culture techniques that have been reported are summarized in Table 2.

PROTOCOLS FOR TISSUE CULTURE OF COLE CROPS

Clonal Regeneration: Cauliflower

A. Shake culture (after Walkey and Woolfitt, 1970).
   1. Divide curds into 40 mm portions.
   2. Surface sterilize for 15 min in 15% sodium hypochlorite solution.
   3. Cut the curds into subportions (3-7 mm diameter).

Table 1. Tissue Culture Methods for the Plant Regeneration of Cole Crops (*Brassica*)

SPECIES	MEDIUM	GROWTH REGULATORS (μM)	EXPLANT	REFERENCE
B. oleracea				
Cauliflower	Linsmaier & Skoog	0.12 KIN + 45.6 IAA	Curd	Pow, 1969
	Margara	5 IBA + 10 BA + 5 GA	Curd	Margara, 1969
	Linsmaier & Skoog	45.6 IAA + 12 KIN	Floret	Walkey & Woolfitt, 1970
	Linsmaier & Skoog	0.9 2,4-D + 14 KIN (callus); 23.2 KIN (plant regeneration)	Leaf vein	Walkey et al., 1974; Crisp & Walkey, 1974; Baroncelli et al., 1973a,b
	MS	5.7 IAA + 2.3 KIN (callus); 0.05-0.5 IAA + 2.3 KIN (plant regeneration)	Leaf	Pareek & Chandra, 1978
	Linsmaier & Skoog	0.9 2,4-D + 13.9 KIN (callus); 0.45 2,4-D + 13.9 KIN (callus)	Leaf vein	Buiatti et al., 1974a,b
	MS	0.44 BA + 5.4 NAA	Curd	Li & Qui, 1981
Broccoli	MS	5.7 IAA + 19.6 2iP (shoot); 1.14 IAA (roots)	Flower buds	Anderson & Carstens, 1977
	MS	5.7-11.4 IAA + 4.7 KIN (callus); 46-51 IAA + 14-28 KIN (shoot)	Leaf	Johnson & Mitchell, 1978a,b
	MS	51-57 IAA + 37-42 KIN (shoot)	Leaf rib	Johnson & Mitchell, 1978a,b
	MS	51-57 IAA + 14-93 KIN	Stem	Hui & Zee, 1980
	MS	2.6 NAA	Cotyledon	Hui & Zee, 1980
	MS	9.8 2iP (plus raw ginseng powder, 1000 mg/l)	Hypocotyl	Hui & Zee, 1980

Table 1. Cont.

SPECIES	MEDIUM	GROWTH REGULATORS (µM)	EXPLANT	REFERENCE
Brussels sprouts	MS	11.4 IAA + 2.3 KIN	Sprout	Clare & Collin, 1974
	MS	4.9 IBA	Petiole	Clare & Collin, 1974
Red cabbage	MS	4.5 2,4-D + 0.5 KIN (callus); 11.4 IAA + 9.3 KIN or 5.7 IAA + 2.3 KIN (shoot)	Hypocotyl	Bajaj & Nietsch, 1975
White cabbage	MS	12 KIN (callus); 60 KIN (shoot)	Meristem tip	Walkey et al., 1980
Leaf-mustard cabbage	MS	2.7 NAA	Hypocotyl	Hui & Zee, 1978
	MS	8.9 BA	Cotyledon	Hui & Zee, 1978
Chinese cabbage	MS	0.44 BA + 5.37 NAA	Axillary bud	Kuo & Tsay, 1977
	Nitsch, WH, Heller	—	Ovary	Inomata, 1976
B. pekinensis x *B. oleracea*	WH	—	Embryo	Nishi et al., 1959; Inomata, 1977c
(Chinese cabbage x common cabbage)	WH, B5	—	Embryo	Feng & Chen, 1981
B. pekinensis x *B. Chinensis*	Nietsch	5.7 IAA + 0.46 KIN + 2.8 GA (10% coconut milk)	Ovary	Inomata, 1968

Table 2. A Summary of Culture Techniques Reported for Cole Crops (*Brassica* spp.)

SPECIES	TECHNIQUE	REFERENCE
B. alboglabra (Chinese kale)	Haploid plants from pollen grains	Kameya & Hinata, 1970
	Plants from hypocotyl and cotyledon explants	Zee & Hui, 1977
	Morphogenesis of hypocotyl explants	Zee et al., 1978
B. campestris (wild yellow mustard)	Relationship between explant hormone content and growth in vitro	Elmsheuser et al., 1978
	Embryo culture: *B. campestris* x *B. oleracea*	Harberd, 1969
	Culture of excised ovaries	Inomata, 1976
	Ovary culture of *B. campestris* x *B. oleracea* crosses	Inomata, 1977
	Ovary culture: 2X x 4X hybrids	Inomata, 1968
	B. campestris x *B. oleracea*: embryo culture	Inomata, 1978a,b
	In vitro fertilization of excised ovules	Kameya & Hinata, 1970a,b,c
	Embryogenesis, plants from anthers	Keller & Armstrong, 1977
	Embryoids and plants from anthers	Keller et al., 1977
	Plants from pollen, embryoids	Keller et al., 1975
	B. campestris x *B. oleracea*: ovary culture	Matsuzawa, 1978
B. chinensis (Chinese cabbage)	*B. chinensis* x *B. pekinensis*: ovary culture	Inomata, 1975
	Aggregation of protoplasts by gelatin	Kameya, 1973
	Haploid plants from pollen grains	Kamey & Hinata, 1970a,b
	Propagation by axillary bud culture	Kuo & Tsay, 1977
	Anther culture	Zhong et al., 1978
B. juncea (Indian mustard)	Plants from hypocotyl and cotyledon explants	Hui & Zee, 1978
	Growth and differentiation in vitro	Pareek & Chandra, 1972
B. napo-brassica (Swede)	Protoplasts regenerate cell walls, no division	Quazi, 1975

Table 2. Cont.

SPECIES	TECHNIQUE	REFERENCE
B. napus (rape)	Nitrogen fixation with *Rhizobium*	Child, 1975
	Inducing effect of cells on nitrogenase activity of *Spirillum* and *Rhizobium*	Child & Kurz, 1978
	Effect of vitamins and hormones on root and shoot formation	Elmsheuser et al., 1978
	Correlation between endogenous and in vitro levels of IAA, GA_3, and ABA	Elmsheuser et al., 1978
	Increased embryoid formation from anthers with temperature shock	Hansson, 1978
	Formation of plants from stem explants	Kartha et al., 1974a
	Protoplast fusion	Kartha et al., 1974b
	Plants from mesophyll protoplasts	Kartha et al., 1974c
	Embryogenesis and plants from anther culture	Keller & Armstrong, 1977
	Microspore-derived plants; high frequencey	Keller & Armstrong, 1978
	Embryogenesis and plants from anther culture	Keller et al., 1977
	Organogenesis from flower stalks	Margara & Leydecker, 1978
	Lipids in tissue culture	Radwan et al., 1978
	Regeneration from stem explants of haploids	Stringham, 1977
	Protoplast regeneration and stem embryogenesis of haploid androgenic plants	Thomas et al., 1976
	Embryogenesis from microspores	Wenzel, 1975
	Protoplasts from leaves of haploids	Wenzel, 1973
	Ploidy level and phenotype of haploids	Wenzel et al., 1977
B. nigra (black mustard)	Tissue differentiation in vitro	Khanna et al., 1971

SPECIES	TECHNIQUE	REFERENCE
B. oleracea acephala (Kale)	Effect of vitamins and hormones on explants	Elmsheuser et al., 1978
	Pollen germination and self incompatability	Ferrari & Wallace, 1975
	Callus from leaf protoplasts; roots	Gatenby & Cocking, 1977
	Ploidy chimeras from stem explants	Horak, 1972
	Polyploid plants from tissue culture	Horak et al., 1971
	Polyploid and diploid plants from stem pith	Horak et al., 1975
	Polyploid plants from tissue culture	Landa & Lustinec, 1971
	Plants regenerated	Lustinec & Horak, 1970
	Dry weight and protein content of explants	Lustinec et al., 1972
B. oleracea botrytis (cauliflower)	Toxicity of An, Cu, Hg, and Pb in vitro	Barker, 1972
	Genetics of in vitro differentiation	Baroncelli et al., 1973a,b, 1974a
	Use of in vitro culture for early selection	Baroncelli et al., 1974b
	In vitro growth and differentiation genetics	Buiatti et al., 1974a,b
	Cloning by meristem culture	Crisp & Walkey, 1974
	Practical aspects of meristem culture	Grout & Crisp, 1977
	Transplanting plants from meristem culture	Grout & Aston, 1978a
	Modified leaf anatomy in regenerated plants	Grout & Aston, 1978b
	Wax on leaves of regenerated plants	Grout, 1975
	Effect of ginseng on regeneration	Hui & Zee, 1980
	Detection and use of incompatable lines	Landa, 1974
	In vitro production of plants	Li & Qiu, 1981
	Auxins and antiauxins; bud formation in vitro	Margara, 1977
	Bud neoformation	Margara, 1969
	In vitro propagation	Murashige et al., 1977
	Three methods of in vitro propagation	North, 1953
	Somatic embryogenesis	Pareek & Chandra, 1978
	Clonal propagation from curd	Pow, 1969

Table 2. Cont.

SPECIES	TECHNIQUE	REFERENCE
B. oleracea botrytis (cauliflower)	Curd fragments cultured in liquid medium	Rahn-Scoher, 1977
	Rooting of cortical buds	Trimboli et al., 1978
	Cloning via shake culture	Walkey & Woolfitt, 1970
	Production of virus free plants	Walkey et al., 1974
	Vegetative propagation	Wenzel et al., 1977
B. oleracea bullata (savoy)	Culture cold hardened tissue—check in vitro response to cold	Saint, 1977
B. oleracea capitata (cabbage)	Root, stem, cotyledon, and leaf explants produce plants	Bajaj & Nietsch, 1975
	Haploid plants from pollen grains	Braak, 1972
	Pollen germination; self incompatibility	Ferrari & Wallace, 1975
	Organogenesis from polyploids following somatic reduction	Maryakhina & Butenko, 1974a
	Somatic reduction in culture	Maryakhina & Butenko, 1974b
	Differentiation of roots, shoots	Mascarenhas et al., 1978
	Roots and shoots from leaf vein callus; cultivar variation important	Miszke & Skucinska, 1976
	Postgametic incompatibility	Neumann, 1973
	Diploid plants from anther culture	Nishi et al., 1974
	Roots and shoots from leaf tissue	Primo-Millo & Harada, 1975
	Amino acid metabolism in leaf callus	Shevyakova & Komizerko, 1969
B. oleracea caulorapa (kohlrabi)	No in vitro report	

SPECIES	TECHNIQUE	REFERENCE
B. oleracea gemmifera (Brussels sprouts)	Meristem culture	Clare & Collin, 1973
	Propagate parent lines from explants	Clare & Collin, 1974
	Propagate inbred parent lines	Dunwell & Davies, 1975
B. oleracea italica (broccoli)	Propagate inbred parent lines	Anderson & Carstens, 1977
	Cost of in vitro propagation	Anderson et al., 1977
	Pollen germination, self incompatibility	Ferrari & Wallace, 1975
	Effect of temperature on in vitro propagation	Johnson & Mitchell, 1978a
	Propagation from leaf, leaf rib, and stem explants	Johnson & Mitchell, 1978b
	Albino haploids and diploids from anthers	Nishi et al., 1974
	Tubular structures in endoplasmic reticulum of cultured broccoli	Quan et al., 1974
	Diploid plants from anthers	Quazi, 1978
	Callus and roots from anthers	Tseng & Lin, 1975
B. pekinensis (Chinese cabbage)	Plants from anther culture	Deng & Guo, 1977
	Artificial amphidiploid in cross with *B. oleracea*	Feng & Chen, 1981
	Haploids from pollen	Kameya & Hinata, 1970
	Axillary bud culture	Lee et al., 1973
	Embryo culture: *B. pekinensis* x cabbage hybrids	Shin, 1972
B. rapa (turnip)	Vitamins and hormones: effects on explants	Elsheuser et al., 1978a
	Protoplasts regenerate cell wall; no division	Quazi, 1975

4. Transfer the curds into liquid Linsmaier and Skoog medium with 12 μM KIN and 45.6 μM IAA (1 piece per 25 ml medium in 100 ml conical flask).
5. Incubate at 24-26 C, 16 hr photoperiod. Continuously shake the culture at 80 cycles/min.
6. Callus will form after a few days; roots and shoots will develop after 3-4 weeks. When the roots and shoots are well developed, the plantlets can be transferred to pots.

B. Solid culture (after Buiatti et al., 1974a,b).
   1. Sterilize the curds with 16% commercial hypochlorite for 25 min.
   2. Cut the curds into smaller pieces.
   3. Place the pieces on a Linsmaier and Skoog solid medium with 0.9 μM 2,4-D and 14 μM KIN, or 0.44 μM BA and 5.4 μM NAA.
   4. Keep at 25 ± 1 C under continuous light.
   5. Culture for 25 days and then transfer into the same medium without 2,4-D but with 23.2 μM KIN for shoot and root differentiation.

C. Somatic embryo formation (after Pareek and Chandra, 1978).
   1. Wash the leaves with 10% solution of a detergent. Cut into small pieces and then sterilize in a 5% solution of sodium hypocholite for 20 min.
   2. Cut the leaf lamina into pieces and then plant on MS agar medium, at 26 ± 2 C and in weak light.
   3. The medium is supplemented with 5.7 μM IAA and 2.3 μM KIN.
   4. After embryoid initiation transfer the callus masses into the same medium with a lower concentration of IAA (0.05-0.5 μM).

Clonal Regeneration: Broccoli

(After Hui and Zee, 1980; Johnson and Mitchell, 1978b.)

1. Cut the sterilized cotyledons, hypocotyls, and leaves into small 3.5 mm pieces.
2. Culture the explants on MS agar medium with 46-51 μM IAA and 14-28 μM KIN. Some cultivars undergo better differentiation with the addition of 500-1000 mg/l raw ginseng power in combination with 2.6 μM NAA + 2.2 μM BA or 2.7 μM NAA + 9.8 μM 2iP.
3. Place the tubes at 27 C under diffused light (about 2.5 Klux) with a 16 hr daylength.
4. Plantlets will form after culturing for about a month.

Clonal Regeneration: Brussels Sprouts

(After Clare and Collin, 1974.)

1. Swab the sprout with ethanol and then remove outer leaves on sterile petri dishes. (The leaf petioles require a vigorous wash with 10% calcium hypochlorite for 5 min.)

2. Section the inner leaves into 1-2 mm thick slices and then place onto MS medium with 0.9 µM 2,4-D and 2.3 µM KIN.
3. After callus has developed transfer the explant onto the same medium with 11.4 µM IAA, 2.3 µM KIN, and 4.4 µM BA for shoot and root initiation.
4. After the shoots have formed, remove the shoot apex to increase the number of plantlets.
5. When the plantlets reach a height of 5 cm and have well developed roots, transfer to soil.

Clonal Regeneration: White Cabbage

(After Walkey et al., 1980.)

1. Excise the meristem into 0.5-2 mm diameter pieces.
2. Culture the meristem tips in MS medium with 12 µM KIN and then transfer onto the same medium with 60 µM KIN for shoot induction.
3. Place the culture tubes at 22 ± 1 C and 16 hr daylength.
4. To induce root development, transfer the shoots to a kinetin-free medium.
5. After root formation, the plantlets can be transferred into soil.

Clonal Regeneration: Leaf Mustard Cabbage

(After Hui and Zee, 1978.)

1. Use hypocotyl and cotyledon raised from sterilized seeds. If leaf explants are used, sterilize the leaves by first immersing the leaves in 70% ethanol for 30 sec and then for 10 min in 20% Clorox + 0.1% Tween 20.
2. Culture the explant (3-5 mm diameter) on MS medium with either 2.7 µM NAA + 8.9 µM KIN or 2.6 µM NAA + 8.8 µM BA.
3. Place the tubes at 27 ± 1 C and 16 hr daylength.
4. Plantlets will form after culturing for 30-40 days; then transfer to soil.

Clonal Regeneration: Chinese Cabbage

(After Kuo and Tsay, 1977.)

1. Surface sterilize the axillary buds with 75% ethyl alcohol for 1.5 min, and for 10 min in 10% sodium hypochlorite.
2. Culture on MS medium with 0.44 µM BA and 5.4 µM NAA.
3. Roots will form after 2 weeks. After plantlets have formed, they can be transferred to soil.

Hybrid Embryo Culture

(After Feng and Cheng, 1981.)

1. Use common cabbage (*B. oleracea*, 2n = 18) as the female parent and Chinese cabbage (*B. pekinensis*, 2n = 20) as the male plant.
2. Remove young embryos from the seed pods 30–40 days after pollination.
3. Culture the young embryos in WH medium to induce callus formation.
4. Transfer the callus tissues onto a B5 medium to induce shoot and root formation.
5. After plantlets have formed transfer the hybrid plantlets to pots.

Anther Culture: Regeneration of Diploid Plants from Broccoli

Recently a number of workers have attempted to regenerate plants from anthers of cole crops with success (Quazi, 1978; Deng and Guo, 1977). The pollen being haploid, is capable of regenerating haploid plants that can then be developed into new varieties.

1. Use 2-4 mm long young flower buds.
2. Sterilize with 70% ethanol for 30 sec and then with 8% sodium hypochlorite for 15–20 min.
3. Place 8–12 anthers on medium in 70 mm petri dish.
4. The medium is composed of MS inorganic salts and B5 amino acids with 2-4 μM 2,4-D and 1-2 μM BA for callus induction, and 1 μM GA_3 + 5 μM BA + 10% v/v CW for differentiation.
5. Culture at 26 C under continuous fluorescent light of low intensity.
6. After the plantlets have developed, transfer them to pots.

Anther Culture: Regeneration of Plants from Chinese Cabbage

(After Deng and Guo, 1977.)

1. Isolate anthers from 0.2-0.4 cm flower buds.
2. Cutlure the anthers on Nitsch medium with 9 μM 2,4-D. Callus will form after 25–30 days.
3. Transfer callus to a differentiation medium using Nitsch basal medium with 2.8 μM IAA and 2.3 M KIN, but reduce the sucrose concentration to 1-2%.

Recently methods have been developed to permit plant regeneration from protoplasts of *B. oleracea* using methods developed for rapeseed protoplast regeneration. Development of protoplast methodology could be very useful for production of novel genetic combinations for release of new plant varieties.

Protoplast Isolation, Culture, and Regeneration of *Brassica* spp.

(After Xu et al., 1982.)

1. *Brassica alba, B. campestris, B. oleracea,* and *B. napus* seeds are surface sterilized in 0.1% w/v mercuric chloride, and 0.1% w/v sodium lauryl sulfate, rinsed three times in sterile distilled water, and germinated in the dark at 25 C on 0.6% agar containing 0.2% sucrose.
2. Within 2 days, 1 cm long apical root sections are excised and sliced transversely if large enough.
3. Root sections are plasmolyzed for 1 hr in a solution of 0.71 M mannitol, 0.19 mM KH_2PO_4, 1.0 mM KNO_3, 10.1 mM $CaCl_2 \cdot 2H_2O$, 0.98 mM $MgSO_4 \cdot 7H_2O$, 0.96 μM KI, 0.01 μM $CuSO_4 \cdot 2H_2O$ at pH 5.6 (CPW solution).
4. After plasmolysis, root sections are incubated 16 hr at 25 C (50 rpm) in the above CPW solution containing 2% Rhozyme, 4% Meicelase, and 0.3% Macerozyme.
5. The protoplast preparation is purified by filtration through a 64 μm nylon filter followed by centrifugation at 100 x g for 10 min.
6. Protoplasts that pellet following centrifugation are resuspended in CPW containing 0.61 M sucrose instead of 0.71 M mannitol and are centrifuged 100 x g for 5 min. Protoplasts that float on sucrose are collected and washed twice with CPW 0.71 M mannitol medium.
7. Protoplasts are resuspended in culture medium (modified Kao and Michayluk, 1975) and incubated in the dark for 14 days.
8. Fresh medium is added at 7-10 day intervals.
9. Callus colonies 30-45 days old are transferred to B5 agar medium containing 0.087 M sucrose, 4.5 μM 2,4-D, and 4.4 μM BA.
10. After 2-3 weeks on B5, callus is transferred to MS containing 11.0 μM IAA and 4.4 μM BA for shoot regeneration.

FUTURE PROSPECTS

Tissue culture techniques have been developed for most of the important cole crops (Table 2). But in our experience, two key variable factors that will affect the success of plant regeneration in these crops need some consideration. First, the cole crops often show large variability in the plant regeneration percentage, and second, the plant regeneration ability among cultivars within a species is usually quite variable. However, recently we have found that both these problems could be partially solved using medium supplemented with ginseng powder (Zee et al., 1980). Table 3 summarizes the effects of ginseng powder in increasing the frequency of plant regeneration for certain cole crops.

Table 3. Effect of Ginseng on Plant Regeneration from Certain Cole Crops (*Brassica* spp.)

SPECIES	EXPLANT	MEDIUM COMPOSITION	PLANT REGENERATION (%)
B. alboglabra	Hypocotyl	MS + 4.65 µM KIN + 5.37 µM NAA	58.3
		Above medium, + 1000 mg/l ginseng powder	98.4
B. chinensis	Cotyledon	MS + 8.88 µM BA + 2.68 µM NAA	43.3
		Above medium, + 250 mg/l ginseng powder	75.0
B. juncea	Cotyledon	MS + 9.83 µM 2iP + 2.68 µM NAA	55.5
		Above medium, + 1000 mg/l ginseng powder	94.4
	Hypocotyl	MS + 8.88 µM BA + 2.68 µM NAA	62.5
		Above medium, + 1000 mg/l ginseng powder	73.5

REFERENCES

Anderson, W.C. and Carstens, J.B. 1977. Tissue culture propagation of broccoli, *Brassica oleracea* italica group, for use in F_1 hybrid seed production. J. Am. Soc. Hortic. Sci. 102:69–73.

Anderson, W.C., Meagher, G.W., and Nelson, A.G. 1977. Cost of propagating broccoli plants through tissue culture. HortScience 12:543–544.

Bajaj, Y.P.S. and Nietsch, P. 1975. In vitro propagation of red cabbage (*Brassica oleracea* L. var. capitata). J. Exp. Bot. 26:883–890.

Barker, W.G. 1972. Toxicity levels of mercury, lead, copper, and zinc in tissue culture systems of cauliflower, lettuce, potato, and carrot. Can. J. Bot. 50:973–976.

Baroncelli, S., Buiatti, M., Bennici, A., and Santoro, A. 1973a. Genetics of differentiation in vitro in *Brassica oleracea* var. botrytis: A preliminary note. Inf. Bot. Ital. 5:104.

Baroncelli, S., Buiatti, M., and Bennici, A. 1973b. Genetics of growth and differentiation "in vitro" of *Brassica oleracea* var. botrytis. I. Differences between inbred lines. Z. Pflanzenzuecht. 90:99-107.

Baroncelli, S., Buiatti, S., Bennici, A., and Pagliai, M. 1974a. Genetics of growth and differentiation "in vitro" of *Brassica oleracea* var. botrytis. III. Genetic correlation and ontogenetic unity. Z. Pflanzenzuecht. 72:275-282.

Baroncelli, S., Buiatti, M., and Tesi, R. 1974b. The use of in vitro culture for early selection in cauliflower. Genet. Agrar. 28:170-176.

Braak, J.P. 1972. The production of haploid plants from pollen grains. Landbouwkd. Tijdschr. 84:50-55.

Buiatti, M., Baroncelli, S., Bennici, A., Pagliai, M., and Tesi, R. 1974a. Genetics of growth and differentiation in vitro of *Brassica oleracea* var. botrytis. II. An in vitro and in vivo analysis of a diallel cross. Z. Pflanzenzuecht. 72:269-274.

Buiatti, M., Baroncelli, S., and Bennici, A. 1974b. Genetics of growth and differentiation in vitro of *Brassica oleracea* var. botrytis. IV. Genotype-hormone interactions. Z. Pflanzenzuecht. 73:298-302.

Child, J.J. 1975. Nitrogen fixation by *Rhizobium* sp. in association with non-leguminous plant cell cultures. Nature (London) 253:350-351.

______ and Kurz, W.G.W. 1978. Inducing effect of plant cells on nitrogenase activity by *Spirillum* and *Rhizobium* in vitro. Can. J. Microbiol. 24:143-148.

Clare, M.V. and Collin, H.A. 1973. Meristem culture of brussels sprouts. Hortic. Res. 13:111-118.

______ 1974. The production of plantlets from tissue cultures of brussels sprout (*Brassica oleracea* var. gemmifera D.C.). Ann. Bot. 38: 1067-1076.

Crisp, P. and Walkey, D.G.A. 1974. The use of antiseptic meristem culture in cauliflower breeding. Euphytica 23:305-313.

Deng, L.P. and Guo, Y.H. 1977. Induction of plants from anther culture of *Brassica pekinensis* Rupr. In: Proceedings of Anther Culture Symposium, Guang Zhou. Science Press, China (in Chinese).

Dunwell, J.M. and Davies, D.R. 1975. Technique for propagating inbred brussels sprout strains by means of tissue culture. Grower 84:105-106.

Elmsheuser, H.A., Lein, C., and Neumann, K.H. 1978. An investigation of the relationship between phytohormone content and growth of 3 *Brassica* spp. in tissue culture. Z. Pflanzenphysiol. 88:25-32.

Feng, W. and Cheng, Y.H. 1981. Artificial amphidiploids obtained in an interspecific cross (*Brassica oleracea* L. x *B. pekinensis* Rupr.). Acta Hortic. Sinica 8:37-40.

Ferrari, T.E. and Wallace, D.H. 1975. Germination of *Brassica* pollen and expression of incompatability in vitro. Euphytica 24:757-765.

Gatenby, A.A. and Cocking, E.C. 1977. Callus formation from protoplasts of marrow stem kale. Plant Sci. Lett. 8:275-280.

Grout, B.W.W. 1975. Wax development on leaf surfaces of *Brassica oleracea* var. Currawong regenerated from meristem culture. Plant Sci. Lett. 5:401-405.

______ and Aston, M.J. 1978a. Transplanting of cauliflower plants regenerated from meristem culture. II. Carbon dioxide fixation and the development of photosynthetic ability. Hortic. Res. 17:65-71.

______ and Aston, M.J. 1978b. Modified leaf anatomy of cauliflower plantlets regenerated from meristem culture. Ann. Bot. 42:993-995.

______ and Crisp, P. 1977. Practical aspects of the propagation of cauliflower by meristem culture. Acta Hortic. 78:289-296.

Hansson, B. 1978. Temperature shock: A method of increasing the frequency of embryoid formation in anther culture of swede rape (*Brassica napus*). Sver. Utsadesfoeren. Tidskr. 89:141-148.

Harberd, D.J. 1969. A simple effective embryo culture technique for *Brassica*. Euphytica 18:425-429.

Horak, J. 1972. Ploidy chimeras in plants regenerated from tissues cultures of *Brassica oleracea* L. Biol. Plant. 14:423-426.

______, Landa, Z., and Lustinec, J. 1971. Production of polyploid plants from tissue cultures of *Brassica oleracea*-D. Phyton Rev. Int. Bot. Exp. 28:7-10.

______, Lustinec, J., Mesicek, J., Kaminek, M., and Polakova, D. 1975. Regeneration of diploid and polyploid plants from the stem pith explants of diploid marrow stem kale (*Brassica oleracea* L.). Ann. Bot. 39:571-577.

Hui, L.H. and Zee, S.-Y. 1978. In vitro plant formation from hypocotyls and cotyledons of leaf-mustard cabbage (*Brassica juncea* Coss.). Z. Pflanzenphysiol. 89:77-80.

______ 1980. The effect of ginseng on the plantlet regeneration percentage of cotyledon and hypocotyl explants of broccoli. Z. Pflanzenphysiol. 96:297-301.

Inomata, N. 1968. In vitro culture of ovaries of *Brassica*-D hybrids between *Brassica chinensis*-D and 4X *Brassica pekinensis*-D. I. Culture medium IAA, auxin, gibberellic acid, kinetin, nutritional requirements. Jpn. J. Breed. 18:139-146.

______ 1975. In vitro culture of ovaries of *Brassica* hybrids between 2X and 4X. Histological studies at certain developmental stages. Jpn. J. Genet. 50:1-18.

______ 1976. Culture in vitro of excised ovaries in *Brassica campestris* L. I. Development of excised ovaries in culture media, temperature, and light. Jpn. J. Breed. 26:229-236.

______ 1977a. Production of interspecific hybrids between *Brassica campestris* and *Brassica oleracea* by culture in vitro of excised ovaries. I. Effects of yeast extract and casein hydrolysate on the culture of excised ovaries. Jpn. J. Breed. 27:295-304.

______ 1977b. Culture in vitro of excised ovaries in *Brassica campestris* L. III. Development of excised ovaries at various days after pollination and in various cultivars. Bull. Univ. Osaka Prefect. 29:1-6.

______ 1977c. In vitro culture of *Brassica* hybrids between 2X and 4X. I. Culture medium. Jpn. J. Breed. 18:17-26.

______ 1978a. Production of interspecific hybrids in *Brassica campestris* x *B. oleracea* by culture in vitro of excised ovaries in the crosses of various cultivars. Jpn. J. Genet. 53:161-173.

______ 1978b. Production of interspecific hybrids between *Brassica campestris* and *Brassica oleracea* by culture in vitro of excised ovaries. II. Effects of coconut milk and casein hydrolysate on the development of excised ovaries. Jpn. J. Genet. 53:1-11.

Johnson, B.B. and Mitchell, E.D., Jr. 1978a. The effect of temperature on in vitro propagation of broccoli from leaf rib and leaf explants. In Vitro 14:334.

______ 1978b. In vitro propagation of broccoli from stem, leaf, and leaf rib explants. HortScience 13:246-247.

Kameya, T. 1973. The effects of gelatin on aggregation of protoplasts from higher plants. Planta 115:77-82.

______ and Hinata, K. 1970a. Test tube fertilization of excised ovules in *Brassica*-D. Jpn. J. Breed. 20:253-260.

______ and Hinata, K. 1970b. Induction of haploid plants from pollen grains of *Brassica*. Jpn. J. Breed. 20:82-87.

Kartha, K.K., Gamborg, O.L., and Constabel, F. 1974a. In vitro plant formation from stem explants of rape (*Brassica napus* cv. Zephyr). Physiol. Plant. 31:217-220.

______, and Kao, K.N. 1974b. Fusion of rapeseed and soybean protoplasts and subsequent division of heterokaryotes. Can. J. Bot. 52: 2435-2439.

Kartha, K.K., Michayluk, M.R., Kao, R.N., Gamborg, O.L., and Constabel, F. 1974c. Callus formation and plant regeneration from mesophyll protoplasts of rape plants (*Brassica napus* cv. Zephyr). Plant Sci. Lett. 3:265-271.

Kao, K.N. and Michayluk, M.R. 1975. Nutritional requirements for growth of *Vicia hajastana* cells and protoplasts at a very low population density in liquid media. Planta 126:105-110.

Keller, W.A. and Armstrong, K.C. 1977. Embryogenesis and plant regeneration in *Brassica napus* anther cultures. Can. J. Bot. 55: 1383-1388.

______ 1978. High frequency production of microspore derived plants from *Brassica napus* anther cultures. Z. Pflanzenzuecht. 80:100-108.

Keller, W.A., Rajhathy, T., and Lacapra, J. 1975. In vitro production of plants from pollen in *Brassica campestris*. Can. J. Genet. Cytol. 17:655-666.

Keller, W.A., Lacapra, J., and Armstrong, K.C. 1977. Embryogenesis and plantlet regeneration in anther culture of *Brassica* spp. In Vitro 13:146.

Khanna, P., Mohan, S., Nag, T.N., and Jain, S.C. 1971. Tissue differentiation in ten plant species in vitro. Indian J. Plant Physiol. 14: 35-43.

Kuo, C.G. and Tsay, J.S. 1977. Propagating Chinese cabbage by axillary bud culture. HortScience 12:456-457.

Landa, Z. 1974. Promising methods for obtaining, detecting, and utilizing incompatible lines of *Brassica oleracea* L. Genetica 5:125-138.

______ and Lustinec, J. 1971. Production of polyploid plants from tissue cultures of *Brassica oleracea* L. Phyton 28:7-10.

Lee, S.S., Chung, J.H., and Han, K.Y. 1973. A study of media for axillary bud culture in Chinese cabbage, *Brassica pekinensis*. Research Reports of the Office of Rural Development, Horticulture 15:9-13.

Li, S.H. and Qiu, W.D. 1981. The production of plants from tissue cultures of cauliflower (*Brassica oleracea* var. botrytis L.). Acta Hortic. Sinica 8:33-36.

Lustinec, J. and Horak, J. 1970. Induced regeneration of plants in tissue cultures of *Brassica oleracea* D. Experientia 26:919-920.

Lustinec, K., Kaminek, M., and Privaratsky, J. 1972. Increase in dry weight and protein content of *Brassica oleracea* and *Nicotiana tabacum* pith explants during short term cultivation on simple media. Biol. Plant. 14:376-378.

Margara, J. 1969. Study of factors in the neoformation of buds in tissue culture of cauliflower D. *Brassica oleracea* var. botrytis D. Ann. Physiol. Veg. 11:95-112.

_____ 1977. The effect of auxins on the in vitro formation of new buds in cauliflower. C.R. Acad. Sci. Ser. D. 284:1883-1885.

_____ and Leydecker, M.T. 1978. Different types of organogenesis oserved in rape, *Brassica napus* var. olifera Metzg. C.R. Acad. Sci. Ser. D. 287:17-20.

Maryakhina, I. and Butenko, R.G. 1974a. The culture of isolated cabbage tissues and the genetic characteristics and development of regenerated plants. Skh. Biol. 9:216-227.

_____ 1974b. Somatic reduction in a cabbage tissue culture. Cytol. Genet. 8:74.

Mascarenhas, A.F., Hendre, R.R., Nadgir, A.L., Barve, D.M., and Jagannathan, V. 1978. Differentiation in tissue culture of cabbage. Indian J. Exp. Bot. 16:122-125.

Matsuzawa, Y. 1978. Studies on interspecific hybridization in *Brassica*. Effect of temperature on the development of hybrid embryos and improvements of the cross success rate by ovary culture in *B. campestris* x *B. oleracea* crosses. Jpn. J. Breed. 28:186-196.

Miszke, W. and Skucinska, B. 1976. In vitro vegetative propagation of *Brassica oleracea* var. capitata L. Z. Pflanzenzuecht. 76:81-82.

Murashige, T. et al., 1977. Symposium on tissue culture for horticultural purposes. Acta Horticult. 78:1-459.

Nishi, S., Kawata, J., and Toda, M. 1959. On the breeding of interspecific hybrids between two genomes, "c" and "a," of *Brassica* through the application of embryo culture techniques. Jpn. J. Breed. 8:7-14.

Nishi, S., Oosawa, K., and Oswa, K. 1974. Studies on the use of anther culture. I. Formation of calluses from anthers of various vegetable crop species and development of plantlets. Bulletin of the Vegetable and Ornamental Crops Research Station, A. No. 1, pp. 1-40.

North, C. 1953. Three methods for vegetative propagation of *Brassica oleracea*. J. R. Hortic. Soc. 78:106-111.

Pareek, L.K. and Chandra, N. 1972. Growth and differentiation in *Brassica juncea* tissues grown in vitro. Indian Sci. Cong. Assoc. Proc. 59(part 3):372-373.

_____ 1978. Somatic embryogenesis in leaf callus from cauliflower (*Brassica oleracea* var. botrytis). Plant Sci. Lett. 11:311-316.

Pow, J.J. 1969. Clonal propagation in vitro from cauliflower curd. Hortic. Res. 9:151-152.

Primo-Millo, E. and Harada, H. 1975. Morphogenesis and vegetative propagation from leaf tissue of red cabbage (*Brassica oleracea* var. Tete de Negre). C.R. Acad. Sci. Ser. D. 280:2845-2847.

Quan, S.G., Chi, E.Y., and Chaplin, S.M. 1974. Tubular structures in endoplasmic reticulum of cultured broccoli. J. Ultrastruct. Res. 48: 92-101.

Quazi, H.M. 1975. Isolation and culture of protoplasts from *Brassica*. N.Z. J. Bot. 13:571-576.

_____ 1978. Regeneration of plants from anthers of broccoli (*Brassica oleracea* L.). Ann. Bot. 42:437-475.

Radwan, S.S., Grosse-Oetringhaus, S., and Mangold, H.K. 1978. Lipids in plant tissue cultures. VI. Effect of temperature on the lipids of *Brassica napus* and *Tropaeolum majus* cultures. Chem. Phys. Lipids 22:177-184.

Rahn-Scoher, M.J. 1977. A preliminary study of the in vitro culture of fragments of cauliflower heads (*Brassica oleracea* var. botrytis). Sci. Agron. Rennes 12:21-23.

Saint, A.M. 1977. The maintenance and loss of cold resistance in leaf lamina fragments from Savoy cabbages, cv. Pontoise, hardened naturally. C.R. Acad. Sci. Ser. D. 285:1435-1437.

Shevyakova, N.I. and Komizerko, E.I. 1969. Amino acid metabolism in a culture of callus tissue from cabbage-D leaves, *Brassica oleracea* var. capitata-D, under salinization conditions. Dokl. Akad. Nauk. USSR 186:1441-1444.

Shin, J.S. 1972. Studies on embryo culture. Breeding interspecific hybrids between *Brassica pekinensis* and cabbage. Bull. Agric. Res. Taiwan 1:27-35.

Stringham, G.R. 1977. Regeneration in stem explants of haploid rapeseed (*Brassica napus* L.). Plant Sci. Lett. 9:115-119.

Thomas, E. and Wenzel, G. 1975. Embryogenesis from microspores of *Brassica napus*. Z. Pflanzenzuecht. 74:77-81.

Thomas, E., Hoffman, F., Potrykus, I., and Wenzel, G. 1976. Protoplast regeneration and stem embryogenesis of haploid androgenic rape. Molec. Gen. Genet. 145:245-247.

Trimboli, D.S., Prakash, N., and DeFossard, A.R. 1978. The initiation rooting and establishment of cortical buds in cauliflower. Acta Hortic. 78:243-248.

Tseng, M.T. and Lin, C.I. 1975. Influence of various concentrations of IAA, 2,4-D and kinetin on anther cultures of radish, broccoli, and sunflower. J. Agric. Assoc. China 91:23-31.

Walkey, D.G.A. and Woolfitt, J.G.M. 1970. Rapid clonal multiplication of cauliflower by shake culture. J. Hortic. Sci. 45:205-206.

Walkey, D.G.A., Cooper, V.C., and Crisp, P. 1974. The production of virus-free cauliflowers by tissue culture. J. Hortic. Sci. 49:273-275.

Walkey, D.G.A., Neeley, H.A., and Crisp, P. 1980. Rapid propagation of white cabbage by tissue culture. Sci. Hortic. 12:99-107.

Watts, L.E. and George, R.A.T. 1963. Vegetative propagation of autumn cauliflower. Euphytica 12:314-345.

Wenzel, H. 1973. Isolation of leaf protoplasts from haploid plants of petunia, rape, and rye. Z. Pflanzenzuecht. 69:58-61.

_____, Hoffman, F., and Thomas, E. 1977. Anther culture as a breeding tool in rape. I. Ploidy level and phenotypes of androgenic plants. Z. Pflanzenzuecht. 78:149-155.

Xu, Z.H., Davey, M.R., and Cocking, E.C. 1982. Plant regeneration from root protoplasts of *Brassica*. Plant Sci. Lett. 24:117-121.

Zee, S.Y. and Hui, L.H. 1977. In vitro plant regeneration from hypocotyl and cotyledons of Chinese kale, *Brassica alboglabra*. Z. Pflanzenphysiol. 82:330-335.

Zee, S.Y., Wu, S.C., and Yue, S.B. 1978. Morphogenesis of the hypocotyl of Chinese kale. Z. Pflanzenphysiol. 90:155-163.

Zee, S.Y., Wu, S.C., and Hui, L.H. 1980. Research on tissue culture explant morphogenesis and the enhancement effect of some Chinese drugs on explants: A summary report. Acta Physiol. Sinica 6:419-428.

Zhong, Z.X., Ren, Y.Y., and Dai, W.P. 1978. Preliminary studies on anther culturing of *Brassica chinensis*. Acta Bot. Sinica 20:180-181.

CHAPTER 10
Tomato

S. A. Kut, J. E. Bravo, and *D. A. Evans*

HISTORY OF CULTIVATION

Lycopersicon species originated in the Andean region of South America. To this date, the only natural habitats of wild species are found from northern Chile to Ecuador. Domestication of the tomato, however, appears to have begun in Mexico where wild populations of the cherry tomato, *L. esculentum* v. cerasiforme, ancestor to the cultivated tomato, are still found. The path of the cultivated tomato from South America to Mexico and on to Europe is documented by the greater similarity in alloenzymes between modern cultivars and Mexican genotypes than between modern cultivars and Andean genotypes. In addition, no signs of early tomato cultivation have been found in the Andean region.

From Mexico, the domesticated tomato was taken to the Mediterranean region of Europe in the sixteenth century and on to North America in the eighteenth century (Rick, 1976). The germplasm available in North America has had, until recently, a very narrow genetic base. However, the infusion of South American wild species genetic information since the 1950s has allowed improvement of the cultivated tomato.

ECONOMIC IMPORTANCE

In 1978, the top U.S. producers of processing tomatoes were California, Ohio, and Indiana. Production amounted to 7,280,588 tons on

333,856 acres, totaling $455,042,000. Fresh market tomatoes, produced mainly in Florida, California, and Alabama totaled 1,054,650 tons for $412,954,000 on 126,503 acres (Lorenz and Maynard, 1980).

World production of tomatoes reached 47,518,000 metric tons in 1979 with daily consumption at 31 g per capita. Table 1 lists the ten largest countries for production of tomatoes worldwide.

The United States, the top world tomato producer, also imports a significant amount of tomatoes (Table 2). Several countries that are large producers also export a large quantity of tomatoes (e.g., United States, Mexico, and Spain). On the other hand, several large exporters are not among the leading producers (e.g., Netherlands, Romania, Jordan, and Morocco), suggesting that tomatoes are grown in these countries principally for export.

Table 1. World Production of Tomatoes[a]

COUNTRY	METRIC TONS (x 1000)
USA	6754
USSR	6210
Italy	5023
China	4130
Turkey	3500
Egypt	2571
Spain	2173
Greece	1666
Brazil	1611
Mexico	1420

[a]Based on FAO statistics, 1980.

Table 2. Tomato Imports and Exports[a]

IMPORT		EXPORT	
Country	Metric Tons (x 1000)	Country	Metric Tons (x 1000)
West Germany	360	Mexico	373
U.S.	296	Netherlands	357
France	183	Spain	273
U.K.	178	Romania	150
Canada	136	U.S.	120
Czechoslovakia	89	Jordan	110
		Morocco	96
		Bulgaria	89

[a]Based on FAO statistics, 1980.

BREEDING ACCOMPLISHMENTS AND GOALS

In recent years selection for new characteristics has resulted in cultivars of tomatoes that are strikingly divergent. For example, processing tomatoes are determinate, resulting in compact growth habit. In addition, they have uniform ripening of fruit, and to facilitate mechanical harvesting, tomatoes are small and firm. Backyard garden tomatoes, on the other hand, have indeterminate vines that must often be staked to maximize yield. In some cases, the fruit ripens over several months and tomatoes are often selected for large, juicy fruit. Fresh and processing tomato varieties also differ in several fruit quality characters, including shape, flavor, and consistency. Greenhouse tomatoes, mostly hybrid varieties, are also strikingly different from fresh market or processing tomatoes.

In recent years tomato breeding has concentrated on increased yields, improved plant quality, alteration of plant growth habit, and pest resistance (Rick, 1976). Several single gene mutations have been valuable for the development of new varieties, including the sp gene for determinate growth habit. Wild species have also been quite useful for introduction of valuable traits, particularly disease resistance. Resistance to several bacterial, fungal, and viral diseases has been introduced from wild *Lycopersicon* species into cultivated tomato (Rick, 1982). These include resistance to bacterial canker, bacterial wilt, leaf mold, anthracnose, fusarium and verticillium wilts, late blight, and tobacco mosaic virus. In several cases, such as *Cladosporium fulvum*, resistance is controlled by a single dominant gene (Kerr et al., 1980). *Lycopersicon peruvianum* and *L. pimpinellifolium* have been particularly useful sources of disease resistance. As well, *L. peruvianum* served as the source of resistance to root-knot nematodes. Jointless pedicel (j-2) originated from *L. cheesmanii*, and high pigment (Ip) from *L. chmielewskii* (Rick, 1982).

It is expected that continued effort will be directed toward use of wild species to incorporate useful traits into cultivated tomato. As the tomato is expanded into new geographical regions, new pests become economically important. For example, *Phytophthora infestans* has evolved as an important pest in Brazil (Maschio and Sampaio, 1982), but it is absent from most U.S. locations. Also, new strains of existing pathogens are beginning to appear. A third race of *Fusarium* wilt has recently been reported in Australia (Grattidge, 1982). As nutrients are depleted from arable land and interest accumulates in using poor lands, breeding for stress tolerance will become increasingly important. Wild species germplasm is again a unique source of stress tolerant characteristics. Several collections have been identified with useful characteristics including salt tolerance in *L. cheesmanii*, drought tolerance in *S. pennellii*, insect resistance in *L. hirsutum*, cold tolerance in *L. hirsutum* and *S. lycopersicoides*, heat tolerance in *L. pimpinellifolium*, and moisture tolerance in *L. esculentum* v. cerasiforme. No new stress tolerant cultivars have yet been developed using these wild species, due to the lack of efficient screening procedures or economic incentives (Rick, 1982). However, effort is currently being directed toward developing new stress tolerant varieties using these wild species (e.g., Epstein et al., 1980).

Several hybrid varieties of tomato have recently been released, but all are produced using hand emasculation. This procedure of cross pollination is tedious and costly. Nevertheless, hybrid seed is widely grown in California. It has been suggested that introduction of a useful male sterile system would result in several new inexpensive hybrid varieties (Rick, 1982). Future research on the tomato will work toward such a male sterile as well as improved color, solids, flavor, and nutrition.

REVIEW OF THE LITERATURE

Callus Culture

Although many laboratories have succeeded in the induction of tomato callus growth in cell culture, continued callus proliferation has resulted in accumulation of chromosomal aberrations and the production of highly polyploid populations of cells (2n = 24, 4n = 48, 6n = 72, ...). Polyploidization is first evident after only one or two subcultures. Aneuploidy also results in a loss of shoot regeneration capacity. Only the wild species *L. peruvianum* (Thomas and Pratt, 1981b) can be used to establish callus cultures which can be maintained with the capacity for shoot regeneration after one year in culture.

Early studies on tomato callus initiation and growth from several carotenoid mutants of cultivated tomato used frozen pea extract as an undefined additive to the basal medium (Fukami and Mackinney, 1967). Ulrich and Mackinney (1969) modified this medium and published a very complex medium that resulted in better proliferation of callus from hypocotyl, stem, and fruit tissue. Since their purpose was to study biochemical pathways of carotenoids, no attempt was made to regenerate the callus. However, two observations from this work continue to hold for most tomato lines in vitro: successful initiation and continued proliferation of tomato callus in vitro depends on (a) genotype specific requirements for certain medium additives (such as growth hormones) and (b) the age and vigor of the donor plants used to initiate the cultures. They also used K_2SO_4 to prevent browning of the callus and thiourea to promote cell division. Most laboratories today employ more defined media such as MS (Murashige and Skoog, 1962) or B5 (Gamborg et al., 1968) for callus culture. However, undefined additives such as CW and casein hydrolysate are still frequently used.

DeLanghe and DeBruijne (1976) induced callus growth from stem internode explants of four CCC (chlormequat- 2 chloroethyl trimethylammonium chloride) -treated *L. esculentum* strains on LS (Linsmaier and Skoog, 1965) medium containing 1.0 µM IBA + 1.0 µM IAA + 1.0 µM KIN + 1.0 µM 2iP. They were able to maintain the shoot regeneration capability of this callus for over two years by subculturing the callus at three week intervals on LS medium supplemented with 1.0 µM KIN. Shoot regeneration was accomplished upon transfer of callus to LS medium containing 1.0 µM ZEA or 20% CW. However, genetic variability was observed in the F_1 progeny of at least one of 30 regenerated plants.

Thomas and Pratt (1981b) were able to induce friable, rapidly growing callus from the wild species *L. peruvianum*. This callus retained the capacity to regenerate diploid plants even after a year in cell culture. However, the proportion of tetraploid regenerated plants increased with time in culture. Subsequently, an interspecific hybrid line (L2) was developed that exhibited the best callus growth rate and plant regeneration efficiency of the *L. peruvianum* strains used in its production. Unfortunately, L2 contains only 25% of cultivated tomato ancestry. The L2 line forms 2 g of callus from each 1 cm leaf disk in only 3 weeks. Suspension cultures composed of well dispersed cells can also be maintained. Diploid shoots can be regenerated from 2-month-old callus 1-2 months after transfer to regeneration medium. Diploid shoots have been regenerated from several-month-old suspension cultures.

APPLICATIONS OF CALLUS CULTURE. One of the first applications of tomato callus culture was for the biochemical study of carotenoid synthesis (Ulrich and Mackinney, 1969). These authors assumed that, since callus cells are totipotent, then cultures of tomato callus should be capable of synthesizing all of the carotenoids encoded in any specific tomato genome. Unfortunately, their culture conditions promoted fruit but not leaf carotenoid production, even though hypocotyl explants were used to initiate callus.

Other studies have been reported in which a function expressed on the whole plant level was not elicited in callus. For example, Ellis (1978) examined tolerance to the herbicide metribuzin whose primary site of action is the photosynthetic electron transport system (Trebst and Wietoska, 1975). Ellis employed nonphotosynthetic cell suspension cultures initated from hypocotyls in an attempt to isolate tomatoes with decreased sensitivity to metribuzin. Although tomato lines employed in this study exhibited a range of sensitivity to metribuzin from relatively sensitive to relatively tolerant, no differential effect of the herbicide was observed on the growth of their respective cell suspension cultures.

Organ cultures (root tips) were compared to callus cultures for uptake of exogenously supplied sugars (Chin et al., 1981). These researchers found significant differences in the ability of the two cultures to take up and grow in different types of sugars as well as differences in the rates of sugar uptake.

Tal et al. (1978) employed callus cultures to study salt tolerance in *L. peruvianum* and *Solanum pennellii*. Tomato callus initiated from leaf, stem, or root tissue had an identical response to salt. Callus from the salt tolerant wild species *L. peruvianum* and *S. pennellii* proliferated to a greater degree under saline cell culture conditions than did salt sensitive cultivated *L. esculentum*. These results suggested that the better osmotic adjustment in the wild species functions on the cellular level and is independent of the organization of the intact plant. Callus cell culture may be uniquely useful for the study of the mechanism of salt tolerance.

Liebisch (1980) demonstrated that tomato cell suspension cultures exhibited identical enzymatic conversions and metabolism of various gibberellins to that observed in intact tomato plants. These cell suspensions provide a simpler in vitro system for the study of these phytohormones.

Tomato callus has also been used to isolate genetic variants. Unlike other crop species such as tobacco and carrot, friable tomato callus cultures necessary for efficient mutant isolation from suspension culture have not been successfully induced to regenerate shoots except for *L. peruvianum*. Thomas and Pratt (1981b, 1982) have succeeded in isolating variants with increased tolerance to the herbicide paraquat using the L2 interspecific hybrid. Regenerated plants expressed a slight increase in paraquat tolerance. It was suggested that these plants were probably heterozygous for the mutated allele(s) as determined by selfing and outcrossing experiments with wild-type, paraquat-sensitive L2, followed by paraquat tolerance tests on callus initiated from the progeny. Although further genetic testing is required, the fact that an increase in tolerance was observed in both R_0 and some R_1 plants suggests the usefulness of the L2 strain in mutant isolation. Agronomic applications of mutants from L2 await the successful use of L2 in a breeding program with cultivated tomato.

Aluminum-resistant variants were also selected from callus cultures as well as from plated suspension cultures of cultivated tomato by Meredith (1978b). These variants were stable, but the cultured cells were highly polyploid and regenerated no plants.

The ultimate value of the isolation of genetic variants from tomato callus culture requires either the development of techniques for establishment and maintenance of normal, diploid callus cultures that retain the capacity to regenerate shoots, or the use of a strain such as L2 to breed these characteristics into agronomically important strains.

An alternative to callus culture for mutant isolation followed by shoot regeneration trials for tomato may prove to be shoot culture itself. Shoot regeneration in tomato is preceded by a period of callus proliferation, both processes occurring in sequence on shoot regeneration medium. Evans and Sharp (1983) demonstrated shoot culture of leaf explants of cultivated tomato to be mutagenic. They recovered twelve nuclear, single gene, morphological mutations and one possible nuclear, single gene mutation rendered diploid by mitotic recombination. All mutations were sexually transmitted to subsequent generations according to simple Mendelian genetics. Since no mosaics were observed among 230 regenerated plants, the mutations apparently occurred in callus cells before shoot formation, with each mutant shoot regenerating from a single mutant cell. These results suggest that plant regeneration from leaf explants could be used to rapidly introduce genetic variability into cultivated tomato.

Tomato callus cultures have also been useful in plant pathology. Callus can be used as an effective tool to study mechanisms of disease resistance and susceptibility. For example, Warren and Routly (1970) have used tomato callus cultures to investigate the nature of single-gene (Ph_1) resistance to the fungal pathogen *Phytophthora infestans*. Cell culture is useful as a contained system which is free of other pathogens prevalent in the greenhouse or the open field.

Finally, Doering and Ahuja (1967) used callus cultures in their studies on a dominant tumor-like condition which occurred in a tomato line established by repeatedly backcrossing an *L. esculentum* x *L. chilense* hybrid to a second *L. esculentum* variety. They found that callus cultures could be used to distinguish tumorous from nontumorous tomato plants based on their growth response to different types and concentrations of hormones, as well as on an overall lower growth rate of the tumorous callus for all combinations of growth regulators tested. Thus callus culture can be used to study metabolic differences between normal and tumorous tissues.

Embryo Culture

Embryo culture is used to rescue hybrid plants from sexual crosses blocked by postzygotic barriers. This method is especially important to achieving interspecific hybridization which is used for the introduction of new and desirable traits from wild species into cultivated tomato. Smith (1944) first used this technique in tomato to isolate interspecific hybrids between cultivated tomato *L. esculentum* and the wild species *L. peruvianum*. Since that time, embryo culture has been used to produce *L. esculentum* x *L. peruvianum* hybrids between various *L. esculentum* and *L. peruvianum* varieties. Rick and Smith (1953) employed both *L. esculentum* x *L. peruvianum* and *L. esculentum* x *L. chilense* embryo culture hybrids for studies on the nature of inheritance in these crosses. De Nettancourt et al. (1974) used hybrids of *L. esculentum* x *L. peruvianum* produced by embryo culture to study both the genetic and ultrastructural anomalies involved in self- and cross-incompatibility. De Nettancourt's hybrids have been backcrossed with limited success to various *L. esculentum* varieties and are being evaluated for disease resistance and agronomic characters (Ancora et al., 1981). In other studies, Hogenboom (1972a-e) succeeded in developing self-compatible *L. peruvianum* varieties that were cross-compatible with *L. esculentum*. Although these self-compatible *L. peruvianum* varieties decrease the need for embryo culture in *L. esculentum* x *L. peruvianum* crosses this technique is still invaluable in effecting other presently incompatible interspecific crosses such as *L. esculentum* x *L. chilense* (Rick, 1963). The embryo culture technique was also employed to recover the intergeneric hybrid *L. esculentum* x *Solanum lycopersicoides* (Rick, 1951).

Thomas and Pratt (1981a) improved on the technique of embryo culture as part of a study to introduce the rapid callus proliferation and shoot regeneration traits found in *L. peruvianum* into *L. esculentum*. These authors were able to recover diploid hybrid plants with a 4% plating efficiency from immature seeds by embryo callus culture. The embryo callus exhibited the rapid callus proliferation and efficient shoot regeneration capacities observed for the *L. peruvianum* parent, thus supporting their efforts to breed these traits into *L. esculentum*.

Root Culture

Initial work on root culture of vegetable crops (including tomato) began in the early 1900s. Robbins (1922) and Kotte (1922) were the first to report successful culture of isolated root tips. White (1934a) later established a culture system using medium with yeast extract in which excised tomato roots could be maintained for indefinite lengths of time. Since that time, most of the work with tomato root culture has centered around determination of the specific nutrients which are essential for prolonged root growth in cell culture (White, 1963; Street and Melhuish, 1965; Street and Shillito, 1977; Chin et al., 1981).

White (1934a) showed that by growing tomato-root cultures at 15 C instead of 25-27 C, the period between subcultures could be increased from 1 to 8 weeks without reducing tissue viability. However, Petru (1965) found that a drop in day/night temperature from 23-24 C/20 C to 19 C/16 C for a 14-day period resulted in gradual senescence and death of root cultures over a 6-week period after return to higher temperatures. Petru also reported that a more acidic initial culture medium (pH 4.9) was most conducive to the culture of excised roots. Norton and Boll (1954) succeeded in regenerating shoots by first initiating callus growth from roots of *L. peruvianum*.

APPLICATIONS OF ROOT CULTURE. White (1934b) maintained tobacco mosaic virus (TMV) in cultured tomato roots. Street and Melhuish (1965) employed tomato root cultures to study the release of metabolites, specifically amino acids, by plant tissue and organ cultures. Chin et al. (1981) employed root culture for the study of sugar uptake and metabolism in excised tomato roots vs. callus culture cells.

Tomato root organogenesis has also been examined using tomato leaf disks in culture. Coleman and Greyson (1977a,b), employing a modified MS agar medium, investigated the effects of various auxins and GA_3 on the initiation of rooting. Coleman et al. (1980) also investigated the interaction of ethylene and auxins on root regeneration.

Shoot Culture

Tomato shoot regeneration occurs as a result of organogenesis as opposed to embryogenesis and requires an initial period of callus proliferation which is stimulated by endogenously synthesized auxin or by the addition of auxin and cytokinin to the regeneration medium. The amount of callus produced and the shoot regeneration efficiency of tomato explants is dependent on the following variables: (a) types and concentrations of growth regulators in the culture medium, (b) growth condition of donor plant, (c) age of donor plant, (d) explant of choice, (e) portion or age of donor tissue, and (f) genotype of the donor plant. Tables 3 and 4 list published culture media for shoot regeneration from several *L. esculentum* varieties and wild tomato species.

Norton and Boll (1954) first reported regeneration of tomato shoots in cell culture. They obtained shoots from callus induced from root cultures of species *L. peruvianum*. DeLanghe (1973) reported regeneration of shoots from the cultivated species *L. esculentum* from callus initiated from stem internodes.

Padmanabhan et al. (1974) investigated the effects of various combinations of IAA and KIN on the regeneration of shoots from leaf explants of one normal and two mutant strains of *L. esculentum*. They

Table 3. Shoot Regeneration from Cultivated Tomato Varieties

VARIETIES	GROWTH REGULATORS[a] (μM)		SHOOT MEDIUM	EXPLANT	REFERENCE
	Callus	Shoot Formation			
Rutgers	11.4–22.8 IAA + 9.3–18.6 KIN	22.8 IAA + 18.6 KIN	MS	Leaf	Padmanabhan et al., 1974
Canatella, King Plus		10 ZEA + 20% CW	LS	Stem internodes	DeLanghe & DeBruijne, 1976
Canatella, Roma, Ventura, King Plus	1 KIN	1 ZEA (or) 20% CW	LS	Stem internodes	DeLanghe & DeBruijne, 1976
Starfire	1 NAA + 1 BA (or) 10 NAA + 1 ZEA (or) 0.1 KIN	0.1–10 IAA + 10 BA (or) 0.1–1 IAA + 1 ZEA (or) 5 BA (or) 5 ZEA	MS salts + B5 vitamins	Leaf	Kartha et al., 1976
Starfire, 15 mutant clones[b]	0.5 or 2.5 NAA[c] + 1–20 BA	0.5 NAA[c] + 10 BA	MS	Leaf	Behki & Lesley, 1976
Starfire, Bush Beefsteak, Xa-ag, T1-, yg-5, pv	0.5 NAA[c] + 10–50 ZEA	0.5 NAA[c] + 50 ZEA	MS	Leaf	Behki & Lesley, 1976
Starfire	5–10 BA (or) 5 ZEA (or)	0.1–10 BA (or) 1–10 ZEA (or)	MS salts + B5 vitamins	Shoot apical meristems	Kartha et al., 1977

Table 3. Cont.

VARIETIES	GROWTH REGULATORS[a] (μM)		SHOOT MEDIUM	EXPLANT	REFERENCE
	Callus	Shoot Formation			
Starfire (cont.)	0.1-1 IAA + 5-10 BA (or) 1-10 NAA + 0.1-1 BA (or) 0.1-10 IAA + 0.1-10 ZEA (or) 0.1-10 NAA + 1-10 ZEA (or) 5-10 IAA	0.1-10 IAA + 0.1-10 BA (or) 0.1 NAA + 10 BA (or) 0.1-10 IAA + 0.1-10 ZEA (or) 0.1-10 NAA + 0.1-10 ZEA (or) 5-10 IAA			
Rheinlands Rhum, Flacca mutant	2.7-32.2 NAA + 2.3-27.9 KIN (or) 1.1-11.4 IAA + 8.9 BA	27.9 KIN (or) 1.1 IAA + 8.9 BA	MS	Leaf	Tal et al., 1977
Bizon	22.8-34.3 IAA + 9.3-18.6 KIN (or) 91.3 IAA + 18.6 KIN[e]	28.5-34.3 IAA + 0-9.8 IBA + 23.2-37.2 KIN[d] (or) 91.3 IAA + 18.6 KIN + 197.8 ADE[f]	MS + 2% sucrose	Cotyledon, leaf, stem	Vnuchkova, 1977

VARIETIES	GROWTH REGULATORS[a] (μM)		SHOOT MEDIUM	EXPLANT	REFERENCE
	Callus	Shoot Formation			
Bizon (Cont.)	(or) 91.3 IAA + 18.6 KIN + 197.8 ADE	(or) 18.6–27.9 KIN + 197.8 ADE			
TH hybrid (Pol x Pusa)	—	1 IAA + 1 ZEA	MS salts + B5 vitamins	Leaf	Dhruva et al., 1978
Pol cultivar	2 IAA + 5 2iP	100 IAA + 10 BA	MS salts + B5 vitamins	Leaf	Dhruva et al., 1978
Apedice (A)	—	5 IAA + 3–5 2iP	MS macros + Fe-EDTA + Nitsch & Nitsch micro & organic supplements	Hypocotyl	Ohki et al., 1978
Porphyre (P)	1 IAA + 5–10 BA	1–3 IAA + 1–5 2iP	See medium for Apedice (A)	Hypocotyl	Ohki et al., 1978
A x P hybrid	—	1–10 IAA + 3–5 2iP	See medium for Apedice (A)	Hypocotyl	Ohki et al., 1978

Table 3. Cont.

VARIETIES	GROWTH REGULATORS[a] (μM)		SHOOT MEDIUM	EXPLANT	REFERENCE
	Callus	Shoot Formation			
P x A hybrid	—	1 IAA + 5 2iP	See medium for Apedice (A)	Hypocotyl	Ohki et al., 1978
A, P, A x P, P x A	—	1 IAA + 15 2iP	See medium for Apedice (A)	Leaf	Ohki et al., 1978
Pixie	22.8 IAA + 18.6 KIN	22.8 IAA + 18.6 KIN	MS	Leaf	Herman & Haas, 1978
VFNT Cherry	10.7 NAA + 4.4 BA	0.6–1.2 IAA + 9.1 ZEA	MS	Established leaf callus	Meredith, 1979
L. esculentum x *L. peruvianum*	0.54 NAA + 2.6 GA + 0.22 BA	0.54 NAA + 2.6 GA + 0.22 BA	MS	Stem internode	Cappadocia & Sree Ramulu, 1980
	0.54 NAA + 2.6 GA_3 + 0.22 BA	247.3 ADE + 20% CW	MS	Callus from stem internodes	Cappadocia & Sree Ramulu, 1980
Karnatak hybrid	2.8 IAA	2.8 IAA + 9 BA	MS	Hypocotyl, cotyledon	Gunay & Rao, 1980
Xa-ag	2.3 2,4-D + 4.4 BA	9.1 ZEA	MS	Leaf	Behki & Lesley, 1980

VARIETIES	GROWTH REGULATORS[a] (μM) Callus	Shoot Formation	SHOOT MEDIUM	EXPLANT	REFERENCE
8 varieties	—	1 IAA + 10 BA	MS	Leaf, hypocotyl, cotyledon	Frankenberger et al., 1981a,b
Xa-2/xa-2 heterozygote	20 NAA + 5 BA (initiation) 5 NAA + 1 BA (maintenance)	5 IAA + 20 BA	MS	Hypocotyl, cotyledon	Seeni & Gnanam, 1981

[a]Concentrations given induced the highest frequency of response.

[b]Various chlorophyll, morphological and auxotrophic mutants: yg-1; yg-2; yg-4; yg-5; yg-6; Xa-1; Xa-2; Xa-ag; suf-aw,d; pe-lg; t ; 1-dcary; Yv-co; pv; Tl⁻.

[c]Concentrations given induced a response in the majority of the strains tested but were not necessarily optimal for any one specific strain.

[d]Best combination for eventual shoot regeneration from callus initiated during the fall and winter months.

[e]Best combinations for eventual shoot regeneration.

[f]Best combinations for eventual shoot regeneration from callus initiated during the spring and summer months.

Table 4. Shoot Regeneration from Wild Tomato and Related Species

SPECIES	GROWTH REGULATORS[a] (µM)		SHOOT MEDIUM	EXPLANT	REFERENCE
	Callus	Shoot Formation			
L. peruvianum	—	198 ADE	Gautheret[b] + 10 mg/l cysteine	Roots	Norton & Boll, 1954
L. peruvianum	1 IAA + 1 IBA + 1 KIN + 1 2iP	1 IAA + 1 IBA + 1 KIN + 1 2iP	LS	Stem internodes	DeLanghe & DeBruijne, 1976
L. peruvianum	2.7–32.2 NAA + 0–27.9 KIN	11.4 IAA + 0 or 9.3 KIN (or) 34.2 IAA + 27.9 NAA (or) 27.9 KIN (or) 0–11.4 IAA + 8.9 BA	MS	Leaf	Tal et al., 1977
Solanum pennellii	2.7–32.2 NAA + 0–27.9 KIN (or) 11.4 IAA + 8.9 BA	0–0.12 IAA + 8.9 BA (or) 9.3 KIN	MS	Leaf	Tal et al., 1977

SPECIES	GROWTH REGULATORS[a] (μM)		SHOOT MEDIUM	EXPLANT	REFERENCE
	Callus	Shoot Formation			
L. peruvianum, *L. pimpinellifolium*, *L. cheesmanii*, *L. glandifolium*, *L. chmielewskii*, *L. parviflorum*, *S. lycopersicoides*, *S. pennellii*	—	5 BA	MS	Leaf	Kut & Evans, 1982
L. chilense, *L. hirsutum*	—	9 ZEA	MS	Leaf	Kut & Evans, 1982

[a]Concentrations given induced the highest frequency of response.

[b]See Street and Shillito, 1977.

succeeded in regenerating shoots from callus initiated on medium containing 22.8 μM IAA and 18.6 μM KIN. Shoots were produced from 8-25% of the explants. In this report, the quantity of callus produced was not directly related to the morphogenetic potential of the explant. The tomato strain that produced the most prolific callus growth did not have the highest efficiency in shoot regeneration. Varietal-specific in vivo plant vigor was also independent of efficiency of callus production or shoot regeneration.

Behki and Lesley (1976) investigated the morphogenetic response of fifteen mutant lines and two commercial varieties of *L. esculentum* in the presence of various combinations of NAA + BA and NAA + ZEA. Callus was produced by all clones using 0.5 or 2.5 μM NAA with 1.0-20 μM BA. The optimal combination of growth regulators for shoot regeneration varied considerably among the genetic lines studied. Shoots were regenerated in 4-6 weeks from 10 of the lines using 0.5 μM NAA and 10 μM BA, while the 2 remaining lines required higher BA or lower NAA for regeneration. The optimal phytohormone combination reported by Padmanabhan et al. (1974) produced shoots in only 3 of the 12 lines. A study of the morphogenetic response of leaf vs. stem explants for one clone indicated a significant difference in the response of these 2 tissues to the same hormone combinations. ZEA was substituted for BA for regeneration of 6 of the tomato clones and was generally more effective than BA. The optimal combination that was sufficient to regenerate all 6 lines was 0.5 μM NAA and 50 μM ZEA.

DeLanghe and DeBruijne (1976) investigated shoot regeneration from four *L. esculentum* cultivars and one *L. peruvianum* strain. Shoots were regenerated from callus initiated from stem internodes. *L. peruvianum* produced the most shoots with the greatest efficiency. The best response was obtained on medium with 2 auxins and 2 cytokinins (i.e., 1.0 μM IBA + 1.0 μM IAA + 1.0 μM KIN + 1.0 μM 2iP). This medium did not produce shoots in the 4 *L. esculentum* cultivars. For *L. esculentum*, optimal regeneration of 20-40% was obtained with 10 μM ZEA. Significant callus proliferation resulted from GA_3 at 1.0 or 10 μM. Shoot formation was stimulated by ABA in some cases, but was inhibited in others. Working on the assumption that the lower efficiency of shoot regeneration obtained for *L. esculentum* might be due to a high level of endogenous GA, DeLanghe and DeBruijne (1976) added CCC (chlormequat: 2-chloroethyl trimethyl ammonium chloride) at 1000 ppm to the medium. This supplement had no effect on shoot regeneration. However, pretreatment of plants serving as sources of stem explants with an aqueous spray containing 2000 ppm CCC stimulated shoot regeneration. Callus from explants of these pretreated plants could be maintained in medium supplemented with 1.0 μM KIN by subculturing every three weeks. Normal shoots capable of rooting could be regenerated from callus maintained for two years on this medium after transfer to medium containing 1.0 μM ZEA or 20% CW. This suggested the maintenance of genetic stability in the long-term callus culture. No phenotypic variability was observed among 234 regenerated plants examined. However, one variant did appear in the F_1 population of 30 of these regenerates. This suggests the induction of at least some genetic variability in the cell culture propagation of crop plants.

Kartha et al. (1976) studied growth and morphogenetic responses of leaf sections of the *L. esculentum* variety Starfire to various individual phytohormones and combinations of phytohormones over a concentration range of 0.1-10 µM. The goal was to establish a culture system with a high shoot regeneration frequency. Of the cytokinins tested, BA and ZEA, each at 5 µM, elicited the best morphogenetic response from leaf sections with multiple shoot formation and rooting. BA was superior to ZEA as shoot development occurred earlier. Kinetin alone induced only erratic and inconsistent shoot formation at 5 µM but at 0.1 µM was the most effective cytokinin for callus induction. As reported earlier, DeLanghe and DeBruijne (1976) observed a similar response of stem internodes to low concentrations of KIN. Both IAA and NAA alone induced callus growth and root formation but no shoot regeneration. The combination of BA + IAA was more efficient in shoot regeneration than BA + NAA. A concentration of 10 µM BA induced multiple shoot formation irrespective of the concentration of IAA. The combination of BA + NAA, on the other hand, induced only erratic shoot formation. Combinations of KIN + IAA and KIN + NAA induced only callus and root growth except for a few shoots recovered at 10 µM KIN and 0.1 µM NAA. As in the case of BA, the combination ZEA + IAA was more efficient in shoot regeneration than ZEA + NAA. A concentration of 1.0 µM ZEA + 0.1 or 1.0 µM IAA induced multiple shoot formation. Thus, the medium containing ZEA + IAA resulted in the regeneration of shoots over a wider range of concentrations than did medium with BA + IAA. The ZEA + NAA medium produced mainly callus except for a few shoots at 1.0 or 10 µM ZEA and 0.1 µM NAA. The effectiveness of the various hormone combinations on shoot regeneration, were as follows (in descending order): ZEA + IAA, BA + IAA, ZEA + NAA, BA + NAA, KIN + NAA, and KIN + IAA. These results suggest that endogenous levels of auxin preclude the need for a strong exogenously supplied auxin like NAA but require the presence of a stronger cytokinin than KIN for shoot regeneration. A specific cytokinin-auxin balance must be established to achieve morphogenesis.

Kartha et al. (1977) also reported high frequency shoot regeneration from shoot apical meristems of seven-day-old seedlings of the same *L. esculentum* variety, Starfire. Apical meristem culture can be used to produce virus-free plants from infected plants or to clone desirable genotypes. Meristem cells are more rapidly growing and more uniformly diploid than those of more mature tissues such as leaves. Thus, Kartha et al. (1977) suggested that meristem culture might provide a more efficient plant propagation system resulting in fewer polyploid or chromosomally aberrant plants. High frequency shoot regeneration was observed for a wider range of horomone concentrations than was observed for leaf explants (Kartha et al., 1976; see Table 3). Shoots could again be regenerated on BA and ZEA alone but with 90-95% efficiency. The highest frequency of shoot regeneration was obtained for several concentrations of BA + IAA. Several concentrations of BA + NAA also produced shoots. The combination of ZEA + IAA was slightly less efficient in regeneration than BA + IAA, and ZEA + NAA was only slightly less efficient than ZEA + IAA. At a concentration of 1.0-10 μM, IAA alone could induce the same high frequency of shoot regeneration, whereas NAA alone at the same concentrations induced only a low frequency of regeneration.

Tal et al. (1977) investigated the regeneration potential of leaf sections from the flacca (flc) mutant of *L. esculentum* which has a lower content of ABA than normal tomatoes and, consequently, higher endogenous KIN- and auxin-like activity. They compared regeneration of flc leaves to leaves of *L. peruvianum* and *Solanum pennellii*. *L. peruvianum* expressed the highest morphogenetic potential of the lines studied, followed by *S. pennellii*, while the flc mutant regenerated at an extremely low frequency in all hormone combinations tested. The endogenous hormonal imbalance in the flc mutant did not affect its regenerative potential as compared to the normal cultivar. Thus, the lower endogenous ABA content of *L. peruvianum* cannot be solely used to explain its higher regeneration efficiency. The authors suggested that the high morphogenetic potential of *L. peruvianum* might be related in some way to the high degree of heterozygosity of this self-incompatible species.

Vnuchkova (1977) tested 150 media variations for callus formation and shoot regeneration from the Bizon variety of *L. esculentum*. She found that media having a high nitrogen content result in rapid blackening and atrophy of explant material. Media such as White's medium (1943) which are nutritionally deficient were also not conducive to growth. Large numbers of different vitamins at high doses, such as ascorbic and nicotinic acids, were also ineffective. Vnuchkova also observed that tomato explants did not respond to casein hydrolysate or yeast extract in the medium.

Although callus could be induced by several concentrations and combinations of hormones, Vnuchkova noted that only the denser callus could be induced to form shoots. Dense callus growth was achieved with higher hormone concentrations. Because it tended to induce friable callus growth, 2,4-D was found to be unsuitable for regeneration. Vnuchkova observed that the types and concentrations of hormones necessary to induce shoot differentiation from explants depended on the season of the year and was, therefore, probably dependent on endogenous concentrations of hormones. Shoot buds developed into plantlets at a higher efficiency when cultured under 5000-10,000 lux.

Many of the regenerated plants obtained by Vnuchkova had reduced fertility. Upon cytogenetic analysis of the callus cultures, a large degree of polyploidization (up to 200 chromosomes per cell) was observed, as well as lagging chromosomes, bridges, and fragments. Polyploidization was the result of endomitosis and steadily increased after the third passage in culture. Optimal media could be developed for callus growth that decreased chromosomal abnormalities. Vnuchkova also found that an increase in callus regeneration potential could be achieved by the addition of adenine to the culture medium. Adenine has been reported to promote growth of diploid cells while inhibiting polyploid callus (Nitsch et al., 1967).

Ohki et al. (1978) conducted a detailed study of the variability of in vitro shoot-regeneration capacity within a particular species as well as the transmission of this capacity through hybridization. They found that efficiency of shoot formation was dependent upon (a) the genotype of the plant, (b) the tissue or organ serving as the source of explant material (varying types and concentrations of endogenous growth regulators), (c) the position within the organ from which the explant was excised (uneven distribution of endogenous growth regulators and inequality of histological constitution), (d) the maturity of the organ (changes in types and concentrations of growth regulators as development proceeds; increasing difficulty in inducing dedifferentiation of cells expressing specialized physiological functions and no longer undergoing active cell division), and (e) the types and concentrations of growth hormones present in the culture medium (critical cytokinin:auxin balance).

Ohki's results on the comparison of the regeneration capacity of hybrids vs. parents depended on the hormone concentrations present in the medium. Hypocotyl explants of one parental line, Porphyre (P), always exhibited a higher regeneration capacity than the other parent, Apedice (A). In one medium, the organogenetic capacity of hypocotyl sections of the two hybrids was intermediate to that of the two parents suggesting an interaction of genomes. In two other media, the P x A hybrid regenerated a greater number of shoots than the A x P hybrid suggesting a maternal effect on shoot regeneration. The P x A hybrid also produced a greater number of shoots than either of its parents suggesting heterosis, while A x P regenerated an intermediate number of shoots in comparison with the parents. The results obtained for a similar experiment conducted with leaf explants were quite different. Shoot regeneration from leaf sections required higher concentrations of cytokinin as well as a ratio of cytokinin to auxin of 10-20:1. Secondly, no clear difference in regeneration capacity was observed between the parental lines and their hybrids (Table 3) due largely to the high degree of variability in regeneration between individual leaf explants. Parent A could be regenerated more efficiently than parent P from leaves. Analysis of plants regenerated from both hypocotyl and leaf explants revealed that approximately 10% were morphologically abnormal with curled, small, thick leaves, short and thick internodes, and/or small sterile flowers. Some of the morphologically abnormal plants as well as several morphologically normal plants were observed to be tetraploid (4n) or mixoploid (2n + 4n). This recovery of genetic variants must be taken into consideration and minimized by employing optimal culture conditions if cell culture is to be used for plant propagation.

Herman and Haas (1978) investigated the origin of shoots regenerated from tomato explants. They repeatedly subcultured leaf-derived callus and observed new shoot formation only from that callus which continued to contain fragments of shoot of leaf tissue. Co-culture of callus pieces not containing shoot or leaf tissue with callus containing growing shoots did not stimulate the unorganized callus to undergo morphogenesis. These results suggested that regenerated tomato shoots only originated as a result of adventitious growth from differentiated tissue of the explant and not as a result of reorganization after callus formation. However, shoots have subsequently been recovered from dissociated or reorganized callus (Meredith, 1979; Cappadocia and Sree Ramulu, 1980; Kut and Evans, 1982).

Meredith (1979) investigated tomato shoot regeneration from established callus cultures. In both her initial experiments employing callus and cell suspension cultures from 4 lines of *L. esculentum* and 1 line of *L. pimpinellifolium* which had been maintained in culture for 17-28 months, and her later experiment employing 4-month-old callus, Meredith was successful in regenerating shoots from only one genotype, VFNT Cherry. Leaf-derived callus of VFNT Cherry expressed better morphogenetic response than anther-derived callus. Although all three cytokinins tested (BA, 2iP, and ZEA) could induce shoot regeneration alone, ZEA produced the greatest number of shoots. Meredith ascertained that a low level of auxin was necessary to prevent loss of callus vigor with successive transfers during shoot induction. Regeneration required 3-5 months from the oldest callus cultures, but only 2 months from young callus. Although young callus produced shoots of normal morphology which rooted easily, the shoots regenerated from old callus were morphologically abnormal and did not form roots.

Cassells (1979) attempted to increase the number of tomato shoots regenerated per explant. The maximum number of shoots per explant had been 6 (DeLanghe and DeBruijne, 1976), with most laboratories reporting only 1-3 shoots per explant. Employing stem explants from plants grown during winter months, Cassells noted that a slower growing, dense callus formed at the apical end of the explant and that is was from this callus that 1 or 2 shoots arose. An apical-dominance effect exerted by these initial shoots could be overcome by excising the shoots and/or by subculturing pieces of the callus to fresh medium. By adding 0.2 μM 2,3,5-triiodobenzoic acid (TIBA) to the medium, callus was induced along the entire length of the explant instead of being limited to the cut ends. Shoots were formed along the stem explant and at a higher frequency (31 vs. 1.7 per explant) using TIBA. However, no difference in shoot regeneration efficiency was observed for TIBA-treated vs. nontreated explants from plants grown during the summer (30 vs. 30 per explant). Cassells' results suggest a seasonal variation in endogenous auxin concentration which influences shoot regeneration capacity.

Cappadocia and Sree Ramulu (1980) succeeded in regenerating additional shoots from stem internode callus by changing the hormone content of the regeneration medium. They added a high level of ADE and CW and omitted auxin. However, this subcultured callus yielded an increasing percentage of tetraploid plants with 50% of the regenerates being tetraploid after 24 months in culture while only 25% of the initial regenerates were tetraploid.

Gunay and Rao (1980) investigated the propagation of hybrid cultivated tomato plants via hypocotyl and cotyledon explants. Hypocotyl explants were more responsive than cotyledon explants producing shoots in a greater number of hormone combinations. The hormones tested were BA, KIN, IAA, and NAA. The best combination was BA + IAA which induced shoot formation in 100% of the cultured explants. BA + NAA and KIN + NAA did not induce shoot formation.

Behki and Lesley (1980) developed an optimal medium for callus induction from leaf explants that would give the highest shoot regeneration frequency. They first investigated the effect of nitrogen source on shoot formation and found that decreasing the total nitrogen content below 80% of MS medium inhibited shoot regeneration as did the presence of an equal or excess amount of NH_4^+ as compared to NO_3^- as the nitrogen source. NO_3^-, however, could be substituted for NH_4^+ without any deleterious effect on plant regeneration. Behki and Lesley subsequently found that successful shoot regeneration from leaf callus depended on the types and concentrations of growth regulators in both the callus induction medium and the shoot differentiation medium, as well as on length of incubation on callus induction medium before transfer to shoot differentiation medium. Preliminary results from auxin experiments for callus induction from Xa-ag were as follows. First, IAA alone induced profuse root growth. A combination of IAA + BA yielded improved callus growth, but root formation was still substantial. Second, NAA alone induced a harder callus plus roots. A combination of NAA + BA inhibited root growth while inducing good growth of white, friable callus. Third, 2,4-D alone quickly stimulated rapid callus proliferation that was white and friable but which could not be induced to regenerate shoots except at 0.18 µM if it was transferred to shoot differentiation medium within 12 days of initial explant. A combination of 2,4-D + BA produced callus that maintained potential for morphogenesis for at least 18 days after initial explant for the Xa-ag clone. Lastly, Behki and Lesley found that only callus induced in the light was capable of subsequent morphogenesis on shoot differentiation medium.

Frankenberger et al. (1981a) examined the effects of explant source and donor plant growth conditions on shoot regeneration for eight different tomato varieties. For plants grown during the fall season, they found significantly higher shoot-forming capacity in leaf explants from younger (6-week-old) plants, leaves closest to the apical meristem, and the proximal (closest to the petiole) portion of leaflets. On the other hand, no effect of leaf maturity on shoot-forming capacity was observed for spring-grown plants. There was an overall three-fold reduction in shoot-regeneration capacity in spring-grown as compared to fall-grown plants confirming earlier results of Vnuchkova (1977) and Cassells (1979). Experiments with hypocotyls and cotyledons from seedlings grown in vitro and in the greenhouse suggest that both explant source and environment of the donor plant significantly affect the total number of shoots produced by each genotype. In contrast to Gunay and Rao (1980), cotyledon explants produced a greater number of shoots than hypocotyl explants. In vitro-grown seedlings produced a greater number of shoots than greenhouse-grown seedlings. All of the above results emphasize the importance of precise reporting of environmental growth conditions and explant source when conducting regeneration experiments.

In further studies on the genetic basis for shoot formation, Frankenberger et al. (1981b) found no correlation between shoot-regeneration capacity and growth habit of the plant from which explant material was taken. A diallel cross was completed with six genotypes displaying a large difference in shoot-forming capacity in an effort to study the heritability of this trait. Shoot-forming capacity was found to be highly heritable. A large degree of the variability in shoot-forming capacity among the hybrids was accounted for by additive gene action, while the high shoot-forming capacity of two of the three high shoot-forming genotypes was associated with the presence of recessive genes.

Kut and Evans (1982) investigated the regeneration potential of leaf explants of eight *Lycopersicon* species and two closely related *Solanum* species as compared to cultivated *L. esculentum* on MS medium supplemented with eight different previously published hormone combinations. We identified that MS medium supplemented with 5 μM BA is generally sufficient for regeneration of most of these species. The best regeneration response on this medium was elicited from *L. peruvianum*, *S. lycopersicoides*, and *L. glandifolium*, while a comparable response was obtained for *L. chilense* in the presence of 9 μM ZEA. *L. hirsutum* proved to be the most recalcitrant species, with only one ecotype producing one shoot in only one of three explants after 13 weeks in culture in the presence of 9 μM ZEA. We have since succeeded in obtaining a shoot from a second ecotype of *L. hirsutum*. Both ecotypes have been maintained in cell culture for extended periods of time and have regenerated additional shoots. All of the wild species studied with the exception of *L. hirsutum* and *L. pimpinellifolium* regenerated more efficiently than *L. esculentum*.

APPLICATIONS OF SHOOT CULTURE. Cappadocia and Sree Ramulu (1980) established shoot cultures of a self-incompatible (SI) *L. esculentum* x *L. peruvianum* hybrid to study the interaction of the alleles of *L. peruvianum* controlling SI and the alleles controlling unilateral interspecific incompatibility (USI). Tetraploidy can lead to self-compatibility (SC) in species such as *L. peruvianum*, which displays a gametophytic monofactorial system of SI. The presence of two different alleles of the SI locus in the same pollen grain results in a breakdown of the incompatibility phenotype. Tetraploid cell culture regenerates of the diploid hybrid proved to be SI and reciprocally cross-incompatible with the diploid hybrid and other tetraploid regenerates. In this case, there was no heteroallelic pollen effect on SI, suggesting that the SI and USI alleles function independently of each other and cannot complement each other. Furthermore, in studies conducted by Ancora and Sree Ramulu (1981) with regenerates of self-incompatible *L. peruvianum*, it was reported that the presence in pollen of heteroalleles for the SI locus was necessary but not sufficient for the establishment of SC.

Seeni and Gnanam (1981) established tomato shoot cultures to study production of plant chimeras. Of the shoots regenerated from yellow-green plants heterozygous for the Xa-2 chlorophyll mutation (Xa-2/xa-2), 10-15% were chimeras. No chimeras were regenerated from homozygous green (xa-2/xa-2) or homozygous albino (Xa-2/Xa-2) plants. Leaves of the chimeras possessed dark green, yellow-green, and white areas. From each of these areas plants could be regenerated that were pure for that color. Although the genotypes of the particular colored area regenerates were not determined, these experiments suggest segregation or recombination occurs in mitotic cells grown under cell culture conditions and implies the usefulness of shoot culture to obtain new pure breeding plant genotypes or phenotypes.

Anther and Pollen Culture

The production of haploid plants from anther or pollen culture would allow for the fixation and analysis of new gene combinations from hybrid plants (including interspecific hybrids) in less time than is required in conventional breeding programs, as well as allowing the establishment of homozygous true, breeding tomato lines. Haploid cell cultures would also prove more amenable to mutant isolation, somatic cell protoplast fusion, and genetic engineering techniques. However, the achievement of these goals in tomato has been frustrated by an extremely low efficiency of haploid/dihaploid plant recovery and the lack of repeatability between laboratories.

The efficiency of haploid plant production depends on at least 5 variables: (a) growth conditions of the donor plant, (b) the stage of development of the pollen cells at the time of anther removal (best results for tomato have been obtained for pollen mother cells in meiosis), (c) cell culture conditions, including medium components, that need to be optimal for haploid callus and shoot production but should minimize stimulation of maternal tissue growth as well as minimizing polyploidization over time, (d) variation in genotypic potential for haploid callus induction, and (e) variation in genotypic potential for haploid/dihaploid shoot regeneration from cultured callus. Published growth conditions for haploid callus and/or haploid shoot production are listed in Table 5.

Apparently, haploid and dihaploid shoot production occurs as a result of organogenesis from an initial proliferation of callus and not via androgenesis. However, Dao et al. (1976) observed embryogenesis as one of three in vitro development pathways followed by pollen cells cultured at the uninucleate microspore stage. Embryos degenerated at the 16-20 cell stage. The other two developmental pathways were callus production and normal microsporogenesis. The latter also eventually degenerated. Pollen cells cultured at the tetrad stage (termination of meiosis) resulted in microsporogenesis. The greatest proliferation of callus was observed from pollen mother cells in early meiosis. Dao and Shamina (1978) also reported embryos from binucleate pollen. These embryoids degenerated at the 16-32 cell stage. Cappadocia and Sree Ramulu (1980) also observed globular embryos from cultured anthers of *L. esculentum* x *L. peruvianum* hybrids which developed up to the 32 cell stage. These embryos and the only callus proliferation observed were induced solely from anthers containing microspores in the late mononucleate mitotic stage.

Most of the research conducted on tomato anther culture has centered on either optimizing medium composition or influencing pollen cell responsiveness by various pretreatments. Sharp et al. (1971) found that the addition of ADE, YE, CH, or CW elicited very little or no growth stimulation of cultured tomato anthers. Increasing the sucrose concentration resulted in an increased percentage of anthers producing callus up to a maximum of approximately 43% at 0.4 M sucrose. Haploid callus was stable maintained in vitro for 7 months. This callus was induced to undergo organogenesis with resulting root but no shoot formation when transferred to medium with a reduced sucrose content of 0.12 M.

Table 5. Anther and Pollen Culture of Tomato

VARIETIES	GROWTH REGULATORS (μM) Callus	Shoot Formation	BASAL MEDIUM	GROWTH CONDITIONS	REFERENCE
Rutgers, Yellow Jubilee	1.45 2,4-D + 0.6 KIN (haploid callus initiation)	2.26 2,4-D + 1.16 KIN (only roots regenerated)	Modified White + 0.4 M sucrose (callus)	12/12 hr pp[a], 25 C	Sharp et al., 1971
	(or) 5.37 NAA + 0.6 KIN (haploid callus initiation)		Modified White + 0.12 M sucrose (shoot)		
	2.26 2,4-D + 1.16 KIN (haploid callus maintenance)				
MHVF145-21-4P, VF145-22-8/ YTMF8BS$_2$-S$_1$S$_6$S$_1$BB, No. 65-B-2	10.74 NAA + 22.23 KIN (haploid callus initiation)	0.54 NAA + 9.3 KIN (haploid shoots)	DBMI[b]	24 hr light then dark, 27 C, 70% RH (callus); 16/8 hr pp, 27 C (shoot)	Gresshoff & Doy, 1972
	(or) 10.74 NAA + 4.65 KIN (haploid callus initiation)		DBMII[b]		
	26.85 NAA + 0.05 KIN (haploid callus maintenance)		DBMIII[b]		

VARIETIES	GROWTH REGULATORS (μM) Callus	Shoot Formation	BASAL MEDIUM	GROWTH CONDITIONS	REFERENCE
Rutgers	27.14 2,4-D (haploid callus)	No shoots regenerated	Modified White + 0.12 M sucrose	12/12 hr pp, 25 C	Sharp et al., 1972 (pollen culture)
Ailsa Craig dl/dl	0.45-4.5 2,4-D (gametophytic and somatic callus)	None; 1/2 sucrose only roots regenerated	Nitsch & Nitsch (1969), LS	12/12 hr pp, 20-25 C	Hender, 1973
50 varieties	0.05-0.5 2,4-D + 4.9-24.7 ADE	No shoots regenerated	MS, Nitsch (1969), DBMII	16/8 hr pp, 24-32 C	Levenko et al., 1977
Gruntovyi, Gribovskii	0-21.5 NAA + 0-18.6 KIN + 0-2 mg/l Sh-12[c]	0.54-21.5 NAA + 0.47-18.6 KIN + 0-0.2 mg/l Sh-12 (uninucleate microspores - embryoids) 0.54 NAA + 0.47 KIN + 0-2 mg/l Sh-12 (binucleate pollen - embryoids)	Modified DBM	16/8 hr pp, 18-25 C, 70% RH	Dao & Shamina, 1978
ms10[35]	2.85 IAA + 1.11 ZEA (gametophytic and somatic callus)	2.85 IAA + 1.11 ZEA (diploid shoots from maternal tissue)	MS	16/8 hr pp, 27 C	Zamir et al., 1980, 1981

Table 5. Cont.

VARIETIES	GROWTH REGULATORS (μM)		BASAL MEDIUM	GROWTH CONDITIONS	REFERENCE
	Callus	Shoot Formation			
4S-101, 4S-102, Montfavet 63-4 x HS-101	10.74 NAA + 4.65 KIN (haploid callus initiation)	0.45–4.5 2,4-D + 8.87 BA (only meristematic buds regenerated)	DBMII[b] (callus)	Dark (callus initiation), 16/8 hr pp (subculturing), 26–30 C	Gulshan & Sharma, 1981
	9.05 2,4-D + 0.44 BA (haploid callus maintenance)		MS (shoot)		
ms10[35]	2.85 IAA + 1.11 ZEA (gametopytic and somatic callus)	2.85 IAA + 1.11 ZEA (about 10% of regenerated shoots were haploid)	MS	16/8 hr pp, 27 C	Ziv et al., 1982

[a]Photoperiod: hours in light/hours in dark.

[b]Gresshoff and Doy basal media (1972).

[c]1-(4-methyl-6-phenyl-3-cyclohexenyl)butanol-1-OH-3; a synthetic biostimulant.

Gresshoff and Doy (1972) developed their own media but obtained haploid callus from anthers of only three of 43 varieties tested. Pollen mother cells in early meiosis elicited the highest frequency of callus production which varied from 50–70%. Haploid callus was observed only after a swollen pollen sac had burst through the anther wall. All other callus growth, especially in the filament area, originated from diploid maternal tissue. Haploid plants with abnormal leaf morphology were regenerated. The totipotency of the haploid callus was further suggested by the development of pseudo-fruits of one centimeter diameter without seeds after transfer of callus which had been incubated on shoot differentiation medium for 10 weeks to DBM II supplemented with 26.85 μM NAA + 9.3 μM KIN. Haploid callus was stably maintained in vitro for one year. Stimulation of the growth rate of this callus was later achieved by the addition of 8.2 μM biotin and 2.1 μM calcium pantothenate.

Devreux et al. (1976) succeeded in producing callus from *L. peruvianum* anthers. However, only diploid, tetraploid, and chimeric plants were regenerated from diploid maternal tissue.

Experiments by Levenko et al. (1977) suggested that ratios and concentrations of hormones play a major role in callus production, while the basal medium mineral salts do not. The frequency of callus induction (4–50%) could be doubled by the addition of extracts of anthers, beech leaves, or phytohemagglutinin but was not affected by plating on media conditioned by previous callus growth. The ratio of polyploid and aneuploid to haploid cells in primary callus cultures varied between genotypes (from 75–100% polyploid + aneuploid) but increased with time in culture, indicating the necessity of inducing haploid shoot formation soon after callus initiation. No shoots were regenerated.

Cappadocia and Sree Ramulu (1980) observed that pretreatment of buds containing microspores in late mononucleate mitosis with 45.2 μM 2,4-D for 6 hr at 4 C enhanced callus proliferation four-fold. However, only diploid plants derived from maternal tissue were regenerated.

Based on the enhanced frequency of haploid recovery from other species such as tobacco by cold pretreatment of anthers which disrupts normal microsporogenesis, Zamir et al. (1980) investigated the use of anthers from 15 nonallelic, recessive, male-sterile (ms) tomato mutants for haploid shoot production. The ms 10^{35} mutant was the only mutant which consistently induced callus proliferation. The introduction of this mutation into several genotypes which did not previously produce callus in anther culture resulted in the induction of significant callus growth. The ms 10^{35} mutation arrests pollen development at the tetrad stage of meiosis. Callus consisting in part of haploid cells was produced only from anthers cultured soon after the initiation of meiotic breakdown (1.5–2.5 mm in length). However, this callus was highly polyploid and contained callus induced from cells of the anther wall. A concentration of 58.4 mM sucrose in the anther culture medium favored callus proliferation while a 29.2 mM sucrose concentration favored shoot formation of unreported ploidy level.

Employing the same culture system, Zamir et al. (1981) then made use of ms 10^{35} mutant plants heterozygous for additional morphological and isoenzyme markers. Based on the analysis of these markers, plants regenerated from their anther-derived, mixoploid callus were shown to have originated from maternal diploid tissue.

Gulshan and Sharma (1981) succeeded in inducing callus proliferation from anthers containing microspores in the early uninucleate stage. They found that medium containing NAA and KIN stimulated development of callus only from sporogenous tissue. No response was elicited from somatic tissue. Combinations of 2,4-D and BA stimulated development of both types of callus. However, MS medium containing 2,4-D and BA was optimal for callus maintenance suggesting that different morphogenetic phases of plant development in cell culture may have very different nutritional requirements. Gresshoff and Doy's DBM II (1972) was found to be the optimal basal medium for callus induction. No shoots could be regenerated, even though a number of meristematic nodules were observed. Cytological studies of microspore development in culture indicated that only about 3% of the microspores were capable of eventual development into callus. These microspores underwent a symmetric ("B" pathway of androgenesis) first mitotic division. Morphogenetic studies on callus transferred to shoot differentiation media suggested the possible requirement for the presence of several different growth regulators, such as 2,4-D + NAA + BA + KIN, to achieve successful shoot regeneration.

Ziv et al. (1982) succeeded in regenerating and confirming the identity of dihaploid homozygous plants employing the ms 10^{35} mutant and culture system, described by Zamir et al. (1980), and a genetic marker. As reported by Zamir et al. (1980), the ms 10^{35} gene exerted a significant positive influence on anther-callus production in all cultivars into which it was introduced, although the quantity of callus produced was dependent upon the cultivar to which the ms 10^{35} gene had been transferred. However, no apparent effect was exerted by the ms 10^{35} gene on shoot regeneration. There also was a gene dosage effect of the ms 10^{35} gene on callus induction frequency as is illustrated in the following example for plants of one particular genotype: homozygous sterile was 25%, heterozygous fertile was 14%, and homozygous fertile was 5%. Sucrose concentrations higher than 29.2 mM or lower than 14.6 mM were also found to result in a drastic reduction in callus and shoot yield. The extent of the stimulatory effect of the lower sucrose concentrations on callus production but not shoot regeneration depended on the genotype.

Pollen culture is another means of obtaining haploid plants. The culture of isolated pollen cells eliminates confusion as to the gametophytic or somatic origin of diploid cells and plants. Sharp et al. (1972) were the first to report successful cultivation of isolated tomato pollen cells by nurse culture. Pollen cells were isolated in liquid culture medium by teasing the walls of anthers with dissecting needles. Nurse cultures were prepared as follows: (a) an intact anther cone from a single 1 cm bud was placed horizontally on agar medium, (b) a sterile filter paper disk was placed over the anther cone, and (c) approximately ten pollen cells were pipetted onto the filter paper. Plating efficiency ranged from 0-60%. No report was made of shoot regeneration, but haploid callus colonies were obtained.

Protoplasts

Protoplasts of *Lycopersicon* species have been readily isolated for many years (Gregory and Cocking, 1965). However, these protoplasts have consistently degenerated in culture. Plants of only two *Lycopersicon* species have been regenerated from mesophyll protoplasts. *L. peruvianum* protoplasts were first regenerated by Zapata et al. (1977), and more recently *L. esculentum* cv. Lukullus, a cultivated tomato, was regenerated by Morgan and Cocking (1982).

Because of the fragile nature of tomato protoplasts, much attention has been given to the manner in which donor plants are grown. Low light intensities (4000–8000 lux) have proven to be optimal for release of viable protoplasts (Morgan and Cocking, 1982; Zapata et al., 1981). Cassells and Barlass (1976, 1978) reported that leaves of plants grown under high light intensities contained ten times as much calcium pectate as those grown under low light intensities. This required higher enzyme concentrations to release protoplasts from leaves of plants grown at high light intensities. Commonly, donor plants are grown at 25 C. Use of young, rapidly growing plants, 3-8 weeks old, has given the highest viable protoplast release. Additionally, Tal and Watts (1979) reported that growing plants under a high relative humidity (82%) enhanced protoplast viability.

Pre-plasmolysis, the incubation of mesophyll tissue in an osmoticum before enzyme treatment, has often been utilized for *Lycopersicon* species (DeWit, 1976). In our laboratory, pre-plasmolysis has been found to increase the percentage of intact protoplasts released from mesophyll tissue. Mannitol alone or a salt solution containing mannitol have been used successfully as pre-plasmolyzing solutions (Morgan and Cocking, 1982; Muhlbach and Thiele, 1981).

The types and propagations of enzymes used for isolation of tomato mesophyll protoplasts has varied. Early reports warned of a decrease in protoplast viability when Driselase (Muhlbach, 1980) or Macerozyme (DeWit, 1976) was included in the enzyme mixture. However, in the reports of plant regeneration from protoplasts these two enzymes have been employed for protoplast isolation. Many groups have combined Meicelase, Macerozyme, and Driselase in a salt-sugar mixture for protoplast release. Generally, 13-18 hr at 25-29 C in darkness is required for this treatment (Zapata et al., 1981; Morgan and Cocking, 1982). A stationary, rather than shaking incubation is used due to the fragile nature of tomato protoplasts.

The fragile nature of these protoplasts also discourages use of filtration as a means of purifying crude protoplast preparations. Rather, floatation of the protoplasts on a sucrose cushion provides a more gentle and effective purification method (Zapata et al., 1977).

Because of the recalcitrant nature of cultured tomato protoplasts, much effort has been given to developing a medium in which these protoplasts will divide rapidly. The B5 medium of Gamborg et al. (1968), modified by Zapata et al. (1977), has commonly been used to culture tomato protoplasts. This medium was used in the two reports in which plants were regenerated from protoplasts of *L. peruvianum*

(Zapata et al., 1977) and *L. esculentum* (Morgan and Cocking, 1982). The balance of salts and vitamins in this medium was found to be essential for callus regeneration from protoplasts. Culture medium solidified with agar has been utilized as often as liquid culture medium. However, in the only reported protoplast regeneration of cultivated tomato, protoplasts were suspended in an agar medium (Morgan and Cocking, 1982).

Protoplast culture conditions include an initial dark period of one week at a temperature of 25-30 C followed by a gradual increase in light intensity. Zapata et al. (1977) reported that no cell division took place in *L. esculentum* and *L. peruvianum* protoplasts cultured at temperatures less than 25 C. Optimum culture temperature for these species was 27-29 C. Commonly employed protoplast culture density has been 0.5-2 x 10^5 protoplasts per ml (Morgan and Cocking, 1982; Tal and Watts, 1979). However, there have been no detailed reports testing this density.

Plant growth conditions, protoplast culture medium, and growth environmental conditions have all been found to influence the stability and viability of tomato mesophyll protoplasts. Variations in the reported conditions for protoplast growth will be needed to allow regeneration of additional species. Indeed, even within one species, *L. esculentum*, only one cultivar, Lukullus, out of 14 cultivars studied was regenerated from protoplasts under the conditions described by Morgan and Cocking (1982). This indicates that if tomato protoplasts are to be useful for somatic hybridization studies and transfer of useful characters from wild to cultivated species, further refinements of current techniques must be made in order to regenerate a wide range of *Lycopersicon* species and cultivars.

PROTOCOLS

Callus Culture

No general isolation and culture procedure has yet been established for tomato callus. This is due partly to the large degree of genetic variability among tomato species and varieties for callus production. Another problem encountered has been the high degree of chromosome instability (aneuploidy and polyploidy) observed in most of the tomato callus cultures established to date. The only exceptions have been certain varieties of *L. peruvianum* and the L2 (*L. esculentum* x *L. peruvianum* backcrossed line containing 25% *L. esculentum*) breeding line (Thomas and Pratt, 1981b).

EXPLANT PREPARATION. A variety of sterilization techniques have been employed depending on the type of tissue to be placed in culture. Leaves, stem pieces, and unopened flower buds (for anther culture) are usually prepared for sterilization by first rinsing the plant material under a gentle stream of tap water. A gentle rinsing in a dilute detergent solution can also be used to remove contaminants and soil particles from soil-grown plants. Tissues and seeds can then be dipped in 70% ethanol followed by immediate rinsing in sterile distilled water. Ethanol is often not used for sterilization as the leaves of many tomato species are very sensitive to ethanol. Many laboratories omit ethanol treatment entirely for all types of tissue due to deleterious effects on the response of the explanted tissue in cell culture. Each of the aforementioned tissues as well as seeds are then placed in a dilute solution of hypochlorite. The most common practice is to employ a 5-10% solution of commercial bleach (ca. 0.25-0.5% sodium hypochlorite). A few grains or drops of detergent can be added to the bleach solution to promote tissue wettability. Sterilization time in the bleach solution ranges from 5 to 15 min depending on the sensitivity of the tissue. Seeds suspected of being heavily contaminated can be sterilized in a 20% commercial bleach solution for 15-20 min. A thorough rinsing of the tissues is then achieved by dipping in two to three successive changes of sterile distilled water.

Leaves are generally cut into 5-10 mm^2 pieces exposing a cut surface on all 4 sides to allow maximum exposure to the culture medium. Thicker stem pieces are peeled to remove the epidermis and cortex, leaving the pith tissue, and then cut into 2-3 mm disks of about 1 mm thickness (Doering and Ahuja, 1967).

Flower buds must be dissected to remove the anthers. Seeds are normally germinated on the callus culture medium to provide a sterile source of explant material. Individual tissues of sterile seedings used as explant material include the hypocotyl, cotyledons, stem sections, and true leaves.

CULTURE MEDIA. The more defined media such as MS (Murashige and Skoog, 1962) and B5 (Gamborg et al., 1968) are used in conventional laboratories. The most commonly used of the two is MS medium. Several modifications of MS medium have been employed to improve the growth of different varieties and species. These modifications include increasing vitamin concentrations (Meredith, 1978a), omitting glycine (Meredith, 1978a), and adding an undefined supplement such as 100-150 ml CW (Thomas and Pratt, 1981a; Warren and Routley, 1970) or 0.05-0.2% CH (Gamborg, 1975). Meredith (1978a) also employed the mineral salts of CS5 medium (Scowcroft and Adamson, 1976) for better growth of the *L. esculentum* variety "Marglobe." Chin et al. (1981) used a modified White's medium (Sheat et al., 1959) for the initiation of callus from stem sections.

Hormone additives, as stated before, vary in type and concentration depending on the requirements of the particular species or variety being studied. Examples of hormones used by a number of laboratories are presented in Table 6. The pH of the medium before the addition of agar is adjusted to between pH 5.5 and 6.0. All media are sterilized by autoclaving. For a detailed description of the medium and culture conditions used for the initiation of callus from leaves and seeds of *L. peruvianum* and the L2 hybrid of Thomas and Pratt (1981a,b), please refer to the protocol section describing embryo culture.

Table 6. Hormones Used in Callus Culture Studies

AUXIN (μM)	CYTOKININ (μM)	REFERENCE
5.71 IAA	4.65 KIN	Doering & Ahuja, 1967
5.71 IAA + 4.52 2,4-D	4.65 KIN + 0.44 BA	Ulrich & Mackinney, 1969
5.37 NAA	4.4 BA	Warren & Routley, 1970
1.0 IBA + 1.0 IAA	1.0 KIN + 1.0 2iP	DeLanghe & DeBruijne, 1976 (callus induction)
—	1.0 KIN	DeLanghe & DeBruijne, 1976 (callus maintenance)
10.74 NAA	4.4 BA	Meredith, 1978a
28.54 IAA + 2.26 2,4-D	1.395 KIN	Meredith, 1978a
9.05 2,4-D	4.92 2iP	Thomas & Pratt, 1981a
5.0 NAA	1.0 BA	Chin et al., 1981

CULTURE CONDITIONS. Callus growth is initiated in the dark to inhibit phenol oxidation or polyvinylpyrrolidone (PVP) can be added to the culture medium at 500 mg/l to disperse phenolics into the medium (Evans et al., 1981). Cultures are incubated at 25-27 C. Once callus proliferation is visible, callus has been subcultured once every 7-28 days depending on the growth rate of the callus line.

Embryo Culture

The following technique for the production of hybrids by embryo callus culture was developed by Thomas and Pratt (1981a).

EXPLANT PREPARATION. Fruits with unbroken skins are harvested 30-40 days after pollination depending on the rate of fruit development of the species used as female in the cross. A range of post-pollination time periods may need to be sampled in order to determine the stage most efficient in embryo callus production. The fruit is surface sterilized by dipping in 95% ethanol, soaking in 5% sodium hypochlorite for 5 min, and then rinsing well in sterile distilled water. Fruit is dissected on sterile absorbent paper which facilitates the removal of the jelly-like coating on the seeds.

CULTURE MEDIA. All media consist of salts, sucrose, inositol, vitamins, and agar as in MS medium, and are supplemented with 3.0 μM thiamine. A solution of salts, vitamins, inositol, and hormones is adjusted to pH 6.0 and autoclaved separately. Solutions of sucrose, agar, and CW are each autoclaved separately. All solutions are mixed after autoclaving.

Callus initiation and maintenance medium is supplemented with 9.05 μM 2,4-D, 4.92 μM 2iP, and 100 ml/l CW. The vitamins pyridoxine•HCl and nicotinic acid are deleted from callus medium. Shoot regeneration medium is supplemented with 9.12 μM ZEA and the sucrose concentration is reduced from 3 to 2%. Rooting medium is devoid of hormones, the agar concentration is reduced to 6 g/l, and the following components are deleted: inositol, thiamine, pyridoxine•HCl, and nicotinic acid.

CULTURE CONDITIONS. Callus cultures are initiated and maintained in the dark at 27 C. Callus should be produced within 2 months. This callus is subcultured onto fresh callus medium as soon as growth is visible. After an additional 2-3 weeks, sufficient callus should be available to be divided into several pieces and placed onto regeneration medium. Shoot regeneration and rooting are conducted at room temperature (20-30 C) with a 16-hr photoperiod of fluorescent light. Shoot regeneration cultures are subcultured every 3 weeks. Shoots should be obtained within 3 months, while rooting takes an additional 2 weeks.

Root Culture

EXPLANT PREPARATION (White, 1963). Young seedlings are the best source of root material for the initiation of root culture. Isolated seeds are sterilized or seeds are aseptically removed from sterilized mature fruit. Seeds are germinated in the dark on moistened sterile filter paper or Hoagland's seed germination agar. The roots are severed from seedlings when they are 1-3 cm long and placed in liquid nutrient medium. Roots are subcultured every week by severing 1-3 cm long root tips from the root mass and placing them in fresh medium.

CULTURE MEDIUM. The most commonly used medium is White's general cell culture medium (White, 1943) supplemented with 1-2% sucrose, thiamine•HCl, pyridoxine•HCl, nicotinamide, and occasionally glycine (Street and Shillito, 1977); or a modification thereof (Sheat et al., 1959). The medium is prepared in liquid form.

CULTURE CONDITIONS. Root cultures are grown in the dark at 25-27 C.

Shoot Culture

EXPLANT PREPARATION. Leaves, stem internodes, hypocotyls, and cotyledons are the most common tissue sources of tomato explants for shoot regeneration. All procedures employed for explant culture are identical to those given for callus induction. Shoots normally regenerate within 4-6 weeks.

CULTURE MEDIA. MS basal medium is the most widely used medium for shoot culture, although Gautheret's medium (Norton and Boll, 1954) and LS medium (DeLanghe and DeBruijne, 1976) have also been employed. Various modifications of the vitamin content and concentrations have been made in many instances with B5 medium vitamins figuring as the most common replacement. Sucrose concentrations vary from 0.058 to 0.088 M. A summary of optimum concentrations and combinations of growth hormones is presented in Table 3 for cultivated *L. esculentum* varieties and in Table 4 for wild tomato species. Shoots are most commonly rooted in half-strength MS medium without growth regulators. For those resistant to rooting in this medium, a small amount of auxin may be added to the medium such as 2.8-17.1 μM IAA.

CULTURE CONDITIONS. Explant cultures are most commonly grown at 24-27 C. A photoperiod of 16 hr of light is provided by a combination of fluorescent and incandescent bulbs supplying 1000-10,000 lux. For cultures grown in a growth chamber, a relative humidity of 50-70% is usually supplied.

Anther Culture

A generalized procedure for production of haploid callus and regeneration of shoots is not currently available. This is due in part to the variability in reported results between tomato genotypes and between laboratories.

EXPLANT PREPARATION. Various concentrations of ethanol and calcium hypochlorite have been used alone or in combination to achieve sterilization of tomato floral buds. Most often the buds are gently removed and placed in 70% ethanol for 10-20 seconds, followed by a treatment of 5-7 min in 0.25-7% calcium hypochlorite. All sterilizations are followed by several rinses in sterile distilled water. The anther cone is then dissected from the bud and the individual anthers carefully separated by dislodging them at the filament end.

CULTURE MEDIA. A variety of basal media have been employed for haploid callus and shoot induction (Table 5). According to Levenko et al. (1977), the mineral salts present in each of these media are equally adequate. The major effects on callus and shoot production appear to be the result of organic supplements. The ratios and concentrations of hormones have the most significant effect. It has been suggested that combinations of NAA and KIN might be critical for the initiation of a pure haploid callus culture (Gulshan and Sharma, 1981). No hormone combination has been identified that produces only haploid shoots. The recovery of diploid or dihaploid shoots has been reported using IAA and ZEA (Zamir et al., 1980; Ziv et al., 1982), or a combination of several hormones such as 2,4-D + NAA + BA + KIN (Gulshan and Sharma, 1981).

CULTURE CONDITIONS. Anthers have most often been incubated under artificial lighting of 2000-20,000 lux for a light/dark period of 12/12 or 16/8 hr. Growth temperatures were either held constant at 25 or 27 C or allowed to vary at room temperatures from 24 to 32 C. Cultures maintained in environmental chambers were supplied with 70% relative humidity.

Protoplasts

PROTOPLAST PREPARATION (after Morgan and Cocking, 1982).

1. Sow *L. esculentum* cv. Lukullus in soil-less compost and grow under 16 hr of light (7000 lux).
2. The first true leaves of 3-4 week old plants are surface sterilized in 8% (v/v) Domestos for 25 min and rinsed 5 times in sterile distilled water.
3. Remove the leaf lower epidermis and float, peeled surface downward, for 1 hr in the following plasmolysis solution: 0.19 mM KH_2PO_4, 1 mM KNO_3, 10.1 mM $CaCl_2 \cdot 2H_2O$, 1 mM $MgSO_4 \cdot 7H_2O$, 0.96 μM KI, 0.1 μM $CuSO_4 \cdot 5H_2O$ (CPW salts), 0.5 M Mannitol pH 5.7.
4. Replace plasmolysis solution with a filter-sterilized enzyme solution consisting of 1.5% Meicelase P (Meiji Seika Kaisha Ltd.), 0.15% Driselase (Kyowa Hakko Kogyo Co., Ltd.), and 0.15% Macerozyme (Yakult Biochemicals) dissolved in CPW salts, 0.49 M Mannitol, pH 5.7, and an antibiotic mixture consisting of 1.1 mM ampicillin, 0.02 mM tetracycline, 10 mg/l gentamycin, and 0.06 mM rifampicin. Incubate 13 hr in darkness at 27 C.
5. Gently squeeze leaf pieces to release protoplasts.
6. Centrifuge crude protoplast preparation 100 x g for 4 min.
7. Resuspend pelleted protoplasts in CPW salts solution containing 0.6 M sucrose. Centrifuge 120 x g for 8 min.
8. Protoplasts which float to the top are removed and resuspended in modified B5 (Gamborg et al., 1968) medium (Zapata et al., 1977) to a density of 1×10^5 per ml.

CULTURE CONDITIONS.

1. 1.5 ml of protoplast suspension is mixed with an equal volume of warm protoplast medium containing 1% agar in a 5 cm petri dish. Petri dishes are sealed with Nescofilm (Nippon Shoji Kaisha, Ltd.) and incubated at 30 C in darkness for 7 days.

2. After 7 days dark treatment, plates are transferred to a light intensity of 800 lux at 27 C.
3. Small colonies of cells regenerate after 28 days. Mannitol concentration is then reduced at 2-week intervals by transferring blocks of agar containing colonies to solid culture medium containing 0.33 M mannitol followed by 0.16 M mannitol.
4. Protoplast derived callus is then placed on shoot regeneration medium containing MS salts and vitamins, 0.058 M sucrose, 0.8% agar, and 4.6 μM ZEA (MSZ).
5. Callus is subcultured monthly to fresh MSZ medium until plantlets have regenerated and grown to a height of 4 cm.
6. Regenerated plantlets are subcultured to MS agar medium containing 0.54 μM NAA for root formation.

FUTURE PROSPECTS

While several isolated reports suggest that cell culture methodology can be applied to tomato, several difficulties still exist that must be circumvented before tomato cell culture can be used as a breeding tool. This is evident as all of the following are presently difficult if not impossible: establishment of cell suspension cultures with stable chromosome number, control of plant regeneration from cell cultures, recovery of plants from isolated protoplasts, or recovery of haploid plants from cultured anthers. Nonetheless, tomato has several characteristics that justify further research. It is an important commercial crop having a well characterized genetic system, and a relatively short generation time. Recent success with closely related wild *Lycopersicon* species for protoplast regeneration (Zapata et al., 1977) and with the VFNT variety of tomato for regeneration from long-term cell cultures (Meredith, 1979) are encouraging.

Establishment of chromosomally stable cell suspension cultures is important for several cell genetic studies. Chromosome instability accumulates shortly after cell cultures of tomato are established (Meredith, 1979). Several authors have reported unstable polyploid and aneuploid cell cultures of tomato. Once stable lines are established they should be subcultured frequently to insure continuous dilution of aberrant cell populations (Evans and Gamborg, 1982). Stable cell lines should retain the capacity for plant regeneration. This characteristic would be extremely useful for mutant isolation and protoplast fusion. Those variants already selected for agriculturallly valuable characters using existing cell lines have been incapable of plant regeneration (cf. Meredith, 1978b). Stable cell lines would also be useful to examine secondary product synthesis. Cultivated tomato or wild species suspension cultures could be useful for production of several compounds. For example, the natural insecticide, 2-tridecanone, produced by *L. hirsutum*, would be an interesting candidate (Williams et al., 1980). Stable callus or cell suspension cultures could also be useful for pathological studies. In other species, callus has been used to screen for resistance to fungal pathogens (Helgeson et al., 1976). Cell lines capable of plant regeneration could also be used to select for resistance to pathogen culture filtrates or specific toxins (e.g. Gengenbach et al., 1977).

Regeneration of plants from long-term cultures may be useful for recovery of variants or for cloning (Sharp et al., 1982). As mentioned above, chromosomally stable cell suspension cultures are important to generate new variants to be integrated into a breeding program. In this respect, mutagenesis or somaclonal variation could be explored to identify novel variants. Alternatively, plant regeneration can be used to clonally propagate certain genotypes. For example, hybrid tomato varieties used for greenhouse production could be cloned using shoot tip propagation. Similarly, male sterile parent plants used for hybrid seed production could be propagated in vitro.

Protoplast regeneration would offer a means of genetically modifying single cells and recovering complete plants. Protoplasts of two species could be fused to create novel hybrids. Several interspecies combinations would be useful if integrated into a breeding program. Most wild species contain characters that would be useful if integrated into cultivated tomatoes, such as salt tolerance in *L. cheesmanii* or cold tolerance in *L. hirsutum*.

While there have been isolated reports of recovery of haploid plants from cultured anthers, there has not yet been a report of consistent isolation and recovery of haploid cultivated tomato plants. The goal in the future will be to produce haploid cell lines which can be maintained and manipulated in vitro for extended periods, while retaining the capacity to regenerate haploid, dihaploid, or, in the case of fusion experiments, hybrid diploid plants. Because of the difficulty of confirming the origin of plants that are diploid or have higher ploidy level, genetic markers should be introduced into any line studied in anther culture. This is relatively easy as several genetic lines are available for tomato. Isoenzyme markers have been proposed (Zamir et al., 1981) as an advantageous means of identifying haploid regenerates. Because of codominant expression of isoenzymes, it is easy to separate heterozygotes from each homozygous haploid-type. Haploids would be particularly useful to isolate recessive mutations using cultured cells, and to obtain somatic hybrids with 2n = 24 chromosomes.

KEY REFERENCES

Evans, D.A. and Sharp, W.R. 1983. Single gene mutations in tomato plants regenerated from tissue culture. Science 221:949-951.

Kartha, K.K., Gamborg, O.L., Shyluk, J.P., and Constabel, F. 1976. Morphogenetic investigations on in vitro leaf culture of tomato (*Lycopersicon esculentum* Mill. cv. Starfire) and high frequency plant regeneration. Z. Pflanzenphysiol. 77:292-301.

Kut, S.A. and Evans, D.A. 1982. Plant regeneration from cultured leaf explants of eight wild tomato species and two related *Solanum* species. In Vitro 18:593-598.

Morgan, A. and Cocking, E.C. 1982. Plant regeneration from protoplasts of *Lycopersicon esculentum* Mill. Z. Pflanzenphysiol. 106:97-104.

Thomas, B.R. and Pratt, D. 1981b. Breeding tomato strains for use in cell culture research. PMB Newsletter 2:102-105.

Ziv, M., Hadary, D., and Kedar, N. 1982. Dihaploid plants regenerated from tomato anthers in vitro. In: Proceedings of the Fifth International Congress on Plant Tissue Culture.

REFERENCES

Ancora, G. and Sree Ramulu, K. 1981. Plant regeneration from in vitro cultures of stem internodes in self-incompatible triploid *Lycopersicon peruvianum* Mill. and cytogenetic analysis of regenerated plants. Plant Sci. Lett. 22:197-204.

Ancora, G., Saccardo, F., Cappadocia, M., and Sree Ramulu, K. 1981. Backcross progenies from *Lycopersicon esculentum* L. x hybrid (*L. esculentum* x *L. peruvianum* Mill.). Z. Pflanzenzuecht. 87:153-157.

Behki, R.M. and Lesley, S.M. 1976. In vitro plant regeneration from leaf explants of *Lycopersicon esculentum* (tomato). Can. J. Bot. 54: 2409-2414.

______ 1980. Shoot regeneration from leaf callus of *Lycopersicon esculentum*. Z. Pflanzenphysiol. 98:83-87.

Cappadocia, M. and Sree Ramulu, K. 1980. Plant regeneration from in vitro cultures of anthers and stem internodes in an interspecific hybrid, *Lycopersicon esculentum* L. x *L. peruvianum* Mill. and cytogenetic analysis of the regenerated plants. Plant Sci. Lett. 20:157-166.

Cassells, A.C. 1979. The effect of 2,3,5-triiodobenzoic acid on caulogenesis in callus cultures of tomato and *Pelargonium*. Physiol. Plant. 46:159-164.

______ and Barlass, M. 1976. Environmentally induced changes in the cell walls of tomato leaves in relation to cell and protoplast release. Physiol. Plant. 37:239-246.

______ and Barlass, M. 1978. A method for the isolation of stable mesophyll protoplasts from tomato leaves throughout the year under standard conditions. Physiol. Plant. 42:236-242.

Chin, C.K., Haas, J.C., and Still, C.C. 1981. Growth and sugar uptake of excised root and callus of tomato. Plant Sci. Lett. 21:229-234.

Coleman, W.K. and Greyson, R.I. 1977a. Analysis of root formation in leaf discs of *Lycopersicon esculentum* Mill. cultured in vitro. Ann. Bot. 41:307-320.

______ 1977b. Promotion of root initiation by gibberellic acid in leaf discs of tomato (*Lycopersicon esculentum*) cultured in vitro. New Phytol. 78:47-54.

Coleman, W.K., Huxter, T.J., Reid, D.M., and Thorpe, T.A. 1980. Ethylene as an endogenous inhibitor of root regeneration in tomato leaf discs cultured in vitro. Physiol. Plant. 48:519-525.

Dao, N.T. and Shamina, Z.B. 1978. Cultivation of isolated tomato anthers. Fiziol. Rast. 25:155-160.

______, and Butenko, R.G. 1976. Tomato anther culture. Haploid Inf. Serv. 16:10-14.

DeLanghe, E. 1973. Vegetative multiplication in vitro. Int. Assoc. Plant Tissue Cult. Newsletter 9:7.

______ and DeBruijne, E. 1976. Continuous propagation of tomato plants by means of callus culture. Sci. Hortic. 4:221-227.

DeNettancourt, D., Devreux, M., Laneri, V., Cresti, M., Pacini, E., and Sarfatti, G. 1974. Genetical and ultrastructural aspects of self and cross incompatibility in interspecific hybrids between self-compatible *Lycopersicon esculentum* and self-incompatible *L. peruvianum*. Theor. Appl. Genet. 44:278-288.

Devreux, M., Sree Ramulu, K., Laneri, V., and Ancora, G. 1976. Anther culture of *Lycopersicon peruvianum*. Haploid Inf. Serv. 15:11-12.

DeWit, P. 1976. Isolation and culture of tomato mesophyll protoplasts. Acta Bot. Neerl. 25:475-480.

Dhruva, B., Ramakrishnan, T., and Vaidyanathan, C.S. 1978. Regeneration of hybrid tomato plants from leaf callus. Curr. Sci. 47:458-460.

Doering, G.R. and Ahuja, M.R. 1967. Morphogenetic studies of a genetically controlled tumor-like condition in *Lycopersicon* hybrids. Planta 75:85-93.

Ellis, B.E. 1978. Non-differential sensitivity to the herbicide Metribuzin in tomato cell suspension cultures. Can. J. Plant Sci. 58:775-778.

Epstein, E., Norlyn, J.D., Rush, D.W., Kingsbury, R.W., Kelley, D.B., Cunningham, G.A., and Wrona, A.F. 1980. Saline culture of crops: A genetic approach. Science 210:399-404.

Evans, D.A. and Gamborg, O.L. 1982. Chromosome stability of cell cultures of *Nicotiana* spp. Plant Cell Reports 1:104-107.

Evans, D.A., Sharp, W.R., and Flick, C.E. 1981. Plant regeneration from cell cultures. In: Horticultural Reviews, Vol. 3 (J. Janick, ed.) pp. 214-314. AVI Pub., Westport, Connecticut.

Frankenberger, E.A., Hasegawa, P.M., and Tigchelaar, E.C. 1981a. Influence of environment and developmental state on the shoot-forming capacity of tomato genotypes. Z. Pflanzenphysiol. 102:221-232.

______ 1981b. Diallel analysis of shoot-forming capacity among selected tomato genotypes. Z. Pflanzenphysiol. 102:233-242.

Fukami, T. and Mackinney, G. 1967. Culture of tomato callus tissue. Nature 213:944-945.

Gamborg, O.L. 1975. Callus and cell culture. In: Plant Tissue Culture Methods (O.L. Gamborg and L.R. Wetter, eds.) pp. 1-10. National Research Council of Canada, Prairie Regional Laboratory, Saskatoon, Saskatchewan.

______, Miller, R.A., and Ojima, K. 1968. Nutrient requirements of suspension cultures of soybean root cells. Exp. Cell Res. 50:151-158.

Gengenbach, B.G., Green, C.E., and Donovan, C.M. 1977. Inheritance of selected pathotoxin resistance in maize plants regenerated from cell cultures. Proc. Nat. Acad. Sci. USA 74:5113-5117.

Grattidge, R. 1982. Occurrence of a third race of *Fusarium* wilt of tomatoes in Queensland. Plant Dis. Rep. 66:165-166.

Gregory, D.N. and Cocking, E.C. 1965. The large scale isolation of protoplasts from immature tomato fruit. J. Cell. Biol. 24:143-146.

Gresshoff, P.M. and Doy, C.H. 1972. Development and differentiation of haploid *Lycopersicon esculentum* (tomato). Planta 107:161-170.

Gulshan, T.M.V. and Sharma, D.R. 1981. Studies on anther cultures of tomato—*Lycopersicon esculentum* Mill. Biol. Plant. 23:414-420.

Gunay, A.L. and Rao, P.S. 1980. In vitro propagation of hybrid tomato plants (*Lycopersicon esculentum* L.) using hypocotyl and cotyledon explants. Ann. Bot. 45:205-207.

Helgeson, J.P., Haberlach, G.T., and Upper, C.D. 1976. A dominant gene conferring disease resistance to tobacco plants is expressed in tissue cultures. Phytopathology 66:91-96.

Hender, A.B. 1973. Anther culture in the cultivated tomato, *Lycopersicon esculentum*. In: Annual Report Glasshouse Crops Research Institute 1972, pp. 130-133. Rustington, Littlehampton, Sussex, England.

Herman, E.B. and Haas, G.J. 1978. Shoot formation in tissue cultures of *Lycopersicon esculentum* Mill. Z. Pflanzenphysiol. 89:467-470.

Hogenboom, N.G. 1972a. Breaking breeding barriers in *Lycopersicon*. 1. The genus *Lycopersicon*, its breeding barriers and the importance of breaking these barriers. Euphytica 21:221-227.

______ 1972b. Breaking breeding barriers in *Lycopersicon*. 2. Breakdown of self-incompatibility in *L. peruvianum* (L.) Mill. Euphytica 21: 228-243.

______ 1972c. Breaking breeding barriers in *Lycopersicon*. 3. Inheritance of self-incompatibility in *L. peruvianum* (L.) Mill. Euphytica 21:244-256.

______ 1972d. Breaking breeding barriers in *Lycopersicon*. 4. Breakdown of unilateral incompatibility between *L. peruvianum* (L.) Mill. and *L. esculentum* (L.) Mill. Euphytica 21:397-404.

______ 1972e. Breaking breeding barriers in *Lycopersicon*. 5. The inheritance of the unilateral incompatability between *L. peruvianum* (L.) Mill. and *L. esculentum* Mill. and the genetics of its breakdown. Euphytica 21:405-414.

Kartha, K.K., Champoux, S., Gamborg, O.L., and Pahl, K. 1977. In vitro propagation of tomato by shoot apical meristem culture. J. Am. Soc. Hortic. Sci. 102:346-349.

Kerr, E.A., Kerr, E., Patrick, Z.A., and Potter, J.W. 1980. Linkage relation of resistance to *Cladosporium* leaf mold (CF-2) and root-knot nematodes (Mi) in tomato and a new gene for leaf mold resistance (cf-11). Can. J. Genet. Cytol. 22:183-186.

Kotte, W. 1922. Wurzelmeristem in Gewebekultur. Ber. Deutsch. Bot. Ges. 40:269-272.

Levenko, B.A., Kunakh, V.A., and Yurkova, G.A. 1977. Studies on callus tissue from anthers. I. Tomato. Phytomorphology 27:377-382.

Liebisch, H.W. 1980. Vergeichende Untersuchungen uber den Stoffwechsel von GA, GA_3, and GA_9 in Zellsuspensionkulturen von *Lycopersicon esculentum* und in vershceidenen intakten pflanzen. Biochem. Physiol. Pflanzen. 175:797-805.

Linsmaier, E. and Skoog, F. 1965. Organic growth factor requirements of tobacco tissue culture. Physiol. Plant. 18:100-127.

Lorenz, O.A. and Maynard, D.N. 1980. Knott's Handbook for Vegetable Growers. Wiley and Sons, New York.

Maschio, L.M.A. and Sampaio, I.B.M. 1982. Epiphytology and control of *Phytophthora infestans*, the late blight fungus, on tomato plants. Pesqui. Agropec. Bras. 17:715-719.

Meredith, C.P. 1978a. Response of cultured tomato cells to aluminum. Plant Sci. Lett. 12:17-24.

_____ 1978b. Selection and characterization of aluminum-resistant variants from tomato cell cultures. Plant Sci. Lett. 12:25-34.

_____ 1979. Shoot development in established callus cultures of cultivated tomato (*Lycopersicon esculentum* Mill.). Z. Pflanzenphysiol. 95: 405-411.

Muhlbach, H.-P. 1980. Different regeneration potentials of mesophyll protoplasts from cultivated and a wild species of tomato. Planta 148:89-96.

_____ and Thiele, H. 1981. Response to chilling of tomato mesophyll protoplasts. Planta 151:399-401.

Murashige, T. and Skoog, F. 1962. A revised medium for rapid growth and bioassays with tobacco tissue cultures. Physiol. Plant. 15:473-497.

Nitsch, J.P. and Nitsch, C. 1969. Haploid plants from pollen grains. Science 163:85-87.

_____, Rossini, L.M.E., and Bui Dang Ha, D. 1967. The role of adenine in bud differentiation. Phytomorphology 17:446-453.

Norton, J.P. and Boll, W.G. 1954. Callus and shoot formation from tomato roots in vitro. Science 119:220-221.

Ohki, S., Bigot, C., and Mousseau, J. 1978. Analysis of shoot-forming capacity in vitro in two lines of tomato (*Lycopersicon esculentum* Mill.) and their hybrids. Plant Cell Physiol. 19:27-42.

Petru, E. 1965. Growth, development, and sudden death of excised tomato roots. In: Proceedings of an International Conference on Plant Tissue Culture (P.R. White and A.R. Grove, eds.) pp. 19-23. McCutchan Pub., Berkeley.

Padmanabhan, V., Paddock, E.F., and Sharp, W.R. 1974. Plantlet formation from *Lycopersicon esculentum* leaf callus. Can. J. Bot. 52: 1429-1432.

Rick, C.M. 1951. Hybrids between *Lycopersicon esculentum* Mill. and *Solanum lycopersicoides* Dun. Proc. Natl. Acad. Sci. USA 37:741-744.

_____ 1963. Differential zygotic lethality in a tomato species hybrid. Genetics 48:1497-1507.

_____ 1976. Tomato. In: Evolution of Crop Plants (N.W. Simmonds, ed.) pp. 268-273. Longman, London.

_____ 1982. The potential of exotic germplasm for tomato improvement. In: Plant Improvement and Somatic Cell Genetics (I.K. Vasil, W.R. Scowcroft, and K.J. Frey, eds.) pp. 1-28. Academic Press, New York.

_____ and Smith, P.G. 1953. Novel variation in tomato species hybrids. Am. Nat. 87:359-373.

Robbins, W.J. 1922. Wurzelmeristem in Gewebekultur. Ber. Deutsch Bot. Ges. 40:269-272.

Scowcroft, W.R. and Adamson, J.A. 1976. Organogenesis from callus cultures of the legume, *Stylosanthes hamata*. Plant Sci. Lett. 7:39-42.

Seeni, S. and Gnanam, A. 1981. In vitro regeneration of chlorophyll chimeras in tomato (*Lycopersicon esculentum*). Can. J. Bot. 59:1941-1943.

Sharp, W.R., Dougall, D.K., and Paddock, E.F. 1971. Haploid plantlets and callus from immature pollen grains of *Nicotiana* and *Lycopersicon*. Bull. Tor. Bot. Club 98:219-222.

Sharp, W.R., Raskin, R.S., and Sommer, H.E. 1972. The use of nurse culture in the development of haploid clones in tomato. Planta 104: 357-361.

Sharp, W.R., Evans, D.A., Flick, C.E., and Sommer, H.E. 1982. Strategies and specifications for management of in vitro propagation. In: USDA Symposium on Plant Propagation (W. Mendt, ed.). Osmund Press, New Jersey.

Sheat, D.E.G., Fletcher, B.H., and Street, H.E. 1959. Studies on the growth of excised roots. VIII. The growth of excised tomato roots supplied with various inorganic sources of nitrogen. New Phytol. 58: 128-141.

Smith, P.G. 1944. Embryo culture of a tomato species hybrid. Proc. Am. Soc. Hortic. Sci. 44:413-416.

Street, H.E. and Melhuish, F.M. 1965. Nitrogen nutrition of excised roots. The release of amino acids by growing root cultures. In: Proceedings of an International Conference on Plant Tissue Culture (P.R. White and A.R. Grove, eds.) pp. 25-43. McCutchan pub., Berkeley.

Street, H.E. and Shillito, R.D. 1977. Nutrient media for plant organ, tissue, and cell culture. In: Handbook Series in Nutrition and Food. Diets, Culture Media, and Food Supplements, Vol. IV (M. Rechcigl, Jr., ed.) pp. 305-359. CRC Press, Cleveland.

Tal, M. and Watts, J.W. 1979. Plant growth conditions and yield of viable protoplasts isolated from leaves of *Lycopersicon esculentum* and *L. peruvianum*. Z. Pflanzenphysiol. 92:207-214.

Tal, M., Dehan, K., and Heiken, H. 1977. Morphogenetic potential of cultured leaf sections of cultivated and wild species of tomato. Ann. Bot. 41:937-941.

Tal, M., Heikin, H. and Dehan, K. 1978. Salt tolerance in the wild relatives of the cultivated tomato: Responses of callus tissue of *Lycopersicon esculentum*, *L. peruvianum*, and *Solanum pennelli* to high salinity. Z. Pflanzenphysiol. 86:231-240.

Thomas, B.R. and Pratt, D. 1981a. Efficient hybridization between *Lycopersicon esculentum* and *L. peruvianum* via embryo callus. Theor. Appl. Genet. 59:215-219.

______ 1982. Isolation of paraquat tolerant mutants from tomato cell cultures. Theor. Appl. Genet. 63:169-176.

Trebst, A. and Wietoska, W. 1975. Hemmung des photosynthetischen electrontransports von chloroplasten durch metribuzin. Z. Naturforsch. 30c:499-504.

Ulrich, J.M. and Mackinney, G. 1969. Callus cultures of tomato mutants. I. Nutritional requirements. Physiol. Plant. 22:1282-1287.

Vnuchkova, V.A. 1977. Development of a method for obtaining regenerate tomato plants under tissue culture conditions. Fiziol. Rast. 24: 1094-1100.

Warren, R.S. and Routley, D.G. 1970. The use of tissue culture in the study of single gene resistance of tomato to *Phytophthora infestans*. J. Am. Soc. Hortic. Sci. 95:266-269.

White, P.R. 1934a. Potentially unlimited growth of excised tomato root tips in a liquid medium. Plant Physiol. 9:585-600.

______ 1934b. Multiplication of viruses of tobacco and aucuba mosaics in growing excised tomato root tips. Phytopathology 24:1003-1011.

______ 1943. A Handbook of Plant Tissue Culture. The Jacques Cattell Press, Lancaster, Pennsylvania.

______ 1963. The Cultivation of Animal and Plant Cells, 2nd ed. Ronald Press, New York.

Williams, W.G., Kennedy, G.G., Yamamoto, R.T., Thacker, J.D., and Bordner, J. 1980. 2-Tridecanone: A naturally occurring insecticide from the wild tomato *Lycopersicon hirsutum* f. *glabratum*. Science 207:888-889.

Zamir, D., Jones, R.A., and Kednar, N. 1980. Anther culture of male-sterile tomato (*Lycopersicon esculentum* Mill.) mutants. Plant Sci. Lett. 17:353-361.

Zamir, D., Tanksley, S.D., and Jones, R.A. 1981. Genetic analysis of the origin of plants regenerated from anther tissues of *Lycopersicon esculentum* Mill. Plant Sci. Lett. 21:223-227.

Zapata, F.J., Evans, P.K., Power, J.B., and Cocking, E.C. 1977. The effect of temperature on the division of leaf protoplasts of *Lycopersicon esculentum* and *Lycopersicon peruvianum*. Plant Sci. Lett. 8:119-124.

Zapata, F.J., Sink, K.C., and Cocking, E.C. 1981. Callus formation from leaf mesophyll protoplasts of three *Lycopersicon* species: *L. esculentum* cv. Walter, *L. pimpinellifolium*, and *L. hirsutum* f. *glabratum*. Plant Sci. Lett. 23:41-46.

SECTION V
Root and Tuber Crops

CHAPTER 11
Potato

S. A. Miller and *L. Lipschutz*

ORIGIN AND HISTORY OF THE POTATO

There are seven species of cultivated potato, including diploids (*Solanum ajanhiuri*, *S. goniocalyx*, *S. phureja*, *S. stenotomum*), triploids (*S.* x *chauca*, *S.* x *juzepczukii*), tetraploids (*S. tuberosum* ssp. *tuberosum*, *S. tuberosum* ssp. *andigena*), and one pentaploid (*S.* x *curtilobum*) (Hawkes, 1978b). *Solanum tuberosum* ssp. *tuberosum* is the potato cultivated in the Northern Hemisphere. It is widely held that ssp. *tuberosum* evolved from ssp. *andigena* through artificial selection after the latter was introduced into Europe in the late sixteenth century. The origin of ssp. *andigena* is the central Andes of South America, in the region of southern Bolivia and northern Peru (Hawkes, 1978a). The andigena potato was adapted to short daylengths and was not productive in the long summer days of the European growing season. Selections for earliness are believed by some to have resulted in the gradual change from ssp. *andigena* to ssp. *tuberosum*. Others (Lewis, 1966) have suggested that the change occurred after the late blight epiphytotic of the 1840s, which decimated the European and North American potato crop and led to intensive selection for resistance to the late blight pathogen. More recently, Grun (1979) suggested that today's cultivated potato may be derived from *S. tuberosum* ssp. *tuberosum* from the coastal region of south-central Chile. He found that cytoplasmic factors (cytoplasmic genes) of potato were very similar to those of Chilean biotypes of ssp. *tuberosum* but strikingly different from those of ssp. *andigena*. The cytoplasmic factors of ssp. *tuberosum* interact with chromosomal genes of ssp. *andigena* to produce both male and female sterility; a change from ssp. *andigena* to ssp. *tuberosum* would require a change in these factors, presumably leading to sterility.

The potato was introduced into the United States from Europe in 1621 via Bermuda; these first introductions were grown in Virginia (Hawkes, 1978a). Another introduction, into New Hampshire, was made in 1719 by a group of Irish immigrants. From this area potato production spread slowly to other parts of New England and the Middle Atlantic states (Smith, 1968). In 1851, the Reverend Chauncey Goodrich of Utica, New York, imported a number of tubers from Chiloe Island off the coast of Chile. One of these produced plants that proved to be well-adapted to the growing conditions of New York, and were named Rough Purple Chile. One of the progeny from selfing this clone, named Garnet Chile, became a popular variety. The ancestry of a number of varieties, including Russet Burbank, can be traced back to Garnet Chile (Howard, 1978). The variety Russet Burbank originated as a somatic mutant (sport) of the cultivar Burbank in the early 1900s. Today it is by far the most utilized variety, comprising about 40% of the total North American potato acreage (Thornton and Sieczka, 1980).

DISTRIBUTION AND ECONOMIC IMPORTANCE

The potato is a crop of worldwide importance and is an integral part of the diet of a large proportion of the world's population. Its production ranks fifth among major food crops, behind wheat, rice, maize, and barley (Hooker, 1981). Potatoes supply at least 12 essential vitamins and minerals including an extremely high density of vitamin C; a single medium-sized potato supplies at least 65% of the U.S. Federal daily recommendation for this vitamin (Thornton and Sieczka, 1980). Potatoes also provide significant amounts of protein, carbohydrates, and iron (Gray and Hughes, 1978). In 1980, the yearly per capita consumption of fresh and processed potatoes in the United States was 116 pounds (Schoenemann, 1982).

The potato is best suited to cool, temperate regions or to high elevations (2000 m) in the tropics. The USSR is the world's largest producer, followed, in order, by Western Europe, China, North America, Latin America, the Far East, the Near East, Africa, and Oceania. World potato production approached 300 million tons in the 1970s (Dalton, 1978). In the United States, the major potato-producing states are Idaho, Washington, Maine, California, Wisconsin, and North Dakota. The value of the 1981 crop was placed at 1.8 billion dollars (Schoenemann, 1982).

BREEDING GOALS

The potato is a tetraploid, vegetatively propagated crop, and as such poses several problems for plant breeders. These include a high level of heterozygosity, the common occurrence of pollen sterility, selection difficulties in the seedling and first clonal years, slow rate of increase, difficulties in germplasm storage and transport, and the build-up of viruses (Howard, 1978).

First and foremost among breeding objectives is to increase yield while maintaining quality. This may be accomplished by altering the date of maturity, increasing the rate of establishment, increasing the number of saleable tubers per plant, and introducing resistance to major diseases and pests. Long oval tubers with white flesh, resistance to mechanical damage and disease, and appropriate flavor and texture are required for consumer acceptance. Processing potatoes have a different set of requirements, including a high dry matter content, a low content of reducing sugars, and resistance to discoloration (Howard, 1978).

LITERATURE REVIEW

Callus and Cell Suspension Cultures

The first successful establishment of a potato callus tissue culture was reported by Steward and Caplin in 1951. Tuber parenchyma cells were induced to proliferate in a basal medium containing 2,4-D and CW; within 5 weeks, actively growing calli, 50-fold larger than the explants, were produced. Similar results were obtained later by Chapman (1955), and in succeeding years callus cultures were produced from many different explant types, including leaves, petioles, ovaries, anthers, stems, roots, and shoot tips (Anstis and Northcote, 1973; Dunwell and Sunderland, 1973; Wang and Huang, 1975; Roest and Bokelmann, 1976; Lam, 1977a; Wang, 1977; Gavinlertvatana and Li, 1980). However, regeneration of plantlets from callus proved to be quite difficult and was not achieved for many years after Steward and Caplin's (1951) initial report. In 1975, Lam reported the induction of embryoid bodies and shoots from potato tuber callus growing on a medium containing MS salts, Nitsch and Nitsch (1969) organic addenda, sucrose, CH, IAA, GA, KIN, and BA. The addition of BA to this medium was essential for shoot induction from somatic embryos. However, the shoots that were produced were abnormal, and the medium was subsequently refined to allow the production of normal plantlets (Lam, 1977a). The modifications primarily involved growth regulators: the concentrations of IAA and KIN were increased, and NAA and ZEA were added. Roest and Bokelmann (1976) regenerated shoots from callus formed from the cut rachises of compound leaves cultured on a modified MS medium containing BA, IAA, and GA as the growth regulators. The addition of GA to the culture medium was required for shoot development and elongation; in its absence only nodule-like structures that usually did not develop further were formed on the callus surface. Shoots produced on medium containing GA were excised and rooted on MS medium containing a low concentration of IAA.

The role of GA appears to be in promoting shoot development from previously formed shoot meristems. Jarret and coworkers (1981) reported that while GA inhibited the initiation of shoot meristems from cultured tuber disks, it was absolutely required for shoot growth and development. Similar results have been reported for other plant species (Murashige, 1964; Lane, 1979). Multiple shoots may be produced from callus derived from shoot tips on media supplemented with

2,4-D, NAA, and KIN (Wang and Huang, 1975), or with NAA alone (Roca et al., 1978). Not surprisingly, in the latter case addition of BA and GA to the NAA-containing medium resulted in dramatic increases in the number of shoots produced per shoot-tip callus. Plants have also been regenerated from leaf calli of dihaploid potato clones (Jacobsen, 1977).

Suspension cultures can be generated readily from friable callus cultures (Bajaj and Dionne, 1967; Anstis and Northcote, 1973; Lam, 1977b) in media containing an auxin, usually 2,4-D, and little or no cytokinin. Suspension cultures of the cv. King Edward exhibited a lag phase of 2 days, followed by a period of exponential growth that lasted for about 2.5 weeks. The growth curve appeared to be relatively independent of the size of the initial inoculum, beyond a fresh weight of 16 mg callus/ml medium (Anstis and Northcote, 1973). Plantlets may be regenerated from single, isolated suspension cells (Lam, 1977b) and from protoplasts derived from cell suspensions (Wenzel et al., 1979), but the regeneration frequency and ploidy of the regenerated plants appears to be dependent, at least in part, on the length of cell culture. A regeneration frequency of only 10% was achieved from isolated cells of the cv. Superior, which had been maintained as a callus line for 12 months prior to establishment of the suspension culture (Lam, 1977b). In dihaploid potatoes, Wenzel et al. (1979) observed morphological and chromosome number variation in plants regenerated from protoplasts isolated from suspension cultures established after a lengthy callus phase. Callus cultures from leaves of dihaploid potato clones have produced cells with 2x, 4x, and 8x ploidy levels, as well as aneuploids, after 4 to 6 weeks of culture. The frequency of polyploidy and aneuploidy varied among different callus genotypes, and the induction of octoploids was also dependent on the source of auxin in the culture medium. Many of the plants regenerated from these calli also exhibited polyploidy or aneuploidy (Jacobsen, 1981). Thus, while callus and suspension cultures are morphogenetically competent and may provide a source of regenerated plantlets for a variety of uses, the possibility that chromosomal aberrations can occur in callus and suspension cells must be taken into consideration.

Although the tendency of ploidy levels to increase during callus culture can be a serious detriment to applications where phenotypic stability is required (e.g. in vitro propagation), this characteristic of callus cultures may also be used to an advantage. In vitro chromosome doubling could be used to double the chromosome number of dihaploid lines, providing an alternative to colchicine treatments (Jacobsen, 1981; Wenzel et al., 1979, 1982). Certain procedures, e.g., limiting the duration of the callus phase, may reduce the frequency of aneuploidy and/or other genetic aberrations (Jacobsen, 1981). Utilizing a relatively short culture period, Roca et al. (1978) regenerated multiple shoots from meristem tip calli that did not vary in leaf, flower, or tuber morphology, or in chromosome number. Plantlets produced through this technique have been used for the international exchange of germplasm from the International Potato Center (CIP) in Lima, Peru (Roca et al., 1979). Regeneration of virus-free plantlets from callus cultures (Wang and Huang, 1975; Wang, 1977) may also serve as an adjunct to the more traditional meristem tip culture methods used for virus elimination from potato. (See the ensuing discussion of meristem tip culture.)

Limited success has been reported for utilizing callus tissue culture systems in breeding for disease resistance. Ingram and Robertson (1965) demonstrated that resistance to race 4 of *Phytophthora infestans* was expressed by tissue culture aggregates of potato varieties resistant to the pathogen, but not by aggregates derived from susceptible varieties. Later, Behnke (1979) selected potato calli resistant to culture filtrates of *P. infestans*; the resistance was not race-specific and could be maintained through plantlet regeneration and subsequent callus induction. Plants regenerated from filtrate-resistant calli were generally somewhat more resistant to *P. infestans* than control plants: lesions on leaves of most of the selected plants were smaller than lesions on control leaves, but the amount of sporulation by *P. infestans* was not affected (Behnke, 1980a). Calli from the same dihaploid potato clones were also screened for resistance to culture filtrates of *Fusarium oxysporum* (Behnke, 1980b). Calli were selected that were somewhat resistant to the filtrates, but none were immune. The response of regenerated plants to *F. oxysporum* was not reported. Thus, there appears to be a potential for selecting calli in vitro and subsequently producing plants with greater quantitative resistance to fungal pathogens. However, these in vitro systems will require further refinements and testing before they become widely applicable to breeding for disease resistance.

Meristem Tip, Shoot Tip, and Axillary Bud Culture

The culture of meristems from potato shoots and tuber sprouts has a long history and several very important applications to potato production. Early attempts to culture potato meristems met with limited success, although in most cases a few plantlets were produced (Norris, 1954; Morel and Martin, 1955; Kassanis, 1957). Growth and development of excised buds was improved by increasing the concentration of inorganic nutrients in the culture medium and adding GA (Morel and Muller, 1964; Goodwin, 1966). Murashige-Skoog (MS) medium, with its high complement of inorganic nutrients, particularly ammonium, potassium, and nitrate ions, has subsequently proven to be the culture medium of choice (Mellor and Stace-Smith, 1977; Henshaw et al., 1980a,b). The medium is normally supplemented with GA, although low concentrations of other growth regulators, e.g. KIN, BA, and IAA may be beneficial to shoot growth and/or multiple shoot formation (Novak et al., 1980). High concentrations of NAA are inhibitory to shoot development and usually induce callus formation. However, low concentrations of the auxin (<0.4 μM) in combination with BA or GA promote the development of vigorous, rooted shoots (Gregorini and Lorenzi, 1974; Pennazio and Vecchiati, 1976).

Excised meristems may be cultured on media with or without agar; meristems grown in liquid medium are often supported by filter paper wicks (Mellor and Stace-Smith, 1977) although this is not always necessary as the buds can float on the surface of the medium (Pennazio and Redolfi, 1973). Pennazio and Redolfi (1973) reported that although the presence of agar did not affect root initiation, root development appeared to be somewhat inhibited compared to plantlets grown on liquid medium.

Other factors that influence shoot development and rooting of potato meristems include the size of the explant, light intensity and composition, temperature, and genotype. In general, the meristem explant consists of the apical dome and one to four leaf primordia. Although the apical dome can be excised and cultured alone, the procedure is a difficult one, and the success rate is usually very low (Kassanis, 1957; Kassanis and Varma, 1967). The larger the explant, the more likely it is to grow and produce roots; however, explants larger than 0.7 mm may be virus-infected. Mellor and Stace-Smith (1977) reported consistently using excised buds between 0.3 and 0.7 mm in length to balance these considerations.

Pennazio and Redolfi (1973) investigated the influence of light intensity and spectrum on the growth and development of meristem tips. They found that high intensity (4000 lux) fluorescent lamps with a spectrum approximating daylight but enriched in red light promoted root development and plantlet vigor. An incubation temperature of 22-23 C is usually the most conducive for shoot development and rooting; higher temperatures may accelerate shoot development in some cultivars but cause abnormal development in others (Mellor and Stace-Smith, 1977). Genotype differences are also reflected in the nutritional requirements for shoot and root growth, growth rates, and multiplication rates (Mellor and Stace-Smith, 1969, 1977).

The applications of meristem-tip culture are primarily of four types: (a) virus eradication, (b) in vitro propagation, (c) germplasm storage, and (d) germplasm exchange.

VIRUS ERADICATION. Meristem-tip culture, either alone or in combination with chemotherapy or thermotherapy, has been used to eradicate one or more viruses from more than 100 potato cultivars. Those eliminated include PVA, PVG, PVM, PVS, PVX, PVY, leaf-roll virus, parakrinkle virus,, and potato spindle tuber viroid (see reviews by Mellor and Stace-Smith, 1977; Wang and Hu, 1980). This technique has become extremely important for potato as a major means of freeing cultivars of virus infection and increasing productivity. Following indexing of meristem-derived plantlets by serological tests, electron microscopy, the inoculation of indicator plants, or other methods (Slack, 1980; Walkey, 1980), those that are determined to be virus-free may be increased by clonal propagation.

Although early workers considered the apical meristem to be free of viruses (Morel and Martin, 1952, 1955), more recent evidence demonstrated that some viruses are present in this tissue. Appiano and Pennazio (1972) observed particles of PVX in the apical dome of potato meristems by electron microscopy. The particles were only found in the cytoplasm and apparently did not cause any ultrastructural abnormalities in these cells. Later, Pennazio and Redolfi (1974) reported a significant correlation between explant size and the percentage of PVX-free plantlets obtained from meristem-tip culture; the smaller the explant, the greater the ability to recover PVX-free plantlets. In addition, although 81% of the smallest (0.12 mm) meristems were infected, 52% of the plantlets obtained were PVX-free. Thus, PVX was eradicated from some of the infected meristems during the culture period. Mellor and Stace-Smith (1977) suggested that virus eradication during meristem-tip culture may be the result of metabolic changes that occur in the meristem due to excision injury. Enzymes necessary for virus replication may not be available in cultured meristem tips for a time period sufficient to allow viral RNA to be degraded. Larger meristems would presumably be injured less by the excision process.

Heat treatment (thermotherapy) and chemotherapy of plants prior to meristem-tip culture, or treatment of meristems during culture have been successful in eradicating a number of viruses that are difficult to eliminate by meristem-tip culture alone (Mellor and Stace-Smith, 1977; Wang and Hu, 1980; Walkey, 1980). Cassells and Long (1982) recently reported the elimination of PVX, PVY, and PVM from several potato clones by culturing 0.1-0.15 mm meristem tips on medium containing 205 μM Virazole (syn. ribavirin). Klein and Livingston (1982) also produced PVX-free plantlets from meristem tips of two potato cultivars (Russet Burbank and Red McClure) cultured in the presence of this compound. However, virazole was phytotoxic and delayed shoot development up to 2 months when present in low concentrations (41 μM); a high concentration (410 μM) killed meristems from both cultivars.

Thermotherapy for virus eradication involves holding the diseased plant at temperatures at or near 37 C for several weeks, long enough to inactivate the virus. Meristem-tip explants from these plants are then cultured and mature plants tested for the presence of the virus. Some viruses are more difficult to eradicate than others; PVS, PVX, and potato spindle tuber viroid are the most difficult (Mellor and Stace-Smith, 1977). Stace-Smith and Mellor (1968) treated PVX- and PVS-infected plants with air temperatures alternating between 33 and 37 C for 6-8 weeks prior to meristem-tip culture to free them from both viruses. MacDonald (1973) cultured meristem tips from sprouts of heat-treated tubers and produced a few plants free of PVX and PVS. Recently, five potato clones were freed of PVS by alternating incubation regimes between temperatures supraoptimal and optimal for host growth prior to meristem isolation and shoot regeneration (Lozoya-Saldana and Dawson, 1982). Heat tolerance, virus inactivation, and bud development were cultivar-dependent, in agreement with the results of other workers (Sip, 1972; MacDonald, 1973).

IN VITRO PROPAGATION. Once virus-free potato clones have been produced by meristem-tip culture, they may be multiplied by in vitro techniques. Several techniques have been developed for the rapid propagation of potatoes from shoot tips, several of which require an intermediate callus phase (Wang, 1977; Roca et al., 1978). At least 50 plantlets may be regenerated from a single shoot tip utilizing the "multi-meristem" technique developed by Roca et al. (1978). Although this involves the regeneration of shoots from shoot-tip callus, varietal characteristics appeared to be stable for the 10 varieties evaluated. However, phenotypic stability should not be assumed for all potato cultivars. In vitro propagation methods that utilize axillary bud development from meristem-tip, shoot-tip, and bud cultures are more likely to produce plantlets identical to the parent since de-differentiation and subsequent organogenesis or embryogenesis, with potential accompanying genetic changes, do not occur (Hu and Wang, 1983). When meristem culture is combined with in vitro layering (Wang, 1977) and in vitro mass tuberization (Wang and Hu, 1982), a large number of propagules (30-50 miniature tubers per explant in 4 months) may be produced (Wang and Hu, 1982). Goodwin et al. (1980a) reported multiplication rates of at least 10-25-fold after 8 weeks of culture, beginning with shoot tips 15-20 mm in length. Multiplication rates generally vary among cultivars (Goodwin, 1980a,b).

GERMPLASM STORAGE. In vitro systems for the storage and preservation of potato germplasm have received considerable attention in recent years and practical alternatives to conventional means of storage of vegetative material are now available (Henshaw et al., 1980a,b; Roca et al., 1982; Schilde-Rentschler et al., 1982). Cryopreservation and minimal-growth storage have both been applied to potatoes, with generally satisfactory results. Bajaj (1981) regenerated plants from meristems frozen at -196 C, stored for 2 years, then thawed at 35-39 C. Survival of the meristems was greatest when dimethyl sulfoxide (DMSO, 5%), glycerol (5%), and sucrose (5%) were combined as the cryoprotectant. Meristems from axillary buds and shoot tips survived better than meristems from tuber sprouts. In some cases, basal callus was produced on the thawed meristems, which could be induced to differentiate shoots. Shoot tips of two other *Solanum* species, *S. etuberosum* (Towill, 1981), and *S. goniocalyx* (Grout and Henshaw, 1978) have also been shown to survive freezing to -196 C and thawing. These positive results indicate the excellent potential of cryopreservation of meristem tips for long-term storage of potato germplasm.

Minimal growth storage has been demonstrated to be applicable to a wide range of *Solanum* species and cultivars (Henshaw et al., 1980a,b; Henshaw, 1982; Roca et al., 1982; Schilde-Rentschler et al., 1982). Meristem-tip cultures are commonly used, primarily due to requirements for genetic stability and maintenance of morphogenetic potential. The objective of this method is to reduce the growth rate of stored cultures so that transfer is only necessary on a yearly, rather than monthly basis. This has been accomplished for potato primarily by reducing incubation temperatures and increasing the osmolarity of the culture medium (Henshaw, 1982; Schilde-Rentschler et al., 1982). Low concentrations of ABA may also be used but may reduce viability in some varieties, depending on the medium composition (Roca et al., 1982). The in vitro culture collection at CIP is routinely maintained at 8-10 C on MS medium containing 14.6 mM sucrose plus 220 mM mannitol (Schilde-Rentschler et al., 1982). Workers at CIAT (Centro International de Agricultura Tropical; Cali, Columbia) reported 92%

viability of potato cultures stored for 10-12 months at 8-10 C on MS medium containing 0.58 µM GA, 19 µM ABA and 88 mM sucrose (Roca et al., 1982). A mean survival rate of 80% for several *Solanum* spp. was achieved using MS medium containing 88 mM sucrose but no hormones, and alternating temperatures between 6 C at night and 12 C during the day (Henshaw et al., 1980a).

INTERNATIONAL EXCHANGE OF GERMPLASM. With techniques at hand for long term storage of pathogen-free material, international distribution of germplasm should become less restrictive. Aseptic cultures of *Solanum* species have been shipped from CIP to 14 countries (Roca et al., 1979) in accordance with the quarantine regulations. Meristem culture techniques provide adequate safeguards to prevent the dissemination of dangerous pests and diseases into countries or regions. The procedures also permit rapid multiplication of disease-free stocks from a few imported cultures; a 5-10-fold increase of plantlets can be achieved every 2-3 weeks.

Anther Culture

Unlike many Solanaceous species, the cultivated potato has not responded well to anther culture. Dihaploid plantlets have been extracted from tetraploid cultivars (Dunwell and Sunderland, 1973), and monohaploids from dihaploid clones (Foroughi-Wehr et al., 1977; Sopory et al., 1978; Wenzel et al., 1979), but the frequency is often very low. Efforts utilizing other *Solanum* spp. have been more successful; Irikura (1975) obtained plantlets from pollen embryos of 17 *Solanum* spp. and 3 interspecific hybrids of *Solanum* in anther culture. These included both tuber-bearing and nontuber-bearing species, but not the cultivated *S. tuberosum* ssp. *tuberosum*.

Despite the difficulties that have been encountered, some progress has been made in the anther culture of cultivated potato. Several factors have emerged that are of critical importance: the components of the culture medium, the stage of development of the microspore, and the genoype of the donor plant. Sopory (1979) reported that the addition of 0.18 M (6%) sucrose, 0.5% activated charcoal, 6 µM IAA and 4.0 µM BA to MS medium resulted in a high frequency (40%) of anthers that produced embryos in a selected dihaploid clone. The high concentration of sucrose was only necessary for the initial stages of embryogenesis; transfer of anthers to MS medium plus 0.06 M sucrose after 6-8 days of culture on high sucrose medium did not inhibit embryogenesis in these anthers. The effect of BA could not be duplicated by other cytokinins, and IAA proved to be the most effective auxin. However, in a more recent report (Wenzel and Uhrig, 1981), auxins were omitted entirely from the anther culture medium used. The beneficial effects of activated charcoal on the induction of embryogenesis from microspores during anther culture are not understood, but it has been suggested that it may absorb inhibitors in the medium or released from the anther wall (Fridborg and Eriksson, 1975).

The developmental stage of the microspore has a significant influence on the development of embryos (Dunwell and Sunderland, 1973; Sopory et al., 1978). The frequency of embryogenesis is very low (<4%) in anthers containing tetrads or binucleate microspores. The optimum stage is generally considered to be the uninucleate (Sorpory et al., 1978) or late uninucleate (Dunwell and Sunderland, 1973; Weatherhead and Henshaw, 1979a,b) stage.

The factor with perhaps the greatest influence on the success of anther culture in potato is the genotype of the donor plant. It has long been noted that some potato cultivars and dihaploid breeding lines respond more readily in anther culture than others (Dunwell and Sunderland, 1973; Foroughi-Wehr et al., 1977; Jacobsen and Sopory, 1978; Wenzel et al., 1979; Wenzel and Uhrig, 1981). The "responsiveness" of certain genotypes appears to be a heritable characteristic, but the mode of inheritance has not been ascertained (Jacobsen and Sopory, 1978; Wenzel and Uhrig, 1981). The failure of some genotypes to produce viable embryos in culture may be the result of the presence of genes that are lethal in monohaploids or in the homozygous state in diploids, but not in the heterozygous dihaploid donor plant (Wenzel, 1980b). In addition, Wenzel and Uhrig (1981) suggested that the differences between genotypes may also result in part from differences in sensitivity to components of the culture medium, primarily the growth regulators. They made several crosses between genotypes that responded well in anther culture and genotypes that responded poorly but contained other valuable traits (i.e., disease resistance). The F_1 families differed for each cross, but for two of the crosses reported, it was possible to isolate several clones that responded well in anther culture. These results indicate that breeding for "regenerative capacity" in anther culture is possible and may be an important part of an in vitro potato improvement program.

One of the major difficulties associated with anther culture of potato is the production of plants with more than the haploid number of chromosomes (Jacobsen and Sopory, 1978). When a dihaploid clone is used as a donor plant, monohaploid regenerates are expected, but dihaploids, polyploids, and aneuploids are also recovered (Sopory et al., 1978; Jacobsen and Sopory, 1978; Wenzel et al., 1979). The direct induction of pollen embryos without an intermediate callus step (Sopory et al., 1978), androgenesis, should reduce the number of polyploids and aneuploids. However, the ploidy level may increase during embryo development, possibly as a result of chromosome doubling and negative selection against monohaploid embryos (Jacobsen and Sopory, 1978). Doubled monohaploids may not be distinguishable from dihaploids arising from sporophytic tissue at an early stage; thus, a dominant marker "embryo spot" controlled by two complementary genes, has been utilized to verify the origin of regenerated plantlets (Jacobsen and Sopory, 1978). Another means of verifying the microspore origin of regenerated plantlets is to culture isolated pollen instead of whole anthers. This has been attempted for both dihaploid clones (Sopory, 1977) and a tetraploid cultivar of *S. tuberosum* (Weatherhead and Henshaw, 1979a), but regeneration beyond embryo formation has not been successful.

Protoplast Culture

Protoplast isolation, culture, and plantlet regeneration have been achieved for numerous tetraploid cultivars (Shepard and Totten, 1977; Butenko and Kuchko, 1978; Shepard, 1980b; Gunn and Shepard, 1981; Thomas, 1981; Shepard, 1982; Karp et al., 1982; Bokelmann and Roest, 1983) and dihaploid breeding lines (Binding et al., 1978; Wenzel et al., 1979) of *Solanum tuberosum* ssp. *tuberosum*, as well as the wild species *S. nigrum* (Nehls, 1978), *S. dulcamara* (Binding and Nehls, 1977), *S. phureja* (Schumann et al., 1980), and the hybrid *S. phureja* x *S. chacoense* f. *gibberulosum* (Grun and Chu, 1978). Protoplasts were first isolated a decade ago from *S. tuberosum* tubers (Lorenzini, 1973; Tomiyama et al., 1974), but they did not divide to form viable callus tissue. In 1975, Upadhya defined culture conditions which allowed the division of protoplasts from leaves of greenhouse-grown plants of the cv. Seiglinde. Viable calli were produced, and in some of these, root morphogenesis occurred. However, shoots never regenerated from any of the callus tissues. Shepard and Totten (1977) were first to report the regeneration of potato plantlets from protoplasts, using the tetraploid cv. Russet Burbank. The following year, Binding et al. (1978) regenerated plants from protoplasts isolated from shoot cultures of four dihaploid potato breeding lines. Plant regeneration was also achieved from protoplasts isolated from suspension cultures of one of these dihaploid clones (Melchers, 1978; Wenzel et al., 1979). Butenko and Kuchko (1978) obtained plantlets from leaf mesophyll protoplasts of potato plants (cv. Priekulsky ranii) grown in vitro, but the regenerates were abnormal in morphology and chromosome number. Since the initial report (Shepard and Totten, 1977), Shepard and coworkers (Shepard, 1980a,b, 1981, 1982; Shepard et al., 1980; Gunn and Shepard, 1981) have refined the original protocols and achieved plant regeneration from protoplasts of a number of potato cultivars, in addition to Russet Burbank, and progressed to the callus stage for several others (Table 1).

Several different protocols have been established for protoplast isolation, culture, and plant regeneration from *S. tuberosum* and some related wild species. In general, for successful results careful attention must be paid to numerous factors, particularly (a) the growth conditions and status of the source plant, (b) the composition of the protoplast isolation solution and culture media, and (c) the culture environment, i.e., temperature and illumination.

SOURCE PLANT. Field-grown potato plants have proven not to be satisfactory for protoplast isolation (Butenko et al., 1977). Greenhouse-grown plants may be slightly better but are often unreliable, depending on the weather and the method of protoplast isolation used (Upadhya, 1975; Nehls, 1978). In order to obtain an acceptable degree of reproducibility, it is usually necessary to grow source plants under strictly controlled environmental and nutritional conditions. Shepard and coworkers (Shepard and Totten, 1977; Shepard, 1980a,b, 1982; Gunn and

Table 1. Production of Callus and Plant Regeneration from Cultivars of Potato (*Solanum tuberosum* ssp. *tuberosum*)

CULTIVAR	CALLUS PRODUCTION	PLANT REGENERATION	REFERENCE
Russet Burbank	+	+	Shepard & Totten, 1977
Priekulsky ranii	+	+[a]	Butenko & Kuchko, 1978
Atlantic	+	+[b]	Shepard, 1980b
Superior	+	+[b]	Shepard, 1982
Katahdin	+	+[b]	Shepard, 1982
Maris Piper	+	+	Gunn & Shepard, 1981
Maris Bard	+	+	Gunn & Shepard, 1981; Thomas, 1981
Fortyfold	+	+	Karp et al., 1982
Feltwell[d]	+	+	Gunn & Shepard, 1981
Foxton[d]	+	+	Gunn & Shepard, 1981
Teal[d]	+	+	Gunn & Shepard, 1981
Kingston[d]	+	-	Gunn & Shepard, 1981
F49/52[d]	+	-	Gunn & Shepard, 1981
Bison	+	-	Shepard, 1982
Butte	+	-	Shepard et al., 1980
Targhee	+	-	Shepard et al., 1980
USDA 41956	+	-	Shepard et al., 1980
Sieglinde	+	-[c]	Upadhya, 1975

[a]Plants were abnormal in morphology and chromosome number.

[b]Only shoot morphogenesis reported; rooting is assumed.

[c]Roots but not shoots were formed.

[d]Experimental cultivars.

Shepard, 1981) grew potato plants from tubers in controlled environment growth rooms under standard conditions for moisture, nutrition, temperature, illumination, and relative humidity. They found that strict adherance to a standard regime was absolutely required for successful protoplast isolation and culture. Other workers have utilized shoot cultures (Table 2) or plants grown in vitro (Butenko et al., 1977) to overcome the problem of variability in field- or greenhouse-grown plants. The use of shoot culture or plants grown in vitro may be somewhat less complicated than the use of growth-room grown plants, since the "preconditioning" and "stabilizing" treatments prior to protoplast isolation (see Protocol section) in the latter may not be necessary. However, cryptic contamination of shoot cultures by bacteria and/or other microorganisms has been reported to be a persistent problem (Thomas, 1981).

PROTOPLAST ISOLATION SOLUTIONS AND CULTURE MEDIA. A variety of protoplast isolation solutions (Table 2) and culture media have been utilized by different workers. Protocols vary with respect to enzymes, osmotic components (Table 2), and environmental conditions. No single enzyme or combination of enzymes has been used by all workers. Macerozyme was utilized to release protoplasts from leaves of *S. tuberosum* cultivars (Shepard and Totten, 1977) and dihaploid clones (Binding et al., 1978), *S. dulcamara* (Binding and Nehls, 1977), and *S. nigrum* (Nehls, 1978). However, Grun and Chu (1978) found this enzyme to be detrimental to the isolation of viable protoplasts from leaves of the hybrid *S. phureja* x *S. chacoense*. Schumann et al. (1980) tested several enzyme combinations (Table 2), all of which were effective in isolating protoplasts from cell suspension cultures of *S. phureja*.

Similarly, the osmotic components of protoplast isolation media vary between protocols (Table 2), although organic components (sucrose, mannitol, and sorbitol) are preferred to inorganic components (e.g., $CaCl_2$). The enzyme treatments were carried out under a range of temperatures (25-30 C) and light regimes (darkness to 5000 lux incandescent illumination). In addition, the time needed to complete protoplast isolation also varied, depending on the concentration of enzymes used.

The procedures designed to complete the cycle from isolated protoplasts to regenerated plants may be relatively simple (Binding et al., 1978; Thomas, 1981), to rather complex (Shepard, 1980b, 1982). Shepard's most recent protocol (Shepard, 1982) utilizes five media with varying types and amounts of major and minor mineral salts, organic addenda, osmotic components, and growth regulators. Minor changes in these media were required to accommodate potato cultivars less "culturable" than Russet Burbank. On the other hand, the procedure developed by Binding and coworkers (1978) for dihaploid potato clones required three media for protoplast culture and plant regeneration. The culture medium was a modification of Kao and Michayluk's (1975) medium 8p (Melchers, 1978), and shoot morphogenesis was induced on MS or B5 media with appropriate growth regulators. As in Shepard's protocol, plants could not be regenerated from all of the potato clones tested using these procedures.

Table 2. Source of Protoplasts, Enzymes, and Osmotic Components for Protoplast Isolation from *Solanum* Species[a]

SPECIES	SOURCE	ENZYMES	OSMOTIC COMPONENTS	REFERENCE
S. tuberosum	Leaves (growth room)	0.1% Macerozyme R-10, 0.5% Cellulase R-10	0.3 M sucrose	Shepard & Totten, 1977
S. tuberosum	Leaves of shoot cultures	0.3–1.9% Macerozyme R-10, 1.5–2.0% Meicellase	0.5 M mannitol	Binding et al., 1978
S. tuberosum	Shoot culture	1.5% Meicellase, 0.1% Pectolyase	0.5 M mannitol	Thomas, 1981
S. dulcamara	Leaves of shoot cultures	0.3% Macerozyme, 1.5% Meicellase	0.5 M mannitol, 5 mM $CaCl_2$	Binding & Nehls, 1977
S. nigrum	Leaves of shoot cultures	0.3% Macerozyme, 1.5% Meicellase	0.25 M mannitol, 0.25 M sorbitol, 5 mM $CaCl_2$	Nehls, 1978
S. phureja x *chacoense*	Leaves (greenhouse)	0.4% Cellulysin	0.3–0.4 M mannitol	Grun & Chu, 1978
S. phureja	Suspension cultures	2.5% Cellulase Onozuka SS, 1.0% Cellulase II, 0.5% Driselase	0.5 M sorbitol	Schumann et al., 1980
	Suspension cultures	2.0% Cellulase II	0.5 M sorbitol	Schumann et al., 1980
	Suspension cultures	5.0% Cellulase Onozuka SS, 0.2% Cellulase II, 0.2% Driselase	0.5 M sorbitol	Schumann et al., 1980
	Suspension cultures	0.5% Cellulase II	0.5 M sorbitol	Schumann et al., 1980

[a] From procedures for which plant regeneration from protoplasts was demonstrated.

CULTURE ENVIRONMENT. Isolated protoplasts of potato have generally been cultured under low light intensities, e.g., 400 lux continuous white fluorescent light (Shepard, 1980a) until cell clusters or small calli have formed. However, Shepard and Totten (1977) reported that higher illumination, at least 4000 lux, was necessary for intense greening of potato calli and subsequent shoot morphogenesis. A photoperiod of 16 hr was as effective as continuous illumination, provided that the light intensity was sufficiently high. Once shoots had formed and elongated on rooting medium, the light intensity was reduced to 200 lux to promote root development. The optimum temperature for growth and development was 24 C; temperatures greater than 28 C were generally detrimental.

Somaclonal Variation

Morphological variation has often been reported among potato plants regenerated by in vitro techniques involving an intermediate callus phase. For example, Van Harten and coworkers (1981) observed mutation frequencies from 50.3% and 12.3% among plants regenerated from potato petiole/rachis and leaf calli respectively, in the absence of intentional mutagenesis treatments. On the other hand, plants propagated via meristem tip or shoot tip culture are usually true to type (Hu and Wang, 1983). An exception is the report of Denton et al. (1977), in which a number of phenotypic variations occurred among plants regenerated from shoot tip cultures. There were no detectable differences in electrophoretic banding patterns for tuber protein, however, and it was suggested that the morphological differences were due to environmental factors. Variation of a number of horticultural traits and disease resistance has been reported among plants regenerated from protoplasts of commercial potato cultivars (Shepard et al., 1980; Ayers and Shepard, 1981; Thomas et al., 1982). Sixty-five clones derived from protoplasts of cv. Russet Burbank were shown to differ in at least one character from the parent cultivar (Shepard et al., 1980; Secor and Shepard, 1981). Some of these variations included (a) growth habit—several protoclones produced vines of a less indeterminate type, and thus a more compact canopy, than the parent; (b) maturity date—three protoclones set tubers earlier and four matured later than Russet Burbank; (c) tuber characteristics—a number of protoclones differed from their parent in size, shape, skin color and smoothness, and total tuber count, although none of the protoclones had higher total tuber weight than the parent clone; and (d) light requirements for flowering—several protoclones flowered with a photoperiod of only 13 hr, in contrast to the 16-hr requirement for Russet Burbank. Protoclones also showed differential sensitivity to two potato pathogens, *Alternaria solani*, which causes early blight of potato, and *Phytophthora infestans*, the causal agent of late blight (Matern et al., 1978; Shepard et al., 1980). Five-hundred protoclones of Russet Burbank were treated with culture filtrates of *A. solani*, and five of the clones proved to be less sensitive than the parent clone to the filtrates. When these protoclones were inoculated with the pathogen, four of the five showed increased resistance. In a separate experiment, 2.5% of 800 protoclones inoculated with race 0 of *P. infestans* were more resistant than their parent Russet Burbank (Shepard et al., 1980). The resistant responses ranged from scattered necrotic flecking on the leaves to a low level of tolerance. Several of the resistant protoclones were inoculated with race 1,2,3,4 of *P. infestans*, and these were also less susceptible than Russet Burbank to this race of the pathogen. The type of resistance that was expressed resembled minor gene or quantitative resistance, rather than major gene resistance; in all cases, some sporulation took place in the lesions on leaves of inoculated plants. There was no relationship between resistance to *A. solani* and to *P. infestans* among the protoclones (Shepard et al., 1980).

Variation was also observed among plants regenerated from protoplasts of in vitro shoot cultures of the tetraploid cv. Maris Bard (Thomas et al., 1982). Ten morphological characters were evaluated, including leaf color, gloss, variegation, anthocyanin pigmentation, leaflet shape, vigor, and plant height. Only one of the 23 clones observed resembled Maris Bard morphologically; most differed from the parent clone and from each other in three or more characteristics, and nearly all were less vigorous than Maris Bard. Shoots derived from the same callus, and presumably the same single protoplast, gave rise to morphologically dissimmilar plants. Thus at least part of the observed variation arose during the callus phase of plantlet regeneration from protoplasts. Cytogenetic analysis of plants regenerated from Maris Bard demonstrated a high incidence of aneuploidy (Karp et al., 1982); all but one of 26 regenerates were aneuploid, with chromosome numbers ranging from 46 to 92. However, about 30% of the regenerates from the tetraploid cv. Fortyfold were euploid, and chromosome numbers of aneuploid plants ranged from 46 to 49. Plants derived from cultured shoots or leaf pieces of Maris Bard all contained 48 chromosomes. The differences in the incidence and degree of aneuploidy between these two cultivars may have been caused by variations in culture technique, since plants were regenerated using different protocols. Protoplasts of Maris Bard were isolated, cultured, and allowed to regenerate shoots by the method of Thomas (1981; see Protocols), while the protocol for plant regeneration from protoplasts of Fortyfold was similar to that developed by Shepard (1980b; see Protocols). The source of protoplasts, and the number and concentration of growth hormones used, differed in the two techniques. However, since the methods were not compared with a single cultivar, the influence of genotype on the incidence of aneuploidy cannot be discounted.

In contrast to the results described above for tetraploid potato cultivars, little variation was observed among plants regenerated from protoplasts of dihaploid potato clones (Wenzel et al., 1979). The incidence of aneuploidy was very low among these regenerates; only 4% of the plants were aneuploid, a result of the gain or loss of one chromosome. Nearly all of the regenerated plants were tetraploid, due to spontaneous chromosome doubling in vitro. The frequency of aneuploids increased, however, when the callus phase of the procedure was lengthened, or when suspension cultures were used as a source of protoplasts.

Thus, while some of the discrepancies between the results obtained using dihaploid clones and tetraploid cultivars may be explained on the basis of karyotype changes (or lack thereof), other factors may also be involved. Shepard and coworkers (1980) reported that 5 of the variant Russet Burbank protoclones examined cytogenetically contained the euploid number of chromosomes. In addition, other workers have observed somaclonal variation in other crops in the absence of alterations in chromosome numbers (Liu and Chen, 1976; Skirvin and Janick, 1976; Evans and Sharp, 1983). In a very recent study, Evans and Sharp (1983) identified 13 putative nuclear gene muations in tomato plants regenerated from cultured leaf explants. Single gene mutations occurring in simplex loci of tetraploid potato cultivars during protoplast culture have good potential for expression in regenerated plants. However, a similar mutation in dihaploid protoplast cultures would have to occur prior to chromosome doubling, thus reducing the chances of expression in regenerated plants.

Several additional mechanisms have been suggested for somaclonal variation (Larkin and Skowcroft, 1981), including cryptic chromosome rearrangement, DNA sequence transposition, mitotic recombination, gene amplification or depletion, and cryptic virus elimination. A considerable amount of additional study is necessary to determine if any of these mechanisms are involved with the observed variation in potato.

Somatic Hybridization

The difficulty of applying conventional techniques to potato breeding, combined with recent successes in regenerating plants from protoplasts of potato and other Solanaceous species, has led to considerable interest in somatic hybridization as an additional means of improving potato cultivars. Wenzel and co-workers (Wenzel et al., 1979, 1982; Wenzel, 1980a) proposed an "analytical synthetic breeding scheme" for potato in which protoplast fusion played an integral role. Dihaploid potato clones derived parthenogenetically—via *S. tuberosum* (4x) x *S. phureja* (2x) crosses (Hougas and Peloquin, 1957)—from superior tetraploid lines would be selected for desirable characteristics, then subjected to haploidization via microspore culture with subsequent spontaneous in vitro chromosome doubling. The resultant homozygous dihaploids could then be crossed with other dihaploids to yield F_1 hybrids with desired characteristics. Selected hybrids would finally be combined by protoplast fusion to yield tetraploid clones with the optimal combination of qualitatively and quantitatively inherited traits.

Progress in the area of somatic hybridization within *Solanum tuberosum* has been hampered by a lack of availability of selectible markers. Thus, of the 2000 plants regenerated from a fusion of protoplasts isolated from two dihaploid *S. tuberosum* clones (Wenzel, 1980b), none were somatic hybrids (Wenzel et al., 1982). Regenerated calli were preselected on the basis of increased vigor, a characteristic of fusion products of *Datura* species (Schieder and Krumbeigel, 1980; Schieder and Vasil, 1980). However, somatic hybrids of potato are apparently no more vigorous than the parent clones, and none of the preselected calli proved to be somatic hybrids (Wenzel et al., 1982). The use of complementary mutations (e.g., chlorophyll deficiency) or different dominant phenotypic markers in the two lines to be fused, or the mechanical isolation of fusion products (Gleba and Hoffman, 1978), may overcome the present difficulties. The latter option may be the more attractive one, as it would not require prior genetic manipulation of the parent material to incorporate the appropriate markers or mutations.

Limited success has been reported for somatic hybridization between *Solanum tuberosum* and related wild *Solanum* species. Butenko and Kuchko (1979, 1980) fused protoplasts of *S. tuberosum* (cv. Priekulsky rannii, 2n = 4x = 48), isolated from callus tissue, with leaf mesophyll protoplasts of *S. chacoense* (2n = 2x = 24). The selection scheme was based on the inability of the parents to either sustain cell division (*S. chacoense*) or regenerate shoots (*S. tuberosum*) in the media in which the fusion products were cultured. Three shoots were regenerated, but two lacked chlorophyll and did not survive. The third was not chlorophyll deficient and could be rooted successfully and vegetatively propagated. A number of morphological characteristics (e.g., leaf morphology, color and shape of the tubers) were intermediate between the two parents, and analysis of the ribulose-1,5 bisphosphate carboxylase (RuBPCase) subunits verified the hybrid nature of the clone. The small subunit of this enzyme (nuclear-encoded) contained a polypeptide pattern corresponding to the expression of nuclear genes from both *S. tuberosum* and *S. chacoense*. In addition, analyses of organelle DNA of the somatic hybrid indicated that the mitochondria were derived from *S. chacoense* and the chloroplasts from *S. tuberosum* (A.A. Kuchko, personal communication). The hybrid contained only 60 chromosomes (expected number: 72), possibly as a result of (a) loss of chromosomes in the hybrid, or (b) the fusion of a *S. chacoense* mesophyll protoplast (2n = 24) with an aneuploid callus protoplast of *S. tuberosum* containing 36 chromosomes instead of the euploid 48. Such aneuploid cells were observed in calli of *S. tuberosum* (Butenko and Kuchko, 1980).

Recently, Binding and coworkers (Binding et al., 1982) carried out fusion experiments with protoplasts isolated from shoot cultures of dihaploid *S. tuberosum* clones and an atrazine-resistant clone of *S. nigrum*. Shoots were regenerated from more than 2000 protoplast clones and selected on the basis of morphology; a total of 11 putative somatic hybrids were identified.

Solanum tuberosum has also proven to be amenable to protoplast fusion and somatic hybridization with other Solanaceous genera. In 1978, Melchers et al. reported the fusion of callus protoplasts of a dihaploid clone of *S. tuberosum* with leaf mesophyll protoplasts of *Lycopersicon esculentum* Mill. var. *cerasiforme* (Dunal) Alef, mutant yellow green 6 (Rick). Four somatic hybrids were isolated and confirmed by analysis of RuBPCase. In subsequent experiments, additional somatic hybrids were produced (Melchers, 1978; Poulsen et al., 1980), bringing the total number to 12. Restriction endonuclease analysis of plastid DNA from these clones demonstrated that eight contained only potato plastid DNA (ptDNA) (pomatoes; Melchers, 1980) and four contained only tomato ptDNA (topatoes; Melchers, 1980) (Schiller et al., 1982).

All of the hybrids were sterile, and no genetic analysis has been possible. Leaf mesophyll protoplasts of dihaploid *S. tuberosum* have been fused with callus protoplasts of another Solanaceous species, *Nicotiana tabacum.* Isozyme analysis of esterase and amylase demonstrated that the plants obtained were somatic hybrids; however, only abnormal shoots were produced (Gleba and Evans, 1983; Vol. 1, Chapter 7).

PROTOCOL FOR SHOOT-TIP CULTURE

The following protocol (after Goodwin et al., 1980a,b) provides a means of propagating potato plants from relatively large (15-20 mm) shoot tips in the absence of an intermediate callus phase. Proliferation results from the development of axillary meristems on the cultured shoot tip. A protocol for potato meristem-tip culture, including in vitro layering and in vitro tuberization, has been recently described (Hu and Wang, 1983; Volume 1, Chapter 5).

Preparation and Excision of Shoot Tips

1. Scrub tubers in running tap water, then soak them for 10 min in a 1% sodium hypochlorite solution.
2. Cut tubers into 2-4 pieces; soak the pieces for 1 hr in 0.29 mM GA.
3. Incubate cut pieces on filter paper moistened with 1.0 mM $CaCl_2$ at 18 ± 2 C in diffuse daylight.
4. Excise sprouts when they reach a length of 15-25 mm, and dip the cut end in molten (55-60 C) paraffin (to prevent uptake of sterilizing solution).
5. Surface-sterilize excised shoot tips in a 1% sodium hypochlorite solution containing 0.1% Tween 20 for 7 min, with gentle shaking.
6. Rinse the shoots 4 times in sterile distilled water.
7. Cut shoots 10-20 mm in length.

Culture Medium

1. Prepare MS medium without casein hydrolysate or agar; include 2.3 µM KIN and 0.29 µM GA (filter sterilize GA solution and add to cooled, sterilized medium).
2. Dispense 10 or 20 ml MS medium into 50 or 100 ml Erlenmeyer flasks, respectively.

Culture Conditions

1. Place one excised shoot tip in each flask, then cover the flasks to within 1-5 mm from the bottom.
2. Incubate the hooded flasks at 18 ± 2 C under a mercury light with an intensity of 28 µE $m^{-2}s^{-1}$ for 3-4 days.

3. Incubate cultures (same light intensity, temperature as above) on a reciprocal shaker (50 oscillations/min) for 1.5 hr each day.
4. After 4 weeks of shake culture, incubate cultures without shaking under a mercury light with an intensity of 47 μE $m^{-2}s^{-1}$ (16-hr photoperiod), at 21 C during the day and 16 C at night (temperature optima differ among cultivars), for an additional 4 weeks.

Harvesting Axillary Shoots

1. Transfer contents of culture flask to a sterile petri dish.
2. Excise 15-25 mm shoots for rooting or subculture.
3. Cover the petri dish and incubate the culture under the same stationary culture conditions as above; a new crop of shoots should develop every 1-2 weeks, enabling 3-4 crops of shoot tips to be harvested.

Subculture

Repeat the procedure (culture medium through harvest of axillary shoots) using freshly harvested axillary shoot tips, except decrease the concentration of GA to 0.03 μM and increase the KIN concentration to 23 μM.

Rooting Excised Shoots

Root excised shoots by placing them in 50 ml Erlenmeyer flasks containing 5 ml liquid MS medium with 0.57 μM IAA and 29 mM sucrose. Incubate at 21 C during the day (16-hr photoperiod), 17 C at night, under a fluorescent light intensity of 48 μE $m^{-2}s^{-1}$. Transfer the plantlets after 2-2.5 weeks to a 1:1 mixture of perlite and peat moss under high humidity and maintain for 2-6 weeks.

Alternatively, plant shoots 10 mm deep in moist potting mix (56 g blood and bone, 5.6 g potassium nitrate, 56 g superphosphate, 56 g calcium carbonate and 168 g dolomite per 36 liter of a 1:1 perlite:peat moss mixture). Cover containers (25 shoots/liter potting mix per container) with a plastic film for 10-14 days. Remove film and place plantlets in 80% shade at 21 C/16 C (12 hr each). Water when needed with a fertilizer solution. Plants may be transplanted to the field in an additional 2-4 weeks.

PROTOCOL FOR MULTI-MERISTEM CULTURE

The multi-meristem technique developed by Roca et al. (1978) permits the production of multiple shoots from a single shoot tip via organogenesis from basal shoot-tip calli. No morphological abnormalities among plants regenerated using this technique have been reported.

Maintenance of Source Material and Shoot-Tip Excision

1. Maintain virus-free stock plants in insect-proof screenhouses.
2. Surface-sterilize stem segments containing one axillary bud in 0.25% calcium hypochlorite for 5 min; rinse thoroughly with distilled water.
3. Dissect shoot tips 0.5-0.6 mm in length from the axillary buds; place buds on induction medium.

Cutlure Media

1. Induction medium—prepare MS medium containing 88 mM sucrose, 2 mg/l calcium pantothenate, 0.7% agar, 1.1-2.2 μM BA, and 1.5 μM GA (optimum hormone concentrations vary between varieties). Dispense 4 ml aliquots of this medium into 15 x 125-mm test tubes.
2. Proliferation medium—prepare medium as above, but omit the agar and add 0.054 μM NAA, 1.2 μM GA, and 2.2 μM BA. Dispense 15-ml aliquots into 125-ml flasks.

Culture Initiation and Proliferation

1. Culture shoot tips on inductiom medium for 4-6 weeks at 22-24 C, under fluorescent lighting with an intensity of 1000 lux, and a 16-hr photoperiod.
2. At the end of the induction period, transfer shoot tips and basal callus (plus adventitious bud primordia) to proliferation medium. Incubate the cultures on a rotary shaker at 90 rpm, with 2000 lux illumintion (16-hr photoperiod) and a temperature of 22-24 C.
3. Subculture healthy tissue to fresh proliferation medium, up to three transfers.

Multiplication by Nodal Cuttings

1. Transfer nodal cuttings from shoots produced in proliferation medium to MS medium containing 58.4 mM sucrose, 0.8% agar, and 0.6 μM GA. Axillary buds develop into complete plantlets within 2-3 weeks.
2. Transfer plantlets to a potting mix containing a 2:1 mixture of coarse sand and peat moss. Incubate under high relative humidity at 22 C, 6000 lux illumination (16-hr photoperiod) for 1 week.

PROTOCOL FOR ANTHER CULTURE

This protocol has been used to produce haploid plants from dihaploid *S. tuberosum* clones (protocol after Sopory et al., 1978; Wenzel and Uhrig, 1981; Wenzel et al., 1982).

Growth of Donor Plants and Preparation of Anthers

1. Grow plants in the greenhouse at 18 ± 2 C with supplemental lighting (10,000 lux, 16-hr photoperiod) during the short day season. In order to prolong flowering, potato shoots may be grafted onto tomato rootstocks. Harvest buds 4-6 mm in length and incubate them in dry test tubes in the dark at 6 C for 3 days.
2. Surface-sterilize the buds with 1% calcium hypochlorite for 5 min, then rinse thoroughly with sterile distilled water.
3. Gently dissect the anthers from the buds, and place 15 horizontally on the surface of agar-solidified culture medium in 6-cm-diameter petri plates. The developmental state of the anthers, and of the microspores contained in them, are of critical importance in microspore culture. Uninucleate microspores, from anthers 2.0-3.5 mm in length, generally have the highest rate of embryo production.

Anther Culture and Plantlet Development

1. Culture anthers on MS basal medium containing 4.0 µM BA, 180 mM sucrose, and 0.5% activated charcoal, at 26 ± 1 C under a 12-hr photoperiod of 2000 lux illumination. Multicellular structures develop within 8-10 days and rupture the exine; 19-22 days after plating, embryoids rupture the anther wall.
2. Remove the embryoids and place them on MS medium plus 10% CW, 1.3 µM ZEA, and 0.09 M sucrose.
3. When morphogenesis commences, transfer the tissue to MS medium with 2.0 µM BA and 90 mM sucrose. Culture conditions are the same as for the culture of anthers.

PROTOCOLS FOR PROTOPLAST ISOLATION, CULTURE, AND PLANT REGENERATION

Protocol after Shepard (1980a, 1982)

The physiological condition of the source plant, which is determined by its age, nutrition, and the environment in which it is grown, is of utmost importance in potato protoplast isolation and culture.

Growth of Source Plants

1. Grow tubers in moist autoclaved sand, until shoots of about 10 cm are produced.
2. Excise the shoots and plant them in vermiculite in 20-cm pots; fertilize them on a biweekly basis with MS mineral salts containing twice the normal concentration of KH_2PO_4 and $CaCl_2$. Grow the plants at 18 C under a 12-hr photoperiod, with 4000 lux white fluorescent light, and 80% relative humidity. Use leaves from plants less than 50-cm tall, with a fully expanded terminal leaflet, as a source of protoplasts.

Leaf Preconditioning and Stabilizing Treatments

1. Place leaves (upper surface down) on a hormone-salts solution containing 10.7 µM NAA, 4.4 µM BA, and 1 mM each $CaCl_2$ and NH_4NO_3 in the dark at room temperature for 48 hr.
2. Surface-sterilize the leaflets for 15 min in 0.525% sodium hypochlorite, rinse several times in sterile distilled water, then brush gently with a nylon brush.
3. Cut the leaflets into 2-cm-diameter squares and incubate them overnight in 200 ml of a salts solution containing one-tenth the concentration of major mineral salts (except NH_4Cl) and one-fifth the concentration of iron and minor elements, of medium C (Table 3) (medium A of Shepard and Totten, 1977).

Protoplast Isolation.

1. Vacuum-infiltrate pretreated leaf tissue with 100 ml enzyme solution containing 0.3 M sucrose, 0.1 g Macerozyme R-10 (Kinki Yakult Co., Hishinomiya, Japan), 0.5 g Cellulase R-10 (Kinki Yakult Co.), 2 g PVP (mol. wt. 10,000, Sigma Chemical Co., St. Louis), the mineral salts of Shepard and Totten's (1977) medium A, and 0.01 M MES buffer (Sigma Chemical Co.); adjust the pH to 5.6. A 4 hr incubation at 28 C on a gyrotary shaker is sufficient to release the protoplasts.
2. Filter the protoplasts through four layers of sterile cheesecloth into Babcock bottles.
3. Centrifuge at 350 x g for 8 min, then collect the viable protoplasts, which have floated to the top, with a pasteur pipet.
4. Place the protoplasts in a rinse medium containing half the mineral salts of CL medium and 0.3 M sucrose (Table 3).
5. Re-centrifuge at 350 x g for 10 min.
6. Collect, count, and dilute the protoplast suspension to 4×10^4 protoplasts/ml with the rinse solution. A short incubation (1 hr) in this medium helps the protoplasts "adjust" to the relatively high mineral salts concentration of CL-medium.

Protoplast Culture.

1. Two media are used for the initial culture of isolated protoplasts: CL (cell layer), and R (reservoir) (Table 3). The high major mineral salts content of CL medium was beneficial for protoplast survival and early cell divisions, but not for growth beyond the first 2 weeks (Shepard, 1980a). Therefore, a system was devised in which CL and R media were dispensed into alternate quadrants of a 10 cm plastic quadrant (x-plate) petri plate. Slow exchange between the two media is accomplished by slitting the base of each septum with three 5-mm slits. Dispense 10 ml of molten R medium into quadrants I and III, and 3 ml CL medium containing a suspension of protoplasts at $1.0\text{-}1.5 \times 10^4$/ml into quadrants II and

Table 3. Media for Protoplast Isolation, Culture, and Plant Regeneration from Potato (*Solanum tuberosum* ssp. *tuberosum*) (after Shepard, 1980b, 1982)

	MEDIUM[a]				
COMPONENT	CL	R	C	D	T
Major salts (mM)					
NH_4Cl	-	0.5	2.0	5.0	5.0
KNO_3	75.2	18.8	18.8	18.8	18.8
$CaCl_2 \cdot 2H_2O$	12.0	3.0	3.0	3.0	3.0
$MgSO_4 \cdot 7H_2O$	6.0	1.5	1.5	1.5	1.5
KH_2PO_4	5.0	1.25	1.25	1.25	1.25
Iron and Minor elements (μM)					
$Na_2 \cdot EDTA$	55.0	55.0	55.0	55.0	55.0
$FeSO_4 \cdot 7H_2O$	50.0	50.0	50.0	50.0	50.0
H_3BO_4	50.0	50.0	50.0	50.0	50.0
$MnCl_2 \cdot 4H_2O$	50.0	50.0	50.0	50.0	50.0
$ZnSO_4 \cdot 7H_2O$	16.0	16.0	16.0	16.0	16.0
KI	2.5	2.5	2.5	2.5	2.5
$Na_2MoO_4 \cdot 2H_2O$	0.54	0.54	0.54	0.54	0.54
$CuSO_4 \cdot 5H_2O$	0.05	0.05	0.05	0.05	0.05
$CoSO_4 \cdot 7H_2O$	0.05	0.05	0.05	0.05	0.05
Organic Addenda (μM)					
Thiamine-HCl	1.5	1.5	1.5	1.5	3.0
Glycine	27.0	27.0	27.0	27.0	27.0
Nicotinic acid	41.0	41.0	41.0	41.0	41.0
Pyridoxine-HCl	2.4	2.4	2.4	2.4	2.4
Folic acid	1.1	1.1	1.1	1.1	1.1
Biotin	0.2	0.2	0.2	0.2	0.2
Casein hydrolysate[b]	50.0	100.0	100.0	100.0	
Myo-inositol		-	555.0	555.0	555.0
Adenine sulfate	-	-	173.0	346.0	173.0
Osmotic Components (mM)					
Sucrose	200.0	100.0	7.0	7.0	170.0
Myo-inositol	25.0	-	-	-	-
D-mannitol	25.0	100.0	300.0	200.0	-
Sorbitol	25.0	-	-	-	-
Xylitol	25.0	-	-	-	-
Others (μM)					
NAA	5.4	5.4	0.54	-	-
IAA	-	-	-	0.57	-
BA	1.8	1.8	2.25	-	-
ZEA	-	-	-	9.1	-

Table 3. Cont.

COMPONENT	MEDIUM[a]				
	CL	R	C	D	T
Others (µM)					
ABA	-	-	-	0.76	-
MES[c]	-	-	5.0	5.0	5.0
Agar	0.4%[d]	0.6%[d]	1.0%[e]	2.0%	0.5%
pH	5.6	5.6	5.6	5.6	5.6

[a]CL = Cell-layer medium; R = Reservoir medium; C = Callus induction medium; D = Shoot morphogenesis medium; T = Shoot elongation and rooting medium.

[b]mg/l.

[c]Z-N-Morpholino-ethane sulfonic acid.

[d]Sigma Type VII Agarose.

[e]Washed Difco purified agar.

IV. The CL medium is only slightly solidified, with 0.2% agar (Sigma Type VII agarose, washed, Sigma Chemical Co.), which facilitates the transfer of small protoplast-derived calli ("p-calli") to the succeeding medium (medium C) after 2 weeks incubation at 24 C under 400 lux continuous fluorescent light. The R medium contains one-fourth the concentration of major mineral salts of CL medium, as well as 0.5 mM NH_4Cl, two times higher casein hydrolysate, and a reduced level of organic osmotic components (Table 3).

2. After 2 weeks of incubation, transfer the small calli to medium C (Table 3) for continued callus proliferation and greening.
3. Spread about 1 ml of CL medium containing many calli over the surface of a 10-cm petri plate containing 12 ml medium C.
4. Incubate the calli at 24 C under 5000 lux continuous white fluorescent light.
5. After 2 weeks, transfer individual small (0.1-1.0 mm) green calli to fresh medium C at the same incubation temperature and light intensity.
6. Two weeks later, transfer calli to medium D for shoot morphogenesis.

Plant Regeneration.

1. In medium D, NAA and BA are replaced by 9.1 µM ZEA, 0.57 µM IAA, and 0.76 µM ABA to promote the formation of shoots, which generally begin to form after about 4-6 weeks of culture.

2. Transfer small shoots to medium T (Table 3), for elongation and root development. The temperature and light intensity for shoot development and growth are the same as for callus growth on medium C, but the light intensity is reduced to 200 lux for root development.
3. Transfer rooted plantlets to small pots containing a 3:1 mixture of vermiculite and sand, moistened with dilute (0.1 g/l) 20-20-20 fertilizer, and maintain them in growth chambers with 12-hr photoperiods of 5000 lux fluorescent light, and 80% relative humidity.

Protocol after Thomas (1981)

This protocol is simpler than the previous one (Shepard, 1980a, 1982), but has been used successfully for fewer commercial potato cultivars. It is a modification of the procedure of Binding et al. (1978), used for dihaploid clones of *S. tuberosum*. One difference between this protocol and that of Shepard (1980a, 1982) is the use of shoot cultures instead of plants grown in controlled-environment chambers as a source of protoplasts.

Growth of the Source Plants

1. Remove axillary buds, together with a small portion of the stem, from potato plants grown in growth chambers under a 16-hr photoperiod of 23,000 lux light, at 20 C during the light period and 16 C in the dark.
2. Surface-sterilize the buds (with stems) for 5 min in 1.0% sodium hypochlorite, then rinse them six times in sterile distilled water.
3. Separate the buds from the large leaves, stem, and necrotic tissue, and transfer them to MS medium containing 0.09 M sucrose, 0.6% agar, and 2.2 µM BA.
4. After 3-5 weeks of incubation at 26 C with a 16 hr photoperiod (400 lux), shoots approximately 3 cm in length have formed. Multiply these by transferring small segments (3-8 mm in length) containing a leaf and axillary bud to the above medium without BA, and incubating them at 26 C under a 16-hr photoperiod of 1000 lux illumination. Protoplasts may be isolated from these shoot cultures after about 3 weeks. Casein hydrolysate (100 mg/l) may be added to the above medium to detect cryptic contamination by bacteria or other microorganisms.

Protoplast Isolation and Culture

1. Cut shoots into 1-cm pieces and incubate them in 100 ml of an enzyme mixture containing 1.5% meicellase (Meiji Seika Kaisha, Japan), 0.1% Pectolyase (Seishin Pharmaceuticals, Japan), and 0.5 M mannitol (pH 5.6) for 3 hr at 25 C in the dark.

2. Shake the tissue-enzyme mixture by hand intermittently during the incubation period to facilitate the release of protoplasts.
3. Filter the protoplast suspension through sterile stainless steel sieves (pore sizes: 100 μm and 50 μm).
4. Centrifuge the protoplast suspension at 100 x g for 5 min; resuspend the pellet in 0.55 M mannitol and centrifuge at 100 x g for 5 min over 50% Percoll (Pharmacia, Sweden) in mannitol (equal volumes of Percoll and 1 M mannitol are mixed) to separate viable protoplasts from debris (Wernicke et al., 1979).
5. Remove the protoplasts that band on the surface of the Percoll using a pasteur pipet; rinse and centrifuge three times at 80 x g in 0.55 M mannitol.
6. Suspend the final pellet in liquid culture medium containing 0.4 M mannitol (Table 4) at a density of 1 x 10^4 protoplasts/ml.
7. Culture the protoplasts in 3.5-cm plastic dishes (1 ml/dish) or in multi-well plates with 0.5 ml/well, under continuous illumination (400 lux) at 25 C.
8. Two days after plating, dilute the protoplast suspensions 1:1 with culture medium containing 0.3 M mannitol and 0.4% agar; replate in plates or wells at the same environmental conditions for 5 days.
9. Incubate the cultures at 20 C with a 16-hr photoperiod of 1000 lux illumination. Each week for the next 2 months, add fresh medium with 0.2% agar to the cultures, which contain numerous small calli.

Plant Regeneration

1. Transfer the calli (1-10 mm diameter) to MS medium containing 0.03 M sucrose, 0.2 M mannitol, 2.3 μM ZEA, and 0.4% agar, and incubate them under a 16-hr photoperiod with 3000 lux illumination. Calli turn green and shoots develop during incubation on this medium.
2. For rapid shoot elongation, excise the shoots and transfer them to MS medium containing 0.09 M sucrose and 0.44 μM BA. Roots should readily form on hormone-free MS medium under reduced light intensity.

FUTURE PROSPECTS

Developments in cell culture technology provide several new avenues for potato improvement, from the eradication of viruses via meristem-tip culture to the introduction of genetic variability through the isolation and culture of protoplasts. The decline of potato varieties due to the buildup of viruses can now be curtailed; seed certification programs routinely utilize meristem-tip culture to eliminate virus diseases in seed stock (Slack, 1980). Meristem-tip culture also provides a means of rapid propagation of elite material, as well as a method of long-term storage of valuable germplasm. With the increasing loss of genetic variability due to human population pressures and loss of habitat, the

Table 4. Culture Medium for Potato (*Solanum tuberosum* ssp. *tuberosum*) Protoplasts Derived from Shoot Cultures (after Thomas, 1981)

COMPONENT	AMOUNT
Major Salts (mM)	
KNO_3	3.0
$CaCl_2 \cdot 2H_2O$	4.0
NH_4NO_3	3.7
$MgSO_4 \cdot 7H_2O$	1.2
KCl	4.0
KH_2PO_4	1.2
Iron and Minor Elements (µM)	
$FeSO_4 \cdot 7H_2O$	100.0
$Na_2 \cdot EDTA$	220.0
KI	4.5
H_3BO_3	48.5
$MnSO_4 \cdot 4H_2O$	44.8
$ZnSO_4 \cdot 7H_2O$	7.0
$Na_2MoO_4 \cdot 2H_2O$	0.1
$CuSO_4 \cdot 5H_2O$	0.1
$CoCl_2 \cdot 6H_2O$	0.1
Vitamins (µM)	
Glycine	26.6
Nicotinic acid	40.6
Thiamine-HCl	1.5
Pyridoxine-HCl	2.4
Folic acid	1.1
Biotin	0.2
Sugars	
Myo-inositol	0.6 µM
Glucose	0.056 M
Sucrose	0.29 M
Mannitol	0.3–0.4 M
Others	
2,4-D	4.5–22.5 µM
ZEA	2.3 µM
Agar[a]	4.0 g/l
pH	6.0

[a]For solid culture.

preservation of unique genetic material becomes more and more critical. Long-term storage by cell culture methods will help to provide for the needs of potato breeders in the future. Cell culture technology should also ease the difficulties involved in the international exchange of germplasm, since cultured plants can reasonably be assured of freedom from pathogens and pests.

Utilizing callus cultures of potato, scientists may be able to study a number of aspects of potato physiology, from respiration (van der Plas and Wagner, 1982) to plant-pathogen interactions (Miller and Maxwell, 1983; Vol. 1, Chapter 31). This technology may also be applied to breeding for disease resistance. Production of dihaploids and haploids through anther culture, with subsequent in vitro chromosome doubling, will help to make possible analytic breeding programs for potato (Wenzel et al., 1979). Finally, the induction of somaclonal variation and other in vitro methods present a new opportunity to correct some of the deficiencies in the more popular varieties, e.g., Russet Burbank. Thus, there is an excellent potential for in vitro techniques to make a significant contribution to the improvement of the potato.

KEY REFERENCES

Harris, P.M. (ed.). 1978. The Potato Crop. Halsted Press, New York.

Hu, C.-Y. and Wang, P.-J. 1983. Meristem, shoot-tip and bud cultures. In: Handbook of Plant Cell Culture, Vol. 1: Techniques for Propagation and Breeding (D.A. Evans, W.R. Sharp, P.V. Ammirato, and Y. Yamada, eds.) pp. 177-227. Macmillan, New York.

Shepard, J.F. 1980a. Mutant selection and plant regeneration from potato mesophyll protoplasts. In: Genetic Improvement of Crops: Emergent Techniques (I. Rubenstein, B. Gengenbach, R.L. Phillips, and C.E. Green, eds.) pp. 185-219. Univ. Minnesota Press, Minneapolis.

REFERENCES

Anstis, P.J.P. and Northcote, D.H. 1973. The initation, growth, and characteristics of a tissue culture from potato tubers. J. Exp. Bot. 24:425-441.

Appiano, A. and Pennazio, S. 1972. Electron microscopy of potato meristem tips infected with Potato Virus X. J. Gen. Virol. 14:273-276.

Ayers, A.R. and Shepard, J.F. 1981. Potato variation. Environ. Exp. Bot. 21:379-381.

Bajaj, Y.P.S. 1981. Regeneration of plants from potato meristems freeze-preserved for 24 months. Euphytica 30:141-146.

______ and Dionne, L.A. 1967. Growth and development of potato callus in suspension cultures. Can. J. Bot. 45:1927-1931.

Behnke, M. 1979. Selection of potato callus for resistance to culture filtrates of *Phytophthora infestans* and regeneration of resistant plants. Theor. Appl. Genet. 55:69-71.

______ 1980a. General resistance to late blight of *Solanum tuberosum* plants regenerated from callus resistant to culture filtrates of *Phytophthora infestans*. Theor. Appl. Genet. 56:151-152.

______ 1980b. Selection of dihaploid potato callus for resistance to the culture filtrate of *Fusarium oxysporum*. Z. Pflanzenzuchtg. 85: 254-258.

Binding, H. and Nehls, R. 1977. Regeneration of isolated protoplasts to plants in *Solanum dulcamara* L. Z. Pflanzenphysiol. 85:279.

______, Schieder, O., Sopory, S.R., and Wenzel, G. 1978. Regeneration of mesophyll protoplasts isolated from dihaploid clones of *Solanum tuberosum*. Physiol. Plant. 43:52-54.

Binding, H., Jain, S.M., Finger, J., Mordhorst, G., Nehls, R., and Gressel, J. 1982. Somatic hybridization of an atrazine resistant biotype of *Solanum nigrum* with *Solanum tuberosum*. Theor. Appl. Genet. 63: 273-277.

Butenko, R.G. and Kucho, A.A. 1978. Cultivation and plant regeneration studies on isolated protoplasts from two potato species. In: Fourth International Congress of Plant Tissue and Cell Culture. Univ. Press, Calgary, Alberta, Canada.

______ 1979. Production of interspecific somatic hybrids of potato by merging isolated protoplasts. Doklady Akad. Nauh 247:491-495.

______ 1980. Somatic hybridization of *Solanum tuberosum* L. and *Solanum chacoense* Bitt. by protoplast fusion. In: Advances in Protoplast Research: Proceedings of the Fifth International Symposium (L. Ferenczy and G.L. Farkas, eds.). Pergammon Press, Oxford.

______, Vitenko, V.A., and Avetison, V.A. 1977. Production and cultivation of isolated protoplasts from the mesophyll of leaves of *Solanum tuberosum* L. and *Solanum chacoense*. Bitt. Sov. Plant Physiol. (Fiziologiya Rastenii) 24:660-665.

Cassells, A.C. and Long, R.D. 1982. The elimination of potato viruses X, Y, and M in meristem and explant cultures of potato in the presence of virazole. Potato Res. 25:165-173.

Chapman, H.W. 1955. Potato tissue cultures. Am. Potato J. 32:207-210.

Dalton, G.E. 1978. Potato production in the context of world and farm economy. In: The Potato Crop (P.M. Harris, ed.) pp. 647-677. Halsted Press, New York.

Denton, I.R., Westcott, R.J., and Ford-Lloyd, B.V. 1977. Phenotypic variation of *Solanum tuberosum* L. cv. Dr. McIntosh directly from shoot-tip culture. Potato Res. 20:131-136.

Dunwell, J.M. and Sunderland, U. 1973. Anther culture of *Solanum tuberosum* L. Euphytica 22:317-323.

Evans, D.A. and Sharp, W.R. 1983. Single gene mutations in tomato plants regenerated from tissue culture. Science 221:949-951.

Foroughi-Wehr, B., Wilson, H.W., Mix, G., and Gaul, H. 1977. Monohaploid plants from anthers of a dihaploid genotype of *Solanum tuberosum* L. Euphytica 26:361-367.

Fridborg, G. and Eriksson, T. 1975. Effects of activated charcoal on growth and morphogenesis in cell cultures. Physiol. Plant. 34:306-308.

Gavinlertvatana, P. and Li, P.H. 1980. The influence of 2,4-D and kinetin on leaf callus formation in different potato species. Potato Res. 23:115-120.

Gleba, Yu.Yu. and Hoffmann, F. 1978. Hybrid cell lines *Arabidopsis thaliana* + *Brassica campestris*: No evidence for specific chromosome elimination. Molec. Gen. Genet. 165:253-264.

Goodwin, P.B. 1966. An improved medium for the rapid growth of isolated potato buds. J. Exp. Bot. 17:590-595.

______, Kim, Y.C., and Adisarwanto, T. 1980a. Propagation of potato by shoot-tip culture. 1. Shoot multiplication. Potato Res. 23:9-18.

______, Kim, Y.C., and Adisarwanto, T. 1980b. Propagation of potato by shoot-tip culture. 2. Rooting proliferated shoots. Potato Res. 23:19-24.

Gray, D. and Hughes, J.C. 1978. Tuber quality. In: The Potato Crop (P.M. Harris, ed.) pp. 504-544. Halsted Press, New York.

Gregorini, G. and Lorenzi, R. 1974. Meristem-tip culture of potato plants as a method of improving productivity. Potato Res. 17:24-33.

Grout, B.W.W. and Henshaw, G.G. 1978. Cryopreservation of potato shoot-tip cultures with a view towards germplasm storage. In: Abstracts of the Fourth International Congress of Plant Cell and Tissue Culture, p. 104. Calgary Univ. Press, Calgary.

Grun, P. 1979. Evolution of cultivated potato: A cytoplasmic analysis. In: The Biology and Taxonomy of the Solanaceae (J.G. Hakes, R.N. Lester, and A.D. Skelding, eds.) pp. 655-665. Linnean Society Symposium, Series No. 7, Academic Press, New York.

______ and Chu, L.J. 1978. Development of plants from protoplasts of *Solanum* (Solanaceae). Am. J. Bot. 65:538-543.

Gunn, R. and Shepard, J. 1981. Regeneration of plants from mesophyll-derived protoplasts of British potato (*Solanum tuberosum*) cultivars. Plant Sci. Lett. 22:97-101.

Hawkes, J.G. 1978a. History of the potato. In: The Potato Crop (P.M. Harris, ed.) pp. 1-14. Halsted Press, New York.

______ 1978b. Biosystematics of the potato. In: The Potato Crop (P.M. Harris, ed.) pp. 15-69. Halsted Press, New York.

Henshaw, G.G. 1982. Tissue culture methods and germplasm storage. In: Plant Tissue Culture 1982. Proceedings Fifth International Congress Plant Tissue and Cell Culture (A. Fujiwara, ed.) pp. 789-792. Japanese Assoc. for Plant Tissue Culture, Tokyo.

______, O'Hara, J.F., and Westcott, R.J. 1980a. Tissue culture methods for the storage and utilization of potato germplasm. In: Tissue Culture Methods for Plant Pathologists (D.S. Ingram and J.P. Helgeson, eds.) pp. 71-76. Blackwell Scientific Pub., Oxford.

______, Stamp, J.A., and Westcott, R.J. 1980b. Tissue culture and germplasm storage. In: Plant Cell Culture: Results and Perspectives (F. Sala, B. Parisi, R. Cella, and O. Ciferri, eds.) pp. 277-282. Elsevier/North-Holland Biomedical Press, Amsterdam.

Hooker, W.J. 1981. The potato. In: Compendium of Potato Diseases (W.J. Hooker, ed.) pp. 1-4. American Phytopathological Society, St. Paul, Minnesota.

Hougas, R.W. and Peloquin, S.J. 1957. A haploid plant of the variety Katahdin. Nature London 180:1209-1210.

Howard, H.W. 1978. The production of new varieties. In: The Potato Crop (P.M. Harris, ed.) pp. 607-646. Halsted Press, New York.

Ingram, D.S. 1967. The expression of R-gene resistance to *Phytophthora infestans* in tissue cultures of Solanum tuberosum. J. Gen. Microbiol. 49:99-108.

______ and Robertson, N.F. 1965. Interaction between *Phytophthora infestans* and tissue cultures of *Solanum tuberosum*. J. Gen. Microbiol. 40:431-437.

Irikura, Y. 1975. Induction of haploid plants by anther culture in tuber-bearing species and interspecific hybrids of *Solanum*. Potato Res. 18:133-140.

Jacobsen, E. 1977. Doubling dihaploid potato clones via leaf tissue culture. Z. Pflanzenzucht. 80:80-82.

______ 1981. Polyploidization in leaf callus tissue and in regenerated plants of dihaploid potato. Plant Cell Tissue Organ Culture 1:77-84.

______ and Sopory, S.K. 1978. The influence and possible recombination of genotypes on the production of microspore embryoids in anther culture of *Solanum tuberosum* and dihaploid hybrids. Theor. Appl. Genet. 52:119-123.

Jarret, R.L., Hasegawa, P.M., and Bressan, R.A. 1981. Gibberellic acid regulation of adventitious shoot formation from tuber discs of potato. In Vitro 17:825-830.

Kao, K.N. and Michayluk, M.R. 1975. Nutritional requirements of growth of *Vicia hajastana* cells and protoplasts at very low density in liquid media. Planta 126:105-110.

Karp, A., Nelson, R.S., Thomas, E., and Bright, S.W.J. 1982. Chromosome variation in protoplast-derived potato plants. Theor. Appl. Genet. 63:265-272.

Kassanis, B. 1957. The use of tissue cultures to produce virus-free clones from infected potato varieties. Ann. Appl. Biol. 45:422-427.

______ and Varma, A. 1967. The production of virus-free clones of some British potato varieties. Ann. Appl. Biol. 59:447-450.

Klein, R.E. and Livingston, C.H. 1982. Eradication of potato virus X from potato by ribavirin treatment of cultured potato shoot-tips. Am. Potato J. 59:359-365.

Lam, S.L. 1975. Shoot formation in potato tuber discs in tissue culture. Am. Potato J. 52:103-106.

______ 1977a. Plantlet formation from potato tuber discs in vitro. Am. Potato J. 54:465-468.

______ 1977b. Regeneration of plantlets from single cells in potatoes. Am. Potato J. 54:575-580.

Lane, W.D. 1979. Influence of growth regulators on root and shoot initiation from flax meristem-tips and hypocotyls in vitro. Physiol. Plant. 45:260-264.

Larkin, P.J. and Skowcroft, W.R. 1981. Somaclonal variation—a novel source of variability from cell cultures for plant improvement. Theor. Appl. Genet. 60:197-214.

Lewis, H. 1966. Speciation in flowering plants. Science 152:167-172.

Liu, M.-C. and Chen, W.-H. 1976. Tissue and cell culture as aids to sugarcane breeding. 1. Creation of genetic variations through callus culture. Euphytica 25:393-403.

Lorenzi, M. 1973. Obtention de protoplastes de tubercule de pomme de terre. C.R. Acad. Sci. 276:1839-1842.

Lozoya-Saldana, H. and Dawson, W.O. 1982. Effect of alternating temperature regimes on reduction or elimination of viruses in plant tissues. Phytopathology 72:1059-1064.

MacDonald, D.M. 1973. Heat treatment and meristem culture as a means of freeing potato varieties from viruses X and S. Potato Res. 16:263-269.

Matern, U., Strobel, G., and Shepard, J. 1978. Reaction to phytotoxins in a potato population derived from mesophyll protoplasts. Proc. Natl. Acad. Sci. USA 75:4935-4939.

Melchers, G. 1978. Potatoes for combined somatic and sexual breeding methods; plants from protoplasts and fusion of protoplasts of potato and tomato. In: Production of Natural Compounds by Cell Culture Methods (A.W. Alfermann and E. Reinhard, eds.) pp. 306-311. GSF Munchen.

______ 1980. Protoplast fusion, mechanism and consequences for potato breeding and production of potatoes + tomatoes. In: Advances in Protoplast Research (L. Ferenczy and G.L. Farkas, eds.) pp. 283-286. Pergammon Press, Oxford.

Mellor, F.C. and Stace-Smith, R. 1969. Development of excised potato buds in nutrient culture. Can. J. Bot. 47:1617-1621.

______ 1977. Virus-free potatoes by tissue culture. In: Applied and Fundamental Aspects of Plant Cell, Tissue, and Organ Culture (J. Reinert and Y.P.S. Bajaj, eds.) pp. 616-646. Springer-Verlag, Berlin.

Miller, S.A. and Maxwell, D.P. 1983. Evaluation of disease resistance. In: Handbook of Plant Cell Culture, Vol. 1: Techniques for Propagation and Breeding (D.A. Evans, W.R. Sharp, P.V. Ammirato, and Y. Yamada, eds.) pp. 853-879. Macmillan Pub., New York.

Morel, G. 1964. Regeneration des varietes virosees par la culture de meristemes apicaux. Rev. Hortic. 261:733-740.

______ and Martin, C. 1952. Guerison de clahlias atteints d'une maladie a virus. C.R. Acad. Sci. 235:1324-1325.

______ and Martin, C. 1955. Guerison de pommes de terre atteintes de maladies a virus. C.R. Seances Acad. Agric. Fr. 41:472-475.

______ and Muller, J.F. 1964. La culture in vitro du meristeme apical de la pomme de terre. C.R. Acad. Sci. 258:5250.

Murashige, T. 1964. Analysis of the inhibition of organ formation in tobacco tissue culture by gibberellin. Physiol. Plant. 17:636-643.

Nehls, R. 1978. Isolation and regeneration of protoplasts from *Solanum nigrum* L. Plant Sci. Lett. 12:183-187.

Nitsch, J.P. and Nitsch, C. 1969. Haploid plants from pollen grains. Science 163:85-87.

Norris, D.O. 1954. Development of virus-free stock of Green Mountain potato by treatment with malachite green. J. Agric. Res. 5:658-663.

Novak, F.J., Zadina, J., Horackova, V., and Maskova, I. 1980. The effect of growth regulators on meristem tip development and in vitro multiplicaton of *Solanum tuberosum* L. plants. Potato Res. 23:155-166.

Pennazio, S. and Redolfi, P. 1973. Factors affecting culture in vitro of potato meristem tips. Potato Res. 16:20-29.

______ 1974. Potato virus X eradication in cultured potato meristem tips. Potato Res. 17:333-335.

Pennazio, S. and Vecchiati, M. 1976. Effects of naphthalenacetic acid on potato meristem tip development. Potato Res. 19:257-261.

Poulsen, C. Porath, D., Sacristan, M.D., and Melchers, G. 1980. Peptide mapping of the ribulose bisphosphate corboxylase small subunit from the somatic hybrid of tomato and potato. Carlsberg Res. Commun. 45:249-267.

Roca, W.M., Espinoza, N.O., Roca, M.R., and Bryan, J.E. 1978. A tissue culture method for the rapid propagation of potatoes. Am. Potato J. 55:691-701.

Roca, W.M., Bryan, J.E., and Roca, M.R. 1979. Tissue culture for the international transfer of potato genetic resources. Am. Potato J. 56: 1-10.

Roca, W.M., Rodriguez, J., Beltran, J., Roa, J., and Mafla, C. 1982. Tissue culture for the conservation and international exchange of germplasm. In: Plant Tissue Culture, 1982, Proceedings of the Fifth International Congress of Plant Tissue and Cell Culture (A. Fujiwara, ed.) pp. 771-772. Japanese Assoc. Plant Tissue Culture, Tokyo.

Roest, S. and Bokelmann, G.S. 1976. Vegetative propagation of *Solanum tuberosum* L. in vitro. Potato Res. 19:173-178.

Schieder, O. and Krumbiegel, G. 1980. Interspecific and intergeneric somatic hybridization between some Solanaceous species. In: Advances in Protoplast Research, Proceedings of the Fifth International Protoplast Symposium (L. Ferenczy and G.L. Farkas, eds.) pp. 301-306. Pergammon Press, Oxford.

Schieder, O. and Vasil, I.K. 1980. Protoplast fusion and somatic hybridization. Int. Rev. Cytol (Suppl.) 11B:21-46.

Schilde-Rentschler, L., Espinoza, N., Estrada, R., and Lizarraga, R. 1982. In vitro storage and distribution of potato germplasm. In: Proceedings of the Fifth International Congress Plant Tissue and Cell Culture (A. Fujiwara, ed.) pp. 781-782. Japanese Assoc. Plant Tissue Culture, Tokyo.

Schiller, B., Herrmann, R.G., and Melchers, G. 1982. Restriction endonuclease analysis of plastid DNA from tomato, potato, and some of their somatic hybrids. Molec. Gen. Genet. 186:453-459.

Schoenemann, J.A. 1982. What kind of year will it be? '82 Potato Outlook. Am. Veg. Grower 30:6.

Schumann, U., Koblitz, H., and Opatrny, Z. 1980. Plant recovery from long-term callus cultures and from suspension culture-derived protoplasts of *Solanum phureja*. Biochem. Physiol. Pflanz. 175:670-675.

Secor, G.A. and Shepard, J.F. 1981. Variability of protoplast-derived potato clones. Crop Sci. 21:102-105.

Shepard, J.F. 1980b. Abscicic acid-enhanced shoot initiation in protoplast-derived calli of potato. Plant Sci. Lett. 18:327-333.

______ 1981. Protoplasts as sources of disease resistance in plants. Annu. Rev. Phytopathol. 19:145-166.

______ 1982. Cultivar-dependent cultural refinements in potato protoplast regeneration. Plant Sci. Lett. 26:127-132.

______ and Totten, R.E. 1977. Mesophyll cell protoplasts of potato. Isolation, proliferation, and plant regeneration. Plant Physiol. 60:313-316.

______, Bidney, D., and Shahin, E. 1980. Potato protoplasts in crop improvement. Science 208:17-24.

Sip, V. 1972. Eradication of potato viruses A + S by thermotherapy and sprout tip culture. Potato Res. 15:270-273.

Skirvin, R.M. and Janick, J. 1976. Tissue culture-induced variation in scented *Pelargonium* spp. J. Am. Soc. Hortic. Sci. 101:281-290.

Slack, S.A. 1980. Pathogen-free plants by meristem-tip culture. Plant Dis. 64:15-17.

Smith, O. 1968. Origin and history of the potato. In: Potatoes: Production, Storing, Processing (O. Smith, ed.) pp. 1-7. AVI Pub., Westport, Connecticut.

Sopory, S.K. 1977. Development of embryoids in isolated pollen culture of dihaploid *Solanum tuberosum*. Z. Pflanzenphysiol. 94S:441-447.

______ 1979. Effect of sucrose, hormones, and metabolic inhibitors on the development of pollen embryoids in anther cultures of dihaploid *Solanum tuberosum*. Can. J. Bot. 57:2691-2694.

______, Jacobsen, E., and Wenzel, G. 1978. Production of monohaploid embryoids and plantlets in cultured anthers of *Solanum tuberosum*. Plant Sci. Lett. 12:47-54.

Stace-Smith, R. and Mellor, F.C. 1968. Eradication of potato virus X and S by thermotherapy and axillary bud culture. Phytopathology 58: 199-203.

Steward, F.C. and Caplin, S.M. 1951. A tissue culture from potato tuber: The synergistic action of 2,4-D and of coconut milk. Science 111:518-520.

Thomas, E. 1981. Plant regeneration from shoot culture-derived protoplasts of tetraploid potato (*Solanum tuberosum* cv. Maris Bard). Plant Sci. Lett. 23:81-88.

______, Bright, S.W.J., Franklin, J., Lancaster, V.A., Miflin, B.I., and Gibson, R. 1982. Variation amongst protoplast-derived potato plants (*Solanum tuberosum* cv. Maris Bard). Theor. Appl. Genet. 62:65-68.

Thornton, R.E. and Sieczka, J.B. 1980 (eds.). Commercial Potato Production in North America. Am. Potato J. (Suppl. to Vol. 57).

Tomiyama, K., Lee, H.S., and Doke, N. 1974. Effect of hyphal components of *Phytophthora infestans* on the potato-tuber protoplasts. Ann. Phytopathol. Soc. Jpn. 40:70-72.

Towill, L.E. 1981. *Solanum etuberosum*: A model for studying the cryobiology of shoot-tips in the tuber-bearing *Solanum* species. Plant Sci. Lett. 20:315-324.

Upadhya, M.D. 1975. Isolation and culture of mesophyll protoplasts of potato (*Solanum tuberosum* L.). Potato Res. 18:438-445.

Van der Plas, L.H.W. and Wagner, M.J. 1982. Respiratory physiology of potato tuber callus cultures. In: Plant Tissue Culture 1982. Proceedings of the Fifth International Congress of Plant Tissue and Cell Culture (A. Fujiwara, ed.) pp. 259-260. Japanese Assoc. Plant Tissue Culture, Tokyo.

Van Harten, A.M., Bouter, H., and Broertjes, C. 1981. In vitro adventitious bud techniques for vegetative propagation and mutation breeding of potato (*Solanum tuberosum* cv. Desiree). 2. Significance for mutation breeding. Euphytica 30:1-8.

Walkey, D.G.A. 1980. Production of virus-free plants by tissue culture. In: Tissue Culture Methods for Plant Pathologists (D.S. Ingram and J.P. Helgeson, eds.) pp. 109-117. Blackwell Scientific Pub., Oxford.

Wang, P.J. 1977. Regeneration of virus-free potato from tissue culture. In: Plant Tissue Culture and Its Bio-technological Application (W. Barz, E. Reinhard, and M.H. Zenk, eds.) pp. 386-391. Springer-Verlag, Berlin.

_____ and Hu, C.Y. 1980. Regeneration of virus-free plants through in vitro culture. In: Advances in Biochemical Engineering (A. Fiechter, ed.) pp. 62-99. Springer-Verlag, Berlin.

_____ and Hu, C.Y. 1982. In vitro mass tuberization and virus-free seed-potato production in Taiwan. Am. Potato J. 59:33-37.

_____ and Huang, L.C. 1975. Callus cultures from potato tissues and the exclusion of potato virus X from plants regenerated from stem tips. Can. J. Bot. 53:2565-2567.

Weatherhead, M.A. and Henshaw, G.G. 1979a. The induction of embryoids in free pollen culture of potatoes. Z. Pflanzenphysiol. 94:441-447.

_____ 1979b. The production of homozygous diploid plants of *Solanum verrucosum* by tissue culture techniques. Euphytica 28:765-768.

Wenzel, G. 1980a. Protoplast techniques incorporated into applied breeding programs. In: Advances in Protoplast Research, Proceedings of the Fifth International Protoplast Symposium (L. Ferenczy and G.L. Farkas, eds.) pp. 327-340. Pergammon Press, Oxford.

_____ 1980b. Recent progress of microspore culture of crop plants. In: The Plant Genome (D.R. Davies and D.A. Hopwood, eds.) pp. 185-196. The John Innes Charity, Norwich, England.

_____ and Uhrig, H. 1981. Breeding for nematode and virus resistance in potato via anther culture. Theor. Appl. Genet. 59:333-340.

_____, Schieder, O., Przewozny, T., Sopory, S.K., and Melchers, G. 1979. Comparison of single cell culture derived *Solanum tuberosum* L. plants and a model for their application in breeding programs. Theor. Appl. Genet. 55:49-55.

_____, Meyer, C., Przewozny, T., Uhrig, H., and Schieder, O. 1982. Incorporation of protoplast techniques into potato breeding programs. In: Variability in Plants Regenerated From Tissue Culture (E.D. Earle and Y. Demarly, eds.) pp. 290-302. Praeger Pub., New York.

Wernicke, W., Larz, H., and Thomas, E. 1979. Plant regeneration from leaf protoplasts of haploid *Hyoscyamus muticus* L. produced via anther culture. Plant Sci. Lett. 15:239-249.

CHAPTER 12
Yams

P. V. Ammirato

Yam plants are members of the genus *Dioscorea* and produce edible tubers, bulbils, or rhizomes that are of considerable economic importance. They are monocots belonging to the family Dioscoreaceae within the order Dioscoreales (Ayensu, 1972). Yams are the staple foodstuff for millions in many tropical and subtropical countries and are a secondary food for many millions more (Onwueme, 1978). They are also a standby for many populations in times of famine. Indeed, they are used in almost all tropical countries, with the exception of the most arid. Their large-scale cultivation, however, is restricted to three main areas: West Africa, Southeast Asia including adjacent parts of China, Japan, and Oceania, and the Caribbean. In addition, a number of wild species have found widespread use as a source of the steroidal sapogenin, diosgenin, which is the precursor in the commercial synthesis of sex hormones and corticosteroids (Coursey, 1967). These are often referred to as the medicinal yams.

The term yam has been widely misapplied to the edible parts of other genera. In the United States, in particular, "yams" usually refer to the sweet potato, *Ipomoea batatas* L. Poir., a dicot belonging to the Convolvulaceae. Yams are also confused with a number of edible aroids including *Colocasia*, *Alocasia*, and *Xanthosoma* spp. These are most properly referred to as cocoyams, taros, dasheen, eddoes, tanias, or yautias. The name has also been applied to *Maranta arundinacea* L., the arrowroot. In fact, at one time or another, almost any edible starchy root, tuber, or rhizome grown in the tropics has been termed a yam. Coursey (1967) recommended that the term yam be reserved for the economically useful plants or tubers or rhizomes of the genus *Dioscorea*. Recognizing this point, many people refer to the tubers of the genus *Dioscorea* as the "true yams."

HISTORY OF THE CROP

The family Dioscoreaceae is a distinct taxonomic grouping. Burkill (1960) believed its closest affinity is with the tribe Asparagoideae within the Liliaceae, the lily family. There is some disagreement as to the number of taxa within the family. Burkill placed six genera with about 650 species in the family. Others, such as Hutchinson (1959), have removed the three genera with hermaphroditic flowers to other families. Most retain the six genera and place the family in its own order, the Dioscoreales (Ayensu, 1972).

According to Coursey (1967), members of the family were present and well diversified in all parts of the southern world at the end of the Cretaceous. After this period, divergent evolution occurred in the Old and New Worlds leading to separate sections of the genus for the two hemispheres. No member of one section is found in the other. At a later time, during the Mioscene, the ancestral groups in Asia and Africa separated. Some sections of the genus are found in both continents and many species from one show affinities to certain species from the other. However, there are no common species, except perhaps for *D. bulbifera*. Some of the more important food tuber species are given in Table 1 according to their sectional affiliations.

It is believed that the ancestral members of the genus were small herbaceous plants that evolved annual twining stems and subterranean perennial rhizomatous stems. The direction of rotation of the twining stems is either dextral or sinistral (right- or left-handed) and is a taxonomic feature, i.e., all members of the section Enantiophyllum twine to the right (Fig. 1a), those of other sections, e.g., Lasiophyton, Combilium, and Macrogynodium to the left (Fig. 1b; cf. Purseglove, 1973, pp. 99-100). The evolution of climbing stems permitted access to better light; their annual nature coupled with perennial rhizomes or stem tubers allowed the plants to withstand drought and seasonal changes. The tubers originated by contraction of the rhizomes. *Dioscorea villosa*, the only indigenous wild species within the United States, has a rhizomatous habit. The tubers are of stem origin in all species and are replaced annually. They shrink during the vegetative period of growth and new ones form and fill during the later part of the growing season (Fig. 1c). In some species, the axillary bud or buds can develop into aerial tubers or bulbils and in one species, *D. bulbifera*, the bulbils provide the edible yams.

The genus has over 600 species (Ayensu, 1972) and is completely dioecious, i.e., flowers are imperfect (either male or female) and borne on separate plants. This condition evolved from hermaphroditic ancestors. In recent breeding work with *D. rotundata* (Sadik and Okereke, 1975) some monoecious plants, bearing both staminate and pistillate flowers, were produced.

Table 1. Major Edible Yam Species

SPECIES	COMMON NAME	LOCATION
SECTION ENANTIOPHYLLUM		
D. alata L.	Greater yam	Orig: Southeast Asia; Cult: tropics
D. rotundata Poir.	White Guinea yam	West Africa
D. cayenensis Lam.	Yellow Guinea yam	West Africa
D. nummularia Lam.		Indonesia, Oceania
D. opposita Thunb.	Chinese yam	China, Korea, Taiwan, Japan
D. japonica Thunb.		China, Japan
SECTION LASIOPHYTON		
D. dumetorium (Kunth) Pax.	African bitter or cluster yam	West Africa
D. hispida Dennst.	Asiatic bitter yam	India, southern China, New Guinea
SECTION COMBILIUM		
D. esculenta Burk.	Lesser yam	Tropics, esp. Asia & the Pacific
SECTION OPSOPHYTON		
D. bulbifera L.	Potato or aerial yam	Orig: Asia & Africa; Cult: tropics
SECTION MACROGYNODIUM		
D. trifida L.	Cush-cush yam	Orig: New World (northern South America); Cult: Caribbean

Coursey (1976) reported that the domestication of the yams in Asia, Africa, and tropical America occurred separately, involving completely different species (Table 1). In Southeast Asia, *Dioscorea alata* and *D. esculenta* were derived from the Indian center of origin. *Dioscorea alata* then had a subsidiary distribution from the Indo-Malaysian center,

A
B
C
D
E
F
G
H

Figure 1. (a) *Dioscorea alata* stem showing right-handed twining. (b) *D. bulbifera* with left-handed twining. An aerial tuber (bulbil) has formed in one leaf axil. (c) Tuber formation in *D. alata* var. Farm Lisbon. The previous year's tuber (arrow) has shrunk during the first months of stem growth and has remained during the formation of the current year's tubers. (d-h) Tubers of yam species (scale = 10 cm); (d) *D. alata* var. Farm Lisbon (immature tubers), (e) *D. rotundata*, (f-h) *D. bulbifera.*

---

a region more to the southeast. From this location, *D. hispida, D. pentaphylla,* and *D. bulbifera* also originated. These yams were probably taken across the Pacific during the Polynesian migrations. The more temperate species of the section Enantiophyllum, *D. opposita* and *D. japonica,* most likely derived from the Chinese center of origin. The domestication of the African yams was independent of external influences, at least initially, and resulted in *D. cayenensis* and *D. rotundata.* The latter is considered by many to be a subspecies of the former. In the New World, yam domestication, and in particular, *D. trifida,* probably originated in the Brazil-Guyana area and spread to the Caribbean.

ECONOMIC IMPORTANCE

The cultivated yams are grown on a field scale for their edible tubers. These provide the staple carbohydrate food in the "yam zone" of West Africa where daily consumption is from 0.5 to 1.0 kg per person. The edible yams are also of some importance in the Melanesian and Pacific regions and in the Caribbean.

Dioscorea alata, the greater yam (Fig. 1d; cf. Table 1), gives the highest yields of tubers of all the cultivated species and is the most widespread (Martin, 1976b). It has become accepted in West Africa where it competes with *D. rotundata.* The two account for most of world yam production. The white guinea yam, *D. rotundata* (Fig. 1e), is used for the production of "fufu," a traditional African dish that cannot easily be made from most other yams (Martin, 1977). It is more tolerant of drought than *D. alata.* The yellow guinea yam, *D. cayenensis,* is the second important African yam and grows in wetter areas. Its tubers do not store well. *Dioscorea esculenta,* the lesser yam, is cultivated mainly in Asia and the Pacific islands. Plants produce clusters of small tubers that are not bitter or toxic, and, with less fiber, are more palatable. However, the tubers bruise and rot easily, and are not easily stored. Stems of the lesser yam plants are heavily covered with spines that make cultivation difficult (cf. Cobley and Steeles, 1976, pp. 118-123). *Dioscorea bulbifera* is one of the most common and widespread yams known in both Asia and Africa. Particularly superior edible varieties are known in India (Martin, 1976a). It ranks a poor fifth after the other yams but is distinguished for the production of aerial tubers called bulbils in the axils of leaves. The aerial tubers of the African races are sharply angled (Fig. 1f) in contrast to the Asian races (Fig. 1g). The aerial tubers may be produced as early as 3 months after planting and can be picked at any time. Some varieties produce bulbils of substantial size (Fig. 1h). The flavor is distinctive but does not have the appeal of the best varieties of *D. alata* and *D. rotundata* (Martin, 1976a). Given the relatively easy harvesting that can be spread over the growing season, it is a pity that they are not the preferred yam.

Surprisingly, there is relatively little reliable information in the literature on either the production or utilization of edible yams. This is due in part to the fact that they are essentially crops of peasant farmers in the tropics. The confusion over nomenclature has also hampered the gathering of reliable production information. In the annual production books of the Food and Agricultural Organization of the United Nations (FAO), all root-tuber crops are summarized; currently, yam production is not distinguished from sweet potato or cassava which are grown on a much larger scale. Some estimates have been made and the global production of yam during the decade 1965-1974 is summarized in Table 2.

There are a number of important features in yam production. For one, little of the yam crop enters into international trade. The vast majority of yams are consumed within the area or country of production. Second, there has been practically no increase in the yield per hectare (Onwueme, 1978). Finally, there has been a dramatic decline in the importance of the crop. For a number of reasons (see the next section), the growing of yams is particularly labor intensive. Consequently, in some regions, they are gradually being replaced by potatoes (*Solanum tuberosum* L.), sweet potatoes [*Ipomoea batatas* (L.) Poir.], the American taro or cocoyams [*Xanthosoma sagittifolium* (L.) Schott.], and, most important, manioc or cassava (*Manihot utilissima* Pohl.) (cf. Coursey, 1967, pp. 19-22).

The approximate composition of edible yam tubers is water (65-75%), carbohydrates, mainly starch (15-25%), protein (1-2.5%), fiber (0.5-1.5%), ash (0.7-2.0%), and fat (0.05-0.20%). Yams are a better source of protein than cassava and its substitution may increase the incidence of "kwashiorkor" (the protein-deficiency disease). Yams contain approximately 8-10 mg/100 g of ascorbic acid. Since most of it is retained during cooking, yams can be a significant source of vitamin C. Because of this, and the relatively easy handling and storage of the tubers, yams were used for their antiscorbutic properties by sailors on long voyages in the Indian and Pacific Oceans in the pre-European era and later by the Portuguese. They were an important foodstuff on the slave ships travelling between West Africa and the New World. This led to their widespread distribution throughout the tropics.

The more important medicinal yams (Coursey, 1967) are *D. composita* Hemsl., *D. floribunda* Mart. & Gal., and *D. mexicana* in Mexico; *D. elephantipes* (L.) Her. Engl. and *D. sylvatica* Eckl. in South Africa; and *D. deltoidea* Wall. and *D. prazeri* Prain and Burk. found at high altitudes in India. All are wild species. There are a number of problems. Tubers from wild plants are difficult to collect and over-harvesting has threatened natural populations. South Africa has controlled exploitation by a licensing arangement. There have been some efforts at cultivation but with limited success (Coursey, 1976).

Table 2. Mean Annual Production of Yams (1965–1974)[a]

LOCATION	YAM AREA (1000 ha)	PERCENT OF TOTAL	PRODUCTION (1000 tons)	PERCENT OF TOTAL	YIELD (kg/ha)
World	1916	100	18037	100	9443
REGIONS					
Africa	1876	98	17701	98	9461
West Indies[b]	13	1	164	1	12124
COUNTRIES					
Nigeria	1334	70	13619	76	9168
Ivory Coast	187	10	2908	16	7860
Ghana	130	7	1046	6	8517
Togo	107	6	803	4	754
Dahomey	54	3	527	3	9626
Sudan	36	2	199	1	3005
Jamaica	6	0.3	87	0.5	140

[a]Adapted from Onwueme (1978).

[b]Includes Jamaica, the Dominican Republic, Haiti, and Puerto Rico.

In the past, the medicinal yams provided the major source of diosgenin. However, in recent years, alternative sources of diosgenin, such as Fenugreek and other raw materials (e.g., stigmasterol and sitosterol) can be converted to steroids by fermentation processing. These now compete with *Dioscorea* diosgenin (Applezweig, 1977).

BREEDING AND CROP IMPROVEMENT

During the thousands of years that yams have been cultivated, the plants were selected for yield, shallow-rooting, palatability, and lack of toxicity. Little systematic work has been performed to breed improved cultivars. Yet there are a number of major problems (Onwueme, 1978).

One problem is the relation of sett size to yield and the yield itself. If small setts are planted, then there are low yields at harvest. Therefore, large quantities of yam setts must be saved, stored, and planted to ensure reasonable yields. Typically, 2.5 tons or more of setts are utilized per ha. Even from such a volume of material, the typical yield is 12.5 tons per ha. Consequently, one-fifth of a crop must be saved and used as planting material.

The relatively low yield in yams is a problem in itself. Some of this is due to the relatively unimproved cultural practices used in production. It has been estimated, for example, that the present average yield of yam per ha in Nigeria is only about 14% of the potential yield (Olayide, 1972). Another major factor is the absence of high-yielding cultivars and the lack of uniformity in any population due to the method of propagation.

Another problem is that, due to the nature of the plant, yam production is laborious. Almost all of the tasks are currently performed by hand or with hand tools and some tasks are not easily mechanized. For one, the growing vines need to be staked. In addition, most crops are double-harvested and the first harvest, performed so as to allow the plants to continue to grow, is particularly difficult.

A further difficulty is the typically long growing season. The typical interval from planting to harvesting is about 10 months. The tubers, once planted, remain dormant for the first 3 months. Following emergence of the shoot, solely vegetative growth occurs for the next 3 months. Tuber initiation and bulking take 3 months. During the final month, tubers mature but gain no fresh or dry weight. There is then a large investment of labor, a long time to be subject to the vagaries of the environment, for a crop that will yield only about 3 tons of dry carbohydrates per ha.

The final problem concerns difficulties in storage and handling. Yams are stored and handled in the form of fresh tubers and two-thirds of the weight is water. They are bulky to manipulate. In addition, the skin or periderm of the tuber has been selected to be thin and easily peeled. Tubers therefore must be moved carefully, often by hand. They are easily bruised and infected and storage rots are common. They also dry out easily. These factors have led to significant storage losses. Coursey (1967) estimated that over one million tons of yam tuber are lost annually during storage in West Africa alone.

According to Coursey (1967), the most common of the storage rots, and the most serious, is *Botryodiploidia theobromae* Pat. *Rosellinia bunodes* (B. & Br.), *Penicillium* spp., and *Fusarium* spp. also cause tuber rot. Diseases of the plant foliage are less serious and include *Cercospora carbonacea* Miles, a leaf spot; *Gloeosporium pestis* Massee, a wilt; Anthracnose, *Glomerella cingulata* (Stonem.) Spauld. & Schrenk; and witches' broom, *Phylleutypa dioscoreae* Cummins. Virus diseases of the mosaic type are also known.

Yam beetles, *Heteroligus meles* (Billb.), are a serious insect pest feeding on both tubers and growing shoots. Scale insects such as *Aspidiella hartii* Ckll, the beetle, *Crioceris livida* Dalm, and the yam weevil, *Palaeopus dioscorae* Pierce, are all serious pests. Nematodes can also do significant damage and include the yam nematode, *Scutellonema bradys* (Steiner & Le Hew) Andrassy, as well as *Meloidogyne* spp. and *Pratylenchus* spp.

The goals of yam improvement, therefore, would include the breeding and selection of plants that are erect or semi-erect, have a relatively short growing season, and produce tubers with high yield, disease resistance, tough skin, and small round or oval shape that would make them amenable to mechanical harvesting. A higher protein content while maintaining good flavor and texture is important. In addition, the development of special propagules to replace the use of setts would be valuable.

Unfortunately, the typical selection scheme that was followed by yam growers for thousands of years, and is still performed today, involves the selection of the best yielding or growing yam plants and their propagation by setts derived from tubers. As a consequence of this asexual method, many important species and cultivars rarely produce viable seeds; others are completely sterile. Most of the better clones of *D. rotundata* are male plants. It is in *D. alata* and *D. esculenta* where infertility is most extreme. The polyploidy of many of the preferred cultivars also probably contributes to reduced fertility. This condition has precluded the development of traditional breeding programs that could alleviate some of the many problems facing this crop. Recently, seedlings have been obtained and a hybridization program begun in Nigeria at the International Institute of Tropical Agriculture (IITA), in particular with *D. rotundata* (Sadik and Okereke, 1975; Akoroda, 1983). For the majority of species that do not produce seed, the techniques of plant cell culture offer a number of important prospects. A tremendous amount of variation exists within any cultivar and the propagation and dissemination of high yielding, elite individuals would be particularly valuable. The induction of increased variability and the selection of lines with new characteristics, such as erect habit or spineless stems, would be equally as important.

REVIEW OF THE LITERATURE

Clonal Propagation from Nodal and Meristem Cultures

A number of species, such as *D. alata*, can be propagated in the greenhouse or field, albeit inefficiently, through nodal cuttings (Coursey, 1967). One goal of investigations of *Dioscorea* cell and tissue culture research has been to utilize the techniques of meristem culture (cf. Hu and Wang, 1983) for large-scale clonal propagation.

Research began in earnest in the mid-1970s with a fair degree of success in both growing plants from axillary buds and in promoting multiple shoot formation to allow for larger scale propagation (Table 3). The greatest number of successes came from the utilization of nodal segments containing the axillary bud or buds (there often are three per node) and some adjacent stem and petiole tissue but with leaves removed. If the segments are relatively large (greater than 1-2 cm), shoot and root growth will usually proceed on an unsupplemented basal medium (usually the MS medium). If the shoot tips, or apical meristems, are excised, then both an auxin and a cytokinin are usually beneficial.

There are differences among species. Chaturvedi et al. (1977) showed that rooting of stem cuttings of *D. deltoidea* was difficult when compared to similar material from *D. floribunda*. In the former, only 50% of the cutting would root on medium containing NAA and IBA while almost all of the latter would root readily. For *D. deltoidea* a combination of NAA, IBA, 2,4-D, and chlorogenic acid improved the incidence of rooting.

The specific form of cytokinin or auxin may be important. Although KIN proved to be an effective cytokinin in nodal cultures of *D. floribunda* (Lakshimi Sita et al., 1976), Mantell et al. (1980) found BA to be preferable to KIN in apical shoot tips of *D. alata*. Both absolute and relative concentrations are also important. As auxin levels increase, there is a tendency for callus formation, e.g., *D. alata* (Ammirato, 1976) and *D. bulbifera* (Uduebo, 1971). Low levels of auxin appear to be important in enhancing rooting. Cytokinin at moderate concentrations enhances shoot development; at higher levels, it promotes multiple shoots through precocious axillary shoot formation (Ammirato, 1976). Chaturvedi (1975), in one of the first investigations, reported that single-node cutting on medium with 0.5 μM NAA and adenine sulphate rooted, then formed some tuberous tissues with the development of one or two shoots. Explants on a medium with 8.8 μM BA produced on the average 5-6 shoot buds in 20 days. Mantell et al. (1978) reported that one excised nodal segment of *D. alata* or *D. rotundata* would produce a complete plant in 3-5 weeks. Each small plant could be split into 3-5 separate nodal segments (each with one leaflet) and transplanted onto fresh medium. After 14-20 days, new multibranched plantlets developed. They estimated that, on a regular 14-20 day cycle, it would be possible to obtain 65,000 plantlets from a single node within 6 months. Although the numbers are impressive and the results useful, the difficulty, of course, is that each nodal segment must be individually cut and transferred to fresh media. It is a laborious process.

One difficulty is the gradual browning of the culture medium from the cut ends of the stem. The intensity of the response is variable from culture replicate to replicate and from variety to variety and often does not affect growth. Interestingly, in studies with *D. alata* var. Gemelos (Ammirato, 1982), after the plants were established in culture from single-node stem segments, further subculturing, using single-node segments, did not exhibit browning of the medium. For primary cultures, allowing the stem segments to sit in a covered petri dish on barely wet filter paper for a half hour prior to transfer to the culture medium seems to help. The use of anti-oxidants in the medium (see the next section) or low light intensity to inhibit phenolic synthesis may be useful.

Mantell et al. (1978) were among the first to notice that nodal segments removed from vines of differing ages differed in their responses in culture. The effect of age variability of yam vine segments on suitability for propagating nodal cuttings had already been described by Martin and Delphin (1969) for *D. floribunda* and *D. bulbifera*. Similarly, since Preston and Haun (1962) had shown that young plants grown under long photoperiods produced new plants from vine segments more readily, they tested the responses of nodal segments removed from plants grown under 16- and 12-hr photoperiods. Those removed from plants grown under the longer photoperiod consistently produced better shoot growth.

The development of complete plants from excised shoot meristems has also occurred in a number of species (Table 3). Using the technique in combination with a high temperature pretreatment, Mantell et al. (1980) successfully eliminated flexous rod viruses from *D. alata*.

Meristem cultures also offer the prospect of germplasm storage by means of either minimal-growth storage or cryopreservation (Chaturvedi et al., 1982). There has been some success with *D. rotundata* (Henshaw, 1982).

Plants clonally produced by nodal or meristem culture do show a high degree of uniformity. Plants of *D. alata* var. Gemelos propagated by means of nodal cultures show uniformity both in their vegetative growth (Fig. 2a) and in tuber formation (Fig. 2b). Chaturvedi et al. (1982) produced ca. 2,560,000 plants from a single-node stem cutting in one year. All were true-to-type.

Tetraploids of *D. floribunda* were obtained by the application of colchicine to the shoot buds developing on the base of a rooted plant prepared by excising all visible shoots (Chaturvedi, 1979). Cotton wool soaked in 0.05% colchicine was applied for 2 days and then removed. After 30 days, about 10 tetraploid shoots developed that were more vigorous, with stout stems and larger leaves than the normal diploids. The tetraploid shoots were rapidly multiplied using single-node leaf cuttings, and a large number of tetraploid plants were obtained. They are to be used for breeding triploid *D. floribunda*, which may produce bigger tubers with higher diosgenin content.

Tuberization in Culture

Uduebo (1971) first reported that nodal cuttings of *D. bulbifera* would form tubers on medium supplemented with low levels of an auxin, IAA. Nodal segments of both *D. bulbifera* and *D. alata* have produced tubers directly in culture (Ammirato, 1976, 1982). In *D. bulbifera*, a single shoot, followed by a root, will form on unsupplemented MS medium (Fig. 3a). Shoot growth is repressed in the presence of NAA with the formation of a tuber (Fig. 3b). In contrast, zeatin promotes multiple shoot growth (Fig. 3c). In the case of *D. alata* var. Farm Lisbon, there was tuber, shoot, and root growth on the MS medium (Fig. 3d)

Table 3. Nodal and Meristem Culture in *Dioscorea*

SPECIES	GROWTH MEDIUM (µM)	EXPLANT SOURCE	REFERENCE
EDIBLE YAMS			
D. alata			
var. Farm Lisbon	MS	Nodal segments	Ammirato, 1976
var. Gemelos	MS or MS, 0.1 NAA, 10 ZEA	Nodal segments, shoot buds	Ammirato, 1982
var. White Lisbon	MS	Nodal segments	Mantell et al., 1978
	MS, 2.7-5.4 NAA, 0.9 KIN or 0.9 BA	Shoot tips (0.6-2.5 mm)	Mantell et al., 1980
	MS, 2.3 2,4-D, 0.9 BA; or 5.4 NAA, 0.9 BA; or 2.7 NAA, 0.4 BA	Shoot meristems (0.2-0.5 mm)	Mantell et al., 1980
D. batatas	K, S, 4.7 KIN	Nodal segments	Asahira & Nitsch, 1968
D. bulbifera var. Sativa	MS or MS, 0.1 NAA, 10 ZEA	Nodal segments, shoot buds	Ammirato, 1976, 1982
	MS, 0.57 IAA	Nodal segments	Uduebo, 1971
D. macrostachya[a]	MS, 10% CW	Nodal segments	Mapes & Urata, 1970
D. rotundata	MS	Nodal segments	Mantell et al., 1978
var. Habanero	MS, 11.4 IAA, 9.3 KIN to 5.4 NAA[b]	Nodal segments	Cortes-Monllor & Liu, 1982
MEDICINAL YAMS			
D. composita	MS or MS, 0.1 NAA, 10 ZEA	Nodal segments, shoot buds	Ammirato, 1982
D. deltoida	MS, 5 NAA, 5 BA	Shoot apices (0.2-0.5 mm)	Grewal et al., 1977
	MS, 9.9 IBA	Shoot apices	Grewal & Atal, 1976
D. floribunda	MS, 8.9 BA to 2.7 NAA[b]	Nodal segments, shoot tips (1-2 cm)	Chaturvedi,, 1975
	MS, 9.3 KIN, ± 0.54 NAA	Nodal segments	Lakshmi et al., 1976

Table 3. Cont.

SPECIES	GROWTH MEDIUM (μM)	EXPLANT SOURCE	REFERENCE
D. floribunda	MS, 2.2 BA, 0.54 NAA or 0.45 2,4-D	Nodal segments	Sinha & Chaturvedi, 1979
	MS or MS, 0.1 NAA, 10 ZEA	Nodal segments, shoot buds	Ammirato, 1982

[a]Authors express some doubt as to species identification.

[b]Change to secondary medium.

and tuber formation and callusing with NAA in the medium (Fig. 3e). However, tuber formation without callusing occurred with the further addition of 1 μM ABA (Fig. 3f). In studies with *D. batatas* (Asahira and Nitsch, 1968), KIN promoted shoot growth in nodal segments if the culture medium contained ammonium ion in addition to nitrate but fostered tuber formation when ammonium was absent.

Many species of *Dioscorea* form small bulbils in their leaf axils; this often occurs during the rooting of nodal cuttings in the greenhouse (Coursey, 1967). The formation of small tubers on plants grown from nodal and meristem cultures has been observed in *D. floribunda* (Chaturvedi, 1975; Grewal et al., 1977).

Ammirato (1982) reported that both *D. alata* var. Gemelos and *D. bulbifera* var. Sativa plants, after growing for 4-5 months in continuous light, developed numerous aerial tubers (Fig. 3g,h). These tubers could be harvested or the cultures allowed to dry and the tubers and dry plants stored for up to 6 months (Fig. 3i). In one experiment, 37 tubers were harvested from 9 flasks. They ranged in weight from 3 to 1486 mg and from 2 to 32 mm in length. When planted in soil, tubers larger than 90 mg and 8 mm in length all sprouted and produced normal plants and tubers.

Microtuber formation has been observed in other genera (Hu and Wang, 1983) and the procedure has become very important in the propagation of potato, *Solanum tuberosum*. In a similar fashion, the micropropagation of elite cultivars and the induction of microtubers could provide an excellent means for both storage and rapid, inexpensive delivery of clonally propagated *Dioscorea* varieties to the field.

Organogenesis and Embryogenesis from Callus and Suspension Cultures

In his classic text on experimental plant morphology, Goebel (1908) described shoot and root regeneration in tuber sections of *D. sinuata* and *D. japonica*. The discussion is in the section on polarity for,

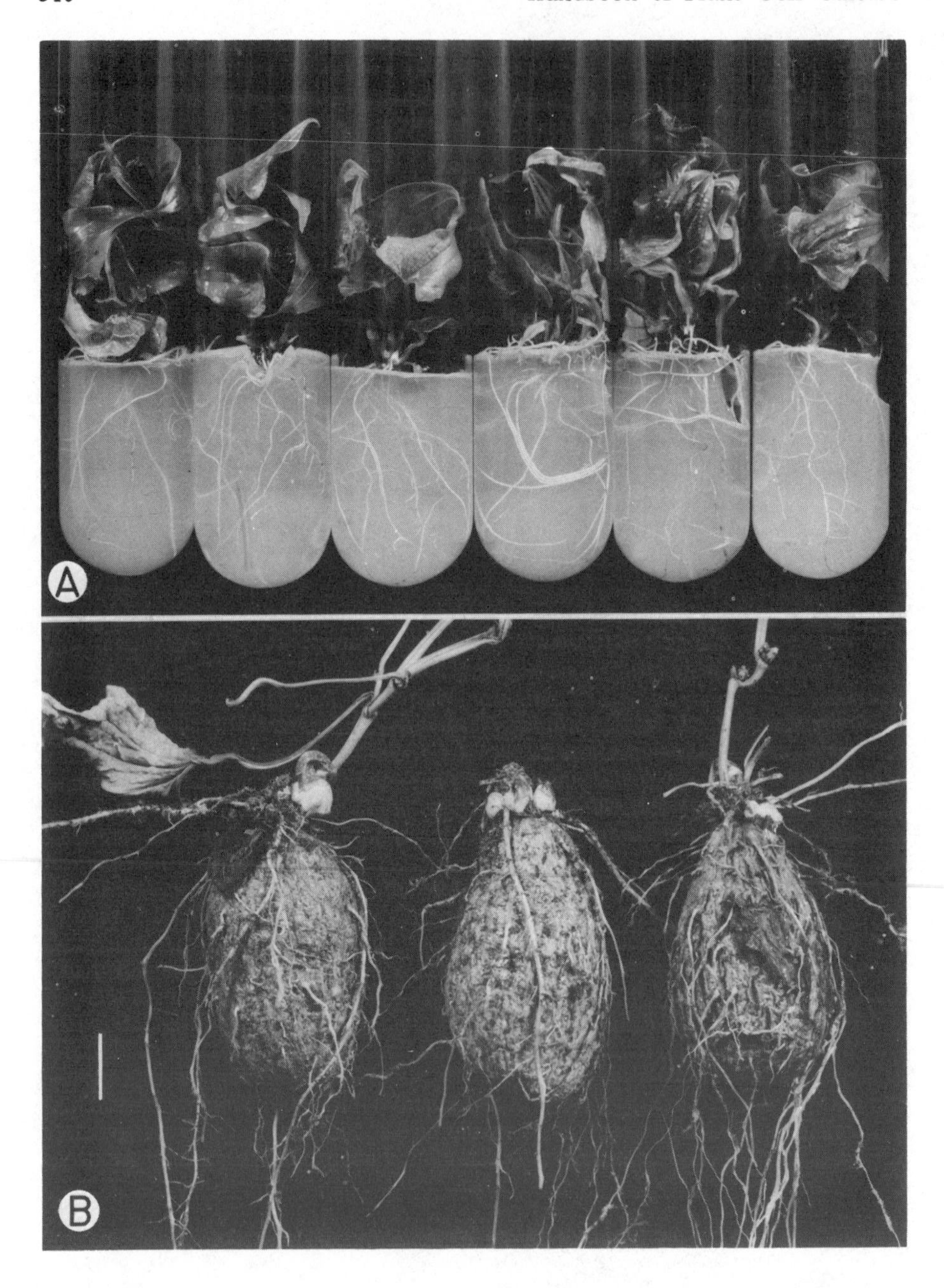
A
B

Figure 2. (a) Clonal propagation of *Dioscorea alata* var. Gemelos from single node stem segments on unsupplemented MS basal medium. (b) Tubers grown after transfer of plants to soil (scale = 1 cm).

---

interestingly, shoot regeneration occurred in the center of the tuber half-slice, while root formation was at the periphery.

Tubers should provide an excellent explant source for organogenesis or embryogenesis in culture. However, tubers collected from the field often contain endogenous fungi (Coursey, 1967) making surface sterilization ineffective. Also, freshly harvested tubers are often dormant, and tuber pieces, when cut, are subject to tissue oxidation and browning.

To circumvent these problems, tubers can be greenhouse grown in sterilized soil or, better still, microtubers may be generated by single-node stem segments or apical meristem cultures. Storage may be used to overcome dormancy, e.g., in the dark at 24 ± 2 C for 4 months (Asokan et al., 1983). Treatment with ethrel has been shown to be effective (Gupta et al., 1979), as has 2-chloroethanol or ethanol (Passum et al., 1982). To minimize tissue oxidation, explants have been immersed in 1.0% ascorbic acid for 10 min after excision and prior to transfer to culture tubes (Asokan et al., 1983). Other anti-oxidants such as cysteine and glutathione may also be effective. Finally, the vegetative parts of the plant, such as the apical meristems and nodal segments, provide fine sources of explant material.

Table 4 lists the yam species that have been regenerated from unorganized callus or suspension cultures and gives some of the conditions. A common element to most of the reports is the use of 2,4-D as the auxin in the culture medium in contrast to IAA, IBA, and NAA used in the nodal and meristem cultures (Table 3). 2,4-D both initiated proliferations and maintained unorganized growth during subcultures. Transfer of the material to a medium (secondary medium) without 2,4-D, with lower levels of 2,4-D, or with another auxin source coincided with the appearance of embryos and/or plants. This sequence is typical of organogenesis (Flick et al., 1983) and especially embryogenesis in plant cell cultures (Ammirato, 1983).

One exception is the direct formation, without conspicuous callus development, of adventitious shoots on the pulvinus base of excised leaves of *D. floribunda* (Sinha and Chaturvedi, 1979). Direct organogenesis and embryogenesis without intervening callus formation has been seen in other genera (Flick et al., 1983; Hu and Wang, 1983; Ammirato, 1983).

It is difficult to explain organogenesis and the development of shoots in some cultures and embryogenesis and somatic embryos in others. The two alternative morphogenetic routes may be triggered by different hormonal or environmental conditions. Or, as has been the case in many cell cultures, somatic embryos initiate but the shoot apex grows out prematurely, thereby obscuring the embryogenic origins.

One feature of most reports is the failure of completely differentiated tissues, such as internodal or mature leaf tissue, to proliferate in culture or to provide callus tissue that could sustain growth. Generally speaking, it was the embryonic, meristematic, or juvenile tissue that generated active cell growth and embryo and plant regeneration.

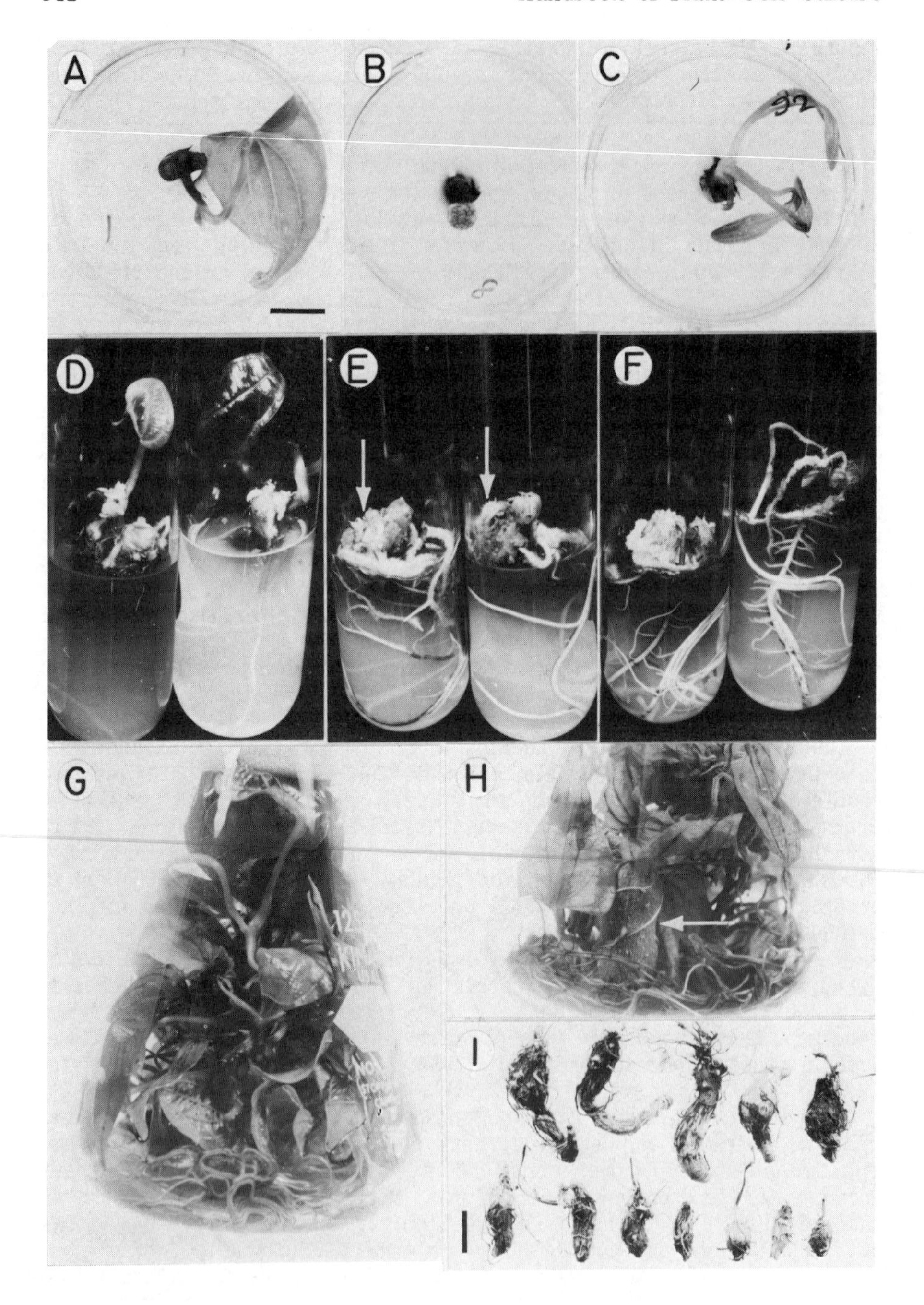
A
B
C
D
E
F
G
H
I

Figure 3. (a–c) Nodal cultures of *D. bulbifera* var. Sativa (scale = 1 cm): (a) shoot formation on MS basal medium, (b) tuber formation of MS with 0.1 μM NAA, (c) multiple shoot formation on MS with 10 μM ZEA. (d–f) Nodal cultures of *D. alata* var. Farm Lisbon: (d) tuber, shoot, and root formation on MS medium, (e) tuber and callus formation (arrows) on MS with 1 μM NAA, (f) tuber formation, free of callusing, on MS with 1 μM NAA and 1 μM ABA. (g–h) Microtuber formation in nodal stem segment cultures of *D. alata* var. Gemelos: (g) plant development after several months growth, (h) small tubers (arrow) have formed as culture has aged. (i) Microtubers harvested from 3 flasks after plants have dried (scale = 1 cm).

Mature cells in *Dioscorea*, as in many monocots, do not for the most part dedifferentiate and reorganize into vascular and cork cambia. Thus, it is the cells that are still dividing that are most easily grown in culture: excised embryos, seedling parts, young petioles and leaves, apical meristems, and tuber tissue. In the latter, there is a prominent cambial layer in the cortex (Martin and Ortiz, 1963).

An additional feature of *Dioscorea* organogenic or embryogenic cultures was the appearance of a mixture of cell types—a pale friable tissue and a more compact tissue with small nodules and often roots (Ammirato, 1978a, 1982; Grewal and Atal, 1976). It is from the latter that regeneration occurs (Ammirato, 1982). Success in regenerating embryos or shoots depends, then, on selectively transferring and growing the correct cell or callus types. This is also true in other monocot cultures and in particular in the grasses (cf. Ammirato, 1983, pp. 97–98).

Almost all the regeneration shown in Table 4 occurred in stationary cultures containing semisolid medium. Somatic embryos were grown, however, in suspension cultures of *D. floribunda* (Ammirato, 1978a, 1982, 1984). Callus was initiated on semisolid medium in the dark and transferred to liquid medium, first in small culture vessels and then in larger culture flasks (Fig. 4a). In earlier studies, 4.5 μM 2,4-D was used to both initiate and maintain growth; later studies showed that 18 μM 2,4-D produced a callus and suspension with a higher percentage of embryogenic to nonembryogenic cell clusters. Somatic embryo maturation could be achieved in two ways. If cultures were allowed to age, watery tissue turned brown and small, conical embryos developed from the dense, white globular tissue (Fig. 4b). Alternatively, the dense masses were selectively transferred to fresh media. For embryo maturation, the small cell clusters or proembryos were transferred to liquid medium containing ZEA and/or ABA. ZEA alone permitted excellent cotyledonary development. The addition of ABA prevented precocious germination.

Developing freely suspended in liquid medium and with the correct formulation, the somatic embryos of *D. floribunda* are remarkable structures (Fig. 4c; Ammirato, 1978b, 1984). They originate as small globular proembryos, then initiate a cotyledonary collar that develops asymmetrically to form the single cotyledon and the sheathing leaf base (Fig. 4d). The single cotyledon has a laminar apical end and a rounded basal end. The zygotic embryo is similar in structure except that the

Table 4. Organogenesis/Embryogenesis in *Dioscorea* Cultures

SPECIES	EXPLANT	1° MEDIUM (μM)	2° MEDIUM (μM)	RESULT	REFERENCE
EDIBLE YAMS					
D. alata var. "purple-fleshed"	Bulbil segments	MS[b], 0.45-0.9 2,4-D, 47-56 KIN	None	Shoots	Asokan et al., 1983
D. macrostachya	Nodal segments	MS, 10% CW, 27 NAA or 23 2,4-D	MS, 10% CW	Shoots	Mapes & Urata, 1970
MEDICINAL YAMS					
D. composita	Nodal segments	RT[a], 4.5 2,4-D	RT, 2.7 NAA, 6.7 BA	Shoots	Datta et al., 1981
D. deltoidea	Seedling tuber	MS, 10% CW	None	Multiple shoots	Mascarenhas et al., 1976
	Seedling tuber	MS, 10% CW, 23 2,4-D or 27 NAA	MS, 4.7 KIN, 4.4-8.9 BA; or 32-55 KIN, 0.05 2,4-D	Multiple shoots	Mascarenhas et al., 1976
	Tuber	MS, 4.5 2,4-D	MS[b], 0.45 2,4-D, 28-55 KIN	Embryos, shoots	Singh, 1978
	Hypocotyl	RT, 4.5 2,4-D	RT, 2.5 IBA	Shoots	Grewal & Atal, 1976
	Tuber or leaf	Not given	17.1 IAA, 1.1-2.2 BA	Embryos	Chaturvedi, 1979

SPECIES	EXPLANT	1° MEDIUM (μM)	2° MEDIUM (μM)	RESULT	REFERENCE
D. floribunda	Excised embryo	MS, 4.5 2,4-D	MS or MS, 0.1 ZEA, ± 0.1 ABA	Embryos	Ammirato, 1978a, 1982
	Leaf (blade + petiole)	MS, 0.54 NAA, 22.2 BA	None	Multiple shoots	Sinha & Chaturvedi, 1979
	Tuber or leaf	Not given	5.7 IAA, 1.1-2.2 BA	Embryos	Chaturvedi, 1979

[a]MS medium modified by Kaul and Staba (1968).

[b]MS modified to contain ammonium nitrate as the sole nitrogen source.

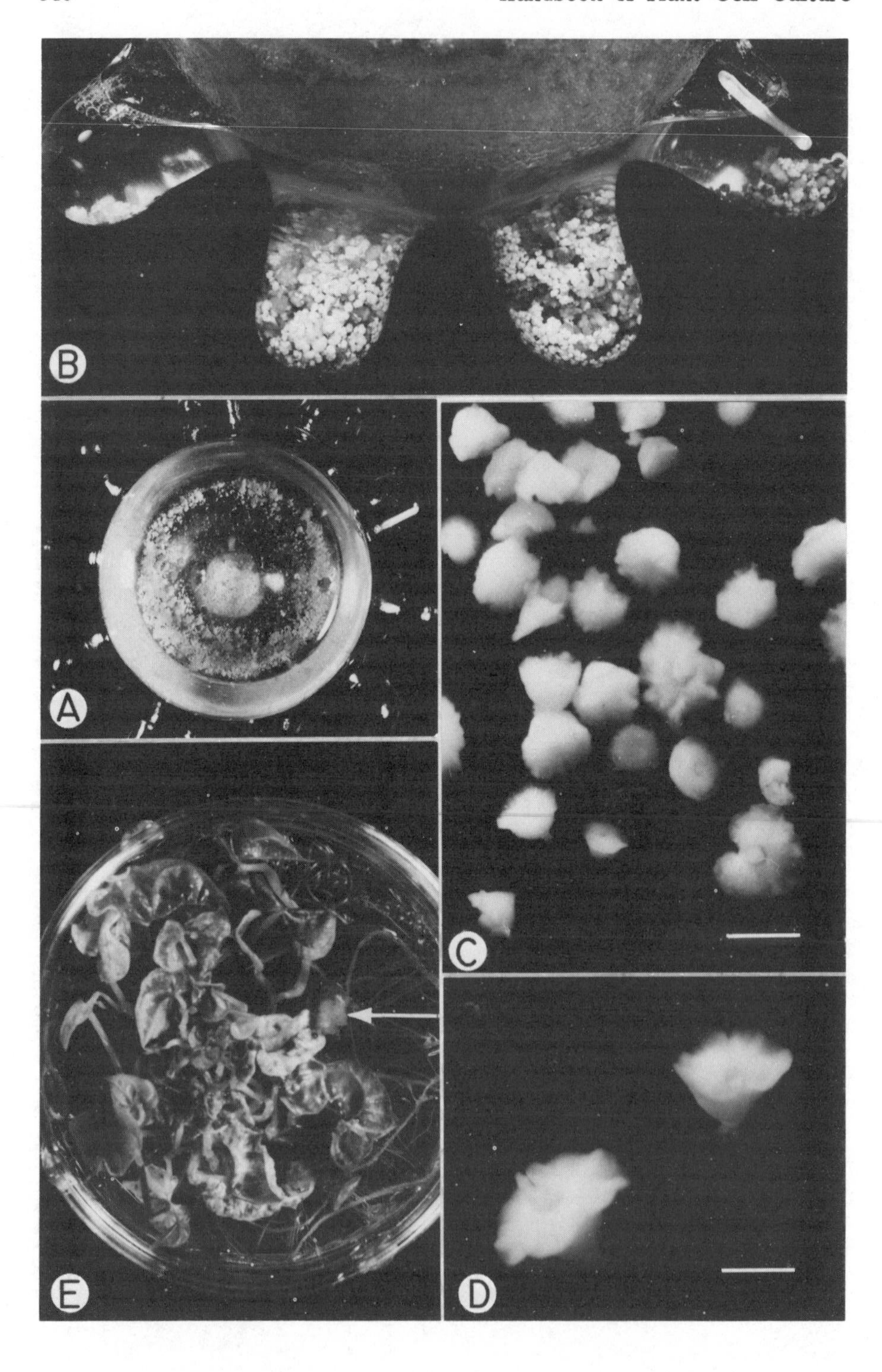
B
A
C
E
D

Figure 4. Somatic embryogenesis in *Dioscorea floribunda*. (a) Special culture flask with a medicinal yam suspension growing on MS with 18 μM 2,4-D. (b) Portion of a flask showing a dense crop of somatic embryos interspersed with darker, watery callus tissue. (c) A portion of the somatic embryo population shown in (b) (scale = 1 mm). (d) Two somatic embryos showing the cotyledonary collar that is a single cotyledon with a sheathing leaf base. The first leaf primordium is visible in the center of each embryo (scale = 0.5 mm). (e) Plantlets that have grown from two somatic embryos on MS with glutamine and 0.2 μM ZEA. The third embryo (arrow) produced roots but the shoot failed to grow out.

---

cotyledon is flattened due to its development in the slit between the two endosperm halves.

A similar response occurred in suspension cultures derived from axillary buds in nodal cultures of *D. bulbifera* (Ammirato, 1982). An excellent suspension culture formed on MS medium supplemented with 4.5 μM 2,4-D and 0.44 μM BA. Somatic embryos developed when the cell clusters were transferred to MS medium with 0.1 μM ZEA. However, they formed plants only weakly.

Dioscorea floribunda somatic embryos grew into plants if the embryos were placed in groups of five on unsupplemented MS basal medium or if placed singly on MS medium supplemented with glutamine and 0.2 μM ZEA (Fig. 4e). Over 40 plants were transferred to soil. All appeared healthy and normal but no further analyses were performed.

The suspension cultures of *D. floribunda*, however, could only be subcultured for several passages after which growth and organized development rapidly declined. Suspension cultures of *D. deltoidea* have been extensively utilized in studies of diosgenin synthesis (Kaul and Staba, 1968; Heble and Staba, 1980). Rarely have organized structures other than roots been observed (e.g., Kaul and Staba, 1968). Analysis of some of these cultures showed that they were very heterogenous with a wide range of chromosome numbers and extensive polyploidy (Karanova and Shamina, 1978). Plants regenerated via somatic embryogenesis in *D. floribunda* by Chaturvedi (1979) showed great variation in vigor, leaf shape, and size and in ploidy level, from diploidy to polyploidy to aneuploidy. Occasionally albino mutants and plants with variegated leaves developed.

The appearance of chromosomal variability in cell cultures is well known (D'Amato, 1978) but only recently has the existence of genetic variability in cell cultures been recognized (Larkin and Scowcroft, 1981). Somaclonal variation offers the potential for generating new types of genetic changes (Evans and Sharp, 1983). There is then the possibility that useful variants can be generated in cell cultures.

PROTOCOLS

Clonal Propagation of *D. alata* by Nodal Culture

1. For greenhouse-grown plants, explants should be removed at the height of the growing season when the aerial shoot is short and

still developing and the days are long. For plants raised in controlled environments, the photoperiod should be set to deliver at least 16 hr of light. Again, plants should be young.

2. Cut the vine at the midpoint of the internode between each node. Quickly take back to the laboratory, trim off the leaf blades by severing at the midpoint of the petiole or pulvinus and shake vigorously for 3 min with a detergent-water mixture. Rinse well many times with tap water.
3. Prepare a 20% (v/v) solution of commercial bleach (final concentration = 1% sodium hypochlorite) and add a few drops of a wetting agent such as Tween 40.
4. Trim each nodal segment so that 2-3 cm remains of the stem above and below each node and also of the pulvinus. Immediately drop each piece into the hypochlorite solution. Place the sealed container on a rotary shaker and agitate for 30-40 min.
5. Quickly transfer the pieces to a sterile petri dish with distilled water and rinse at least 12 times.
6. Trim each nodal segment so that 5 mm remains of each stem end and pulvinus.
7. Place the segments on MS medium solidified with 0.8% agar contained in French square bottles or test tubes with closures that allow good gas exchange. The segments should be oriented upright and pushed into the agar so that the axillary buds are at the level of the medium.
8. Incubate the cultures at 25-27 C either in continuous light or a 16/8 light/dark photoperiod. Shoots and roots should develop in 3-5 weeks.
9. New cultures can be easily established by aseptically removing nodal stem segments and transferring to fresh MS medium.
10. In varieties with some natural bulbil formation, microtubers can be grown by allowing the cultures to age. A long-day photoperiod or continuous light is important.

Somatic Embryogenesis in Suspension Cultures of *D. floribunda*

1. Surface-sterilize seeds for 20 min with 20% (v/v) commercial bleach (1% sodium hypochlorite). Rinse at least 5 times with sterile distilled water.
2. Transfer seeds to sterile petri dishes containing filter paper wetted with distilled water. The seed coats are particularly hard and cannot readily be cut until they have imbibed water.
3. After 24 hr, remove each seed in turn to a second petri dish, remove the circular wing from around each seed, and make two tangential cuts through the seed. The embryo is extremely small and is located in a small depression at one side of the seed. The seed coat often comes to a point just over the embryo. The cuts should be at the end away from the embryo and should intersect each other. About one-half of the seed can be removed leaving a wedge with the embryo enclosed within the base.

4. With a pair of sterile forceps, grasp the upper half of the seed coat and endosperm; with a second pair, grasp the lower half. Lift the top half back and away. The embryo should be intact and attached to one of the two halves.
5. With a sterile transfer needle or loop, the embryo can be scooped up and moved to the culture medium.
6. For *D. floribunda*, the appropriate culture medium is the MS salt and vitamin solution supplemented with 18 µM 2,4-D and solidified with 0.8% agar. Cultures are placed in the dark at 25-27 C. A callus should develop in 4-6 weeks. It is slow growing and you may need to wait as long as 12 weeks for sufficient material to subculture.
7. The small callus is transferred intact to small culture tubes (Steward et al., 1952) containing 10 ml of MS and 18 µM 2,4-D and rotated at 1 rpm in complete darkness. Alternatively, small Erlenmeyer flasks placed on a gyrotary shaker may be used. A highly variable cell suspension develops in 4-8 weeks.
8. Under the dissecting microscope, the small, dense, nearly white, globular clusters are selectively removed with a sterile Pasteur pipette and transferred to fresh medium. After several subculture regimes, there should be enough material to permit transfer to large culture flasks (Steward and Shantz, 1956) or to larger Erlenmeyer flasks. The transfer of cultures to fresh medium during the log phase of growth and the maintenance of a high density appear to be especially critical with *Dioscorea* cultures.
9. For somatic embryo development, the suspension is filtered through a large mesh (250 µm) stainless steel sieve and the cells washed of old medium with MS basal medium.
10. The washed suspension is transferred to culture tubes containing MS and 0.1 µM ZEA. Small conical embryos should appear in 4-6 weeks. Embryos may germinate precociously or develop abnormally. The further addition of 1 µM ABA is often helpful in promoting normal somatic embryo maturation.
11. For plantlet development, transfer the embryos to semisolid MS medium with 0.1 µM ZEA and 500 mg/l glutamine. If the somatic embryos are placed in groups of five or more, unsupplemented basal medium will suffice. Place the bottles or flasks in the light. Plants will develop and can be removed to soil in about 4 weeks. Care must be taken to transfer them early, before they develop too extensive a rooting system. In addition, all medium must be thoroughly removed.

FUTURE PROSPECTS

The ability to take single-node or single-leaf cuttings and regenerate many plants that are genetically identical has opened the door to enhanced propagation of elite cultivars. That microtubers can be generated from this material offers the chance for inexpensive propagules, ones that can be readily available and provide an alternative source of planting material. Farmers may not need to reserve a goodly part of their harvest to use as setts the next year. That suspension cultures can be established and somatic embryos grown from them raises the possibility of large-scale clonal propagation for the future. However, one must be absolutely certain of the genetic fidelity of the regenerates. There are some indications here. Tobacco cultures that were subcultured frequently and during the log phase of growth retained their genetic integrity (Evans and Gamborg, 1982). Carrot cultures that were maintained for only a few months regenerated normal looking plants (Krikorian, 1982). More research is needed to elucidate the conditions for clonal fidelity.

As discussed briefly in an earlier section, somaclonal variation in cell culture can produce beneficial mutations. Somaclonal variants in potato have been isolated with resistance to early blight disease, and with altered growth habit, tuber shape and color, and maturity date (Shepard, 1982). The altered traits were attributed to changes in chromosome number and structure (Karp et al., 1982). Such traits can be stably propagated asexually. The variants that have appeared and will continue to arise in *Dioscorea* cell cultures may very well provide the plants with erect stems that no longer require staking, the absence of spines that makes cultivation difficult, and the small, ovate, hard-skinned tubers that can be mechanically harvested. In any cell culture system, the stumbling block is whether or not plants can be regenerated. If they can, as they are now being done in some *Dioscorea* spp., then variants can be induced and/or selected.

The successes with yam culture are recent enough that the more sophisticated techniques such as protoplast isolation and fusion and genetic manipulation have yet to be tried. One intriguing possibility is a hybrid between *D. alata* and *D. bulbifera*, the resulting plant producing aerial tubers that are delicious and need only be plucked off the vine rather than dug out of the ground. That is also a possibility in the realm of selection via somaclonal variation.

Fusing cells, moving genes around, or selecting variants that have improved disease resistance, greater photosynthetic efficiency, or higher yields are all intriguing prospects. Clonal propagation of elite varieties using nodal, apical, and meristem culture is available right now. There are millions of people who depend on true yams and are either reluctant or unable to change to other crops. The application of cell culture techniques to crop improvement in *Dioscorea* is a valuable undertaking.

ACKNOWLEDGMENTS

This chapter is dedicated to the memory of Marion O. Mapes, a fine scientist, an excellent teacher, and a kind, generous, remarkable friend. I would like to express my thanks to Abraham D. Krikorian, who originally suggested working with this fascinating genus and supplied the first tubers, and Franklin W. Martin, who has patiently provided material and answered queries for a good many years. A special note of thanks to my students of past years, Rhonda Lubka and Ellen Hammer.

This work was supported in part by grants from Barnard College to me and the Merck Foundation to my students.

KEY REFERENCES

Ayensu, E.S. 1972. Anatomy of the Monocotyledons. VI. Dioscoreales. Oxford Univ. Press, New York.

Burkill, I.H. 1960. The organography and the evolution of Dioscoreaceae, the family of yams. J. Linn. Soc. 56:319-412.

Coursey, D.G. 1967. Yams. Longmans, Green and Co., London.

Onwueme, I.C. 1978. Tropical Tuber Crops. Yams, Cassava, Sweet Potato and Cocoyams. John Wiley and Sons, New York.

Purseglove, J.W. 1973. Tropical Crops. Monocotyledons. Longmans, Green and Co., London.

REFERENCES

Akoroda, M.O. 1983. Variation, heritability and genetic advance of eight characteristics in white yam. Theor. Appl. Genet. 66:51-54.

Ammirato, P.V. 1976. Hormonal control of tuber formation in cultured axillary buds of *Dioscorea bulbifera* and *D. alata*. Plant Physiol. 57(suppl):66.

______ 1978a. Somatic embryogenesis and plantlet development in suspension cultures of the medicinal yam, *Dioscorea floribunda*. Am. J. Bot. 65(suppl):27.

______ 1978b. Ontogeny and anatomy of somatic embryos grown from cultured cells of the monocot yam, *Dioscorea floribunda*. Am. J. Bot. 57(suppl):89.

______ 1982. Growth and morphogenesis in cultures of the monocot yam, *Dioscorea*. In: Plant Tissue Culture (A. Fugiwara, ed.) pp. 169-170. Maruzen, Tokyo.

______ 1983. Embryogenesis. In: Handbook of Plant Cell Culture, Vol. 1, Techniques for Propagation and Breeding (D.A. Evans, W.R. Sharp, P.V. Ammirato, and Y. Yamada, eds.) pp. 82-123. Macmillan, New York.

______ 1984. Induction, maintenance, and manipulation of development in embryogenic cell suspension cultures. In: Cell Culture and Somatic Cell Genetics of Plants, Vol. 1 (I.K. Vasil, ed.) pp. 139-151. Academic Press, New York.

Applezweig, N. 1977. *Dioscorea*—The pill crop. In: Crop Resources (D.S. Seigler, ed.) pp. 149-163. Academic Press, New York.

Asahira, T. and Nitsch, J.P. 1968. Tubérisation in vitro: *Ullucus tuberosus* et *Dioscorea*. Bull. Soc. Bot. Fr. 115:345-352.

Asokan, M.P., O'Hair, S.K., and Litz, R.E. 1983. In vitro plant development from bulbil explants of two *Dioscorea* species. Hortic. Sci. 18:702-703.

Chaturvedi, H.C. 1975. Propagation of *Dioscorea floribunda* from in vitro culture of single-node stem segments. Curr. Sci. 44:839-841.

______ 1979. Tissue culture of economic plants. In: Progress in Plant Research, Vol. 1, Applied Morphology and Allied Subjects (T.N. Khoshoo and P.K.K. Nair, eds.) pp. 265-288. Today and Tomorrow's Printers and Publishers, Karolbagh, New Delhi.

______ and Sinha, M. 1979. Mass propagation of *Dioscorea floribunda* by tissue culture. Extension Bulletin of the NBRI, Lucknow, India.

______, Sharma, A.K., Sharma, M., and Prasad, R.N. 1982. Morphogenesis, micropropagation, and germplasm preservation of some economic plants by tissue cultures. In: Plant Tissue Culture 1982 (A. Fugiwara, ed.) pp. 687-688. Maruzen, Tokyo.

Cobley, L.S. and Steele, W.M. 1976. An Introduction to the Botany of Tropical Crops, 2nd ed. Longman, London.

Cortes-Monllor, A. and Liu, L.J. 1982. Tissue culture propagation of yam in Puerto Rico. J. Agric. Univ. Puerto Rico, pp. 419-428.

Coursey, D.G. 1976. Yams. *Dioscorea* spp. (Dioscoreaceae). In: Evolution of Crop Plants (N.W. Simmonds, ed.) pp. 70-74. Longman, London.

D'Amato, F. 1978. Chromosome number variation in cultured cells and regenerated plants. In: Frontiers of Plant Tissue Culture (T.A. Thorpe, ed.) pp. 287-295. Univ. Calgary Press, Calgary.

Datta, S.K., Datta, K., and Datta, P.C. 1981. Propagation of yam, *Dioscorea composita*, through tissue culture. In: Tissue Culture of Economically Important Plants (A.N. Rao, ed.) pp. 90-93. COSTED and ANBS (Asian Network for Biological Sciences) Pub., Singapore.

Evans, D.A. and Gamborg, O.L. 1982. Chromosome stability of cell suspension cultures of *Nicotiana* species. Plant Cell Rep. 1:104-107.

Evans, D.A. and Sharp, W.R. 1983. Single gene mutations in tomato plants regenerated from tissue culture. Science 221:949-951.

Flick, C.E., Evans, D.A., and Sharp, W.R. 1983. Organogenesis. In: Handbook of Plant Cell Culture, Vol. 1, Techniques for Propagation and Breeding (D.A. Evans, W.R. Sharp, P.V. Ammirato, and Y. Yamada, eds.) pp. 13-81. Macmillan, New York.

Goebel, K. 1908. Einleitung in die Experimentelle Morphologie der Pflanzen. B.G. Teubner, Leipzig and Berlin.

Grewal, S. and Atal, C.K. 1976. Plantlet formation in callus cultures of *Dioscorea deltoidea* Wall. Indian J. Exp. Biol. 14:352-353.

Grewal, S., Koul, S., Sachdeva, U., and Atal, C.K. 1977. Regeneration of plants of *Dioscorea deltoidea* Wall. by apical meristem cultures. Indian J. Exp. Biol. 15:201-302.

Gupta, S., Sobti, S.N., and Atal, C.K. 1979. Tuber dormancy in *Dioscorea composita* Hemsl.: Effect of growth regulators. Indian J. Exp. Biol. 17:612-614.

Heble, M.R. and Staba, E.J. 1980. Steroid metabolism in stationary phase cell suspensions of *Dioscorea deltoidea.* Planta Med. Suppl. 124-128.

Henshaw, G.G. 1982. Tissue culture methods and germplasm storage. In: Plant Tissue Culture 1982 (A. Fujiwara, ed.) pp. 789-792. Maruzen, Tokyo.

Hu, C.Y. and Wang, P.J. 1983. Meristem, shoot tip, and bud cultures. In: Handbook of Plant Cell Culture, Vol. 1, Techniques for Propagation and Breeding (D.A. Evans, W.R. Sharp, P.V. Ammirato, and Y. Yamada, eds.) pp. 177-227. Macmillan, New York.

Hutchinson, J. 1959. The Families of Flowering Plants. II. Monocotyledons, 2nd ed. Clarendon Press, Oxford.

Karanova, S.L. and Shamina, Z.B. 1978. The direction of selection of genetically heterogeneous cell populations of *Dioscorea deltoidea* Wall. Biol. Plant. 20:86-92.

Karp, A., Nelson, R.S., Thomas, E., and Bright, S.W.J. 1982. Chromosome variation in protoplast-derived potato plants. Theor. Appl. Genet. 63:265-272.

Kaul, B. and Staba, E.J. 1968. *Dioscorea* tissue cultures. 1. Biosynthesis and isolation of diosgenin from *Dioscorea deltoidea* callus and suspension cultures. Lloydia 32:347-359.

Krikorian, A.D. 1982. Cloning higher plants from aseptically cultured tissues and cells. Biol. Rev. 57:151-218.

Lakshmi, S., Bammi, R.K., and Randhawa, G.S. 1976. Clonal propagation of *Dioscorea floribunda* by tissue culture. J. Hortic. Sci. 51: 551-554.

Larkin, P.J. and Scowcroft, J.M. 1981. Somaclonal variation—a novel source of variability from cell cultures for plant improvement. Theor. Appl. Genet. 60:197-214.

Mantell, S.H., Haque, S.Q., and Whitehall, A.P. 1978. Clonal multiplication of *Dioscorea alata* L. and *Dioscorea rotundata* Poir. yams by tissue culture. J. Hortic. Sci. 53:95-98.

______ 1980. Apical meristem tip culture for eradication of flexous rod viruses in yams (*Dioscorea alata*). Tropical Pest Management 26: 170-179.

Mapes, M.O. and Urata, U. 1970. Aseptic stem culture of a *Dioscorea* clone. In: Tropical Root and Tuber Crops Tomorrow, Vol. II, Proceedings of the Second International Symposium on Tropical Root and Tuber Crops, pp. 25-27.

Martin, F.W. 1976a. Tropical Yams and their Potential, Part 2, *Dioscorea bulbifera*. USDA Agriculture Handbook No. 466, Agricultural Research Service, Washington, D.C.

______ 1976b. Tropical Yams and their Potential, Part 3, *D. alata*. USDA Agriculture Handbook No. 495, Agricultural Research Service, Washington, D.C.

______ 1977. Tropical Yams and their Potential, Part 2, *Dioscorea rotundata* and *Dioscorea cayenensis*. USDA Agriculture Handbook No. 502, Agricultural Research Service, Washington, D.C.

______ and Delphin, H. 1969. Techniques and problems in the propagation of sapogenin-bearing yams from stem cuttings. J. Agric. Univ. Puerto Rico 53:191-198.

______ and Ortiz, S. 1963. Origin and anatomy of tubers of *Dioscorea floribunda* and *D. spiculiflora*. Bot. Gaz. 124:416-421.

Mascarenhas, A.F., Hendre, R.R., Nadgir, A.L., Ghugale, D.D., Godbole, D.A., Prabhu, R.A., and Jagannathan, V. 1976. Development of plantlets from cultured tissues of *Dioscorea deltoidea* Wall. Indian J. Exp. Biol. 14:604-606.

Olayide, S.O. 1972. Agricultural productivity and increased food production under economic development in Nigeria. Proceedings of Annual Conference of Nigerian Economic Society, Ibadan, Nigeria.

Passam, H.C., Wickham, L.D., and Wilson, L.A. 1982. Comparative observations on the polarity of sprouting of bulbils of *Dioscorea bulbifera* L. and *Dioscorea alata* L. Ann. Bot. 49:359-366.

Preston, W.B. and Huan, T.R. 1962. Factors involved in the vegetative propagation of *Dioscorea spiculifera* Seml. Proc. Am. Soc. Hortic. Sci. 80:417-429.

Sadik, S. and Okereke, O.U. 1975. A new approach to improvement of yam *Dioscorea rotundata*. Nature 254:134-135.

Shepard, J.F. 1982. The regeneration of potato plants from leaf cell protoplasts. Sci. Am. 246:154-166.

Singh, J.P. 1978. Effect of nitrogen on shoot bud differentiation of *Dioscorea deltoidea* Wall. callus cultures. Biol. Plant. 20:436-439.

Sinha, M. and Chaturvedi, H.C. 1979. Rapid clonal propagation of *Dioscorea floribunda* by in vitro culture of excised leaves. Curr. Sci. 48:176-177.

Steward, F.C. and Shantz, E.M. 1956. The chemical induction of growth in plant tissue cultures. In: The Chemistry and Mode of Action of Plant Growth Substances (R.L. Wain and F. Wightman, ed.) pp. 165-187. Academic Press, New York.

Steward, F.C., Caplin, S.M., and Miller, F.K. 1952. Investigations on growth and metabolism of plant cells. I. New techniques for the investigation of metabolism, nutrition and growth in undifferentiated cells. Ann. Bot. 16:58-77.

Uduebo, A.E. 1971. Effect of external supply of growth substances on axillary proliferation and development in *Dioscorea bulbifera*. Ann. Bot. 35:159-163.

SECTION VI
Tropical and Subtropical Fruits

CHAPTER 13
Citrus

P. Spiegel-Roy and *A. Vardi*

CITRUS SPECIES AND TAXONOMY

The genus *Citrus* and its wild relatives are members of the Rutaceae, subfamily Aurantioideae, subtribe Citrinae ("True Citrus Fruit Trees"). The "true" *Citrus* group includes six genera: *Eremocitrus*, *Poncirus*, *Clymenia* (all with one species), *Fortunella* (4 spp.), *Microcitrus* (6 spp.), and *Citrus*. The genus *Citrus* is divided into two distinct subgenera—*Eucitrus*, which includes all commonly cultivated species of *Citrus* and *Papeda*. None of the species belonging to the latter subgenus bear edible fruits. All the genera have persistent unifoliate leaves except the monotypic genus *Poncirus*, which has trifoliate deciduous leaves. All the six genera are closely related and all (except the little-studied genus *Clymenia*) have been grafted on one another and interhybridized (Swingle and Reece, 1967). Of economic importance are *Citrus*, *Fortunella* (kumquat), and *Poncirus* (as a rootstock).

Classification of *Citrus* is difficult because most citrus cultivars have been subjected to natural hybridization since ancient times, and because of the wide occurrence of mutants and nucellar embryony. Tanaka (1954) recognized 159 species, according species status to many garden varieties. The Swingle and Reece (1967) system, which comprises 16 species, is the most widely respected. It accords species rank, among others, to mandarin (*C. reticulata* Blanco), Shaddock known also as pummelo (*C. grandis* Osbeck), sour orange (*C. aurantium* L.), sweet orange (*C. sinensis* Osbeck), lime (*C. aurantifolia* Christm.), citron known also as ethrog (*C. medica* L.), lemon (*C. limon* L. Burn. f.), and grapefruit (*C. paradisi* Osbeck).

A more recent classification based on 146 characters of the tree, leaf, flower, and fruit was suggested by Barrett and Rhodes (1976). They recognize only three valid species within edible citrus: mandarin (*C. reticulata* Blanco), pummelo (*C. grandis* Osbeck), and citron (*C. medica* L.). They also speculate about interrelationships of other species of cultivated citrus. They suggest that lime and lemon are tri-hybrids involving citron, pummelo, and *Microcitrus*, with lemon carrying a greater proportion of citron genes; sour orange and sweet orange are hybrids between mandarin and pummelo having a parallel but separate origin within the polytypic *C. reticulata*.

HISTORY OF THE CROP

The true citrus fruits are native to a large Asiatic area extending from the Himalayan foothills of India to north central China and the Philippines in the east; Burma, Thailand, Indonesia, New Guinea, northeastern Australia, and New Caledonia in the southeast. Centers of origin of the main commercial *Citrus* species are difficult to ascertain as they have been subjected to natural hybridization and probably cultivation since ancient times. Cooper and Chapot (1977), using evidence from old Chinese writings, support the contention that almost all known citrus fruit cultivars originated in China, with the exception of lemon and grapefruit. *Poncirus* is native to central and northern China, *Fortunella* to southern China. The mandarin (*C. reticulata*) and pummelo (*C. grandis*) have been mentioned in Chinese writings as far back as during the reign of Ta Ya (2205-2197 B.C.). The citron (*C. medica*) is probably native to southern China and India; it was the first *Citrus* with which the Europeans got acquainted, and for many years the only one known. The citron was probably cultivated in Babylonia 4000 B.C., in Egypt 1500 B.C., and in Media, from which it was introduced to the Near East and Greece during the time of Alexander the Great (Webber et al., 1976). The sweet orange (*C. sinensis*) probably originated in southern China, while the sour orange (*C. aurantium*) is probably indigenous to Southeast Asia. The sour orange was known to the Romans and later was carried by the Arabs in the tenth century to many countries, including Spain. There is no written evidence of actual culture of sweet orange in Europe until the fifteenth century. Certainly introductions were made by Genoese traders, from China (around 1425) and later, by the Portuguese. Lemon followed the same route as sweet orange, however little is known about its center of origin and early distribution. The lime (*C. aurantifolia*) is probably native to the East Indian Archipelago. The grapefruit (*C. paradisi*) appeared in the West Indies, presumably, during the end of the eighteenth century as a natural hybrid between pummelo and sweet orange (Swingle, 1943; Robinson, 1952; Albach and Redman, 1969; Barrett and Rhodes, 1976). Columbus introduced *Citrus* into America probably at the end of the fifteenth century, and at the beginning of the sixteenth century it started to spread in the Americas. Orange, lemon, and pummelo reached South Africa in the mid-seventeenth century and Australia at the end of the eighteenth century. The famous Japanese variety Unshu has been in existence for over 300 years (Webber et al., 1976).

Citrus world trade was already important during the second half of the nineteenth century and has grown considerably in volume, number of varieties, and countries of production since that time.

GEOGRAPHIC LIMITATIONS

At present citrus is grown in tropical and subtropical climates, roughly in a belt extending around the world that follows the equator and spreads on either side of it approximately to latitudes 40° north and south.

Extreme frost limits culture (temperatures below 4 C for a few hours are considered dangerous), and only hardy sour orange types of Unshu mandarin grafted on *Poncirus* can be grown successfully in harsh climates. In tropical regions, near the equator, oranges are grown mainly for local consumption because of poor fruit color and short harvest period. The three major regions in the subtropical latitudes of the northern hemisphere are North America, the Mediterranean, and East Asia including India. Large producers in the southern hemisphere include Brazil, Argentina, South Africa, and Australia.

ECONOMIC IMPORTANCE

Citrus fruits rank second to grapes in the total world fruit production. Production of citrus in 1979 amounted to over 50 million metric tons (FAO, 1980).

The most important *Citrus* varieties are: (a) orange—Navel, Shamouti, Valencia, Hamlin, Pera, Blood oranges (Tarocco and Moro); (b) grapefruit—Marsh seedless, colored grapefruit (Redblush, Ruby, Ray); (c) tangerines and mandarins—Clementine, Satsuma (Unshu), Ponkan, Avana, Ellendale, Dancy; and (d) lemon—Femminello, Eureka, Lisbon, Verna, Fino.

Orange is the most important citrus fruit in world trade, both for fresh fruit and fruit by-products. It is followed by grapefruit, mandarins (recently increased, especially in European markets) and lemons.

There are about 90 citrus-producing countries, but only 17 countries produce more than 1% of the total world tonnage. The major producing countries are: United States (22.3%); Brazil (19%); Japan (7.7%); Mexico, Spain, and Italy (around 5%); India and Israel (around 3%); China and Argentina (2.5%); Turkey and Egypt (2%).

The most important citrus exporting countries are Brazil, Spain, Israel, the United States, Italy, Greece, Morocco, and South Africa. Those eight countries account for 85% of all citrus trade.

Export value in 1980 has reached $1,952,343,000 for oranges including tangerines and clementine, $466,490,000 for lemon and limes, and $302,175,000 for other citrus fruit, including grapefruit.

Very large quantities of sweet orange, grapefruit, and mandarin are processed into juice, frozen concentrates, squashes, citrus-based beverages, and often other products. Smaller proportions are used for manufacture of marmalade, jam, and essential oil.

KEY BREEDING AND CROP IMPROVEMENT PROBLEMS

Diploidy is the general rule in *Citrus* and its related genera. The somatic chromosome number is 2n = 18. Natural tetraploids and triploids occur in small frequency. Tetraploids are of no economic value. They are slow growing and generally arise spontaneously as nucellar seedlings in all major groups of cultivars (Cameron and Frost, 1968). Triploids, on the other hand, regularly arise as sexually produced seedlings. They can also be systematically produced by controlled crossing of tetraploid with diploid (Esen and Soost, 1972a,b). In spite of low seed content or seedlessness only the Tahiti lime and recently Oroblanco have attained commercial significance among triploids. Genetic studies in *Citrus* are complicated because of nucellar embryony, heterozygosity, self-incompatibility, sterility, and long generation cycle. Nucellar embryony in *Citrus* (Frost and Soost, 1968) is an important obstacle for hybridization, but is of great value in producing vigorous, uniform, virus-free rootstocks. Many forms of *Citrus* may have been reproduced and introduced since ancient times, almost entirely by seed. Self-incompatibility has been demonstrated in *C. grandis* (Soost, 1964), Clementine mandarin, tangelos, in some *C. limon* cultivars, and in *C. limettoides* and *C. sinensis* (Soost and Cameron, 1975). Sterility factors in ovule and pollen are widespread in *Citrus* (Iwamasa, 1966). These coupled with parthenocarpic tendency of fruit setting have contributed to the selection for seedless citrus fruit. The long juvenile period, the high heterozygosity, and the paucity of knowledge on mode of inheritance of desirable characters (Vardi and Spiegel-Roy, 1978) are also serious obstacles to breeding programs.

While it is relatively easy to produce new varieties of the mandarin or mandarin type by hybridization, it is very difficult to improve in the same way sweet oranges, grapefruit, and lemon. In such cases search for mutation might provide a better way. Spontaneous mutations occur frequently in *Citrus* and many of the world's most important cultivars have arisen via mutation. Valuable mutations have been found more recently in the Satsuma and Clementine mandarins, Navel and Shamouti oranges, and Marsh grapefruit. Variation between nucellar progeny has also given rise to selections such as the very early ripening Miho and Okitsu clones of Unshu (Iwasaki et al., 1966). Little work has been done on the artificial induction of genetic changes. Using thermal neutrons with seeds of the seeded cultivar Hudson Red, a very highly colored seedless red grapefruit, the Star Ruby, has been obtained (Hensz, 1971).

In rootstock breeding the ultimate goal is the production of rootstocks resistant to diseases (e.g., tristeza, *Phytophthora*), nematodes, and possibly environmental stress (cold, salinity, and drought). For a rootstock to be successful it should result in longevity, high yield, and good fruit quality in the scion. *Poncirus trifoliata* is often employed in most recent rootstock breeding programs (Bitters et al., 1973). Not only does it carry a genetic marker, it also carries resistance factors to tristeza virus, *Phytophthora* rot, nematodes, and has cold-tolerance (Hearn et al., 1974). Intergeneric hybridization offers the possibility of incorporating multiple desirable traits found in different genera into the germplasm of new improved citrus rootstock (Barrett, 1977).

REVIEW OF THE LITERATURE

Somatic Embryogenesis

Culture of developing ovules and seeds at various periods before and after fertilization was initiated in 1958 by Maheshwari and Rangaswamy. In further studies Rangaswamy (1961) and Subharwal (1963) cultured explants containing fully developed nucellar embryos to produce additional embryos and callus.

Mature *Citrus* embryos, probably of nucellar origin, were reported by Ohta and Furusato (1957). Rangan et al. (1968) and Juarez et al. (1976) induced development of nucellar embryos from nucelli of monoembryonic *Citrus* cultivars, with the aim of producing virus-free types from monoembryonic cultivars.

While Maheshwari and Rangaswamy (1958) assumed that only fertilized ovules respond to culture conditions, Bitters et al. (1970), Button and Bornman (1971), Kochba et al. (1972), and Kochba and Spiegel-Roy (1973) demonstrated that nucellar embryos can also be obtained from nucellar explants from unfertilized ovules. Kochba and Spiegel-Roy (1973), as well as Mitra and Chaturvedi (1972), obtained embryogenic callus from unfertilized ovules and nucelli. While Mitra and Chaturvedi (1972) observed a decline and cessation of embryogenesis with prolonged subculture, the Shamouti orange callus obtained 11 years ago by Kochba et al. (1972) still maintains its embryogenic competence.

Unlike *Coffea* leaf tissue in which specific growth regulators have been used in order to induce somatic embryogenesis, nucellar cells in *Citrus* are predetermined as embryogenic cells and hormones are usually not required to induce somatic embryogenesis. However, while plants have been freely induced from nucellar and ovular explants of Eureka and Villafranca lemon (*C. limon*) and the closely related *C. volckameriana*, the addition of 2,4-D was required in order to obtain an embryogenic callus in *C. limon* (Saad, 1975). Nucellar and ovular explants of various *Citrus* species and cultivars have given rise to callus upon culturing on MS medium, without phytohormones, but with the addition of 500 mg/l malt extract (Kochba and Spiegel-Roy, 1973). A recent review (Tisserat et al., 1979) summarizes the available data on asexual embryogenesis in vitro among *Citrus*.

Thus far embryogenic callus has been developed from nine *Citrus* species and cultivars in our laboratory (Vardi et al., 1982). Callus and protoplast cultures have been initiated from a genus closely related to *Citrus*, *Microcitrus* (Vardi et al., 1983). Callus initiated from nucellar and ovular tissue (Kochba et al., 1972; Kochba and Spiegel-Roy, 1973), suspension cultures derived from callus (Kochba et al., 1982a), single cells from callus (Button and Botha, 1975) and protoplasts isolated from nucellar callus (Vardi et al., 1975, 1982) can be regenerated in vitro. Phytohormone autotrophy is a regular phenomenon in *Citrus* cultures of nucellar origin and has been observed also in cultures initially induced with the aid of 2,4-D (Villafranca and Eureka lemon).

The origin of embryos from Shamouti nucellar callus has been studied by Button et al. (1974). The callus seems to be composed of roughly spherical cell clusters delimited by an outer cell wall. Inner cells were found to be joined by numerous plasmodesmata. Later these cells enlarge and the cluster dissociates into new meristematic centers. Callus proliferation continues by budding off new globular bodies from existing ones. Under embryogenic conditions, globular bodies are small, and cells on the periphery as well as within form embryos (Spiegel-Roy and Kochba, 1980).

High internal concentrations of auxin seem to limit the frequency of somatic embryogenesis. The autonomous growth habit of the *Citrus* callus indicate an adequate endogenous level of auxin and possibly of other growth substances. Embryogenesis is inhibited by exogenous application of IAA, NAA, or cytokinins such as KIN, BA, and 2iP (Kochba and Spiegel-Roy, 1977a). On the other hand, somatic embryogenesis was stimulated by inhibitors of auxin synthesis (Kochba and Spiegel-Roy, 1977a) such as HNB (5-hydroxynitrobenzyl-bromide) and AZI (7-aza-indole); and by inhibitors of GA_3 synthesis (Kochba et al., 1978a) such as CCC (2-chloro-ethyltrimethylammonium chloride) and ALAR (succinic acid 2,2,-methyl-hydrazide).

Embryogenic callus converted a large proportion of IAA rapidly into IAA aspartate (Epstein et al., 1977), considered a stable conjugate, thus removing excess IAA from tissues (Davies, 1976). In contrast, a nonembryogenic line formed very little IAA aspartate.

A marked increase in embryogenesis was obtained by low ABA (abscisic acid) 0.04-4.0 μM and by low ethephon (2-chloroethyl-phosphonic acid) at 0.07-7.0 μM. Higher concentrations of ethylene may suppress embryogenesis (P. Spiegel-Roy and H. Neumann, unpublished).

While orange juice has been found to stimulate growth of *Citrus* callus from fruit albedo (Erner et al., 1975), and globular embryo formation from nucellar tissue of cv. Navel (*C. sinensis*) was reported to be stimulated by ascorbic acid (Button and Bornman, 1971), 5.6-56 mM ascorbic acid proved toxic for habituated *Citrus* ovular callus in our work.

While malt extract (ME) was reported to be beneficial to embryogenesis in nucellus cultures (Kochba et al., 1972), habituated nucellar *Citrus* callus lost its dependence on ME for embryogenesis (Kochba and Spiegel-Roy, 1973).

The effect of various sugars on embryogenesis and development of large embryos was studied in detail with Shamouti (Kochba et al., 1978b) and also with five other citrus lines (Kochba et al., 1982b). Galactose and also galactose-yielding sugars (lactose, and to a lesser extent, raffinose) stimulated embryogenesis considerably. These test sugars also supported very good growth in the absence of sucrose. It is not yet known whether the effect of galactose is due to an inhibition of auxin synthesis, or to an effect on ethylene evolution. Suppression of embryogenesis in *Citrus* and carrot by high ethephon levels has been observed by Tisserat and Murashige (1977). Of the sugars tested glucose and fructose were least effective in stimulating embryogenesis.

Addition of IAA, KIN, or GA_3 to galactose, especially above 32 mM sugar, resulted in varying degrees of inhibition of embryogenesis. After 10 weeks in culture inhibition by IAA diminished significantly (compared to 5 weeks) (Kochba et al., 1982b).

The effect of age of callus, omission of sucrose and of irradiation (gamma-rays) on embryogenesis has also been studied (Kochba and Spiegel-Roy, 1977b). Omission of sucrose from the medium for a single culture passage greatly stimulated embryogenesis when the following subculture was performed on a sucrose containing medium (Kochba and Button, 1974). Maximal stimulation of embryogenesis in Shamouti callus was obtained with gamma-irradiation of 12-16 KR (Spiegel-Roy and Kochba, 1973). With habituated callus only callus irradiation stimulated embryogenesis (Kochba and Spiegel-Roy, 1977b). Radiation treatments have been reported to lower endogenous auxin (Chourey et al., 1973). With very high, near-lethal radiation doses (30 KR), *Citrus* cultures survived upon the addition of IAA (Kochba and Spiegel-Roy, 1977b).

The subcultured nucellar *Citrus* callus has maintained its embryogenic potential, manifested in the original cells of the nucellus, for protracted periods. Certain culture and trophic conditions could suppress this ability, others reexpress it. This is also evident from studies on selection of callus lines for tolerance to NaCl (Kochba et al., 1982a) and 2,4-D (Kochba et al., 1980; Spiegel-Roy and Kochba, 1980). Several callus cell lines of cv. Shamouti (*C. sinensis*) and one cell line of sour orange (*C. aurantium*) capable of growing in the presence of up to 0.17 M NaCl were obtained. Some of the salt-tolerant lines regenerate in the presence of galactose, others require the combination of galactose and NaCl in the medium. Three lines that tolerate up to 2×10^{-4} M 2,4-D were also selected. One line has become 2,4-D-dependent, i.e., it retained its embryogenic capacity when cultured on 2,4-D.

Extensive work has been carried out to ensure development of embryos and establishment of plants from *Citrus* nucellar callus and protoplasts. Details are given in subsequent sections on protocol. Four stages of embryos were defined (Kochba et al., 1974), and rooting was promoted in all stages of embryonal development by GA_3 alone or in conjunction with ADE (Kochba and Spiegel-Roy, 1976).

Organogenesis

Citrus tissue culture and callus formation can be initiated, in addition to ovules, from a number of organs; however, embryogenic capacity is dependent primarily on the source of the explant.

ALBEDO AND JUICE VESICLES. Nonembryogenic callus has been obtained from citrus albedo (the white inner portion of citrus peel) by several authors (Murashige and Tucker, 1969; Erner et al., 1975), and from the juice vesicles of lemon (Kordan, 1959, 1977).

STEM AND LEAF. Callus formation was obtained at the abscission zone between the petiole and the branch from citrus bud explants of old Shamouti orange trees (*C. sinensis* L. Osbeck). As the aim of the study (Altman and Goren, 1971, 1974) was to elucidate the role of growth regulators and environmental conditions in bud development, no further experiments were carried out to determine the capacity of such callus to differentiate. The first report on shoot morphogenesis from organs other than ovules and seeds was by Grinblat (1972). Callus, buds, and plants were developed from in vitro cultured explants of stem segments derived from seedlings of *C. madurensis* L. cv. Calamondin. Embryogenic callus has been also obtained from stem and leaf explants from the following *Citrus* species: Shaddock (*C. grandis* L. Osbeck), sweet orange (*C. sinensis* L. Osbeck by Chaturvedi and Mitra (1974), and stem segments of grapefruit (*C. paradisi* Macf.) and sweet lime (*C. limettoides* Tanaka) by Raj Bhansali and Arya (1978, 1979).

ENDOSPERM. Callus tissue, embryos, and triploid (3n = 27) plantlets have been obtained from young endosperm at the cellular stage (about 2 months after anthesis). The extracted endosperm from Bei pei pummelo (*C. grandis* L. Osbeck) and Chin-cheng orange (*C. sinensis* Osbeck) were cultured on 2MT medium containing GA (5.77-43.3 µM) (Wang and Chang, 1978).

ANTHERS. Haploid plantlets (n = 9) were obtained from in vitro anther culture, at the uninucleate stage, of *Poncirus trifoliata* L. Raf. (Hidaka et al., 1979). Cultured anthers of sour orange (*C. aurantium* L.) resulted in diploid (2n = 18) plantlets only (Hidaka et al., 1981).

Protoplast-Derived Plants

One of the recent approaches to plant cell genetics and crop improvement is based on the utilization of protoplasts. In spite of considerable efforts during the last decade, protoplast systems have been reported in only a dozen genera (see Evans and Bravo, 1983). Moreover, *Citrus* is the only woody plant among these genera. The *Citrus* protoplast system was developed in our laboratories (Vardi et al., 1975, 1982; Vardi and Raveh, 1976; Galun et al., 1977; Vardi, 1977). The initial study was based on Shamouti orange (*C. sinensis* L. Osbeck) nucellar calli that was obtained by Kochba et al. (1972) and Kochba and Spiegel-Roy (1973). The aims of the study, at the earliest stage were (a) development of a method for protoplast isolation and culture, (b) accomplishment of the full sequence from protoplasts into plants, and (c) exposure of protoplasts to mutagenic treatment such as X-ray irradiation and EMS in order to obtain plants originating from treated single cells (Vardi et al., 1975; Galun et al., 1977; Vardi, 1977). To obtain self-sustained growth at lower minimal seeding densities the feeder layer technique was utilized (Vardi and Raveh, 1976). X-ray irradiated protoplasts as well as nonirradiated protoplasts seeded above a feeder layer enhanced the regeneration of somatic embryos. This conforms with a previous finding (Spiegel-Roy and Kochba, 1973) that irradiation of Shamouti ovular callus promotes embryogenesis. Plants regenerated from those embryos were diploid.

Six of them were grafted on sour orange and have now reached the fruiting stage (Vardi et al., 1982). Since the beginning of our work we extended this method to other *Citrus* species to permit protoplast isolation and subsequent culture in vitro and regeneration into functional plants.

The full sequence of plant regeneration has been achieved with protoplasts of new callus lines from the following: (a) *C. sinensis* L. Osbeck (orange) cv. Nucellar Shamouti, cv. Shamouti Landau; (b) *C. aurantium* L. (sour orange); (c) *C. reticulata* Blanco (mandarin) cv. Murcott, cv. Dancy, cv. Ponkan; (d) *C. paradisi* Macf. (grapefruit) cv. Duncan; and (e) *C. limon* L. Burn. f. cv. Villafranca. In all but the Villafranca lemon the in vitro culture of callus did not cause polyploidization or aneuploidy. Such maintenance of diploidy is noteworthy since polyploidization was frequently observed in protoplast derived plants (e.g. *Nicotiana sylvestris*; Zelcer et al., 1978). Among the existing protoplast systems (Vasil and Vasil, 1980) *Citrus* is outstanding in another important aspect: in all the systems but *Citrus* auxin is an obligatory requirement for protoplast division. The same procedure for protoplast isolation, optimal growth conditions, and plant regeneration can be applied to nucellar callus from all citrus species (Vardi et al., 1982). However, choice of pectinase source may affect the viability of protoplasts depending on the species used. The reaction to osmotic stabilizers was often crucial; while orange and sour orange protoplasts showed no sensitivity to sorbitol, lemon indicated some sensitivity and mandarin (Murcott, Ponkan, Dancy) protoplasts proved very sensitive (Vardi et al., 1982).

A protoplast system has so far been achieved in citrus only with protoplasts derived from nucellar callus. A system for the release of protoplasts from cotyledons, seedling leaves, and mature leaves was reported by Burger and Hackett (1981). Of these, only protoplasts derived from cotyledons underwent a few cell divisions.

PROTOCOLS

Callus and Embryo Formation

1. Immerse young fruits about 4 weeks after anthesis (optimal stage for each cultivar has to be determined separately) in 1% sodium hypochlorite for 10-15 min and then rinse 3 times in sterilized, distilled water.
2. Excise axenically the ovules from the sterilized young fruits. Culture about 20 ovules in each 5 cm petri dish containing Murashige and Tucker (1969) basal medium (MT) with 0.15 M sucrose, 500 mg/l malt extract (ME), and 1% agar, then seal dishes with Parafilm.
3. Incubate the petri dishes at 26 ± 1 C, 16 hr dim light each day (nucellar callus starts to emerge after 4-12 weeks).
4. For the first three passages subculture the callus on the same culture medium (MT + ME) and then on MT (ME omitted), at identical temperature and light regimes.

5. For lemon callus only—when embryos start to appear from the cultured ovules, subculture ovules with embryos on MT + ME supplemented with 2.3 µM 2,4-D until the formation of a nucellar callus and then follow as in step 4.
6. For callus maintenance subculture every 4-5 weeks.

Embryo Formation, Development, and Plant Regeneration

1. To stimulate embryogenesis, subculture the callus on an embryogenic medium. (The embryogenic medium used by us is MT containing 0.055 M galactose instead of 0.15 M sucrose.)
2. In order to increase size of embryos, transfer the embryos to MT suspension culture medium containing 0.055 M galactose, 0.03 mM GA_3, and 500 mg/l ME for 5 weeks.
3. Transfer the embryos from suspension to a 10-cm petri dish containing MT medium with 0.06 M sucrose, 0.03 mM GA_3, and 500 mg/l ME for 5 weeks.
4. Subculture the cotyledonary embryos, for further development of shoots and roots, into test tubes containing 15 ml each of MT medium plus 0.15 M sucrose, 0.1 mM ADE, 0.003 mM GA_3, and 0.7% agar, for about 5-7 weeks.
5. To improve the root system, transfer plantlets into test tubes, on a filter paper bridge (Whatman 1), containing liquid medium with the inorganic salts of MT and 0.06 M sucrose.
6. Upon development of the root system transfer plants to a Jiffy pot (no. 7), in a closed plastic bag to maintain high humidity.
7. Upon further development plants are transferred to a greenhouse (25 ± 1 C) on a sterilized potting mixture. High humidity is maintained during first 7-10 days with gradual removal of cover.

Note: During steps 1-6 incubate at 26 ± 1 C, 16 hr light each day.

In Vitro Selection

FROM CALLUS.

1. Choose the best growing cultures to start the selection procedure. Subculture 40 mg fresh weight explants on marginal concentration of selecting agent (10 replicates).
2. After 5 weeks best growing cultures are selected to start cell lines. Each line is subcultured in 10 replicates.
3. Recurrent selection over 10 passages every 5 weeks.
4. Best lines selected on basis of growth or embryogenesis and subjected to higher concentration of selected agent.
5. Tolerant lines subjected to stability test (3 subculture periods in absence of selective agent).
6. Reculture on chosen concentration of selective agent.
7. Stimulation of embryogenesis and regeneration of plants from resistant or tolerant lines tested as in previous protocol, steps 3-7.

8. Plants from resistant (or tolerant) and control lines tested.
9. Clonal propagation of resistant types.

FROM SUSPENSION.

1. Subculture explants (250 mg fresh weight) in 100 ml Erlenmeyer flask containing 25 ml MT medium. Erlenmeyer flasks are placed on a gyratory shaker, 100 cycles/min (25 ± 1 C, 16 hr light daily).
2. For determination of fresh weight, cultured cells are collected by vacuum filtration, through Whatman 1 filter paper, after 3 weeks of growth.
3. Selection performed in liquid medium, containing selection agent, for at least 10 passages, every 3 weeks.
4. Best lines tested for stability: 3 passages of growth in the absence of selection agent.
5. Resuspension on chosen concentration of selective agent.
6. As in previous protocol, steps 7-9.

Protoplast Isolation and Culture

1. Use nucellar callus, subcultured on MT every 2-3 weeks.
2. Place about 0.5 g callus in 10 cm petri dish containing 10 ml maceration medium for overnight incubation at 26 ± 1 C. The maceration medium consists of 0.3% pectinase (fungal, Koch Light Laboratories) or macerozyme (R-10), 0.2% cellulase (Onozuka RIO, Kinki Yakult Mfg. Co.), and 0.1% driselase (Kyowa Hakko Kogyo Co.) dissolved in 0.3 M sucrose, 0.4 M mannitol, and half the amount of MT macroelements. The maceration solution is filter sterilized (0.2 μm) before use.
3. Isolate protoplasts by sequential filtering through nylon screen of 50 and 30 μm pore openings.
4. Collect filtrate in centrifuge tube and centrifuge at 100 g for 5 min (3 times) with liquid culture medium (MT containing 0.3 M sucrose and 0.3 M mannitol).
5. Suspend protoplasts at final density of 10^5 cells/ml in culture medium, MT containing soft agar (0.6%), and plate as 4 ml aliquots in 5 cm, Parafilm sealed, petri dishes.
6. Incubate in continuous dim light, 26 ± 1 C.
7. Embryo culture and plant regeneration as in plant regeneration protocol, steps 3-7, described earlier.

Callus and Plant Differentiation from Stem and Leaf

FROM CALAMONDIN SEEDLINGS GROWN IN VITRO.

1. Extract seeds under aseptic conditions and remove seed coat.
2. Place each seed in culture tubes containing modified MS medium [medium composition in mM: mineral solution as in MS with the

addition of sucrose (146.0), myo-inositol (0.55), niacinamide (4 x 10^{-3}), pyridoxine·HCl (3 x 10^{-3}), thiamine·HCl (3 x 10^{-4}), glycine (0.026), ME (500 mg/l), and agar (1%)]. Incubate in growth chamber 16/8 hr light/dark, 27 C/21 C.

3. After 6 weeks, when seedling attains 10–15 cm length, excise stem segment (0.5 cm long) from the region between the cotyledon and the first leaves. Cut the stem segment longitudinally to stimulate callus formation. Place 3 explants in each culture tube [containing modified MS supplemented with ME (500 mg/l), BA (4.4 x 10^{-4}), NAA (5.4 x 10^{-4}), agar (1%)], and incubate as in step 2 (based on Grinblat, 1972).

FROM SEEDLINGS GROWN IN VITRO OF *C. GRANDIS* (L.) AND *C. SINENSIS* (L.) OSBECK.

1. Place small segment of stem (or leaf), from in vitro grown seedlings, in culture tubes containing modified MS medium [medium composition in mM: NH_4NO_3 (18.75), KNO_3 (14.85), $CaCl_2$ (2.72), KH_2PO_4 (1.1), $MgSO_4 \cdot 7H_2O$ (1.46), thiamine·HCl (0.03), pyridoxine·HCl (0.012), nicotinic acid (0.02), folic acid (2.2 x 10^{-4}), riboflavin (2.6 x 10^{-4}), biotin (4 x 10^{-4}) ascorbic acid (0.03), sucrose (146.0), agar (7%) supplemented with KIN (1.16 µM), NAA (13.44 µM), and 2,4-D (1.13 µM)] and incubate in 3000 lux fluorescent light for 14 hr daily at 26 C.
2. For callus maintenance use the same modified MS but with a lower NAA concentration (2.26 x 10^{-3}).
3. To induce shoot formation, from stem or leaf callus, use modified MS supplemented with BA (1.1 µM), NAA (0.5 µM), and ME (500 mg/l).
4. For rooting the callus regenerated shoots use modified MS supplemented with NAA (2.7 µM) or NAA and IBA (1.23 µM each) (based on Chaturvedi and Mitra, 1974).

FUTURE PROSPECTS

Breeding by recombination and selection in *Citrus* is limited in application because of nucellar embryony, heterozygosity, and a long juvenile period. A very large proportion of the improved citrus cultivars originated as bud mutations. The common procedure of budwood or seed treatment with mutagens also causes loss of mutations because of diplontic selection and formation of chimeras. Procedures employed for recurrent budwood selection are cumbersome and require large field space. Use of cell cultures for induction of mutants is therefore of great interest. *Citrus* nucellar callus offers many of the properties for mutant selection in vitro. Embryogenesis and the diploid cell number are maintained for prolonged periods. Nucellar embryos reproduce the maternal genotype. Once a favorable variant plant has been obtained, it can be readily propagated clonally. Selection of lines from nucellar callus have been so far performed, and cell lines and embryos selected, with Shamouti orange (*C. sinensis*) and sour orange (*C. aurantium*) for tolerance to NaCl, with Shamouti for tolerance to 2,4-D, and with a cell line of Villafranca (*C. limon*) for tolerance to the toxin generated by *Phoma tracheiphila* (Nadel and Spiegel-Roy, in preparation). Tolerance to NaCl is of great importance for utilizing saline water and increasing tolerance of citrus rootstocks and stionic combinations to adverse soil conditions due to salt.

Possibilities for plant improvement exist in *Citrus* also by making use of protoplasts for mutagenesis and especially for somatic hybridization through protoplast fusion. Hybridization between *Citrus* and *Citrus* relatives from related genera is hampered because of incompatibilities, poor germination of hybrid seed, and inviability of hybrids in some species combinations. Moreover, many crosses in *Citrus* cannot be performed because both parents are nearly 100% apomictic. Protoplasts from nucellar callus have been recovered from eight *Citrus* species and cultivars (Vardi et al., 1982) and variability in nutritional requirements determined (Vardi, 1982). *Citrus* relatives may harbor important genes for stress and disease tolerance, this being of importance especially for rootstock improvement in the future.

Plants have been successfully raised from anthers in *Poncirus trifoliata* (Hidaka et al., 1978) an important *Citrus* rootstock. Development of haploid plants from anthers of *Citrus* could be important for genetic studies and mutagenesis. Further development of methods for bud culture (Chaturvedi and Mitra, 1974) could be of importance for the maintenance and exchange of *Citrus* germplasm. Excision of sexual hybrids of polyembryonic seed parents and their subsequent growth in tissue culture could prove helpful in breeding, especially with crosses of special interest.

In vitro recovery of nucellar plants from monoembryonic cultivars (*C. reticulata*) has been accomplished (Rangan et al., 1968). However, plants of nucellar origin from the monoembryonic Clementine mandarin (*C. reticulata*) have phenotypic variation (Juarez et al., 1976). Moreover, since the development of in vitro shoot-tip grafting in *Citrus* (Navarro et al., 1975), a more dependable method of obtaining true to type, virus-free material from established clones in general, and especially in monoembryonic cultivars, would be useful. Recent developments of the technique and uses of shoot-tip (meristem) grafting in *Citrus* have been reviewed by Navarro (1981). Rejuvenation of established *Citrus* clones and obtaining virus-free material without using nucellar seed (Cameron and Frost, 1968) is of great practical importance.

ACKNOWLEDGMENT

This work is dedicated to our late colleague, Jehoshua Kochba.

KEY REFERENCES

Hidaka, T., Yamada, Y., and Shichijo, T. 1979. In vitro differentiation of haploid plants by anther culture in *Poncirus trifoliata* (L.) Raf. Jpn. J. Breed. 29:248-254.

Maheshwari, P. and Rangaswamy, N.S. 1958. Polyembryony and in vitro culture of embryos of *Citrus* and *Mangifera*. Indian J. Hortic. 15: 275-282.

Navarro, L., Roistacher, C.N., and Murashige, T. 1975. Improvement of shoot-tip grafting in vitro for virus free clones. J. Am. Soc. Hortic. Sci. 100:475-479.

Spiegel-Roy, P. and Kochba, J. 1980. Embryogenesis in *Citrus* tissue cultures. In: Advances in Biochemical Engineering (A. Fiechter, ed.) Vol. 16, pp. 27-48. Springer-Verlag, Berlin, Heidelberg.

Vardi, A. 1982. Protoplast derived plants from different *Citrus* species and cultivars. Proc. Int. Soc. Citriculture (1981) 1:149-152.

REFERENCES

Albach, R.F. and Redman, G.H. 1969. Composition and inheritance of flavanones in citrus fruit. Phytochemistry 8:127-143.

Altman, A. and Goren, R. 1971. Promotion of callus formation by abscisic acid in citrus bud cultures. Physiol. Plant. 47:844-846.

______ 1974. Interrelationship of abscisic acid and gibberellic acid in the promotion of callus formation in the abscission zone of citrus bud cultures. Physiol. Plant. 32:55-61.

Barrett, H.C. 1977. Intergeneric hybridization of *Citrus* and other genera in citrus variety improvement. Proc. Int. Soc. Citriculture (1977) 2:586-589.

______ and Rhodes, A.M. 1976. A numerical taxonomic study of affinity relationships in cultivated *Citrus* and its close relatives. Syst. Bot. 1:105-136.

Bitters, W.P., Murashige, T., Rangan, T.S., and Nauer, E. 1970. Investigations on established virus-free citrus plants through tissue culture. Calif. Citrus Nurserymen's Soc. 9:27-30.

Bitters, W.P., McCarty, C.D., and Cole, D.A. 1973. Evaluation of trifoliate orange. I Congreso Mundial de Citricultura, Muracia-Valencia Spain, Vol. 2, pp. 557-563.

Burger, D.W. and Hackett, W.P. 1981. Protoplast culture of several citrus tissues. HortScience 16:417.

Button, J. and Bornman, C.H. 1971. Development of nucellar plants from unpollinated and unfertilized ovules of the Washington naval orange in vitro. J. S. Afr. Bot. 37:127-134.

Button, J. and Botha, C.E.J. 1975. Enzymic maceration of *Citrus* callus and regeneration of plants from single cells. J. Exp. Bot. 26:723-729.

Button, J., Kochba, J., and Bornman, C.H. 1974. Fine structure of and embryoid development from embryogenic ovular callus of 'Shamouti' orange (*Citrus sinensis* Osb.). J. Exp. Bot. 25:446-453.

Cameron, J.W. and Frost, H.B. 1968. Genetics, breeding and nucellar embryony. In: The Citrus Industry (W. Reuther, L.D. Batchelor, and H.J. Webber, eds.) Vol. II, pp. 325-381. Univ. Calif. Div. Agric. Sci., Univ. of California Press, Berkeley.

Chaturvedi, H.C. and Mitra, G.C. 1974. Clonal propagation of *Citrus* from somatic callus cultures. HortScience 9:118-120.
Chourey, P.S., Smith, H.H., and Combatti, N.C. 1973. Effects of X-irradiation and indole acetic acid on specific peroxidase isozymes in pith tissue of a *Nicotiana* amphiploid. Am. J. Bot. 60:853-857.
Cooper, C. and Chapot, H. 1977. Fruit production with special emphasis on fruit for processing. In: Citrus Science and Technology (S. Nagy, P.E. Show, and M.K. Veldhuis, eds.) Vol. II, pp. 1-27. AVI Publ., Westport, Connecticut.
Davies, P. 1976. Bound auxin formation in growing stems. Plant Physiol. 57:197-202.
Epstein, E., Kochba, J., and Neumann, H. 1977. Metabolism of indoleacetic acid by embryogenic and non-embryogenic callus lines of Shamouti orange (*Citrus sinensis* Osb.). Z. Pflanzenphysiol. 85:263-268.
Erner, Y., Reuveni, O., and Goldschmidt, E.E. 1975. Partial purification of growth factors from orange juice which affects citrus tissue culture and its replacement by citric acid. Plant Physiol. 56:279-282.
Esen, E. and Soost, R.K. 1972a. Tetraploid progenies from 2X and 4X crosses of *Citrus* and their origin. J. Am. Soc. Hortic. Sci. 97:410-414.
______ 1972b. Unexpected triploids in *Citrus*: Their origin, identification and possible use. J. Hered. 62:329-333.
Evans, D.A. and Bravo, J.E. 1983. Protoplast isolation and culture. In: Handbook of Plant Cell Culture, Vol. 1 (D.A. Evans, W.R. Sharp, P.V. Ammirato, and Y. Yamada, eds.) pp. 124-176. Macmillan, New York.
Food and Agriculture Organization. 1980. Tabulated information 1969, 1979, Vol. 34. Production Yearbook 1980, FAO, Rome.
Frost, H.B. and Soost, R.K. 1968. Seed reproduction: Development of gametes and embryos. In: The Citrus Industry (W. Reuther, L.D. Butchelor, and H.J. Webber, eds.) Vol. II, pp. 292-334. Univ. Calif. Div. Agric. Sci., Univ. of California Press, Berkeley.
Galun, E., Aviv, D., Raveh, D., Vardi, A., and Zelcer, A. 1977. Protoplasts in studies of cell genetics and morphogenesis. In: Proceedings in Life Science (E. Reinhard and A.W. Alfermann, eds.) pp. 301-312. Springer-Verlag, Berlin.
Grinblat, U. 1972. Differentiation of citrus stem in vitro. J. Am. Soc. Hortic. Sci. 97:599-603.
Hearn, C.J., Hutchinson, D.J., and Barrett, H.C. 1974. Breeding citrus rootstock. HortScience 9:357-358.
Hensz, R.A. 1971. Star Ruby, a new deep-red-fleshed grapefruit variety with distinct tree characteristics. J. Rio Grande Val. Hortic. Soc. 25:54-58.
Hidaka, T., Yamada, Y., and Shichijo, T. 1981. Plantlet formation from anthers of sour orange (*Citrus aurantium* L.). In: Proc. Int. Soc. Citriculture (1981) 1:153-155.
Iwamasa, M. 1966. Study on the sterility in genus *Citrus* with special reference to the seedlessness. Bull. Hortic. Res. Sta. Jpn. (Ser. B) 6:1-77.

Iwasaki, T., Nishiura, M., and Okudai, N. 1966. New citrus varieties Okitsu-Wase and Miho-Wase. Bull. Hortic. Res. Sta. Jpn. (Ser. B) 6: 83-93.

Juarez, J., Navarro, L., and Guardiola, J.L. 1976. Obtention des plants nucellaires de divers cultivars de clementiniers au moyen de la culture de nucelle in vitro. Fruits 31:751-762.

Kochba, J. and Button, J. 1974. The stimulation of embryogenesis and embryoid development in habituated ovular callus from the 'Shamouti' orange (*Citrus sinensis*) as affected by tissue age and sucrose concentration. Z. Pflanzenphysiol. 73:415-421.

Kochba, J. and Spiegel-Roy, P. 1973. Effects of culture media on embryoid formation from ovular callus of 'Shamouti' orange (*Citrus sinensis*). Z. Pflanzenzuecht. 69:156-162.

_______ 1976. Improvement of plant production in *Citrus* by tissue culture. Plant Prop. 22:11-12.

______ 1977a. The effects of auxins, cytokinins and inhibitors on embryogenesis in habituated ovular callus of the 'Shamouti' orange (*Citrus sinensis*). Z. Pflanzenphysiol. 81:283-288.

______ 1977b. Embryogenesis in gamma-irradiated habituated ovular callus of the 'Shamouti' orange as affected by auxin and by tissue age. Environ. Exp. Bot. 17:151-159.

______, and Safran, H. 1972. Adventive plants from ovules and nucelli in *Citrus*. Planta 106:237-245.

Kochba, J., Button, J., Spiegel-Roy, P., Bornman, C.H., and Kochba, M. 1974. Stimulation of rooting of *Citrus* embryoids by gibberellic acid and adenine sulphate. Ann. Bot. 38:795-802.

Kochba, J., Spiegel-Roy, P., Neuman, H., and Saad, S. 1978a. Stimulation of embryogenesis in *Citrus* ovular callus by ABA, ethephon, CCC and Alar and its suppression by GA_3. Z. Pflanzenphysiol. 89:427-432.

Kochba, J., Spiegel-Roy, P., Saad, S., and Neuman, H. 1978b. Stimulation of embryogenesis in citrus tissue culture by galactose. Naturwissenschaften 65:261.

Kochba, J., Spiegel-Roy, P., and Saad, S. 1980. Selection for tolerance to sodium chloride (NaCl) and 2,4-Dichlorophenoxyacetic acid (2,4-D) in ovular callus lines of *Citrus sinensis*. In: Plant Cell Culture: Results and Perspectives (F. Sala, B. Parisi, R. Cella, and O. Ciferi, eds.) pp. 187-192. Elsevier North-Holland Biomedical Press, Amsterdam.

Kochba, J., Ben Hayyim, G., Spiegel-Roy, P., Neuman, H., and Saad, S. 1982a. Selection of stable salt-tolerant callus cell lines and embryos in *C. sinensis* and *C. aurantium*. Z. Pflanzenphysiol. 106:111-118.

Kochba, J., Spiegel-Roy, P., Neuman, H., and Saad, S. 1982b. Effect of carbohydrates on somatic embryogenesis in subcultured nucellar callus of *Citrus* cultivars. Z. Pflanzenphysiol. 105:359-362.

Kordan, H.A. 1959. Proliferation of excised juice vesicles of lemon in vitro. Science 129:779-780.

______ 1977. Mitosis and cell proliferation in lemon fruit explants incubated on attenuated nutrient solutions. New Phytol. 79:673-677.

Mitra, G.C. and Chaturvedi, H.C. 1972. Embryoids and complete plants from unpollinated ovaries and from ovules of in vivo-grown emasculated flower buds of *Citrus* spp. Bull. Torrey Bot. Club 99:184-189.

Murashige, T. and Tucker, D.P.H. 1969. Growth factor requirements of citrus tissue culture. Proc. First Int. Citrus Symp. 3:1155-1161.

Navarro, L. 1981. *Citrus* shoot-tip grafting in vitro (STG) and its applications: A review. Proc. Int. Soc. Citriculture (1981) 1:453-456.

Ohta, Y. and Furusato, K. 1957. Embryo culture in *Citrus*. Seiken Jiho 8:49-54.

Raj Bhansali, R. and Arya, H.C. 1978. Differentiation in explants of *Citrus paradisi* Macf. (grapefruit) grown in culture. Indian J. Exp. Biol. 16:409-410.

_____ 1979. Organogenesis in *Citrus limettiodes* (sweet lime) callus culture. Phytomorphology 29:79-100.

Rangan, T.S., Murashige, T., and Bitters, W.P. 1968. In vitro initiation of nucellar embryos in monoembryonic citrus. Hortic. Sci. 3:226-227.

Rangaswamy, N.S. 1961. Experimental studies on female reproductive structures of *Citrus microcarpa* Bunge. Phytomorphology 11:109-127.

Robinson, T.R. 1952. Grapefruit and pommelo. Econ. Bot. 6:228-245.

Saad, S. 1975. Factor affecting the growth of lemon (*C. limon*) callus in vitro culture. M.S. Thesis, Hebrew Univ., Jerusalem.

Sabharwal, P.S. 1963. In vitro culture of ovules, nucelli and embryos of *Citrus aurantifolia* Swingle. In: Plant Tissue and Organ Culture (P. Maheshwari and N.S. Rangaswamy, eds.) pp. 265-274. Univ. of Delhi, India.

Soost, R.K. 1964. Self-incompatability in *Citrus grandis* (Linn.) Osbeck. Proc. Am. Soc. Hortic. Sci. 84:137-140.

_____ and Cameron, J.W. 1975. *Citrus*. In: Advances in Fruit Breeding (J. Janick and J.N. Moore, eds.) pp. 507-540. Purdue Univ. Press, West Lafayette, Indiana.

Spiegel-Roy, P. and Kochba, J. 1973. Stimulation of differentiation in orange (*Citrus sinensis*) ovular callus in relation to irradiation of the media. Rad. Bot. 13:97-103.

Swingle, W.T. 1943. The botany of *Citrus* and its wild relatives of the orange. In: The Citrus Industry (H.J. Webber and L.D. Batchelor, eds.) Vol. I, pp. 129-474. Univ. Calif. Div. Agric. Sci., Univ. of California Press, Berkeley.

_____ and Reece, P.C. 1967. The botany of *Citrus* and its wild relatives. In: The Citrus Industry (W. Reuther, L.D. Batchelor, and H.J. Webber, eds.) Vol. I, pp. 190-430. Univ. Calif. Div. Agric. Sci., Univ. of California Press, Berkeley.

Tanaka, T. 1954. Species problem in *Citrus* (Revisio aurantiacearum IX) pp. 152. Japanese Society Promotion Science, Ueno, Tokyo.

Tisserat, B. and Murashige, T. 1977. Effect of ethephon, ethylene and 2,4-dichlorophenoxyacetic acid on asexual embryogenesis in vitro. Plant Physiol. 60:437-439.

Tisserat, B., Esan, E.B., and Murashige, T. 1979. Somatic embryogenesis in Angiosperms. In: Horticultural Reviews (J. Janick, ed.) Vol. I, pp. 1-78. Purdue University. AVI Pub., Westport, Connecticut.

Vardi, A. 1977. Isolation of protoplasts in *Citrus*. Proc. Int. Soc. Citriculture 2:575-578.

_____ and Raveh, D. 1976. Cross-feeder experiments between tobacco and orange protoplasts. Z. Pflanzenphysiol. 78:350-359.

______ and Spiegel-Roy, P. 1978. *Citrus* breeding, taxonomy and the species problem. Proc. Int. Soc. Citriculture 1:51-57.

______, Spiegel-Roy, P., and Galun, E. 1975. *Citrus* cell culture: Isolation of protoplasts, plating densities, effect of mutagens and regeneration of embryos. Plant Sci. Lett. 4:231-236.

______, Spiegel-Roy, P., and Galun, E. 1982. Plant regeneration from citrus protoplasts: Variability in methodological requirements among cultivars and species. Theor. Appl. Genet. 62:171-176.

______, Spiegel-Roy, P., and Galun, E. 1983. Protoplast isolation, plant regeneration and somatic hybridization in different *Citrus* species and *Microcitrus*. In: Proc. 8th International Protoplast Symposium (I. Potrykus, C.T. Harms, A. Hinnen, R. Huetter, J. King, and R.D. Shillito, eds.) pp. 284-285. Birkhauser Verlag, Basel.

Vasil, I.K. and Vasil, V. 1980. Isolation and culture of protoplasts. In: Perspectives in Plant Cell and Tissue Culture (I.K. Vasil, ed.) Int. Rev. Cytol. 11b:1-19.

Wang, T.-Y. and Chang, C.-J. 1978. Triploid citrus plantlet from endosperm culture. In: Proceedings of Symposium on Plant Tissue Culture, pp. 463-467. Science Press, Peking.

Webber, H.J., Reuther, W., and Lawton, H.W. 1967. History and development of the Citrus industry. In: The Citrus Industry (W. Reuther, L.D. Butchelor, and H.J. Webber, eds.) Vol. I, pp. 1-39. Univ. Calif. Div. Agric. Sci., Univ. of California Press, Berkeley.

Zelcer, A., Aviv, D., and Galun, E. 1978. Interspecific transfer of cytoplasmic male sterility by fusion between protoplasts of normal *Nicotiana sylvestris* and X-ray irradiated protoplasts of male-sterile *N. tabacum*. Z. Pflanzenphysiol. 90:397-407.

CHAPTER 14
Pineapple

T. S. Rangan

The pineapple shares with other major food plants the distinction of having been selected, developed, and domesticated by people of prehistoric times and passed on to us through earlier civilizations. Like the tomato, avocado, cocoa bean, potato, and maize, it is a gift of the New World to the Old. Before the discovery of America, the Indians of tropical America had selected a number of varieties of pineapple for fruit size, quality, and seedlessness. The wild pineapples that are still found in certain parts of tropical America are the stocks from which these early domesticated varieties were derived.

HISTORY AND GEOGRAPHICAL DISTRIBUTION

It is the resemblance to a pine cone that gives the fruit its Spanish name Pina and the English name pineapple. The generic name *Ananas* is derived from the Tupi-Guarani Indian name *Nana* meaning excellent fragrance. The pineapple was unknown to people of the Old World before Columbus made his second voyage to the New World. Columbus discovered the plant and its fruit in 1493 when his expedition landed on an island in the lesser Antilles of the West Indies to which Columbus gave the name Guadeloupe. The records of early explorers and colonists show that pineapple was widely distributed throughout most of tropical America. After the discovery of the New World, pineapple spread rapidly throughout the tropics. It was introduced or reported growing in the Old World as follows: Madagascar, 1548; southern India, 1550; the Philippines, 1558, where the first pina cloth was made about 1570; West Africa, 1602; South Africa, 1660; Mauritius, 1661; and Australia, 1830. They were introduced into Hawaii early in the nineteenth century.

Although all the known members of the Bromeliaceae [except for one species, *Pitcairnia feliciana* (A. Chev.) Harms & Mildbr., which was discovered in Guinea in West Africa] are indigenous to the New World, there are some puzzling references suggesting that pineapple might have been known to some of the ancient civilizations. Among the numerous products of India found in ancient Egypt was the pineapple, carvings of which were seen in the tombs. It was therefore suggested that the pineapple was native to India or had existed there thousands of years ago. Merrill (1954) suggested that the Romans might have known the pineapple about 2000 years ago at the time of the destruction of Pompei.

The place, the manner of origin and domestication of pineapple is shrouded in secrets of antiquity and it is not possible to determine with certainty the time and place of these events. Baker and Collins (1939) believed the place of origin to be somewhere in the area including central and southern Brazil, northern Argentina, and Paraguay (15-30° south latitude and 40-60° west longitude). In some regions of this area, three species were found to be growing under natural conditions and in some localities in close proximity to each other. The Parana-Paraguay river drainage, suggested as the possible region of origin, was also the home of Tupi-Guarani Indians. These people are believed to have migrated northward and westward in prehistoric times taking the pineapple with them and introducing it to other tribes.

CLASSIFICATION, VARIETIES, AND PRODUCTION

The Bromeliaceae is divided into two distinct habitat groups, the terrestrial and the epiphytic. Pineapple has a terrestrial habitat but shows some features of the epiphytes as being able to store small quantities of water in the axils, the presence of special water storage tissue in the leaves, and the ability to endure considerable periods of drought. The common term pineapple is limited to the genera *Ananas* and *Pseudoananas*. The genus *Pseudoananas* is monotypic (*P. sagenarius*), while *Ananas* consists of five species, namely, *A. bracteatus*, *A. fritzmuelleri*, *A. comosus*, *A. erectifolius*, and *A. ananassoides*.

The modern pineapple is a cultigen and, although a large number of cultivars have been recorded, the cultivated varieties (from the viewpoint of commercial production) may be classified according to their characteristics in the following groups (Samuels, 1970; Knight, 1980).

1. Spanish group—Leaves spiny, shape of fruit globose with large deep-set eyes; fruit weighs from 0.9 to 1.8 kg; rind is deep reddish-orange. Flesh pale yellow to white with spicy-acid taste and fibrous texture. Members of this group are grown primarily for export and local consumption in fresh condition. Cultivars: Red Spanish, Singapore Spanish, Green Selangor, Castilla, and Cabezona.
2. Queen group—Leaves spiny, fruit conical with deep eyes, fruit weighs about 0.5-1.1 kg; rind is yellow, and flesh deep yellow.

Sweet and less acidic than Cayenne and low in fiber. Grown mostly for fresh consumption. Cultivars: Queen, MacGregor, Z(James), Natal, Ripley, and Alexandria.

3. Cayenne group—Leaves are smooth with a few spines near the tip (spiny tip). Fruit cylindrical with a slight taper and with flat eyes, weighing 2.3 kg on the average; rind is dark orange and the flesh is pale yellow to yellow, sweet, mildly acidic, with a low fiber and a tender juicy texture. In addition to canning, the fruit is good for fresh consumption. Cultivars: Cayenne (Smooth Cayenne, Cayenne lisse), Baron Rothschild, Smooth Guatemalan, Typhone, St. Michael, and Esmeralda.
4. Abacaxi group—Leaves spiny, fruit conical in shape and weighs 1.4 kg on the average. Rind is yellow and flesh is pale yellow or white, sweet, tender, and juicy. The fruits do not process well or survive export well but are grown in large quantities in Brazil for domestic consumption. Cultivars: Perola, Sugar Loaf, Papelon, Abakka, Venezolana, and Amarella.
5. Maipure group—Leaves are completely smooth with "piping" (margins folded over). Fruit cylindrical-ovoid to cylindrical weighing 0.8–2.9 kg, and has a yellow to dark orange or red rind with flesh that is white or deep yellow. It is sweeter than Cayenne, fibrous but tender and juicy. Cultivars: Maipure, Bumunguesa, Piamba di Marquita, Rondon, Perolera, and Monte Lirio.

The pineapple is best suited to a mild tropical climate, with temperatures between 16 and 32 C. The plants are injured by temperatures below freezing. Very strong sunshine and dense shade are harmful but the plants can tolerate light shade. The planting density varies and in recent years densities as high as 60,000/ha have been used to raise yields, by producing larger number of fruits that are individually smaller than those with lesser densities. For example, a density of 43,036 plants/ha produced a yield of 87 MT/ha while a density of 63,738 plants/ha yielded 118 MT/ha of fruit in Bangalore, India (Chadha et al., 1974).

The world production of pineapple has shown a steady increase over the years. Much of the increase is due to the expansion of pineapple industry in the developing countries of the Far East, Africa, and Latin America. The major pineapple-producing countries with their production figures in million metric tons are China (including Taiwan), 848; United States, 626; Brazil, 501; Thailand, 500; Philippines, 382; Mexico, 300; Malay peninsula, 264; Ivory Coast, 240; Ecuador, 234; South Africa, 190; Bangladesh, 124; Australia, 112; India, 105; Kenya, 100; and Colombia, 96 (Knight, 1980).

ECONOMIC IMPORTANCE

The edible portion that constitutes about 60% of the fresh fruit contains approximately 85% water, 0.4% protein, 14% sugar, 0.1% fat, and 0.5% fiber. The fruit is a good source of vitamins A and B, and the mill juice contains on dry weight basis 75–83% sugar and 7–9% citric acid. The fruit also contains bromelain, a proteolytic enzyme.

The 3-methylpropionate esters comprise a significant fraction of the pineapple volatile components and have been adapted for use in pineapple flavors. Several odorous lactones have also been found in pineapple, in particular, the gamma- and delta-octalactones and gamma-nonalactones (Flath, 1980).

Pineapples are eaten as dessert fruits throughout the tropics and subtropics and most of the commercial crop of pineapple is canned in the producer countries. The fruit is also made into jam and used as crystallized and glacé fruit. The leaves yield a strong white silky fiber that is used for making a fine fabric called "pina" cloth in the Philippines and Taiwan and is also used for cordage. In southeastern Asia young immature fruits are used as an abortifacient. A variegated form with green, yellow, and pink stripes is grown as an ornamental.

PINEAPPLE MORPHOLOGY

The pineapple is a perennial monocot having a terminal inflorescence and fruit and continues growth after fruiting by means of one or more axillary buds growing into vegetative branches with a new apical meristem. The original plant grown from a separated vegetative shoot is commonly called the plant crop and the term first ratoon plant or fruit is applied to the plant or fruit developed from a lateral axillary branch attached to the axis of the first crop plant. The plant is an herb, 90-100 cm in height, with 70-80 leaves forming a dense rosette attached to the short, thick stem. The root system is shallow and rather limited. A seedling produces a primary root which soon disappears and is replaced by adventitious roots, which are the only roots produced in vegetatively propagated plants. The stem is short and thick with short internodes. When a fruit is developing a few axillary buds in leaf axils elongate to form lateral branches called shoots, which if left intact grow into ratoons, but if removed may be used for propagation. Vegetative branches called suckers, which are more slender and have longer leaves than the shoots, arise from buds on the stem below the soil. Below the inflorescence the buds in axils of short leaves of the peduncle grow out to form "slips." Occasionally buds at the point of junction of peduncle and stem produce "hapas."

The leaves are arranged in a crowded fashion on the stem in a right- or left-handed spiral with a phyllotaxy of 5/13. Leaves are sessile with the lamina shaped like a shallow trough which conducts water to the base of the plant. The upper surface of the leaf is smooth and dark green, while the lower surface is silvery white and scurfy. The leaves also have a water storage tissue below the upper epidermis which when filled with water occupy about half the thickness of the main body of the leaf.

The inflorescence consists of a thick peduncle with about 100-200 hermaphrodite, trimerous flowers, each subtended by a bract. The ovary has three nector secreting glands. The petals, stamens, and style wither following anthesis and remain at the bottom of the blossom cup. The blossom cup, which is a characteristic feature of the mature fruit, is an oval cavity in the shell or rind of each fruitlet formed by the thick and fleshy calyx bending horizontally over the circular ridge formed by the petal and stamen bases. In the depressed center of this blossom cup is the base of the style.

During vegetative growth the stem is slightly dome-shaped, producing leaves, and the first visible evidence of the formation of an inflorescence is the rapid increase in the dome area of the meristem. Following this the floral structures are produced in rapid succession instead of leaves, until the meristem area is gradually reduced in size when it produces only floral bracts and then reverts to the production of leaves which form the crown. The sterile floral bracts at the top of the fruit mark the transition zone between the fruit and the crown. The growth of the crown continues during the development of the fruit but ceases with fruit maturity. The crown can be separated and is used for propagation. The crown is important taxonomically for it is the principal character that separates the genus *Ananas* from the other genera in Bromeliaceae.

The fruit is a parthenocarpic multiple fruit or syncarp formed by the almost complete fusion of 100–200 berry-like fruitlets and their subtending leafy bracts to each other and to the central fibrous peduncle. Generally the fruitlets are seedless but when natural or hand pollination occurs between cultivars and species, 2000–3000 seeds may be produced. All diploid cultivars produce functional pollen and ovules but are self-incompatible. Most of them are cross compatible and set seed when cross pollinated (Purseglove, 1972).

PROPAGATION OF PINEAPPLE

Except in breeding work, pineapple is always vegetatively propagated using the following material: (a) suckers, arising from buds below the ground level; (b) shoots, which are leafy branches arising from buds in leaf axils; (c) slips, which are borne on the peduncle just below or at the base of the fruit; (d) hapas, which are shoots produced at the base of the peduncle; (e) crowns, from the top of the fruit; and (f) butts or stumps, consisting of entire plants after the fruits have been harvested and from which the base of the stem, roots, leaves, and peduncle have been removed.

All of these forms of planting material have considerable resistance to desiccation and may be stored for several weeks before planting. The time taken from planting to harvesting depends on the type of propagules used and is 15-18 months for shoots, about 20 months for slips, and 22-24 months for crowns. However, crowns produce a more uniform crop than shoots. Larger planting material produce larger plants, earlier fruiting, and higher yields.

The average rate of production of propagules in cv. Cayenne is about two per year. It thus takes about 30 years to produce enough planting material for one hectare starting with a single plant. One of the methods devised to speed up the rate of asexual propagation is to excise the dormant buds (in the axil of the leaves on the main stem) with some adjacent stem tissue and plant them in sterilized shaded beds. Some of the buds grow into plantlets with roots and produce mature plants in 2.5–3 years. Other methods of rapid multiplication include planting of pieces of stumps with buds and lateral shoot induction in adult plants (Macluskie, 1939; Purseglove, 1972).

The successful rapid multiplication through in vitro micropropagation of orchids and other horticultural species suggests that this technique is highly advantageous when numerous plants of a desirable quality are required. Consequently several attempts have been made to utilize the in vitro micropropagation method for rapid clonal multiplication of pineapple. Aghion and Beauchesne (1960) obtained a few plantlets from segments of crown cultured in vitro. Mapes (1973) reported partial success in shoot cultures of pineapple. Protocorm-like bodies and plantlets were produced from preshaken shoot tips on MS medium supplemented with adenine or adenosine. Lakshmi Sita et al. (1974) were able to obtain plantlets from the terminal buds of the crown in vitro. However, only one plant was obtained from each bud.

Mathews and Rangan (1979) reported the formation of multiple plantlets in lateral bud cultures of pineapple. Plantlets were obtained from usually dormant axillary buds excised from the crown and grown on MS medium supplemented with NAA, IBA, and KIN (see protocol section). Shaking the culture flasks during growth significantly increased the number of multiple shoots formed. A three-fold increase in the number of multiple shoots per culture as compared to stationary cultures was observed. An average of 62 multiple shoots per culture were produced in shake cultures as compared to 26 per culture in stationary cultures. The multiple shoots, when isolated and transferred to a fresh medium of the same composition with continued agitation, again developed multiple shoots; this cycle could be repeated several times resulting in a large number of plants. The shoots, when transferred to MS medium containing low concentrations of NAA and IBA, developed roots and grew into plantlets that could be transplanted to soil (Mathews et al., 1976; Rangan and Mathews, 1980). A detailed protocol for callus and shoot regeneration in pineapple is summarized in the following section.

Callus cultures were also established from the basal region of in vitro-obtained shoots grown on filter paper discs on MS medium fortified with CH, CW, and NAA. Such callus cultures when transferred to a medium devoid of any growth regulators regenerated plants (Mathews and Rangan, 1981). Leaf explants excised from in vitro plantlets also developed a callus capable of plantlet regeneration (Mathews and Rangan, 1979).

The observations of Wakasa et al. (1978) are somewhat similar to those of Mapes (1973). Various explants such as young syncarp, axillary buds from suckers or slips, small crown and small slip in culture produced a mass similar to a protocorm that grew vigorously and subsequently differentiated plantlets. Wakasa (1979) observed many variants among the regenerated plants with regard to spine, leaf color, wax secretion on leaf surface, and foliage density. The frequency and distribution of these variants appeared to be related to the explant source. While the syncarp and slip developed variants in high frequencies, the crown and axillary buds did so only to a small extent.

Zepeda and Sagawa (1981) have also reported multiple shoot formation in lateral bud cultures of pineapple. They suggest that at least

5000 plants could be produced in 12 months from a single crown. Pannetier and Lanaud (1976) estimated that it might be possible to produce 2 million plants from a single bud in 2 years.

In pineapple, fruit development is parthenocarpic and there is no natural fruit set. However, when intercultivar cross pollinations are made there is abundant seed set. Srinivasa Rao et al. (1981) obtained the regeneration of plantlets in hybrid embryo ('Kew' x 'Queen') callus of pineapple. On average, 20-25 plants were obtained from the callus of one single hybrid embryo.

PROTOCOLS

Multiple Plantlet Formation in Lateral Bud Cultures

1. Dormant lateral buds removed from the crown and cultured on MS medium supplemented with NAA (9.7 μM) + IBA (9.8 μM) + KIN (9.3 μM) develop a number of multiple buds. These regenerated buds, when individually isolated and transferred to a fresh medium, again develop multiple buds. This cycle can be repeated several times.
2. Shaking the culture flasks during growth increases the number of multiple buds formed as compared to stationary liquid cultures.
3. The regenerated shoots can be induced to form roots on MS medium supplemented with NAA (0.96 μM) + IBA (1.96 μM), and subsequently can be transplanted to soil.

Regeneration in Callus Cultures

1. Multiple shoots obtained from lateral bud cultures and maintained on MS medium containing 9.7 μM NAA + 9.8 μM IBA + 9.8 μM KIN are utilized for callus initiation. When such shoots are transferred to MS medium supplemented with 29.0 μM NAA + 29.7 μM IAA and 9.8 μM KIN, callus regenerates from the base of the explant.
2. The callus thus obtained is maintained by regular subculture on MS medium with 57.0 μM NAA + 15% CW + 400 mg/l of CH.
3. When pieces of calli maintained on MS + NAA + CW + CH medium are transferred to MS medium with CW (5%) + CH (400 mg/l), shoot regeneration occurs.
4. The regenerated shoots can either be induced to form multiple shoots by transfer onto MS medium with 9.7 μM NAA + 9.8 μM IBA + 9.8 μM KIN, or induced to form roots on modified WH medium (Rangaswamy, 1961) supplemented with 0.27 μM NAA + 1.9 μM IBA. Subsequently, they can be transplanted to soil.

CONCLUSIONS AND FUTURE PROSPECTS

These observations suggest that it is possible to obtain a large number of plantlets by in vitro micropropagation (Fig. 1). Utilizing the

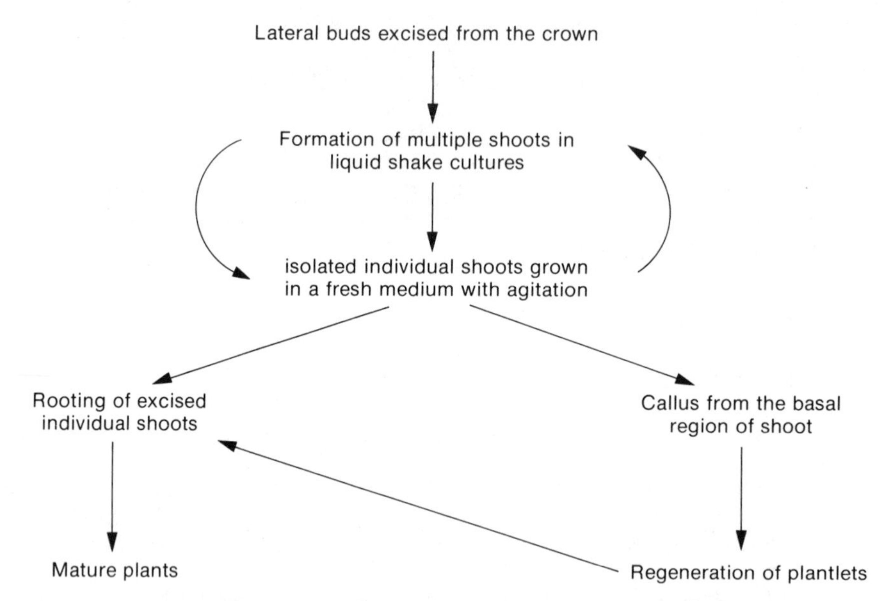

Figure 1. Protocol for micropropagation of pineapple (*Ananas comosus* L. Merr.) through culture of lateral buds from the crown.

shake culture method it would be possible to establish a large population in a fairly reasonable time, which could be used as propagules. The fact that syncarp and slip produced variants among the regenerants while the plants regenerated from the crown and axillary buds maintained the original phenotype to a large extent (Wakasa, 1979) could be exploited by pineapple breeders. It may be possible to mass produce plants similar to the parent by using the axillary bud culture and use the syncarp and slip culture as a means to obtain as many variants as possible. Similarly the regeneration of plants in hybrid embryo callus is also of interest. Although plantlets propagated via the callus phase may not be genetically homogeneous due to chromosome instability (Murashige, 1974) this technique provides a method to establish a large number of hybrid plants from one single hybrid embryo.

The success obtained in rapid multiplication through axillary bud culture of pineapple as well as other bromeliads (Mapes, 1973; Jones and Murashige, 1974; Teo, 1974; Hosoki and Asahira, 1980) suggests that this approach is an efficient method for rapid in vitro clonal propagation.

KEY REFERENCES

Flath, R.A. 1980. Pineapple. In: Tropical and Subtropical Fruits: Composition, Properties and Uses (S. Nagy and P.E. Shaw, eds.) pp. 157-183. AVI Pub., Westport, Connecticut.

Jones, J.B. and Murashige, T. 1974. Tissue culture propagation of *Aechmea fasciata* Baker and other bromeliads. Proc. Int. Plant Prop. Soc. 24:117-126.

Mathews, V.H. and Rangan, T.S. 1979. Multiple plantlets in lateral bud and leaf explant in vitro culture of pineapple. Sci. Hortic. 11:319-328.

______ 1981. Growth and regeneration of plantlets in callus culture of pineapple. Sci. Hortic. 14:227-234.

Purseglove, J.W. 1972. Tropical Crops. Monocotyledons 1. John Wiley and Sons, New York.

Wakasa, K., Koga, Y., and Kudo, M. 1978. Differentiation from in vitro culture of *Ananas comosus*. Jpn. J. Breed. 28:113-121.

REFERENCES

Aghion, D. and Beauchesne, G. 1960. Utilisation de la technique de culture sterile d'organes pour obtenir des clones d'*Ananas*. Fruits 15: 464-466.

Baker, K. and Collins, J.L. 1939. Notes on the distribution and ecology of *Ananas* and *Pseudoananas* in South America. Am. J. Bot. 26: 697-702.

Chadha, K.L., Melanta, K.R., and Shikamany, S.D. 1974. High density planting increases pineapple yields. Indian Hortic. 18:3-5.

Hosoki, T. and Asahira, T. 1980. In vitro propagation of bromeliads in liquid culture. HortScience 15:603-604.

Knight, R., Jr. 1980. Origin and world importance of tropical and subtropical fruit crops. In: Tropical and Subtropical Fruits: Composition, Properties and Uses (S. Nagy and P.E. Shaw, eds.) pp. 1-120. AVI Pub., Westport, Connecticut.

Lakshmi Sita, G., Singh, R., and Iyer, C.P.A. 1974. Plantlets through shoot tip culture in pineapple. Curr. Sci. 43:724.

Macluskie, H. 1939. Pineapple propagation: A new method in Sierra Leone. Trop. Agric. (Trinidad) 16:192-193.

Mapes, M.O. 1973. Tissue culture of bromeliads. Proc. Int. Plant Prop. Soc. 23:47-55.

Mathews, V.H., Rangan, T.S., and Narayanaswamy, S. 1976. Micropropagation of *Ananas sativus* in vitro. Z. Pflanzenphysiol. 79:450-454.

Merrill, E.D. 1954. The Botany of Cook's Voyages. Chronica Botanica 14, Waltham, Massachusetts.

Murashige, T. 1974. Plant propagation through tissue cultures. Annu. Rev. Plant Physiol. 25:135-166.

Pannetier, C. and Lanaud, C. 1976. Divers aspects de l'utilisation possible des cultures "in vitro" par la multiplication végétative de l'*Ananas comosus* L. Merr. variété "Cayenne lisse." Fruits 31:739-750.

Rangan, T.S. and Mathews, V.H. 1980. Growth and multiple plantlet formation in lateral bud, leaf explant and callus culture of pineapple (*Ananas comosus* L. Merr.). In: Plant Cell Cultures: Results and Perspectives (F. Sala, B. Parisi, R. Cella, and O. Ciferri, eds.) pp. 301-304. Elsevier/North Holland Biomedical Press, Amsterdam.

Rangaswamy, N.S. 1961. Experimental studies on female reproductive structures of *Citrus microcarpa* Bunge. Phytomorphology 11:109-127.

Samuels, G. 1970. Pineapple cultivars 1970. Proc. Trop. Reg. Am. Soc. Hortic. Sci. 14:13-24.

Srinivasa Rao, N.K., Dore Swamy, R., and Chacko, E.K. 1981. Differentiation of plantlets in hybrid embryo callus of pineapple. Sci. Hortic. 15:235-238.

Teo, C.H.K. 1974. Clonal propagation of pineapple by tissue culture. Planter (Malaya) 50:58-59.

Wakasa, K. 1979. Variation in plants differentiated from the tissue culture of pineapple. Jpn. J. Breed. 29:13-22.

Zepeda, C. and Sagawa, Y. 1981. In vitro propagation of pineapple. HortScience 16:495.

SECTION VII
Temperate Fruits

CHAPTER 15
Blueberry

J. M. Smagula and *P. M. Lyrene*

HISTORY AND ECONOMIC IMPORTANCE

Blueberries are the edible fruit of plants in the genus *Vaccinium*. This highly variable genus has worldwide distribution, and many of its species have been domesticated as crop plants, although some are exploited primarily or exclusively as wild stands. Blueberry cultivation has received greatest attention in the United States and in Canada. Three distinct types of blueberries are cultivated and breeding programs have produced improved varieties of each of the three types. Lowbush, highbush, and rabbiteye blueberries differ from each other in cultural requirements, area of adaptation, and species from which they were domesticated, and in some cases they differ in chromosome number.

The native lowbush blueberry is an important fruit crop in Maine, the leading producer in the United States, with an annual yield of about 18-20 million pounds. The crop serves as a major source of income for growers, pickers, processors, and distributors in many rural townships in Maine. The lowbush blueberry industry started on the "barrens," a wild, desolate, treeless tract of burned-over land, many thousand acres in extent in Washington County, Maine (Day, 1960). Recorded history of the blueberry industry suggests that Indians living in Washington County burned-over the barrens and forest near their dwellings, partly for protection from surprise attacks and partly to increase the production of various kinds of wild berries which were an important part of their diet. The Indians also dried the berries and sold them to the English settlers by the bushel.

Maine holds the distinction of canning the first blueberries, as early as 1866, a whole generation before blueberries were canned commercially anywhere else in the United States. A current trend in processing is freezing the crop after harvest. Maine produces an average of 18 million pounds annually, with about 90% of the crop frozen and the rest canned.

Highbush blueberries, bred primarily from *V. corymbosum*, the wild highbush blueberry of the eastern United States, are cultivated in Michigan, New Jersey, North Carolina, and Washington state. The breeding of improved highbush cultivars was initiated by the USDA in 1906 (Galletta, 1975). The first varieties were tested and grown in New Jersey. Later breeding programs to develop highbush cultivars or highbush x lowbush hybrid cultivars were begun in Michigan, North Carolina, and Maine. The entire highbush blueberry industry in the United States currently is based in selected cultivars clonally propagated by hardwood or softwood stem cuttings.

Rabbiteye blueberries (*V. ashei* Reade) are the primary cultivated blueberry in the southeastern United States. *Vaccinium ashei* is a vigorous hexaploid species with a native range that seems originally to have been limited to a few river valleys in northwestern Florida and southern Georgia. Between 1900 and 1930, however, over a million native rabbiteye seedlings were dug from the woods and swamps of northwest Florida and were used to establish 3000 acres of cultivated blueberries in Florida (Lyrene and Sherman, 1979). Rabbiteye blueberries have spread from these plantings to surrounding forests and roadsides, and the species is probably more widespread and abundant today than it was a century ago. The breeding of rabbiteye blueberries began in Georgia about 1940 (Galletta, 1975), and more than 12 clonally propagated cultivars have been released.

The production of cultivated blueberries has increased rapidly from 1939 to the present (Table 1). Blueberries are one of the most recently domesticated crop plants, and many problems associated with blueberry breeding, cultivation, and utilization have only recently received attention.

The blueberry industry seems likely to continue its growth during the next few decades. Markets appear available to accommodate increased production. Fresh blueberries are not currently available in many potential markets. Harvest of fresh blueberries could extend from May 1 in Florida to September 15 in Michigan, but fresh blueberries are currently available for only a fraction of this period, even in major marketing areas. In recent years, there has been interest in the establishment of blueberry farms in Australia and New Zealand for production of berries that could be marketed in December, January, and February. The amenability of blueberries to mechanized culture is an important factor favoring growth for the industry. Improved mechanical harvesters and better equipment for cleaning and packing blueberries should reduce picking costs in the future and lower the cost of fresh fruit. The expansion of blueberry culture into the southeastern United States, made possible by the development of improved cultivars of rabbiteye blueberry, promises to add to the growth of the blueberry industry.

GEOGRAPHICAL DISTRIBUTION

Like most members of the heath family, blueberries grow well only in acid soils (pH 5.5 or below). In other respects blueberries are highly

Table 1. Yield of Cultivated Blueberries in the United States and Canada, 1939–1982[a,b]

YEAR	LOWBUSH	HIGHBUSH	RABBITEYE
1939	5[c]	2	0
1949	15[c]	9	0
1959	42	26	1
1969	46	77	-
1979	43	82	-
1980	40	85	5
1981	50	100	-
1982	74	95	6

[a]Approximate, in millions of pounds.

[b]Data are from annual reports of "Blueberry Pack," Division of Markets, Maine Department of Agriculture; and from Eck and Childer (1966) and Lyrene (1983).

[c]Production in Maine only.

diverse in their geographical adaptation, and *Vaccinium* species can be found all the way from the tropics to the arctic circle. Two blueberry species native to the eastern United States, *V. darrowi* and *V. myrsinites*, are evergreen and have essentially no winter chilling requirements. Both extend southward almost to the southern tip of Florida (Sharpe, 1954). Native southern highbush blueberries (*V. corymbosum*) extend southward to Lake Okeechobee in Florida. None of these southern species is of cultivar quality, but all can readily be crossed with northern highbush cultivars, and development of commercial blueberry cultivars for acid-soil areas of south Florida should be possible. Rabbiteye blueberries, although native in northern Florida, are deciduous, and require at least some cold weather (probably 250–500 hr below 7 C for most) to enable heavy fruiting (Spiers and Drapner, 1974; Sharpe and Sherman, 1971). The practical northern limit for cultivation of highbush blueberries is established by minimum winter temperatures. Temperatures below about -30 C, particularly when accompanied by strong winds, kill flowering twigs and reduce or eliminate yields. However, north of this area, lowbush blueberries can be grown because they are protected from extreme temperatures by snow.

UTILIZATION OF THE CROP

Blueberries can be eaten fresh or used in many processed foods. Highbush and rabbiteye blueberries serve as sources for both markets, while lowbush fruit are almost exclusively processed. Lowbush berries are smaller than highbush; lowbush blueberries average 450 per cup contrasted with highbush berries that average 100 per cup. Because of the high count per unit weight or volume, fewer lowbush blueberries are needed to suggest high berry contents in foods. Blueberry species and blueberry cultivars within species show much variation in flavor components: levels of acidity, percent soluble solids, and presence and concentrations of various aromatic compounds. These components affect berry flavor and longevity in storage. Processed blueberries may be frozen, canned, dried, or made into syrups and jellies. Possibilities exist for blueberry juices and blueberry wines, but use of blueberries for these new products awaits increased production. A major use of blueberries is in the baking of pies, muffins, and pancakes.

KEY BREEDING AND CROP IMPROVEMENT PROBLEMS

Each of the three major classes of cultivated blueberries (lowbush, highbush, and rabbiteye) and each production area has its own unique problems and its own cultivar needs. The following breeding objectives apply, in most cases, to all types of blueberries, but the urgency of various objectives varies with area of production, type of blueberry, and purpose for which the fruit is intended.

Consistent and High Yields

This important trait is complex and depends on many factors including plant vigor, which allows newly planted fields to come into early production, late flowering which avoids frost, heavy flowering, attractiveness of flowers to pollinating insects, high percent of flowers retained to fruit maturity, and berry qualities that minimize fruit loss during and after ripening.

Plant Vigor

High vigor makes it easier to get new plantings established, allows them to come into full production at an early age, and reduces losses of plants from producing fields.

Resistance to Diseases and Insects

Like most crops whose primary area of cultivation coincides with its center of origin, blueberries are beset by many disease and insect problems. Potentially serious diseases include a number incited by fungi and several others caused by viruses. For most of these diseases, pronounced variations in cultivar susceptibility have been noted. The highbush blueberry cultivar development program at North Carolina State University provides an excellent example of how breeding for cultivar resistance to disease, in this case, canker incited by *Botryosphaeria corticis*, can have a major payoff (Milholland and Galetta, 1969). Lowbush blueberry fields consist of numerous naturally established clones interspaced with woody and herbaceous weeds, grasses, and sedges. Two species, the sweet lowbush blueberry, *Vaccinium angustifolium Ait.*, and the sourtop or velvet leaf blueberry, *V. myrtilloides Michv.*, dominate most fields (Hall and Aalders, 1961). The diversity of genetic material (clones) in native lowbush blueberry fields and pruning by fire (Eck and Childers, 1966) have kept insects and diseases to a minimum. However, a change to mechanical pruning (Ismail and Yarborough, 1981), and more intensive management through planting selected high yielding clones into existing fields or through the establishment of new fields may increase the incidence of insect and disease problems. Lowbush cultivars have been released (Aalders et al., 1978) but there is no evidence that selection was based on cultivar resistance to disease.

Plant Architecture

Highbush and rabbiteye blueberries range from colonial to almost monopodial, depending on the genetic characteristics of the variety. A limited amount of basal tillering may be desirable in blueberries to increase the bearing surface of the bush and to replace dead or weak canes that are removed by pruning. Excessive tillering, however, makes pruning and mechanical harvesting difficult.

Unlike the highbush species, the production of lowbush blueberries involves extensive rather than intensive management practices. Growers have relied primarily on the management of existing fields after clearing forest land to encourage development of natural stands. Increased plant populations have come about by the slow natural spread of existing species through seed dispersal and rhizomatous growth. The rhizome, which accounts for approximately 70% of the plant volume, is also important in the absorption and translocation of water and nutrients, retention of food reserves and production of new aerial shoots following periodic burning.

Plant height is an important varietal characteristic. Lowbush blueberries must be short enough to benefit from snow protection in extremely cold weather. Some rabbiteye cultivars are so vigorous that extensive pruning for height control is necessary.

Time of Berry Ripening

Prices for fresh blueberries vary widely throughout the year, with the first berries usually bringing the highest prices. This puts rabbiteye blueberries at a disadvantage relative to highbush, since most rabbiteye cultivars have a long interval from bloom to fruit ripening. Length of the bloom-to-ripening interval is under genetic control and is highly variable within each species of blueberry. This variability has been exploited to obtain both early and late ripening varieties that will provide fresh blueberries for a long period in each producing area. In northern Florida, the earliest rabbiteye cultivars reach 50% ripening about June 1, whereas late rabbiteyes do not begin to ripen fruit before September 1. Both earlier- and later-ripening cultivars can undoubtedly be developed by recurrent selection within *V. ashei*. For any one cultivar, it is desirable for mechanical harvesting that all berries ripen nearly simultaneously. This, too, is a highly variable trait among varieties. The rabbiteye cultivar Climax is noted for ripening its berries simultaneously. Lowbush blueberries are harvested almost entirely by metal hand rakes beginning in the last week of July and extending into September. A portion of the crop is lost from early ripening clones in fields that are raked toward the end of the harvest period. A modified Darlington cranberry harvester has recently been shown to be effective in relatively level blueberry fields and would speed up harvest and reduce these losses.

Avoidance of Frost Damage

Blueberry flowers tolerate lower temperatures without injury than do flowers of many fruit crops. Still, crop reduction due to spring frosts sometimes occurs. The best defense is provided by cultivars that flower later than usual. Selection for later flowering may have two disadvantageous side effects: it tends to select varieties that ripen later and varieties with higher chilling requirements which may yield poorly after mild winters in the South. Some uncultivated blueberry species, including *V. arboreum* and *V. stamineum*, avoid frost damage because they bloom a month or more after leafing out in the spring, with the flowers arising on spring growth rather than on wood produced the previous growing season. This flowering habit may someday be transferred to cultivated blueberries.

Self-Fruitfulness

Blueberry varieties range in self-fertility from near 0 to near 100%. Most lowbush blueberry clones are not self-fruitful and honeybees are used to supplement native pollinations in most commercial fields. Existing clones are not necessarily high yielding and the incompatibility of pollen among these clones may be further reducing yields. In general, highbush cultivars are self-fruitful, and blocks planted to single cultivars sometimes give yields as high as blocks interplanted with cultivars to facilitate cross-pollination. By contrast, no available rabbiteye cultivars can safely be planted in single-cultivar blocks. Yet variation in degree of self-fertility among rabbiteye varieties is enough to indicate the possibility of developing highly self-fruitful cultivars (Meader and Darrow, 1944). Availability of such cultivars would eliminate the need for interplanting two cultivars with similar flowering dates in the same field. Even in mixed-variety plantations, most pollinations made by bees are self-pollinations, since a bee may visit many flowers on one plant before moving to the next. It seems likely from this that self-fertile varieties would have better fruit set than self-incompatible ones in years when poor pollinating weather or low bee populations reduce pollen movement.

Berry Quality

This complex trait includes flavor (itself complex), color (blue is usually preferred over black), low seediness, thin skin, firmness, a dry stem-separation scar, and freedom from grit or stone cells. The preferred berry size may range from small for mechanically-harvested berries intended for processing to large for hand-picked fruit. Because fruit size is a yield component, selection for small berry size may tend to isolate varieties with reduced yield potential. Storage life of fresh blueberries varies with variety and is related to scar, skin thickness, and berry chemical constituents. Carefully picked berries of some rabbiteye cultivars can be stored a month or more at 4 C with little loss of quality.

Amenability to Mechanical Harvest

High labor costs dictate that most blueberries will be mechanically harvested in the future. This puts a premium on cultivars whose bush architecture and berry characteristics best suit them to machine harvest. Desirable traits include simultaneous ripening, loose berry clusters, dry scars, reasonably thick skins, and limited suckering.

Ease of Propagation

Blueberry cultivars are clonally propagated by hardwood cuttings, softwood cuttings, and in vitro shoot-tip culture. Varieties differ considerably in how easily they can be propagated by each of these methods. Cultivars that are particularly hard to propagate may be discriminated against by blueberry propagators.

Greater Tolerance for Diverse Soil Types

Highbush blueberries grow best in acid soils that are high in organic matter and moist but not water-logged. Such soils are unavailable in many potential producing areas. Both rabbiteye and lowbush blueberries grow well on acid mineral soils if moisture levels are suitable. One goal in breeding highbush blueberries is to improve their performance on mineral soils. Increased drought tolerance would be desirable in all species. Improved tolerance to root-rotting fungi would improve survival of blueberries on wet soils in warm areas. Tolerance for higher soil pH would be highly desirable in blueberries. Tolerance for high soil pH has been markedly enhanced in several agronomic crops through breeding, and genetic variability for this trait seems to be present in blueberries (Brown and Draper, 1980).

GENETICS AND BREEDING OF BLUEBERRIES

Blueberries are heterozygous, cross-pollinating plants. Each *Vaccinium* species consists of a highly variable population of inter-pollinating plants. No formal studies of inbreeding depression in blueberries have been made, but observations on several small S_1 populations obtained from self-pollination of rabbiteye cultivars indicated that inbred seedlings were much less vigorous than outcrossed progeny (Lyrene, unpublished). It seems reasonable to expect that development of highly vigorous blueberry cultivars will continue to depend upon maintaining high heterozygosity.

Because blueberries are clonally propagated, a new cultivar can originate from a single superb seedling. The standard breeding procedure with highbush, lowbush, and rabbiteye blueberries has been to cross superior selection and cultivars and grow large seedling populations. Because parents are highly heterozygous, F_1 seedling populations are variable, and the best seedlings are selected, propagated, and tested for cultivar potential. Observations of seedlings from various crosses indicate that some parental combinations produce many seedlings with high cultivar potential while other combinations produce few good seedlings. A popular procedure in breeding blueberries and other clonally propagated plants has been to identify parental combinations that produce seedlings of high quality and then repeat these superior crosses to obtain very large populations of superior seedlings from which cultivars can be selected.

Continued progress in breeding requires, however, that old, proven parental combinations eventually be abandoned in favor of new and better parents. The best seedlings obtained from each generation are selected as parents for intercrossing to produce the next generation of seedlings for selection. This procedure is called recurrent selection. It mimics evolution under natural selection and is the fundamental procedure in the breeding of clonally propagated plants. The power of recurrent selection derives from the fact that genetic progress is cumulative over cycles of selection. Because a large number of genes affect variety performance, it is impossible to obtain anything near the best possible genotype in a single generation as a result of an unusually favorable genetic segregation. Since genetic progress is cumulative from generation to generation, simultaneous mild selection for a number of favorable horticultural traits in a series of cycles can produce cultivars far better than could be obtained from "one-shot" selection (Lyrene, 1981b).

Induced polyploidy, induced mutations, and interspecific hybridization are breeding techniques that supplement recurrent selection in blueberries. The primary purpose of induced polyploidy in blueberries is to enable interspecific crosses to be made between species differing in ploidy. Among cultivated blueberries and closely related *Vaccinium* species, most interspecific hybrids are easy to make if both species have the same ploidy. By contrast, crosses between diploid and tetraploid species are difficult, and most of the resulting hybrids are tetraploid rather than triploid, having resulted from fusion of a normal gamete from the tetraploid parent and an unreduced gamete from the diploid parent. Low frequency of unreduced gamete formation makes such interspecific tetraploid hybrids hard to obtain. It is expected that doubling the chromosome number of the diploid species would enhance its ability to form hybrids with tetraploid species.

The objectives of interspecific hybridization are to combine desirable characteristics from two or more species, to exploit heterosis in new ways, and to create new amphidiploid combinations for use in breeding.

Mutation induction has not been used in blueberry breeding, although it has potential for the future. Because the gene pool available to blueberry breeders is wide and diverse, mutation induction appears to have little practical value if it is designed merely to enrich the germplasm pool available for breeding cultivars. The peculiar value of induced mutagenesis in breeding clonal cultivars lies in the fact that established cultivars of known value can be exposed to mutagens to obtain slightly modified forms. Mutations in ripening date, chilling requirement, or in bush architecture are examples of what might be looked for. Precedents for the use of mutation breeding in the improvement of woody clonal cultivars are readily available in apples and in citrus, crops in which many important cultivars have arisen from mutations of older varieties.

REVIEW OF THE LITERATURE

Clonal Propagation

Anderson's (1975) work with rhododendrons was probably the first report of in vitro propagation of an economically important plant in the family Ericaceae. Anderson's rhododendron formulation was the first medium used in Zimmerman's experiments on propagation of highbush blueberries which were initiated in 1977 (Zimmerman and Broome, 1980). Lyrene (1978) reported successful in vitro propagation of rabbiteye blueberry seedlings. Nickerson (1978) reported in vitro shoot production from lowbush blueberry seedlings, but rooting and establishment of plants from these shoots was not successful. Frett and Smagula (1982a) cultured lowbush blueberry single bud explants and found subsequent rooting of shoots produced was affected by growth regulator concentrations in the media. Although the media varied somewhat from one laboratory to another, the cytokinin 2iP was used to stimulate multiple shoot production in each case. The intensity, duration, and quality of light as well as incubation temperatures used in blueberry shoot-tip propagation have been typical of levels commonly used for in vitro propagation of other plant species.

Explants from juvenile stock plants have proved much easier to establish in culture than nonjuvenile explants of similar genotype (Lyrene, 1980). Frett and Smagula (1982b) attempted to overcome variation in clonal response of lowbush blueberries to culture by using stock plant treatments which had the potential of inducing juvenility. These treatments, etiolation at two temperatures, and application of glyphosate and ethophon, did not overcome clonal variation but did improve culture success for specific clones. Cultures established from nonjuvenile explants produced shoots with small leaves, thin stems, and short internodes that resembled young blueberry seedlings grown from seed. In blueberries, as in many plants, the rate of adventitious root formation on softwood cuttings declines as a plant passes from juvenility into adulthood. When nonjuvenile explants were used to establish shoot-tip colonies and the resulting shoots were used as softwood cuttings, the juvenile rooting rate was obtained (Lyrene, 1981a). Temporary phenotypic changes that apparently were not the result of genetic mutations have been observed following in vitro propagation of other crops (Swartz et al., 1981).

Use of colchicine-containing media to double the chromosome number has been reported in 1982 (Lyrene and Perry, 1982). Aside from the advantage of providing an optimum environment in which treated plants can recover from the toxic effects of colchicine, the in vitro colchicine treatments had the further advantage of allowing rapid separation of doubled from nondoubled shoots because of the increased stem diameters of colonies in which chromosome numbers had been doubled.

Table 2 summarizes the key contributions of in vitro techniques for blueberries.

Table 2. Progress in Development of In Vitro Techniques for Blueberries

CONTRIBUTION	REFERENCE
In vitro propagation of a related ericaceous species--Rhododendron	Anderson, 1975
In vitro propagation of highbush and rabbit-eye blueberry seedlings	Lyrene, 1978
In vitro shoot formation of lowbush blueberry seedlings	Nickerson, 1978
In vitro propagation of adult highbush blueberry clones	Zimmerman & Broome, 1980
Use of in vitro culture to produce fast-rooting juvenile phase of adult blueberry clones	Lyrene, 1981
In vitro propagation of adult lowbush blueberry clones	Frett & Smagula, 1982a,b
Production and screening of blueberry polyploids in vitro	Lyrene and Perry, 1982

Applications

Rapid in vitro propagation of disease-free blueberry shoots followed by in vivo rooting is now commercially practical, and at least two large blueberry propagators have established tissue culture laboratories. Experienced blueberry propagators have little trouble propagating most blueberry cultivars by softwood or hardwood cuttings, and the practicability of shoot-tip propagation was not at first obvious. Nonetheless, several advantages of in vitro shoot-tip propagation have become apparent, and will probably lead to increased use of in vitro propagation for blueberries despite the cost of establishing the laboratory.

There are many advantages. First, the use of pathogen-free explants can allow production of plants free of virus and fungus pathogens. Second, shoots from shoot-tip cultures root faster and at higher percentages than the best softwood or hardwood cuttings from the field. Third, shoots from culture can be produced for rooting any month of the year, whereas hardwood and softwood cuttings from field plantings can be made only during certain months of the year. Fourth, conventional propagation by cuttings cut from the field makes desirable the maintenance of field plantings of stock plants near the location where cuttings are to be rooted. This restricts the location of blueberry propagation facilities to blueberry-growing areas and increases the likelihood that newly produced plants will be exposed to blueberry pathogens. Fifth, there are indications (Lyrene, 1981a) that blueberry plants propagated in vitro are more juvenile than plants of the same variety propagated conventionally. This appears to give tissue-cultured ramets increased vigor, and may delay flowering after the plant is set in the field. The duration of the non-fruiting juvenile period in seedling blueberries is not more than two years, and this delay in flowering of tissue-cultured plants would probably be advantageous rather than disadvantageous because it would allow the plants to grow more rapidly. Lastly, newly released cultivars for which cutting wood is in short supply could be expanded much faster using in vitro techniques.

Tissue, Embryo, and Anther Culture

Aside from propagation of pathogen-free plants, culture techniques that could be of value in blueberry breeding include induced mutagenesis and polyploidy, embryo rescue of wide hybrids, dihaploid extraction from tetraploid species through anther culture, and rapid interspecific introgression by wide hybridization and in vitro chromosome elimination. The tissue culture techniques needed to enable use of these breeding tools are already available in some cases and are being developed in others. Some possibilities for using each of these techniques in blueberry breeding are discussed briefly below.

INDUCED MUTAGENESIS. The optimum strategy for inducing mutations in heterozygous clones differs from the optimum for use with homozygous, seed-propagated crops. Mutations are usually recessive, and with homozygous plants that have been exposed to mutagens, few mutations can be detected in the mutagenized or M_0 generation. M_0 plants must be selfed to produce an M_1 generation in which recessive mutant alleles can become homozygous and thus visible. With blueberries and other heterozygous species, mutations should be looked for in the original mutagenized clone rather than in its selfed progeny. The results of induced mutations can be seen in heterozygous clones without growing selfed progeny, even if the mutations are highly recessive. This is because mutations which quiet the dominant allele at heterozygous loci will unmask the effects of the originally recessive allele and produce a phenotype change. Selfing a mutagenized blueberry clone would unleash enormous genetic variation through segregation at heterozygous loci, and any induced mutations would be difficult to identify.

Two problems with induced mutagenesis in clonally propagated plants have been chimera formation and diplontic selection (Dermen and Bain, 1944). In conventional mutagenesis of clonal cultivars, apical and axillary buds have most often been treated. These buds are multicellular and contain more than one initial cell. Because each mutation is a single-cell event, mutation induction results in buds that are chimeras in which some initial cells are mutated and some are unaffected. In such cases, diplontic selection may occur in which normal cells have greater vigor than mutated cells and replace them as initial cells within growing points, with resultant loss of the induced mutation, even before it is detected.

The adventitious buds technique for clonal mutagenesis (Broertjes et al., 1968) offers a powerful tool for circumventing the undesirable effects of diplontic selection in chimeras after mutagenesis of clones. The effectiveness of the technique derives from the fact that most adventitious buds arise from single cells. Thus, mutagenic treatments of tissues that subsequently can be induced to form adventitious shoots give rise to nonchimeral shoots that will be mutant if the cell of origin was mutant.

A limitation of the adventitious bud technique is the current inability to induce adventitious buds in many species. In blueberries, adventitious shoots can be obtained from leaves harvested from established shoot-tip cultures and planted on proliferation medium. The explant source can be irradiated prior to culture initiation and the regenerated plants can be screened for mutations.

The primary goal of mutation induction in blueberries should be the production of modified forms of established cultivars or test selections. Subclones with modified plant architecture, ripening dates, or berry quality could be valuable.

INDUCED POLYPLOIDS. Chimeras and diplontic selection are as much a problem in polyploid induction by colchicine treatment as they are in mutation induction by irradiation; the adventitious bud technique can be as useful in obtaining polyploids as it is in obtaining mutants. Tetraploid derivatives of the diploid blueberry *V. elliottii* and 12-ploid forms of two rabbiteye cultivars have been obtained by treating shoot-tip cultures with colchicine and then using these cultures as explant sources (Lyrene and Perry, 1982). An unexpected advantage of in vitro polyploidization was that polyploid shoots could readily be separated from normal shoots because of their increased vigor and shoot diameter in vitro.

EMBRYO RESCUE IN WIDE HYBRIDIZATION. This technique has not been explored with blueberries but has potential usefulness because of the importance of wide hybridization in blueberry breeding. The triploid block, which results in absence of triploid seedlings following diploid x tetraploid crosses, is strongly expressed in all 2x-4x combinations tested in *Vaccinium*; it prevents the synthesis of artificial hexaploids by triploid doubling, a technique that would enable transfer of genes from diploid and tetraploid species into hexaploid *V. ashei*. Studies of the triploid block on other genera (Marks, 1966) reveal that triploid zygotes ordinarily form but perish because of endosperm malfunction. Thus, triploid embryos might be candidates for in vitro embryo rescue techniques.

DIHAPLOIDS EXTRACTED FROM TETRAPLOID. This technique has not been used with blueberries but could be useful if it could be accomplished. Dihaploid (2x) plants could be derived from highbush (4x) cultivars. These diploids, if sufficiently fertile, could then be crossed and selected, or they could be crossed with normally diploid *Vaccinium* species. Breeding at the diploid level would provide the advantages of simpler gene segregation ratios and less extensive masking of recessive alleles. The final cultivars would probably be developed at the tetraploid level, the tetraploids being produced by means of 2n gametes from the diploid parents. The use of extracted dihaploids in breeding tetraploid cultivars has been called synthetic breeding and has been tested more extensively with *Solanum tuberosum* (Peloquin et al., 1966).

SOMATIC CHROMOSOME ELIMINATION IN WIDE HYBRIDS TO PRODUCE INTROGRESSION LINES. Considerable evidence has accumulated indicating that plants that are wide hybrids may show somatic instability when propagated vegetatively by means other than apical or axillary buds. Somatic cell hybrids in both plants and animals sometimes lose chromosomes of one species component during mitosis. Hybrid zygotes resulting from crosses of *Hordeum vulgare* x *H. bulbosum*, when grown in embryo-rescue medium, usually eliminate all the *H. bulbosum* chromosomes and give rise to haploid embryos with chromosomes derived exclusively from *H. vulgare* (Kasha, 1975). Numerous seedlings from crosses of polyploid *Rubus* had variable chromosome numbers in root tips (Britton and Hull, 1957; Hull and Britton, 1958). When propagated by stem cuttings, these mitotically unstable clones maintained their phenotypes and their mitotic instability as indicated by variability in root-tip chromosome counts. When propagated by adventitious shoots on root cuttings, the unstable clones showed radical changes in phenotype. The "mutant" derivatives were much lower in chromosome number than the original clones and were usually mitotically stable.

Chromosomal instability is enhanced in many tissue culture systems, particularly where growth occurs from callus or from adventitious shoots (Larkin and Scowcroft, 1981). Since wide hybrids have, at least in some cases, inherent mitotic instability, tissue culture propagation of wide hybrids might produce unusually high levels of cytogenetic changes.

Wide hybridization followed by tissue culture propagation of the hybrid could provide a method for expeditiously introgressing one or a few chromosomes from one blueberry species into another. For example, if it were desired to transfer useful traits from the diploid *V. elliottii* into the cultivated tetraploid *V. corymbosum*, an octoploid line of *V. corymbosum* could be obtained with colchicine, and this could be hybridized with *V. elliottii* to produce pentaploid hybrids. These could be propagated in vitro by adventitious budding, and a search could be made for derivatives with reduced chromosome numbers and the desired combination of *V. corymbosum* and *V. elliottii* traits. The technology for using this technique is available now, but whether it would produce the desired result has not been determined.

Key Variables and Factors for Success with In Vitro Techniques

CULTURE MEDIA. The first three media listed in Table 3 all give satisfactory shoot-tip growth with most rabbiteye blueberry seedlings. Zimmerman media has been successfully used with adult highbush (Zimmerman and Broome, 1980) and lowbush blueberries (Frett and Smagula, 1982a,b). The medium based on WPM (McCown and Lloyd, 1981) seems superior with many highbush seedlings (Lyrene, unpublished). The cytokinin, 2iP, appears essential for obtaining good shoot proliferation. Elimination of 2iP typically gives single-stalked cultures. Increasing concentrations of 2iP give increasing numbers of shorter and shorter shoots (Zimmerman and Broome, 1980; Frett and Smagula, 1982a). Auxins (IAA, IBA, NAA, 2,4-D) added to the medium either induce callus which has low differentiation potential or have no effect (Nickerson and Hall, 1976; Frett and Smagula, 1982a). A soft agar (0.4%) gives consistently better shoot-tip growth than a firm agar (0.8%).

JUVENILE VS. ADULT STATE OF EXPLANTS. Juvenile blueberry explants give markedly higher survival and growth rates than adult explants. Shoot tip colonies established from adult explants assume a juvenile-like growth pattern with small leaves, short internodes, and thin stems (Lyrene, 1981a). These "habituated" colonies are as easy to subculture as are colonies initiated from juvenile stock.

EXPLANT GENOTYPE. A number of different blueberry species have been propagated by in vitro shoot culture, including *V. ashei*, *V. corymbosum*, *V. angustifolium*, *V. elliottii*, *V. darrowi*, *V. arboreum*, and *V. stamineum*. These seven species show some differences in growth habit, but overall, their growth patterns seem more remarkable for their similarity than for their differences. Yet within individual species, different clones show striking differences in ease of culture and in in vitro growth habit. One such contrast is provided by NC 1688, a North Carolina highbush test selection which grows readily on either Knops, Anderson, or WPM media described in Table 3, and the highbush cultivar Flordablue, which barely survives on these media. Many similar examples could be cited. The reasons for these differences are unknown.

Table 3. Blueberry Shoot-Tip Culture Media and Their Compositions[a,b]

INGREDIENTS	MEDIUM MODIFIED FROM			
	Knops	Anderson	WPM	Zimmerman
Macronutrients (mM)				
$Ca(NO_3)_2 \cdot 4H_2O$	4.8	2.9	2.4	3.0
K_2SO_4	—	—	5.7	—
KH_2PO_4	1.25	2.7	1.25	3.0
KNO_3	1.88	1.88	—	2.0
$MgSO_4 \cdot 7H_2O$	1.5	1.5	1.5	1.5
NH_4NO_3	—	5.0	5.0	2.0
$(NH_4)_2SO_4$	—	—	—	1.5
Iron (mM)				
$FeSO_4 \cdot 7H_2O$	0.2	0.2	0.2	0.2
$Na_2 \cdot EDTA$	0.2	0.2	0.2	0.2
Micronutrients				
$CoCl_2 \cdot 6H_2O$ (µM)	0.1	0.1	0.1	0.1
$CuSO_4 \cdot 5H_2O$ (µM)	0.1	0.1	1.0	0.1
H_3BO_3 (mM)	0.1	0.1	0.1	0.1
KI (µM)	5.0	5.0	5.0	5.0
$MnSO_4 \cdot 4H_2O$ (mM)	0.1	0.1	0.1	0.075
$Na_2MoO_4 \cdot 2H_2O$ (µM)	1.0	1.0	1.0	1.0
$ZnSO_4 \cdot 7H_2O$ (mM)	0.03	0.03	0.03	0.03
Vitamins-Purine				
Adenine Sulfate (mM)	—	—	—	0.35
Casein hydrolysate (mg/l)	1000.0	1000.0	1000.0	—
Glycine (mM)	0.03	0.03	0.03	—
Myo-inositol (mM)	0.55	0.55	0.55	0.55
Nicotinic acid (µM)	4.1	4.1	4.1	—
Pyridoxine·HCl (µM)	2.4	2.4	2.4	—
Thiamine·HCl (µM)	0.3	0.3	0.3	1.2
Growth Regulators (µM)				
IAA	—	—	—	22.0
2iP	25.0	25.0	25.0	75.0
Carbohydrates and Gelling Agent				
Sucrose (M)	0.088	0.088	0.088	0.088
Agar (g/l)	4.0	4.0	4.0	10.0

[a]pH adjusted to 5.7 (5.2 for Zimmerman) before adding agar.

[b]Medium autoclaved for 15 min at 121 C and 1.1 kg/cm^2.

PROTOCOLS

Establishment of Shoot-Tip Cultures

Several methods have been successfully used to establish proliferating shoot-tip cultures. A major difference involves implanting the original explant in a semisolid medium or first placing it in a liquid medium and then transferring it to a semisolid medium. Blueberry shoot tips and shoot segments with axillary buds have been successfully cultured by placing them directly into a semisolid medium (Zimmerman and Broome, 1980; Cohen and Elliott, 1979; Frett and Smagula, 1982a). One system successfully employed involves autoclaving screw-top vials (inside diameters 1-2 cm) which have been filled to a level of about 2 cm with liquid WPM medium (Table 3). Rapidly growing succulent shoots from greenhouse-grown plants are stripped of their leaves without damaging axillary buds and cut into 3-4 cm long segments, providing two or more explants per shoot. Explants are surface sterilized by gently stirring them for 15 min in a mixture of 1 part Clorox to 3 parts water with 1 drop of Tween-20 per 20 ml of sterilant and rinsed 3 times in sterile water. Each explant is aseptically transferred to a vial so that it leans against the side and approximately half its length is submerged in the medium. Within 2 months, some explants should produce shoots from axillary buds. When shoots have two or more elongated internodes, they should be cut into two-node segments and placed horizontally onto the surface of fresh WPM made semisolid with 0.4% agar. Establishment of cultures of nonjuvenile explants give rather low success rates for most clones due to failure of many stem explants to produce shoots from axillary buds or death of the two-node segments when transferred onto agar medium. If the initial explants are juvenile, neither problem is severe. Even with nonjuvenile explants, however, vigorous colonies will result from a fraction of the two-node cuttings placed on agar-WPM medium. Once a clone is established in culture, it can be increased by subculturing two-node stem segments onto fresh medium. When rooted plants are desired, shoots 1 cm long or longer are stuck to a depth of 0.5 cm in firmly packed peatmoss and placed in a misthouse. Rooting will be complete in 3-8 weeks depending on species.

Seed Germination In Vitro

Blueberry seedlings have responded quite well as a source of plant tissue (Nickerson, 1978; Lyrene, 1980). Germination of seed in vitro provides a convenient source of sterile tissue. Seeds may be extracted by macerating a cup of ripe fruit for approximately 25 sec in a 1000 ml food blender container with 400 ml of water. Skins and pulp can be floated off by adding fresh water, and the seeds will be left in the bottom of the blender. Cleaned lowbush blueberry seeds have been effectively stored dry in paper envelopes at 1 to -23 C without loss of viability. Seeds can be surface sterilized by immersing them for twenty minutes in a mixture of 1 part Clorox to 3 parts water with 1 drop Tween-20 per 20 ml of sterilant. After a rinse in sterile water seeds are placed on water-agar gel made with about 6 g agar/l. Blueberry seeds will not germinate in the dark (Smagula et al., 1980; Scott and Draper, 1967; Austin and Cundiff, 1978); therefore a source of light must be provided. Blueberry seeds will germinate at room temperature, but germination is somewhat faster at low temperatures (10 C) or fluctuating day/night temperatures (25 C/15 C).

Doubling Chromosome Numbers with Colchicine

Colchicine-induced chromosome doubling in vitro may be an important tool for the blueberry breeder. The procedure described by Lyrene and Perry (1982) uses 0.1% colchicine in liquid WPM media. About 10 vigorous shoots are placed in vials containing enough WPM-colchicine to submerge them and rotated or agitated for 24 hr. Shoots are then removed, washed in sterile water, cut into two-node explants and placed on WPM agar medium. Shoots developing in these cultures with atypically large stem diameter have a high probability of being the desired polyploids and should be subcultured, tested, and rooted for use in breeding.

FUTURE PROSPECTS

Blueberry shoot-tip culture will probably become increasingly important in the rapid propagation of disease-free plants. Colchicine-induced chromosome doubling in vitro will be useful in interspecific hybridization. If embryo-rescue techniques can be developed to enable routine rescue of triploid embryos from tetraploid x diploid crosses, it would make possible new ways of combining the traits of diploid, tetraploid, and hexaploid blueberry species. Work is needed to find ways to induce differentiation of blueberry callus. This might increase the efficiency of tissue-culture-induced mutagenesis in blueberries. It might also be a key to production of haploids from anther culture. If media could be found that would allow accurate in vitro selection for tolerance to higher soil pH, tissue culture could be used in developing blueberries with broader soil adaptation. More information may be needed on how high soil pH retards blueberry growth before an effective screening medium can be devised.

Because so many horticultural traits in blueberries are quantitative, it seems likely that recurrent selection and field testing will continue to be the central procedures in blueberry breeding for many years. In vitro mutation breeding that starts with high-quality cultivars and ends with improved variants of these cultivars will probably become an important subsidiary breeding method, and the other techniques discussed in this chapter will also contribute to the efficiency of future blueberry breeding efforts.

REFERENCES

Aalders, L.E. and Hall, I.V. 1975. Germination of lowbush blueberry seeds stored dry in fruit at different temperatures. HortScience 10: 525-526.

______, and Brydon, A.C. 1978. Yields of native clones of lowbush blueberry under cultivation. Fruit Var. J. 32.

Anderson, W.C. 1975. Propagation of rhododendrons by tissue culture: Part 1. Development of a culture medium for multiplication of shoots. Proceedings of the International Plant Propagators Society, Boulder, Colorado. 25:129-135.

Austin, M.E. and Cundiff, J.S. 1978. Factors affecting rabbiteye blueberry seed germination. J. Am. Soc. Hortic. Sci. 103:530-533.

Britton, D.M. and Hull, J.W. 1957. Mitotic instability in *Rubus*. J. Hered. 48:11-20.

Broertjes, C., Haccius, B., and Weidlich, S. 1968. Adventitious bud formation on isolated leaves and its significance for mutation breeding. Euphytica 17:321-344.

Brown, J.C. and Draper, A.D. 1980. Differential response of blueberry (*Vaccinium*) progenies to pH and subsequent use of iron. J. Am. Soc. Hortic. Sci. 105:20-24.

Cohen, D. 1980. Application of micropropagation methods for blueberries and tamarillo. Proc. Int. Plant Prop. Soc. 30:144-146.

______ and Elliott, D. 1979. Micropropagation methods for blueberries and tamarillos. Proc. Int. Plant Prop. Soc. 29:177-179.

Day, C. 1960. A History of Blueberry Industry in Washington County. Special Collections, Folger Library, University of Maine, Orono.

Dermen, H. and Bain, H.F. 1944. A general cytohistological study of colchicine polyploidy in cranberry. Am. J. Bot. 31:451-463.

Eck, P. and Childers, N.F. (eds.) 1966. Blueberry Culture. Rutgers Univ. Press, New Brunswick, New Jersey.

Frett, J.J. and Smagula, J.M. 1982a. In vitro shoot production of lowbush blueberry. Can. J. Plant Sci. 63:467-472.

______ 1982b. Effect of stock plant pretreatment on explant shoot multiplication of lowbush blueberry. Crop Res. (in press).

Galletta, G.J. 1975. Blueberries and cranberries. In: Advances in Fruit Breeding (J. Janick and J.N. Moore, eds.) pp. 154-196. Purdue Univ. Press, West Lafayette, Indiana.

Hall, I.V. and Aalders, L.E. 1961. Cytotaxonomy of lowbush blueberries in Eastern Canada. Am. J. Bot. 48:199-201.

Hull, J.W. and Britton, D.M. 1958. Development of colchicine-induced and natural polyploid breeding lines in the genus *Rubus*. Univ. Maryland Agric. Exp. Sta. Bull. A-91.

Ismail, A.A. and Yarborough, D.E. 1981. A comparison between flail mowing and burning for pruning lowbush blueberries. HortScience 16: 318-319.

Kasha, K.J. 1975. Haploids from somatic cells. In: Haploids in Higher Plants: Advances and Potential (K.J. Kasha, ed.) pp. 67-90. Univ. of Guelph, Guelph, Ontario.

Larkin, P.J. and Scowcroft, W.R. 1981. Somaclonal variation--A novel source of variability from cell cultures for plant improvement. Theor. Appl. Genet. 60:197-214.

Lyrene, P.M. 1978. Blueberry callus and shoot-tip culture. Proc. Fla. State Hortic. Soc. 91:171-172.

______ 1980. Micropropagation of rabbiteye blueberries. HortScience 15:80-81.

______ 1981a. Juvenility and production of fast-rooting cuttings from blueberry shoot cultures. J. Am. Soc. Hortic. Sci. 106:396-398.

______ 1981b. Recurrent selection in breeding rabbiteye blueberries (*Vaccinium ashei* Reade). Euphytica 30:505-511.

______ and Perry, J.L. 1982. Production and selection of blueberry polyploids in vitro. J. Hered. 73:377-378.

______ and Sherman, W.B. 1979. The rabbiteye blueberry industry in Florida—1887 to 1930—with notes on the current status of abandoned plantations. Econ. Bot. 33:237-243.

Marks, G.E. 1966. The enigma of triploid potatoes. Euphytica 15:285-290.

McCown, B.H. and Lloyd, G. 1981. Woody plant medium (WPM)—A mineral nutrient formulation for microculture of woody plant species. HortScience 16:453.

Meader, E.M. and Darrow, G.M. 1944. Pollination of the rabbiteye blueberry and related species. Proc. Am. Soc. Hortic. Sci. 45:267-274.

Milholland, R.D. and Galletta, G.J. 1969. Pathogenic variation among isolates of *Botryosphaeria corticis* on blueberry. Phytopathology 59: 1540-1543.

Nickerson, N.L. 1978. In vitro shoot formation in lowbush blueberry seedling explants. HortScience 13:698.

______ and Hall, I.V. 1976. Callus formation in stem internode sections of lowbush blueberry (*Vaccinium angustifolium* Ait.) cultured on a medium containing plant growth regulators. Hortic. Res. 16:29-35.

Peloquin, S.J., Hougas, R.W., and Gabert, A.C. 1966. Haploidy as a new approach to the cytogenetics and breeding of *Solanum tuberosum*. In: Chromosome Manipulations and Plant Genetics (R. Riley and K.R. Lewis, eds.) pp. 21-28. Oliver and Boyd, Edinburgh.

Scott, D.H. and Draper, A.D. 1967. Light in relation to seed germination of blueberries, strawberries and *Rubus*. HortScience 2:107-108.

Sharpe, R.H. 1954. Horticultural development of Florida blueberries. Proc. Fla. State Hortic. Soc. 66:188-190.

______ and Sherman, W.B. 1971. Breeding blueberries for low-chilling requirement. HortScience 6:145-147.

Smagula, J.M., Michaud, M., and Hepler, P.R. 1980. Light and gibberellic acid enhancement of lowbush blueberry seed germination. J. Am. Soc. Hortic. Sci. 105:816-818.

Spiers, J.M. and Draper, A.D. 1974. Effect of chilling on bud break in rabbiteye blueberry. J. Am. Soc. Hortic. Sci. 99:398-399.

Swartz, H.J., Galletta, G.J., and Zimmerman, R.H. 1981. Field performance and phenotypic stability of tissue culture-propagated strawberries. J. Am. Soc. Hortic. Sci. 106:667-673.

Zimmerman, R.H. and Broome, O.C. 1980. Blueberry micropropagation. Proceedings of the Conference on Nursery Production of Fruit Plants through Tissue Culture: Applications and Feasibility. U.S. Dept. Agric. Pub. ARR-NE-11, Beltsville, Maryland.

CHAPTER 16
Stone Fruits

R. M. Skirvin

The stone fruits (*Prunus* spp.) are an extremely diverse group of woody plants that includes over 400 species of trees and shrubs. The plants are grown worldwide. Some species are grown for their edible fruits and nuts, others are grown for their ornamental interest, while others are of value as rootstocks (Tables 1 and 2).

Economically the most important stone fruit types are as follows: peaches and nectarines (*P. persica*), European and Japanese plums (*P. domestica* and *P. salicina*, respectively), sweet and sour cherries (*P. avium* and *P. cerasus*, respectively), and almond (*P. amygdalus*). Commercial production of the stone fruits is found in many regions of the world. The success of commercial production is limited by the severity of the winter and the number of hours of chilling (temperatures less than 7.2 C). In North America the majority of the stone fruits are produced on the western coast (California, Oregon, Washington, and British Columbia), but they are also grown in important quantities in Michigan and certain areas of the South.

Many of the stone fruits do not propagate well from cuttings so they are grafted onto various types of rootstocks. Some rootstocks are seedlings and others are maintained clonally. Each important stone fruit type has its own series of rootstocks which are used for proliferation. Cummins (1979a,b), for instance, has reported a large number of species that may be suitable rootstocks for cherry.

LITERATURE REVIEW

In order to insure cultivar stability most stone fruits are propagated asexually by grafting. Stone fruits are seldomly propagated by cuttings since many of them are difficult to root from cuttings. The techniques of grafting are very ancient and date back to at least the Roman times (Janick, 1979). The practice of grafting is as much an art as it is a science, requiring skilled labor and considerable time, and it therefore limits the speed of plant propagation. However, it is carried out because until quite recently it was not possible to obtain satisfactory numbers of propagules of many cultivars by cuttings. Even when self-rooted plants are available, many commercial growers and propagators prefer grafted plants because hardiness, disease resistance, and certain growth characteristics can be predicted when cultivars are on rootstocks.

The rootstocks that are used for grafting are either clonal or seedlings. There are a number of good clonal rootstocks for stone fruits, but many of these are difficult to root from cuttings. For this reason most of the stone fruit rootstocks are seedlings. Since it is difficult to obtain highly uniform seed stocks, grafted cultivars are variable. If clonal rootstocks could be propagated easily, then they could be used for grafting and the need for seedling stocks could be reduced or eliminated. There have been outstanding efforts made to improve rootstock and scion rootability (Hartmann and Hansen, 1958; Farmer and Besemann, 1974; Robitaille and Yu, 1980) but none of these systems have proved to be useful enough to become a standard propagation method.

However, even when cultivars can be propagated readily by asexual means, the number of propagules that can be increased is limited by the number of stock plants available. Micropropagation or in vitro propagation can be utilized to substantially increase the rate of multiplication.

There are a number of different ways that a scientist could approach the problem of obtaining large numbers of plants in vitro (Boxus, 1978). Theoretically, the most efficient method would be to stimulate embryogenesis directly from callus. Via embryogenesis, one could obtain thousands of plants in very little time from continuous cultures. However, the only woody fruit crop that will consistently undergo embryogenesis is *Citrus* nucellar callus (Kochba and Spiegel-Roy, 1977).

A second method to propagate large numbers of plants would be to stimulate adventitious bud formation directly from plant organs such as leaves and stems. This could also be used to produce large numbers of plants, but efficient shoot organogenesis has not been accomplished for any fruit crop (Skirvin, 1981). There have been reports of adventitious shoot formation from embryonic almond pieces (Mehra and Mehra, 1974) and sour cherry anthers (Seirlis et al., 1979), but even these published results were inconsistent. There is however a recent report that adventitious shoots could be stimulated from injured roots of three different *Prunus* species (Druart, 1980).

A third method of in vitro propagation involves the stimulation of axillary bud growth. Using shoot cultures, the expansion of dormant axils is stimulated because each of these buds can become a single

Table 1. A Survey of *Prunus* Tissue Culture

SPECIES	EXPLANT SOURCE[a]	RE-SPONSE[b]	PRINCIPLE COMPONENTS OF MEDIUM: Salts: Macro	Salts: Micro	Vitamins[c]	Hormones[d] (mg/l)	Sugar[e]	Other[f]	REFERENCE
ANTHER CULTURE									
P. avium									
'Limburger Vogelkirche'	A	C	See Nitsch, 1972			0 or 1.0 NAA + 1.0 BA	20 S	Light	Jordan, 1974, 1975
		R	MS	MS	MS	1.0 NAA + 0.01 BA	30 S	—	Jordan, 1975
—	Seedling hypo-cotyls	R	See original article	MS	MS + 5.0 CaP + 100 myo	1.0 NAA + 0.1 KIN	30 S	—	Jordan et al., 1980
'Dragena Zheltaya'	A	C	MS	MS	MS		30 S	—	Legeida et al., 1976
'V1629', 'V1021', 'V978', 'V811', 'V340'	A	C	See original article			5 or 10 BA + 2,4-D	30 S	250 myo	Seirlis et al., 1979
P. cerasifera									
'P2175'	A	C	See *P. avium* (Seirlis et al., 1979)						
P. cerasus									
'V603', 'V1280',	A	C	See original article			5 or 10 BA + 2,4-D	30 S	250 myo	Seirlis et al., 1979
'Ferracida'	A	Ad	See original article			10.0 BA + 1.0 IAA	30 S	100 myo	Seirlis et al., 1979

SPECIES	EXPLANT SOURCE[a]	RE-SPONSE[b]	PRINCIPLE COMPONENTS OF MEDIUM						REFERENCE
			Salts						
			Macro	Micro	Vitamins[c]	Hormones[d] (mg/l)	Sugar[e]	Other[f]	
	A	R	See original article			5.0 2,4-D	30 S	250 myo	Seirlis et al., 1979
P. domestica 'P107', 'P149', 'P248', 'P707', 'P994', 'P244'			See *P. avium* (Seirlis et al., 1979)						
P. x *pandora*	—	—	—	—	—	—	—	—	Boxus, 1971
—	A	C + R	Nitsch & Nitsch (1969) + 37.3 Na_2EDTA + 27.8 $FeSO_4 \cdot 7H_2O$			1.0 NAA + 1.0 BA	20 S	Liquid medium	Jordan & Feucht, 1977
P. persica 'Springtime'	A	C	See original article			5 or 10 BA + 1.0 IAA	30 S	100 myo	Seirlis et al., 1979
CALLUS CULTURE									
P. amygdalus	S	C	MS	MS	MS	—	30 S	100 CH	Kester et al., 1977
—	Se (with-out testa)	C	1/2 MS	1/2 MS	1/2 MS	—	30 S	—	Mehra & Mehra, 1974
	R	C	MS	MS	MS	1-3 NAA	30 S	10% CW	Mehra & Mehra, 1974

Table 1. Cont.

SPECIES	EXPLANT SOURCE[a]	RESPONSE[b]	PRINCIPLE COMPONENTS OF MEDIUM: Salts: Macro	Micro	Vitamins[c]	Hormones[d] (mg/l)	Sugar[e]	Other[f]	REFERENCE
	S, L, St	C	MS	MS	MS	4–8 NAA or 4–8 2,4-D	30 S	10% CW	Mehra & Mehra, 1974
	Cotyledons, Em	C	MS	MS	MS	4–10 NAA	30 S	10% CW	Mehra & Mehra, 1974
	Callus from L, Em, or cotyledons	Ad	MS	MS	MS	5.0 NAA + 0.5–1.0 KIN	30 S	1000 CH	Mehra & Mehra, 1974
		R	MS	MS	MS	5.0 NAA + 0.5–1.0 KIN	30 S	1000 CH	Mehra & Mehra, 1974
—	A	C	See original article			0.1 IAA + 1.0 2,4-D + 0.2 KIN	30 S	—	Michellon et al., 1974
P. armeniaca	A	C	MS	MS	MS	2.0 NAA + 2.0 KIN	30 S	—	Harn & Kim, 1972
P. avium 'Limburger Vogelkirsche'	S	C	MS	MS	MS	5.0 NAA + 1.0 BA	30 S	—	Feucht, 1974, 1975
—	C	R	MS	MS	MS	1.0 NAA + 1.0 ABA	30 S	—	Feucht & Dausend, 1976

SPECIES	EXPLANT SOURCE[a]	RE-SPONSE[b]	PRINCIPLE COMPONENTS OF MEDIUM						REFERENCE
			Salts						
			Macro	Micro	Vitamins[c]	Hormones[d] (mg/l)	Sugar[e]	Other[f]	
—	S	C	1/2 MS	1/2 MS	1/2 MS	0.1 BA	20 S	15 chlorogenic acid + 5 tryptamine	Feucht & Johal, 1977
—	S	C	1/2 MS	1/2 MS	1/2 MS	10.0 IAA + 0.1 BA	20 S + 5 sorbitol	Catechines 16 hr day, 5000 lux	Feucht & Nachit, 1977
'F12/1'	S	C	1/2 MS	1/2 MS	1/2 MS	5.0 IAA + 0.1 BA		10–20 rutin + 10–20 dihydroxquercitin	Feucht & Nachit, 1978
P. avium x *P. fruticosa*	C	C	See *P. avium* (Feucht, 1975)						
P. besseyi	C	C	MS	MS	MS + 0.5 CaP + 100 myo	1.0 2,4-D + 1.0 KIN	30 S	1000 CH	Heuser, 1972

Table 1. Cont.

SPECIES	EXPLANT SOURCE[a]	RE-SPONSE[b]	PRINCIPLE COMPONENTS OF MEDIUM						REFERENCE
			Salts						
			Macro	Micro	Vitamins[c]	Hormones[d] (mg/l)	Sugar[e]	Other[f]	
P. canescens	Ax with roots	C	See *P. canescens* (in shoot tip and meristem section)						Druart, 1980
P. cerasus 'Schatten-morelle'	C	C	See *P. avium* (Feucht, 1974, 1975)						
P. dawyckensis	Ax with roots	C	See *P. canescens* (in shoot tip and meristem section)						Druart, 1980
P. fruticosa	C	C	See *P. avium* (Feucht, 1975)						
P. incisa x *P. serrula*	Ax with roots	C	See *P. canescens* (in shoot tip and meristem section)						Druart, 1980
P. mahaleb	C	C	MS	MS	MS	0.25 2,4-D + 0.2 KIN	30 S or 30 Sorbitol	3000 lux, continuous light	Coffin et al., 1976
—	L	C, R	1/2 MS	MS	MS	4.0 NAA + 1.0 BA	20 S	Light	Hedtrich, 1977
P. nigra			See *P. mahaleb* (Coffin et al., 1976)						
P. padus (common bird cherry)	—	—	—	—	—	—	—	—	Bychenkova, 1963
P. padus			See *P. mahaleb* (Coffin et al., 1976)						

SPECIES	EXPLANT SOURCE[a]	RE-SPONSE[b]	PRINCIPLE COMPONENTS OF MEDIUM: Salts						REFERENCE
			Macro	Micro	Vitamins[c]	Hormones[d] (mg/l)	Sugar[e]	Other[f]	
P. persica 'Frau Anneliese Rudolf', 'South Haven', 'Mme Roignat', 'Elberta'	C	C	MS	MS	0.2 thia	0.2–0.5 2,4-D + 0.2 KIN	30 S	—	Caporali & Weltzien, 1974
'Polly'	C	C	See *P. besseyi* (Heuser, 1972)						
—	M	C	MS	MS		0.5 2,4-D + 1.0–2.0 IAA + 0.1–1.0 KIN + 1.0 GA_3	30–40	—	Maroti & Levi, 1977
'Dixired', 'Nectared IV'			See *P. amygdalus* (Michellon et al., 1974)						
—	St	C	WH	WH + Heller (1953)	WH	0.5 NAA	20 S	—	Nekrosova, 1964
'Miller's Late'	—	C	Modified WH		—	1 or 10 NAA + 2.0 KIN	S	pH 7.0	Sommer et al., 1962

Table 1. Cont.

SPECIES	EXPLANT SOURCE[a]	RE-SPONSE[b]	PRINCIPLE COMPONENTS OF MEDIUM						REFERENCE
			Salts						
			Macro	Micro	Vitamins[c]	Hormones[d] (mg/l)	Sugar[e]	Other[f]	
P. pseudocerasus (not *P. pseudocerasus* Lindl., but belonging to section *pseudocerasus*)	C	R	MS	MS	MS	10.0 IBA + 0.1 BA + 0.1 ABA	30 S	—	Feucht & Dausend, 1976
	S	R	MS	MS	MS	10.0 IBA + 1.0 ABA	30 S	—	Feucht & Dausend, 1976
P. serotina	Cambial zone tissue	C	Wetmore & Rier, 1963			1.0 2,4-D	40 S	15% CW, 1g CH, dark, 27–32 C	Caponetti et al., 1971
P. tenella			See *P. mahaleb* (Coffin et al., 1976)						
P. tomentosa	C		See *P. besseyi* (Heuser, 1972)						
EMBRYO CULTURE									
"Cherries"	Em	G	MS	MS	MS	10 GA_3	20 S	pH 5.6	Ivanicka & Pretova, 1980

SPECIES	EXPLANT SOURCE[a]	RE-SPONSE[b]	PRINCIPLE COMPONENTS OF MEDIUM						REFERENCE
			Salts						
			Macro	Micro	Vitamins[c]	Hormones[d] (mg/l)	Sugar[e]	Other[f]	
		Ax	MS	MS with 20 Fe-NaEDTA	0.4 thia + 100 myo	1.0 IBA + 1.0 BA	30 S	pH 5.2, 162 PG	Ivanicka & Pretova, 1980
		Ax	MS	MS with 20 Fe-NaEDTA	0.4 thia + 100 myo	0.2 NAA	30 S	pH 5.6, 2.0 As	Ivanicka & Pretova, 1980
P. avium	Em	G	See original article				20 G	20 Asparagine	Tukey, 1933
P. cerasus 'Montmorency'	Fr	G	See original article						Wittenbach & Bukovac, 1980
PROTOPLAST CULTURE									
P. brigantica									Salesses & Mouras, 1977
P. cerasifera									Mouras et al., 1978
P. persica									Salesses & Mouras, 1977; Mouras et al., 1978

Table 1. Cont.

			PRINCIPLE COMPONENTS OF MEDIUM						
			Salts						
SPECIES	EXPLANT SOURCE[a]	RE-SPONSE[b]	Macro	Micro	Vitamins[c]	Hormones[d] (mg/l)	Sugar[e]	Other[f]	REFERENCE
SHOOT TIP AND MERISTEM CULTURE									
P. accolade	St	Ax	MS	Heller (1953) (x 100)	50 myo + 0.1 bio + 0.5 CaP	0.1 GA_3 + 1.0 BA + 0.22 2,4-D	20 S	pH 5.0, 16-hr day, 28 C night/ 24 C day	Boxus, 1975; Boxus & Quoirin, 1974, 1977; Quoirin et al., 1974, 1975
		R	MS	Heller (1953) (x 100)	50 myo + 0.1 bio + 0.5 CaP	1.0 IBA	20 S	pH 5.0, 16-hr day, 28 C night/ 24 C day	Boxus, 1975; Boxus & Quoirin, 1974, 1977; Quoirin et al., 1974, 1975
P. amygdalus	St	Ax	See original article						Seirlis et al., 1979
'Nonpareil'	St	Ax	See original article			1.0 BA	20 S	—	Tabachnik & Kester, 1977
		R	See original article			1.0 NAA or IBA	20 S	—	Tabachnik & Kester, 1977

SPECIES	EXPLANT SOURCE[a]	RE-SPONSE[b]	PRINCIPLE COMPONENTS OF MEDIUM						REFERENCE
			Salts						
			Macro	Micro	Vitamins[c]	Hormones[d] (mg/l)	Sugar[e]	Other[f]	
P. armeniaca	St	Ax	W	W + Heller (1953)	W	?	20 S	—	Nekrosova, 1964
'Stella'	St	Ax	MS	MS	1.5 μg cyano-cobal-min + 0.5 Fo + 0.5 ribo + 1.0 bio + 1.0 cho-line clo-ride + 1.0 CaP + 1.0 thia + 2.0 nia + 2.0 pyr + 0.5 paba + 100 myo	2.0 BA + 0.1 NAA	30 S	50 As	Skirvin & Chu, 1977; Skirvin et al., 1980
P. austera			See *P. accolade* (Boxus and Quoirin, 1977)						
P. avium									
'F 12/1', 'Early Rivers', 'Redel-finger',	St	Ax, R	See *P. accolade* (Boxus and Quoirin, 1977)						

Table 1. Cont.

SPECIES	EXPLANT SOURCE[a]	RE-SPONSE[b]	PRINCIPLE COMPONENTS OF MEDIUM						REFERENCE
			Salts						
			Macro	Micro	Vitamins[c]	Hormones[d] (mg/l)	Sugar[e]	Other[f]	
Schneider sp. Knorp. Many mature forest	St + root suckers	Ax	MS with 800 in-stead of 1650 NH_4NO_3	MS	0.01 bio + 1 nia + 1 pyr + 1 thia + 1 CaP + 100 myo	5.0 IBA + 4.0 BA + 0.3 GA_3	20 S	1.0 cyst, 200 glut	Chaix, 1981
		R	20% MS + full iron	MS	0.01 bio + 1 nia + 1 pyr + 1 thia + 1 CaP + 100 myo	5.0 IBA	20 S	1.0 cyst, 200 glut	Chaix, 1981
'F 12/1'	Ax	Ax	0.25–0.5 MS salts		?	1.12 BA	21 or 28 S	—	Dunstan, 1981
		R	0.25–0.5 MS salts		?	0.75 NAA	21 or 28 S	—	Dunstan, 1981
'F12/1'	St	Ax	MS	MS with 20 Fe-NaEDTA	0.4 thia + 100 myo	0.1 or 1.0 IBA + 1.0 BA + 0.1 GA	30 S	pH 5.2, 162 PG	Jones & Hop-good, 1979; Jones et al., 1977

SPECIES	EXPLANT SOURCE[a]	RESPONSE[b]	PRINCIPLE COMPONENTS OF MEDIUM: Salts						REFERENCE
			Macro	Micro	Vitamins[c]	Hormones[d] (mg/l)	Sugar[e]	Other[f]	
		R	MS	MS with 20 Fe-NaEDTA	0.4 thia + 100 myo	3.0 IBA	30 S	—	Jones & Hopgood, 1979; Jones et al., 1977
'F 12/1'	St	Ax, R	MS	—	100 myo + 5 gly + .4 thia + .5 nia + .5 pyr	1 μM BA	30 S	15% CW, 5 g/l PVP, pH 5.0, 2000 lux, continuous light	Nemeth, 1979
'F 12/1'	Ax	R	1/2 MS	MS	MS	?	30 S	—	Quoirin et al., 1977
—	St	Ax	MS with 1/2 NH_4NO_3	MS	0.01 bio + 1 CaP + 1 thia + 1 nia + 1 pyr + 100 myo	1.0 IBA + 1.0 BA + 0.1 GA_3	20 S	1 cyst, 200 glut	Riffaud & Cornu, 1981

Table 1. Cont.

SPECIES	EXPLANT SOURCE[a]	RE-SPONSE[b]	PRINCIPLE COMPONENTS OF MEDIUM: Salts: Macro	Salts: Micro	Vitamins[c]	Hormones[d] (mg/l)	Sugar[e]	Other[f]	REFERENCE
		R	20% MS	MS	0.01 bio + 1 CaP + 1 thia + 1 nia + 1 pyr + 100 myo	Quick dip in 5 mM IBA	20 S	16-hr photo-period low light (200 ft-c)	Riffaud & Cornu, 1981
'Early Rivers'	St	Ax	See *P. armeniaca* (Skirvin & Chu, 1977; Skirvin et al., 1980; Miller et al., 1982)						
P. canescens	Ax with roots	R	See original article		0.4 thia + 100 myo	2.0 IBA	20 S	—	Druart, 1980
		C	See original article		0.4 thia + 100 myo	1.0 BA + 0.1 GA_3	20 S	—	Druart, 1980
		Ax + Ad	See original article		0.4 thia + 100 myo	1.0 BA + 0.1 GA_3	20 S	—	Druart, 1980
P. cerasifera Ehrh. 'Newport'			See *P. insititia* (Cheng, 1979)					100 ft-c light (better than 200 ft-c)	

SPECIES	EXPLANT SOURCE[a]	RE-SPONSE[b]	PRINCIPLE COMPONENTS OF MEDIUM						REFERENCE
			Salts						
			Macro	Micro	Vitamins[c]	Hormones[d] (mg/l)	Sugar[e]	Other[f]	
—	St	Ax	MS	MS	0.1 paba + 0.4 thia + 0.5 nia + 0.5 pyr + 100 myo	0.01 IBA + 0.2 BA	30 S	—	Hammer-schlag, 1982a
		R	MS	MS	0.1 paba + 0.4 thia + 0.5 nia + 0.5 pyr + 100 myo	2.5–5.0 IAA	30 S	Dark	Hammer-schlag, 1982a
'Thundercloud'	St	Ax	MS	MS	MS	1.0 BA + 0.1 NAA	30 S	—	Zwagerman & Zilis, 1979
		Ax (sub-culture)	MS	MS	MS	1.0 BA	60 S	—	Zwagerman & Zilis, 1979
		R	MS	MS	MS	0.5 IBA	?	—	Zwagerman & Zilis, 1979
P. cerasus 'Apetala', gr 'Schaerbeek', gr 'Vise'			See *P. accolade* (Boxus and Quoirin, 1977)						

Table 1. Cont.

SPECIES	EXPLANT SOURCE[a]	RE-SPONSE[b]	PRINCIPLE COMPONENTS OF MEDIUM: Salts: Macro	Micro	Vitamins[c]	Hormones[d] (mg/l)	Sugar[e]	Other[f]	REFERENCE
P. cerasus	St	Ax	See *P. armeniaca* (Nekrosova, 1964)						
		R	WH	WH+ Heller (1953)	WH	0.1 IAA or 0.1 NAA	20 S	—	Nekrosova, 1964
'Shubinka'	St	Ax	MS	MS	0.5 thia + 0.5 nia + 0.5 pyr	0.1–0.5 BA	30 S	1.0 As	Popov et al., 1976
		R	Soak shoots for 18 hr in IBA (50 mg/l) then move to rooting medium below						Popov et al., 1976
		R	WH	WH	WH	—	20 S	—	Popov et al., 1976
'Montmorency'	St	Ax	See *P. armeniaca* (Skirvin & Chu, 1977; Skirvin et al., 1980; Miller et al., 1982)						
P. cistena	St	Ax	MS	MS	MS	5.0 µM BA	30 S	—	Lane, 1979
		R	MS	MS	MS	1.0 µM NAA	30 S	—	Lane, 1979
P. dawyckensis	—	Ax	See *P. accolade* (Boxus, 1975; Boxus & Quoirin, 1977)						
	—	R + Ax + Ad	See *P. canescens* (Druart, 1980)						
P. dawyckensis fastigiata			See *P. accolade* (Boxus & Quoirin, 1975)						
P. domestica									
'Brompton', 'St. Julien'			See *P. accolade* (Boxus & Quoirin, 1975)						
'Victoria'	St	Ax	See *P. avium* (Jones & Hopgood, 1979)						

SPECIES	EXPLANT SOURCE[a]	RE-SPONSE[b]	PRINCIPLE COMPONENTS OF MEDIUM: Salts: Macro	Salts: Micro	Vitamins[c]	Hormones[d] (mg/l)	Sugar[e]	Other[f]	REFERENCE
'Stanley'	Ax	R	MS	MS	200 myo + 2 gly + .05 bio + 2 thia + 1.0 choline chloride + 2.5 nia + 0.25 pyr + 1.0 paba + 0.25 fo + 0.5 CaP	3.0 IBA + 1.0 NAA + 0.1 GA_3 + 0.04 KIN 0.01 BA	20 S	6% agar, 1000 CH	Skirvin et al., 1982
P. hally jolivette			See *P. accolade* (Boxus & Quoirin, 1977)						
P. hillieri			See *P. accolade* (Boxus, 1975; Boxus & Quoirin, 1977)						
P. incisa x *serrula*		Ax, R	See *P. accolade* (Boxus, 1975; Boxus & Quoirin, 1977)						
		R, Ax, Ad	See *P. canescens* (Druart, 1980)						
P. insititia 'Pixy'	St	Ax	See original article		2.5 thia + 250 myo	5.0 μM BA + 0.5–5.0 μM IBA or 0.005–0.05 μM 2,4-D or 0.005–0.5 μM NAA	30 S	—	Cheng, 1979

Table 1. Cont.

SPECIES	EXPLANT SOURCE[a]	RE-SPONSE[b]	PRINCIPLE COMPONENTS OF MEDIUM						REFERENCE
			Salts						
			Macro	Micro	Vitamins[c]	Hormones[d] (mg/l)	Sugar[e]	Other[f]	
		R	See original article		2.5 thia + 250 myo	0.25–5.0 μM IBA		—	Cheng, 1979
'Pixy'	St	Ax	MS	MS with 20 Fe-NaEDTA	0.4 thia + 100 myo	0.1 IBA + 1.0 BA	30 S	162 PG, pH 5.2	Howard & Oehl, 1981
		R	MS	MS with 20 Fe-NaEDTA	0.4 thia + 100 myo	3.0 IBA	30 S	162 PG, pH 5.2	Howard & Oehl, 1981
'Pixy'			See *P. avium* (Jones, 1979; Jones & Hopgood, 1979; Negueroles & Jones, 1979)						
P. kursar			See *P. accolade* (Boxus & Quoirin, 1977)						
P. mariana									
'GF 8-1'	St	Ax	MS	MS	MS	0.35 IAA + 0.04 KIN	30 S	—	Monison & Dunez, 1971
'P 2736', 'GF 8-1'	St	Ax	See original article			BA (?)	30 S	—	Seirlis et al., 1979
		R	See original article			BA (0)	30 S	—	Seirlis et al., 1979
P. myrobalan									
'GF 31'			See *P. avium* (Nemeth, 1979)						
P. nipponica			See *P. accolade* (Boxus, 1975; Boxus & Quoirin, 1977)						
—	Ax	R	See *P. avium* (Quoirin et al., 1975)						

SPECIES	EXPLANT SOURCE[a]	RE-SPONSE[b]	PRINCIPLE COMPONENTS OF MEDIUM: Salts: Macro	Micro	Vitamins[c]	Hormones[d] (mg/l)	Sugar[e]	Other[f]	REFERENCE
P. okame			See *P. accolade* (Boxus & Quoirin, 1977)						
P. x *pandora*	Ax	R	See *P. accolade* (Boxus, 1975; Boxus & Quoirin, 1977)						
P. persica Batsch									
'Sunhigh', 'Compact Redhaven'	St	Ax	MS	MS	0.1 paba + 0.4 thia + 0.5 nia + 0.5 pyr + 100 myo	0.01 IBA + 0.2 BA	20 S	Liquid medium on wicks, 21-24 C	Hammerschlag, 1982b
'Nemagard', 'Redhaven', 'Madison', 'Harbelle', 'Harbrite'	St		See *P. armeniaca* (Skirvin & Chu, 1977; Skirvin et al., 1980; Miller et al., 1982)						
'Nemagard'	St	R	MS	MS	—	0.1 NAA	20 S	50 As	Miller et al., 1982
'GF 305'	Ax (for 4 days)	R (induction)	See original article			0.5 IAA or 0.5 NAA	40 S	—	Mosella & Macheix, 1979; Mosella et al., 1980a
	Move to darkness on fresh medium	R (initiation)	See original article			0.1 GA_3 + 0.5 IAA or 0.5 NAA	40 S	24 C dark for 3-4	Mosella & Macheix, 1979; Mosella

Table 1. Cont.

SPECIES	EXPLANT SOURCE[a]	RE-SPONSE[b]	PRINCIPLE COMPONENTS OF MEDIUM						REFERENCE
			Salts						
			Macro	Micro	Vitamins[c]	Hormones[d] (mg/l)	Sugar[e]	Other[f]	
								weeks, 10^{-3} M quer-citin or rutin	et al., 1980a
'GF 305'	St	Ax	See *P. avium* (Hopgood & Jones, 1979)						Negueroles & Jones, 1979
	Ax	R	1/2 Kester et al., 1977			3.0 IBA	30 S	162 PG	Negueroles & Jones, 1979
'Harbrite'	Ax	R	See *P. domestica* (Skirvin et al., 1982)						
P. persica x *P. amygdalus*	St	Ax							
'GF 557'			See *P. avium* (Nemeth, 1979)						
—	St	Ax	See *P. amygdalus* (Tabachnik & Kester, 1977)						
P. rhexii	St	Ax	See *P. accolade* (Boxus & Quoirin, 1977)						
P. salicina									
'Calita'	St	Ax	MS + 20 NaFe-EDTA	MS	—	0.1 IBA	30 S	—	Rosati et al., 1980
		Ax (after 6 weeks)	MS + 20 NaFe-EDTA	MS	100 myo + 0.4 thia	0.1 IBA	30 S	—	Rosati et al., 1980

SPECIES	EXPLANT SOURCE[a]	RE-SPONSE[b]	PRINCIPLE COMPONENTS OF MEDIUM						REFERENCE
			Salts						
			Macro	Micro	Vitamins[c]	Hormones[d] (mg/l)	Sugar[e]	Other[f]	
		R	MS + 20 NaFe-EDTA	MS	100 myo + 0.4 thia	0.1 GA_3 + 4.0 IBA	30 S	26 C	Rosati et al., 1980
P. serrulata 'Kanzan', 'Yedosakura'	St	Ax	See *P. accolade* (Boxus, 1974, 1975; Boxus & Quoirin, 1977)						
P. subhirtella autumnalis	St	Ax	See *P. accolade (Boxus, 1974, 1975; Boxus & Quoirin, 1977)*						
P. yedoensis moerheimii 'Shidare Yoshina'	St	Ax	See *P. accolade* (Boxus & Quoirin, 1977)						
—	Ax	R	See *P. avium* (Quoirin et al., 1977)						

[a]Explant source: A = anthers, Ax = subcultured axillary buds, C = callus, Em = embryo, Fr = fruit, L = leaf, R = roots, S = stem, Se = seed, St = shoot tip.

[b]Response: Ad = adventitious shoot formation, Ax = axillary bud development, C = callus, G = growth in a normal manner, R = roots.

[c]Vitamins: bio = biotin, CaP = calcium pantothenate, Fo = folic acid, gly = glycine, myo = myo-inositol, nia = nicotinic acid, paba = para-aminobenzoic acid, pyr = pyridoxine, ribo = riboflavin, thia = thiamine.

[d]Concentration of growth regulator in mg/l unless otherwise noted.

[e]Sugar: G = glucose, S = sucrose

[f]Other: As = ascorbic acid, CH = casein hydrolysate, CW = coconut water, cyst = cysteine·HCl, glut = glutamine, myo = myo-inositol, PG = phloroglucinol, PVP = polyvinylpyrrolidone.

shoot that can be rooted and grown to a whole plant. Theoretically, this production system is not as fast as the other two, but the method works well for many fruit crops and has been utilized widely for proliferation purposes. It is not unrealistic to expect a millionfold increase in planting stock in a single year by using such a propagation system.

In order to develop a propagation system for a particular stone fruit species or cultivar, one must first develop a culture medium that will supply the excised plant part with all the food, nutrients, vitamins, hormones, etc. that it will need.

Media and Media Components

INORGANIC COMPONENTS. Stone fruits have been grown on many different culture media. Some media are specific for callus cultures while others are designed for shoot proliferation and/or rooting. Most investigators working with fruit crops now utilize some modification of the Murashige and Skoog (MS; 1962) medium (Zimmerman, unpublished survey). In spite of this popularity, other media should not be ignored and researchers should strive for an optimal medium for each crop.

Among the various media that have been reported in the literature the most common ingredients include the 16 basic elements for plant culture as well as a few miscellaneous substances like aluminum and nickel. Some media are very simple, such as Knop's medium which contains only four components, while others are very complex (e.g., MS medium). Kester et al. (1977) reported that almond callus grew poorly on full strength MS medium, but grew better on Knop's medium (which has a very low salt strength). They found that when MS medium was diluted to one-third to one-half strength the callus grew even better. Caponetti et al. (1971) made similar observations when they found that black cherry callus cultures grew best on Wetmore and Rier (1963) medium (a low salt medium) and performed poorly on MS medium. It may be that one of the most important factors concerning media may be related to its overall salt concentration. When one examines various media, one can find that there are three basic types: high salt (MS), intermediate salt (Nitsch and Nitsch, 1969), and low salt (White, WH, 1963) media.

It has been demonstrated that root development is a function of hormones and salt concentration. Many rooting media are actually low salt media (e.g., WH). Quoirin et al. (1977) reported that half-strength MS gave good rooting for four species of *Prunus*. Feucht and Nachit (1978) found that root development was enhanced on half-strength MS macroelements. Riffaud and Cornu (1981) have rooted *P. avium* at 20% MS salt concentration; Dunstan (1981) uses one-quarter to one-half strength MS for rooting *P. avium*. Similar results have been reported by Lane (1979). Hammerschlag (1982a), on the other hand, found that *P. cerasifera* shoots taken from one-quarter strength medium sometimes were stunted and necrotic

Boxus, Quoirin and co-workers have had considerable success propagating many rosaceous plants (including numerous *Prunus* species and strawberry) on their basic culture medium. The most unique feature of their media was the very high level of microelements that they used in comparison to most tissue culture media; they used 100 times the recommended concentrations of Heller (1953). Quoirin and Lepoivre (1977) examined this phenomenon in greater detail and found that manganese was limiting the growth of *Prunus* shoot tips and the formation of axillary buds. The main advantage of the increased microelements had been to provide additional manganese. The "improved" medium that resulted from these investigations reflects this higher rate of manganese.

Another common problem associated with fruit tissue cultures is chlorosis or loss of green pigmentation in vitro. Researchers have assumed that the problem is due to iron deficiency and iron has been added to media in a number of different ways. In some cases the problem of chlorosis may persist despite additional iron. Seirlis et al. (1979) reported that they grew their *Prunus* cultures on medium supplemented with 4x iron levels for several subcultures with little effect. They suggested that the chlorosis could be related to high cytokinin in the medium.

ORGANIC COMPONENTS. **Carbohydrates.** Carbohydrates serve two principle functions in the tissue culture medium: (a) they provide an energy source to the tissues, and (b) they maintain an osmotic potential within the medium. There are a number of carbohydrates that have been used in fruit tissue culture, but the most popular and versatile of these has been sucrose.

The concentration of carbohydrates in most media is 20–30 g/l, but in some instances the level of carbohydrates is much higher. Zwagerman and Zilis (1979), for instance, found that axillary buds of 'Thundercloud' (an ornamental plum) proliferated well on a modified MS medium supplemented with 4.4 μM benzyladenine (BA) and 0.54 μM NAA, but when the sucrose level was raised to 0.175 M the young shoots no longer required NAA for proliferation, and the shoots developed the purple-red coloration characteristic of the cultivar.

Coffin et al. (1976) reported that sorbitol (D-glucitol) is often used with the Rosaceae. They also determined that callus of many rosaceous plants can grow well on medium supplemented with either sucrose or sorbitol. Some plants grew even better with sorbitol than sucrose (e.g., 'Reliance' peach).

Nitrogenous Compounds. Several *Prunus* species have been grown on media supplemented with amino acids or casein hydrolysate (CH), but it remains unclear whether these compounds are absolutely essential for growth.

Vitamins. Almost every medium that is used for plant tissue culture contains at least some vitamins. The most complex vitamin mix for the Rosaceae is that used by Skirvin and Chu (1977), referred to as the "Staba" vitamins (vitamin formulation provided by Staba). The mix contains 10 different vitamins. At the other extreme, the Linsmaier and Skoog (1965) medium has only thiamine·HCl in its formulation. In most cases, the optimal number of vitamins lies between these two extremes. It is quite likely that some vitamins inhibit growth; for instance, Miller et al. (1982) found that when Staba vitamins were eliminated from their cultures of 'Nemagard' peach rootstock, good rooting occurred. They eventually found that riboflavin was the specific vitamin that had inhibited rooting. Almost all researchers report that thiamine, nicotinamide (niacin), and pyridoxine are essential.

Agar and Liquid Medium. One of the most common organic complexes used in tissue culture is agar. Since agar is produced from red algae harvested from the ocean, each batch may be different and may therefore introduce a source of variation between experiments. For this reason many researchers use more purified forms of agar (e.g., "phytagar") and/or partially purify their agar prior to use.

The amount of agar used in the medium varies considerably. MS medium includes 10 g/l (1%), but many researchers have reduced agar concentration to 8 g/l or less. The osmotic potential of high agar concentration, and the concomitant reduction in nutrient and organic matter availability to the tissue is the common reason for reducing the agar concentration.

Many researchers have completely eliminated the use of agar for their cultures by using shake or roller drum cultures and/or by using the so-called Heller (1949) bridges made of filter papers and supported in liquid medium.

Lane (1979) reported that cultures of *P. cistena* did not root well when agar was present in the medium, so he rooted his cultures on Heller bridges. The cause for this inhibition is unknown, but Kohlenbach and Wernicke (1978) have reported that agar can be toxic to certain plant cultures.

Hammerschlag (1982b) reported that 'Sunhigh' and 'Compact Redhaven' peach scions proliferated at 20-30x when grown on liquid medium that was rotated at 7 rpm, while only 8x proliferation rates were achieved on agar medium. Unfortunately, when shoot tips were grown continuously on rotating cultures, they became water-soaked and brittle, therefore her cultures are now incubated on rollers for only 2-3 weeks or are grown on Heller bridges. She has suggested that part of the inhibition could have been due to agar toxicity. She also felt that the liquid medium facilitated increased absorption of hormones and nutrients through the entire shoot surface which facilitated breakage of apical dominance. She went on to explain that another advantage of the liquid medium was that one could identify cultures that contained slow-growing bacteria or fungi, since these cultures would develop a milky appearance. In contrast, Tabachnik and Kester (1977) reported that both rotated liquid cultures and paper wick cultures reduced the rate of almond and almond-peach shoot growth.

Hormones. An examination of the various media and hormone combinations that have been utilized for tissue culture of the stone fruits reveals some interesting facts. First, callus is ubiquitous in cultures and it will develop on almost any medium so long as both auxins and cytokinins are present (unless the callus is habituated). Second, shoot proliferation (both as adventitious buds as well as axillary buds) generally requires the presence of both auxins and cytokinins, but in subculture, a cytokinin alone may be sufficient. Finally, relatively high levels of auxins promote rooting and cytokinins frequently inhibit rooting.

Ghugale et al. (1971) studied the effects of several auxins on the rooting of mulberry (*Morus alba*) tissues. Kinetin (4.6 μM) was combined with each of four auxins, with the following results: (a) KIN + NAA, 20% rooting; (b) KIN + IAA, 50% rooting; (c) KIN + IBA, 70% rooting; and (d) KIN + IPA, 80% rooting.

Dunstan (1981) found that 'F 12/1' (*P. avium*) rootstocks would root readily on a medium with a high concentration of NAA (8 μM). However, these shoots callused and formed roots that were fused and thick. In addition these plants usually died in soil. On the other hand, they found that it was better to root their shoots in lower NAA concentration (4.0 μM) medium which had been relatively high in BA (4.96 μM) in the last transfer. Under these conditions rooting occurred within 4 days. When roots were 0.25-0.5 inch long, the shoot was transferred to soil. If the roots elongated more, before subculture, the survival rate in soil was decreased.

The hormones that are used to stimulate rooting are normally mixed in the culture medium, but some researchers have found value in exposing the young shoot to an auxin treatment and then moving it to an auxin-free medium. Popov et al. (1976) reported that soaking 'Shubinka' sour cherry shoots in a solution of IBA (0.24 mM) for 18 hr improved rootability. Riffaud and Cornu (1981) used either a fast dip in IBA (5.0 mM) or else they incorporated IBA into their medium (5.0 μM) and they obtained similar rooting success from *P. avium*.

In spite of the classical hormone relationships of Miller and Skoog (1953), Nemeth (1979) found that BA (a cytokinin) can stimulate rooting of *P. myrobalan* and *P. avium* when shoots are grown under relatively high light intensity (about 2000 lux). Incidentally, Nemeth also stated that *P. avium* shoots always developed brown lignified tissue at their base prior to root emergence.

It is generally agreed that callus formation is antagonistic to root and shoot proliferation. For this reason most fruit tissue culturalists strive to minimize callus formation. It is also well known that the use of callus cultures increases the likelihood of cellular variability (Skirvin, 1978) and, hence, to ensure clonal stability and maximize shoot and root proliferation, callus should be minimized.

Rooting Cofactors. In 1964 Nekrosova reported that a "leaf extract" from cherry promoted rooting in his cultures. Nekrosova had no idea what his unknown factors were. Since *Prunus* wood contains large quantities of phenolic substances Feucht and his colleagues hypothesized that some of these compounds might serve as rooting co-factors. In a series of experiments (Feucht, 1974, 1975; Feucht and Nachit, 1977) it was found that these compounds were primarily chlorogenic acids, and that there were quantitative differences in the concentration as a function of growth rate of parent plants (i.e., the concentration was low in dwarfed trees and high in vigorous types) (Feucht and

Forche, 1975). Eventually some of these substances were identified and added exogenously to tissue culture medium (Feucht and Nachit, 1977). The authors later found that the addition of rutin and/or dihydroxyquercitin could enhance callus growth for *Prunus* species from the *eucerasus* section (Feucht and Nachit, 1978). They hypothesized that the effect of the chlorogenic acids in vitro was to inhibit IAA oxidases that would normally degrade IAA; the undegraded auxins could then stimulate extra callus (Feucht and Johal, 1977).

Other researchers have made similar observations with other species. Mosella et al. (1980a) reported improved rooting with rutin or quercitin and improved plant quality and rootability with phloridizin (Mosella et al., 1979). Negueroles and Jones (1979) reported that 100% rooting of peach rootstocks occurred on media with phloroglucinol (a breakdown product of phloridizin) compared to 60% rooting without it. They also reported that phloroglucinol improved the average number of roots per culture (7.2 with and 1.7 without phloroglucinol). Chlorogenic acid is also reported to be useful for rooting *P. cerasifera* in the light (Hammerschlag, 1982). On the other hand, some researchers have reported absolutely no benefit of these phenolic substances and Jordan et al. (1980) has reported that chlorogenic acid actually inhibited rooting of his *P. avium*.

Explants for Tissue Culture

Tissue cultures can be established from any portion of a plant, but experience has shown that certain regions of a plant are more amenable to tissue culture conditions than others. Murashige (1977) suggests that this fact should be exploited by the researcher and different portions of the plant should be compared for suitability to culture.

TISSUE AGE. In stone fruits (as with other woody plants) the most responsive explant material is juvenile (Murashige, 1977). Callus derived from almond embryos, for instance, can differentiate to form roots, shoots, and intact plants (Mehra and Mehra, 1974), but callus derived from mature plants does not (Kester et al., 1977; Tabachnik and Kester, 1977).

In spite of the advantages of juvenile wood for tissue culture, juvenile tissue may have limited usefulness. The reason is quite simple—most of the stone fruits have been selected for mature characteristics such as growth habit, flowers, fruits, and foliage color, which cannot be fully assessed until the plant has passed its juvenile phase. As every plant propagator knows, once a plant becomes mature (fruitful), the ability to root may be diminished (Hartmann and Kester, 1975). The ability of the explant from a mature parent to grow well in tissue culture also is diminished.

This returns us to the problem of maturity and juvenility. Since almost any plant part of a cultivar is probably mature and therefore not well suited to tissue culture, the researcher should try to find and utilize other tissues that might be amenable to tissue cultures. Sommer et al. (1962) found that peach mesocarp tissue that contained a section of vascular tissue formed the best callus in vitro. Hedtrich (1977) was able to form callus from leaf lamina, midribs, and petioles, as well as internode segments and roots, but root differentiation only occurred on callus from leaf lamina. In addition he reported that roots which developed directy from leaf tissue were geotropic and had hairs while roots that developed from callus were nongeotropic and had no hairs. Nozeran and Bancilon (1972) suggest that proximal portions of roots and floral organs might be good choices for tissue culture since these regions are often relatively juvenile.

While surveying a prospective tree for explant material, one should note succulent growth, water sprouts, and root suckers (provided that these do not arise from the rootstock). If no juvenile wood is found, and other tissues do not prove suitable, it may be possible to stimulate the appearance of juvenile sprouts by severe pruning (Farmer and Besemann, 1974), cutting off the plant at its base (be wary of grafted plants), hormone treatments, or grafting. Whenever possible, however, tissue of various ages from many organs should be compared to find the most suitable explant.

Although mature tissues typically do not do as well as juvenile tissues in vitro, these differences in growth rate may be negated with time as the explants slowly (or rapidly) revert to juvenility in vitro. Such reversion has been noted by Nozeran and Bancilon (1972) for grape (*Vitis* spp.). They found that with successive subcultures, shoot tips became more juvenile and grew better. Boxus and Quoirin (1972) have reported similar observations for *Prunus*.

EXPLANT-HORMONE INTERACTIONS. Different organs can respond to the ingredients in the medium in various ways. Mehra and Mehra (1974) were the first individuals to achieve adventitious shoot formation from callus of *Prunus*. They germinated seeds of almond (*P. amygdalus*) in vitro and then used the various organs of the sterile seedling to establish cultures. The chemical requirements for differentiation and optimal callusing of the various plant parts differed. For instance, basal medium, NAA (11.0 µM), and CW (10%) were required for root and hypocotyl tissue to develop callus while leaf, stem, cotyledon, embryo, and shoot tips required basal medium, NAA (27.0 µM), and CH (1 g/l). Interestingly, after 2-3 subcultures all calli grew well on medium with 11.0 µM NAA and 10% CW. Subcultured callus of leaves, cotyledons, and embryos formed plants in vitro when a cytokinin was added to the medium. The yield of plants from these organs was very low (8, 5, and 2%, respectively). All regenerated plants were of 2n constitution, however. Extreme variability, based on medium composition, was noted for cell shape, size, and cellular inclusions.

Nitsch (1965) grew peach fruit tissue in vitro. He found that he could culture both mesocarp and endocarp tissue but that they grew differently as a function of the medium components. He reported that 1-naphthalene acid amide had a strong growth regulator effect on mesocarp tissue, while it only moderately affected endocarp tissue. On the other hand, KIN strongly affected endocarp while it hardly affected the mesocarp. Similar observations have been reported by Feucht (1974) who found that cortical parenchyma cells were more inhibited by ABA than cambial tissue. He also found that he could differentially establish callus lines from particular tissues by varying hormone concentrations: high concentrations of NAA (54.0 μM) stimulated cortex proliferation while low concentrations (5.4 μM) stimulated cambial cells.

PREPARING THE EXPLANT FOR CULTURE. Stone fruits of major economic interest are scion- or rootstock-types that have either been developed in breeding programs or located in the "wild." Since these plants have been growing in open conditions for extended periods of time, their organs (stems, buds, bud scales, etc.) are contaminated both externally and internally by various microorganisms. Most of the organisms that are encountered are of no particular importance to the plant in vivo, but result in contamination when cultures are initiated.

The usual method of tissue sterilization is to soak plant parts in a known concentration of sodium hypochlorite (usually supplied as diluted commercial bleach e.g., Clorox). A dilution of 10% v/v bleach is normally sufficient for surface-disinfecting the tissue, particularly when it is mixed with a surfactant (e.g., 0.1% Triton X-100, Tween 20). However, even careful use of bleach followed by good rinsing can produce damage: Fisher and Tsai (1978) reported that growth of their coconut cultures was inhibited by normal levels of bleach. Only when the bleach concentration was reduced to 0.5-1.5% was good growth restored. Tissue that is pubescent presents particular problems, and it is sometimes of benefit to use a vacuum system to ensure good disinfection. Since bleach is toxic to plant cells, it is necessary to wash the tissue 2-3 times with sterile, distilled water to ensure dilution of bleach.

Many researchers have found that a single bleach treatment is not sufficient to ensure contaminant-free cultures. In our laboratory, for instance, we have found that it is very difficult to obtain clean cultures of peach from field-grown plants during the summer. Our normal disinfestation procedures include a 10-min soak in 10% Clorox + 0.1% Triton X-100 followed by two 5-min rinses in sterile distilled water. This treatment works well with most plants, but peaches frequently have 100% contamination with this treatment. We have developed an improved method. First, peach tips are placed into 25% Clorox + 0.1% Triton X-100 for 10 min, followed by two 5-min water rinses. Then the tips are soaked in 70% ethanol for 5 min followed by two more 5-min water rinses. By this procedure we have found that the contamination rate can be reduced to 50% with only a 10% kill rate. Maroti and Levi (1977) reported that it was better to rinse with ethanol (45%) for 3 min followed by a 10-min bleach treatment (5-10%) and 3 rinses with sterile water.

Nekrosova (1964) reported high contamination rates with several fruit crops until optimal sterilization times were determined using 0.1% $HgCl_2$. His results are summarized for the following species: *Prunus armeniaca* (apricot), 2 min; *P. persica* (peach) shoot tips, 6 min, lateral buds, 20-25 min; *P. cerasus* (sour cherry), 6 min; *P. cerasus* var.

Besseyi, 6 min; *Malus sylvestris*, 8 min; *Citrus limon* (lemon) shoot tips, 4 min, lateral buds, 35-40 min; and *Ribes nigrum* (black currant), 18 min.

Jones et al. (1977) have developed a complicated sterilization procedure that is summarized in the following sections.

First Stage of Sterilization.

1. Tissue was dipped in 0.01-0.1% surfactant for 20 sec.
2. Tissue was soaked in 0.14% available Cl solution for 1 min.
3. Tissue was rinsed 3 times in sterile, distilled water.
4. Tissue was transferred to media with no hormones or vitamins for 24 hr.

Second Stage of Sterilization.

1. Tissue was removed from medium and dipped in 0.01-0.1% surfactant for 20 sec.
2. Tissue was soaked in 0.1% benomyl (fungicide) for 15 min.
3. Tissue was soaked in 0.42-0.50% available Cl for 30-40 min.
4. Tissue was rinsed 3 times in sterile distilled water.
5. Tissue was transferred to fresh culture medium.

Hammerschlag (1980, 1982b) has used peach (*P. persica*) in tissue culture to develop sensible methods to minimize fungal and bacterial contamination. She found that untreated control had 100% contamination, while the Skirvin and Chu (1977) method gave 70% contamination. After a series of experiments she finally devised the following method that can give as little as 3% contamination:

1. During the winter, dormant, but fully vernalized, scion wood is gathered from parent trees and placed in water in the laboratory.
2. When shoots begin to emerge, they are removed.
3. The outer leaves are removed from each shoot.
4. The shoot is then reduced in size to 0.5 from 1.0 cm.
5. The shoots are then soaked in 10% Clorox + 0.01% Tween 20 for 15-20 min.
6. The shoots are then placed into a 100 ppm penicillin-streptomycin solution (from Grand Island Biological Company, Grand Island, New York) for 5 min.
7. The shoots are rinsed 3 times in sterile distilled water.
8. The young shoots are then transferred to liquid medium on filter paper wicks for 1-2 weeks.
9. The shoots are then grown normally.

In addition to the surface disinfestation processes described earlier, woody plants can also be plagued by internal contaminants. According to James and Thurbon (1978), even after several weeks in tissue culture, 'M9' apple rootstocks can develop bacterial contamination that is seen as a halo of material oozing near the base of the young explant. Pathologists determined that these apple contaminants consisted of at least five bacterial types. Because of the devasting effect of contamination on established cultures, it is a regular practice at the East Malling Experiment Station (England) to screen all cultures on enriched bacterial medium at one month intervals. Under their system, cultures are transferred to bacterial medium and grown at 25 C for 3 days. Any cultures that show contamination are discarded. James and Thurbon reported that one culture survived 5 screens but it developed contamination on the sixth screen and over 25% of 300 cultures were contaminated. They also report that "screened" cultures proliferate at about 7x per month while "unscreened" shoots grow about 2-3x per month. Jones and Hopgood (1979) reported that they routinely screen their cultures of *P. insititia* and *P. avium* on enriched bacterial medium which consists of yeast extract (3 g/l), bacterial peptone (5 g/l), agar (12 g/l), and pH 6.8-7.0. Lane (1979) cultures his *P. cistena* shoot tips on a medium that is supplemented with CH to encourage the growth of bacteria and fungi. These cultures that survive this treatment are transferred to identical media without CH but with BA.

Hammerschlag (1982a) screens her *P. cerasifera* cultures for contamination while they are growing on liquid medium since contamination will appear as a milky color. Bacterial contamination in strawberry cultures has been successfully controlled using Difco brand antibiotic discs according to Skirvin and Larson (1978). Unfortunately, certain antibiotics also inhibit the growth of the strawberry cultures. In our laboratory we also have been able to reduce the percentage of bacterial contamination by using the commercial antibiotic preparation garamycin (Schering Corp, Kenilworth, New Jersey). Wittenbach and Bukovac (1980) drastically reduced fungal contamination in their cherry cultures by adding benomyl at 0.34 mM.

Cheng (1978) reported a slightly different procedure for sterilizing all of her fruit tree cultures. She routinely washes her explants with water containing Alconox, and then the explants are disinfected with 6-20% Clorox + 0.2% Alconox. The tissue is then placed onto a basal medium without hormones for one week to "condition" the cultures; whereupon they are transferred to normal media. Rosati et al. (1980) also use a conditioning medium that has no thiamine, hormones, or myo-inositol. Their cultures of *P. salicina* are on this medium for 6 weeks after which they are transferred to complete medium.

Differentiation

SHOOTS. The ability of woody plant callus to differentiate organs and whole plants is very limited (Durzan and Campbell, 1974). In spite of this fact there have been a few isolated reports of success. Mehra and Mehra (1974) recovered whole plants from embryo callus of almond. Seirlis et al. (1979) and Gayne et al. (1980) were able to regenerate plants from cherry (*P. cerasus*) anther callus and root callus, respectively. Druart (1980) has been able to stimulate the formation of plants on roots of several *Prunus* species. His system involves the following steps: (a) axillary buds are moved from proliferation medium to rooting medium, (b) roots are developed and allowed to grow until

2-4 cm long, (c) entire plant is lifted from the medium and returned to proliferation medium, (d) after 20-30 days callus is observed on the roots where they were injured during subculture, and (e) the callus is proliferated and then transferred to fresh culture regeneration medium to induce shoot formation.

The degree of proliferation that occurred by this system varied by species. Although the recovery of adventitious plants in vitro has not been reported for many species, it will undoubtedly become more important as other investigators develop modifications of Druart's system.

Shoot tip culture proliferation does not seem to be a particular problem for any of the *Prunus* species investigated to date. There are differences in proliferation rates among species and some cultivars certainly proliferate faster than others. Lane (1979) reports that cultures of *P. cistena* have a 2-month lag period before they begin to grow in vitro. Eventually his cultures proliferate at the rate of about 35x per 3-4 week cycle. He also reports that using this method he can proliferate *Spirea bumalda* at the amazing rate of 300-400x per 3-4 week cycle. Many researchers have reported problems with particular species and cultivars, but most species can be propagated in vitro.

ROOTS. Although it is a rather routine procedure to proliferate most types of stone fruit shoot tips in vitro, the development of roots on these shoots is not easy. Many of the *Prunus* species and cultivars are difficult to root in vivo and their ability to root does not necessarily increase in vitro. For these reasons, much early work has been done with seedlings and easy-to-root species. Successful work with difficult-to-root species has been more sparse. Feucht and Dausend (1976) studied *P. avium* (a difficult-to-root species) in tissue culture. They found that rooting of stem sections in vitro was proportional to the in vivo response. They also found that callus of both species rooted readily on appropriate media.

Cheng (1978) has reported that the use of 2,4-D in her regeneration medium predisposed the shoots of 'Pixy' and 'St. Julien X' plum rootstocks to root on medium without auxin. Seirlis et al. (1979) have reported that the removal of BA from their medium was enough to stimulate roots on various plum rootstocks.

Mosella and his colleagues (Mosella and Macheix, 1979; Mosella et al., 1980a) have developed a whole series of steps that are required in order to stimulate rooting of 'GF 305' peach:

1. Induction phase—cultures are grown in the dark at 24 C for 5 days on medium containing either IAA (2.9 μM) or NAA (2.7 μM).
2. Initiation phase—shoots are transferred to medium that contains either rutin or quercitin and are grown in dark for 3-4 more weeks.
3. Root elongation phase—cultures are transferred to medium with GA (0.3 μM), grown for 1 week in the dark, then grown 2-3 weeks under 16-hr-days at about 1000 lux.
4. Plants are moved to soil.

Many researchers have found that cytokinins, especially BA, will stimulate axillary bud development, but at high concentrations shoot elongation is suppressed (Chaix, 1981; Kester et al., 1977). It may be necessary, therefore, to increase shoot length prior to rooting since larger shoots frequently root better than small shoots. Popov et al. (1976) reported that shoots of 'Shubinka' sour cherry rooted best (about 60%) if they were longer than 5 mm while short shoots rooted at lower rates (10–20%). Quoirin et al. (1974) found that short shoots of *P. accolade* usually died on rooting medium while larger shoots rooted easily and could be easily transferred to the soil. Howard and Oehl (1981) have reported that both small and tall plants of *P. institia* rooted well in vitro, but only the taller plants survived and grew well in the field. They also reported that transplant size was largely a function of the length of time on rooting medium. They reported that cultures that had been on medium for 3, 6, or 9 weeks survived at rates of 50, 80, and 90%, respectively.

GRAFTING. It has been difficult for scientists to decide whether it is more sensible to produce certain woody crops on their own root system or to provide them to growers and researchers in the normal grafted manner. Since rooting remains a difficulty for many of the *Prunus* species and cultivars, grafting is a sensible alternative. There are a number of reasons for grafting: (a) grafted plants bypass the juvenile phase that is associated with nucellar and sexual seedlings; (b) almost all stone fruit cultivars are traditionally propagated by grafting, and when a cultivar is grafted onto a particular rootstock the grower can predict which pathogens he will encounter, the growth habit of the plant, and the hardiness associated with the rootstocks; and (c) shoot tips and meristems grown in tissue culture in conjunction with heat treatments can be useful for the production of virus-free plants.

The method for grafting in vitro seems to be quite simple and is based on the work of Navarro et al. (1975) for grafting *Citrus*. Seedlings that have either been germinated in vitro or grown in a clean greenhouse usually are used as rootstocks. Alskief and Gautheret (1977) described their method of obtaining rootstocks. They took seeds of 'GF 305' peach, disinfested them, soaked them in sterile water for 24 hr, and then moved them to a basal medium where they were stored at 3 C for 90 days to stratify the seeds. The seeds were then moved to the laboratory at 27 C in darkness where they germinated. About 8-10 days after the radicle had emerged, they were ready for grafting. Following grafting the new plant was moved to a medium that was high in sugar. In this case they germinated their seeds on medium with 0.15 M sucrose and the grafted shoots were grown on medium with 0.22 M sucrose. Shoots may be grafted in vitro (Negueroles and Jones, 1979) or in vivo (Mosella et al., 1980a,b; Negueroles and Jones, 1979).

The success rate for grafting is not always high and seems to be a function of experience, the condition and age of the stock and scion, environment, and time of the year (Mosella and Macheix, 1979; Mosella et al., 1980a; Poessel et al., 1980). Mosella et al. (1979) claimed that when their peach scions were pretreated on "Medium B," which contained phloridizin and BA or ZEA, healthy plants developed and these healthy scions grafted quite easily (these plants would also root easily). Another factor in the success of grafting may be related to graft incompatibilities. Martinez et al. (1979, 1981) have reported that incompatibilities between various peach and apricot and peach and myrobalan types will be visible quickly in vitro, whereas such incompatibility may not develop until years after the trees are grafted and grown in vivo. The phenomenon of graft incompatibility also has been explored by Fujii and Nito (1972).

MOVEMENT OF PLANTS TO SOIL. The literature contains many reports of rooted cultures that were lost upon transfer to soil. Since the young tissue culture-derived plant is grown in a sheltered environment at 100% relative humidity, the shock of transfer to unfavorable atmosphere (such as in the greenhouse or laboratory) results in immediate desiccation. Dunstan (1981) has reported that 85% of his cultures have survived transplanting by moving rooted plants to 2-inch pots that are placed under mist for 7-10 days. The mist treatment helps the plant harden and increases its chances for survival in soil. Many researchers pot their rooted cultures in sterile soil or vermiculite. The cultures are covered by a glass vessel, plastic bag, or plastic sheet to maintain humidity. At the end of 10-21 days, many plants can stand the natural atmospheric conditions.

In spite of survival, tissue-culture-propagated plants (self-rooted plants especially) may not be as vigorous as those conventionally propagated due to shock. Howard and Oehl (1981) showed that 'Pixy' plum rootstock micropropagules were quite variable. Two or more seasons after potting, many plants produced weak bush growth instead of strong single-stemmed plants which are most suitable for grafting. Due to these problems the authors developed a system whereby good survival and uniformity could be insured. Their system is outlined as follows:

1. Preplanting acclimatization. Shoots were rooted on Jones and Hopgood medium (1979) for 6 weeks. The closed test tubes were then moved to the greenhouse under normal light for 1 week (16-hr photoperiod).
2. Potting. 2-cm-long shoots were removed from the culture tubes. They were potted into soil that contained 7 parts peat:3 parts sand:1 part loam in 8-cm plots.
3. "Weaning." For 10 days the pots were held at high humidity under clear polyethylene which was gradually removed. The pots themselves were placed on sand benches at 20 C. The plants were grown using 16-hr days under 500 W tungsten bulbs.
4. "Growing on." The pots were moved to raised gravel-filled benches with 400 W mercury discharge lamps. The temperature was not allowed to go below 15 C.
5. GA treatments. GA (0.58 mM) + Tween 20 was sprayed onto the plants until runoff.

The authors found that by using these procedures good growth was ensured. Even plants that had previously been classified as bushy or dwarf now grew in normal fashion.

Jones and Hopgood (1979) have a simpler system for introducing their rootstock shoots to the outside environment. Young shoots are removed from proliferation medium and transferred to rooting medium. After extensive development (usually about 3 weeks), the culture tube and rooted plants are moved to a shaded greenhouse for 10 days. Then the plants are removed and planted in 8-cm pots with "compost." The potted plants are then grown in humid propagating boxes for 14 days after which they grow "normally" in the greenhouse.

The soil mixes used for initial potting vary with the researcher. Stone (1963) reported that carnation shoots grew well in a 1:1:1 mixture of peat:sand:loam covered with a layer of peat or sand. They claimed that this mix was superior to perlite or vermiculite mixes. Rosati et al. (1980) potted their Japanese plum plants in a 1:1:1 mix of peat:perlite:sand. Popov et al. (1976) reported that either a 1:1:1 soil:peat:sand or 3:1 peat:sand mix was satisfactory for 'Shubinka' sour cherry. Dunstan (1981) has used a soil mix that consists of 3:4:1 mix of peat:perlite:sand.

Dunstan (1981) claims to have transferred over 156,000 shoots of apple and peach to soil with an 85% success record. His guidelines for success are as follows: (a) while the plants are young and growing in flats in the greenhouse one should avoid watering the plants with too much pressure since the plants will tend to grow crooked, (b) the young plants are quite sensitive to damping off and they should be treated with fungicides as needed and the scientist should be careful with his sanitation practices, and (c) when plants are to be transplanted to the field in the spring one should harden the plants; when transplanting during the summer, it is a good idea to grow them under 50% shade for awhile before moving to the field to prevent burning of the foliage.

Environmental Factors

TIME OF YEAR. It has been fairly well documented that tissues taken from field-grown plants are not equally amenable to tissue culture conditions throughout the year. Monsion and Dunez (1971) determined that the optimum time to gather *P. mariana* shoot tips and buds for explants in France was between May 25 and August 14 to ensure shoot and root formation.

Skirvin (1981) reported that peach scions gathered in the very early spring and grown in vases in the laboratory represented a good source of explants that were consistently low in contamination. As the season progresses and outside temperatures increase, the rate of contamination showed a concomitant increase.

It would seem likely that on the basis of this information a good source of tissue culture explants would be stored scion wood that could remain refrigerated in ethylene-free rooms until needed. The dormant bud sticks could then be placed in water and allowed to grow. This system has been used in our laboratory and by Hammerschlag (1982b). However, even this system may have problems. Quoirin et al. (1975) have shown that *P. accolade* twigs stored at 4 C from December until April produced abnormal rosetted shoot tip cultures in vitro.

LIGHT. The role that environmental factors play in rooting of *Prunus* remains unclear. Mosella et al. (1980) feel that root induction is very complex and requires several distinct steps in order to yield a whole plant. For peach he has given this scheme: (a) Induction--cultures are grown in dark at 24 C with IAA (2.9 μM) or NAA (2.7 μM); (b) initiation--cultures are grown 3-4 weeks in darkness with rutin or quercitin; (c) elongation--the plants are moved to media with GA (0.29 μM) where they are left in darkness for 1 week after which they are moved to 16-hr days at 2000 lux for 2-3 more weeks; and (d) the rooted plants are then moved to soil. Feucht and Dausend (1976) found that darkness was important for rooting stem segments (without buds) of *P. avium* and *P. pseudocerasus*. Hammerschlag (1982a) reported that she could achieve 100% rooting of *P. cerasifera* axillary shoots when they were incubated on media with IAA with or without GA if they were given a 2-week period in the darkness. Percentage rooting in the light on half-strength MS medium with either IAA or IAA plus chlorogenic acid was 30 and 100%, respectively. It appears that the chlorogenic acid helped to protect the IAA from photodegradation (probably by inhibiting IAA oxidase enzymes). Hammerschlag discusses the significance of this observation in considerable detail in her paper.

Nemeth (1979) has reported the importance of light for the rooting of *P. avium* and *P. myrobalan*. In his studies he reported that the best rooting occurred on medium with BA and continuous high light intensities. For comparison, at 2000, 1000, and 800 lux, the percentage rooting observed was 46.7, 36.5, and 24.6, respectively. Jordan et al. (1980) have made similar observations for *P. avium* and they found that 95% rooting occurred in the light. Therefore, while no specific conclusions can be stated, it is obvious that the investigator should certainly consider comparing rootability under lighted and darkened conditions.

TEMPERATURE. The role of temperature in relation to *Prunus* tissue cultures has only been partially investigated. Hammerschlag (1981a) has reported that peach shoot tips survived better at 21-24 C than they did at 26-30 C. Riffaud and Cornu (1981) have reported that *P. avium* cultures rooted 100% under continuous 19-20 C temperatures with reduced light (200 lux), compared with 33 and 67% rooting under 2500 and 200 lux, respectively, with temperatures of 25 C days and 19 C nights. Hammerschlag (1981d) reported that *P. cerasifera* cultures rooted more quickly at 26 C than 21 C.

Chaix (1981) found that the *P. avium* cultures which had been stored at 2 C for up to 6 months proliferated at an improved rate. In fact the longer the cultures had been stored, the faster they proliferated. Howard and Oehl (1981) found that they could force *P. insititia* cultures to grow more uniformly in the field if they stored their rooted cultures at 0 C for up to 4 months. To prevent desiccation and chlorosis, the cultures were sealed prior to storage, covered with a polyethylene sheet to protect the cultures from dust, and a 100 W tungsten bulb was provided 25 cm above the racks.

Applications

GRAFT COMPATIBILITY IN VITRO. When compatible scions are grafted to rootstocks, the first step in the completion of the graft union is callus formation. If the union is strong and complete, the likelihood of a successful graft is high. The development of the graft union has been studied in vitro. Fujii and Nito (1972) studied the compatibility of callus derived from grape, peach, chestnut, persimmon, pear, and apple in all combinations. In this interesting study it was found that even members of different families could grow together on medium (e.g., persimmon, chestnut, apple, and grape). Some combinations would grow side-by-side without ever touching, and some combinations inhibited each other. The development of graft incompatibilities between peach and myrobalan and peach and apricot was discussed earlier (Martinez et al., 1979). A well known type of graft incompatibility is due to cyanide production by either the scion or the stock. Heuser (1972) has studied this phenomenon with three dwarfing rootstocks of *Prunus* by adding cyanide to his medium and comparing growth rates.

Feucht (1975) cultured stem sections of dwarf, semidwarf, and normal *Prunus* species. The callus fresh weights showed no relationship with the vigor of original trees. The growth potential, therefore, is quite similar in vigorously growing and dwarfing trees when cultured on the same hormone concentrations. It was found, however, that the beginning of cell division of the explants is in relationship with the growth vigor of the *Prunus* species: the more vigorous the growth, the shorter the lag phase is for cell division. The difference between vigorous *P. avium* and dwarf *P. fruticosa* was about 2 days.

DISEASE DEVELOPMENT AND RESISTANCE. Callus cultures have been used to study disease development in *Prunus*. Caporali and Weltzien (1974) infected peach callus cultures with *Taphrina deformans* (Berk.) Tul. in order to study its development. Hammerschlag (1981c) has developed a bioassay whereby peach mesophyll cells (not necessarily callus) are exposed to filtrates of virulent cultures of *Xanthomonas pruni* and valinmycin, stained with mergoganine 540, and examined by fluorescence to determine membrane potential. She found that the susceptible cultivar 'Sunhigh' showed pronounced membrane changes with virulent cultures while only slight changes occurred with avirulent cultures. When the cells were infected with *X. pelargonii* (which is nonpathogenic) only transient modifications were noted. On the other hand, the resistant variety 'Compact Redhaven' showed only slight changes even with virulent cultures.

Within fruit and nut cultivars, there exist certain propagation-related disorders that are listed in virus handbooks as "genetic disorders that resemble virus diseases" (Kester et al., 1977). These occur sporadically within cultivars and are not transmitted to healthy plants by grafting. One of these diseases, "noninfectious bud-failure" (BF) occurs in almond (*P. amygdalus*). The disease expresses itself as necrosis of the shoot buds in midsummer when plants are exposed to higher temperature. Since the disease appears among individuals within a clone, it seems that the disease is induced by high temperatures and once the induction has occurred, the condition is irreversible. Unfortunately, in actual field practice the BF condition may not appear until the trees are quite established. If scion wood could be screened at an early date for its BF state, then induced wood could be eliminated from propagation stock. Kester et al. (1977) were able to grow almond callus on half-strength MS. They found that callus from BF wood will grow well at 25 C, but it would be inhibited at higher temperatures (35 C). In contrast, callus from normal wood grew best at 35 C. In this way almond wood could be screened for BF.

DEVELOPMENTAL INVESTIGATIONS. Because the development of organs in vitro frequently proceeds at a slower rate than in vivo, tissue culture has been of value to scientists interested in detailed developmental studies. Wittenbach and Bukovac (1980) utilized in vitro techniques to culture cherry fruits at various developmental stages. They found that fruits cultured prior to "Stage II" of development did not grow while "Stage III" fruits increased in fresh weight and ripened as indexed by pigment formation and softening.

DEVELOPMENT OF VIRUS-FREE PLANTS. Virus diseases are a very serious problem for stone fruits (Hesse, 1975). As with other plant species, it should be possible to eliminate viruses through a combination of heat treatment and shoot tip culture followed by grafting onto disease-free stock. Surprisingly, very little has been written on this topic for *Prunus*. The general techniques and theory have been discussed by Fridlund (1980), but only Mosella et al. (1980) have actually tested the system. They infected some peach plants with 2 viruses (Plum pox virus and Necrotic ring spot virus), removed the shoot apices, and cultured them onto media for 12-14 days. The shoots were then micro-grafted onto seedling stocks and indexed for virus infection. Well over 50% of the cultures proved to be free of the inoculated virus.

In any case, as virus-free forms of stone fruits become available, they should be maintained in vitro in germplasm repositories to ensure their availability for the future (Fridlund, 1980).

ANTHER CULTURES. The use of anthers and pollen grains in an attempt to produce haploid plants has been investigated by several authors. Anthers have yielded haploid cells of *P. avium* (Legeida et al., 1976), apricot (Harn and Kim, 1972; Michellon et al., 1974), peach (Michellon et al., 1974), and cherry (Jordan, 1974, 1975; Jordan and Feucht, 1977). The ability of anther cultures and pollen grains to regenerate has been limited, but there have been reports of success. Jordan (1975) and Jordan and Feucht (1977) produced multicellular pollen grains that burst the exine from *P. X pandora* and *P. avium*, respectively; in addition, some of their anthers differentiated roots. The only plants that have been regenerated from anther cultures were those reported by Seirlis et al. (1979) for cherry. All of these plants had a diploid chromosome number (not haploid).

PROPAGATION. Almost all tissue cultures which have been established from *Prunus* have been aimed at developing a faster and more economical method of propagating particular cultivars and species. Pierik (1975) has reviewed the use of tissue culture for tree and shrub propagation and concluded that there are four principal reasons for using tissue culture to propagate crops: (a) when the existing methods of vegetative propagation in vivo are too slow or unprofitable, (b) when vegetative propagation in vivo is not possible, (c) when plant breeders require a rapid increase of a new plant type, or (d) when production of plant material that is free from pathogens (viruses, bacteria, fungi, nematodes, etc.) is necessary.

The use of tissue culture for propagation will become more important, according to Faust and Fogle (1980) who predict an increasing development of high- and ultra-high density orchards. Tree requirements for such orchards are high (up to 4000 trees per hectare), since trees in such orchards are planted closely, machine-pruned, and relatively short-lived. Tissue culture has the potential to provide inexpensive, high-quality trees.

According to Zimmerman (1979) and Hammerschlag (1981b), the very largest commercial fruit tissue culture laboratories are in Europe. One of these laboratories is in Cesena, Italy, where a group of fruit growers have organized a cooperative laboratory for propagating a number of fruit and vegetable crops. They have established cultures of several peach rootstocks that are grown in pint canning jars. Their culture room has facilities for about 20,000 of these jars under controlled atmosphere and lighting conditions. Subcultured peach shoots also are rooted in jars and then transferred to soil. The grower can purchase rooted plants fresh from the bottle for about 18¢ each. Peach stocks also can be hand potted in paper pots and transferred to a greenhouse (the Cesena facilities have about 10,000 sq. ft. of greenhouse space) where they are covered with a sheet of plastic. The plastic sheet is gradually raised over several days to harden the plants. The potted greenhouse grown plants are ready to be sold in about 6 weeks at about 30¢ each (1979 prices).

At the INRA Center for Forest Research at Orleans, France, researchers are using tissue culture to propagate desirable clones of cherry (*P. avium*) for forest purposes (Chaix, 1981; Riffaud and Cornu, 1981). They have developed several useful methods for propagating these plants both in vitro and in vivo.

EMBRYO CULTURES. One of the earliest uses for tissue culture was growing immature embryos to maturity in vitro. Tukey (1933) did the very first tissue culture work with stone fruits when he cultured immature aborting embryos of *P. avium* (sweet cherry) on a very simple medium of his own design. The use of his techniques have subsequently been exploited by many stone fruit breeders (Hesse, 1975). Most of the interest in embryo culture has been focused on the rescue of embryos that would normally abort prior to maturity. Some of the media constituents for *Prunus* embryo culture have been discussed by Abou-Zeid (1973). Ivanicka and Pretova (1980) reported that their embryo cultures grew and rooted on culture medium, but some of the plants had such weak root systems that they died when transferred to soil. To overcome this problem, the authors transferred the weaker shoots from embryo culture medium to shoot proliferation medium. Proliferated shoots were removed from the medium and then explanted onto rooting medium where good roots developed. In this manner the percentage of embryo survival was increased considerably.

CLONAL STABILITY IN VITRO. Variation has been an ubiquitous phenomenon associated with the tissue culture of plants (Skirvin, 1978). Variants associated with in vitro plant tissue cultures may be classified as follows: (a) physical and morphological changes in undifferentiated callus, (b) variability in organogenesis, (c) changes manifested in differentiated plantlets, and (d) chromosomal changes. [A very complete discussion of this area has been published by Skirvin (1977, 1978, 1981).]

The amount of variation that has occurred in *Prunus* tissue cultures has been minor to date, mainly because most researchers have confined their investigations to the use of shoot tip cultures which usually are quite stable in vitro. Mehra and Mehra (1974) reported many different types of cells growing in their suspension cultures; some were very elongated, some were highly vacuolated, others had a spiral growth habit, and some cells contained oil droplets. They also reported that only some of their cultures were organogenic. Chromosome counts among the callus cultures revealed 2n, 3n, and 4n cells that appeared after 2 or more subcultures. Legeida et al. (1976) found considerable variation for chromosome number among cultures of *P. avium*. Interestingly, there were differences between parental clones with respect to the percentages of cells of different ploidy level.

Of all of the stone fruit plants that have been regenerated, rooted, and grown in soil, there has been very little variation within a cultivar. Boxus and Quoirin (1977) found no variation among many *Prunus* species and cultivars planted over an 11-year period in Belgium. Jones and Hopgood (1979) reported that 23 of 70 'Pixy' (*P. insititia*) plum rootstocks grew as a rosette instead of the normal upright growth of the parent plant. Howard and Oehl (1981) investigated this phenomenon and found that the variance was probably not due to mutation but was due to lack of chilling. They found that either chilling the young plants or spraying them with GA overcame the rosetted condition.

ASEXUALLY PROPAGATED CULTIVARS. Many horticultural cultivars are propagated by cuttings rather than by seeds because the plants derived from seeds are not "true" or identical to the parental clone. For instance, if a person planted seeds from 'Red Delicious' apple, they would probably never find an offspring that was identical in all respects to the parent. However, with the normal cuttage and graftage procedures that are used to propagate apples (or any *Prunus*), an individual will rarely vary from its parent. Plants produced in tissue culture of many asexually propagated crops have been identical to the parental cultivar except for a small percentage of plants that are similar but not identical. The differences are usually minor, but they may benefit varietal improvement programs. Typical changes have been in leaf shape, fruit color, disease resistance, and growth habit (Skirvin, 1977).

PROTOCOLS FOR TISSUE CULTURE OF *PRUNUS*

Establishment of *Prunus* Cell Cultures

1. Gather dormant scion wood from the field during the middle of winter, bundle it, and take it to an ethylene-free cold storage area where it can be stored at about 0–2 C.
2. When you are ready to establish tissue cultures, remove some of the stored wood, cut off the end of the stick under water (to facilitate movement of water through the xylem), and leave the stick in water in the laboratory or greenhouse at room temperature. If the room is very dry, you may wish to cover the wood with a plastic bag to prevent desiccation.
3. Change the water at least every other day to prevent the growth of microorganisms. It may be necessary to cut the end of your stick at various times to prevent the xylem from becoming blocked by fungal mycelia.
4. When buds begin to break and leaves expand, then they can be removed and used as explant sources. These shoots will have minimal contamination.
5. To initiate callus cultures use young leaves, petioles, or stem tissue. For shoot tip cultures shorten the shoots to 0.5 from 1.0 cm using the techniques of Hammerschlag (1982b) (see previous discussion).
6. The tissue can be disinfested using the techniques of Hammerschlag (1982b), described earlier.
7. Explants should be transferred to an appropriate medium. Fungicides (benomyl) and/or antibiotics may be added to the medium to control contamination.

Maintenance

1. Cultures should be subcultured to fresh medium at 3- to 6-week intervals depending upon the requirements of your own cultures. We try to move our subculture to fresh medium before the cultures begin to turn pale.
2. At 2- to 3-month intervals, cultures should be screened for bacterial contaminants using the techniques of James and Thurbon (1978), summarized earlier. Those cultures with contaminants should be discarded.

Differentiation

1. *Shoot proliferation from axillary bud growth*—use the medium that is best suited to your species and cultivar (see Table 1).
2. *Shoot differentiation from callus*—the methods of Druart (1980), summarized earlier, are certainly the most interesting and should be attempted for other species.
3. *Root development on shoots*—this area is extremely variable and there is no generalized technique. Each researcher has his own system and the new investigator should try several systems until he finds one that works well for his species or variety.

Movement to Soil and Growing in the Greenhouse and Field

1. Rooted shoots should be moved to small pots and grown under mist. The best method seems to be system proposed by Dunstan (1981) and summarized earlier.
2. If regenerated plants are stunted or bushy instead of growing normally, it would be appropriate to give cultures a cold treatment prior to potting plants in soil. The method of Howard and Oehl (1981) is appropriate.

FUTURE PROSPECTS

The major objective of stone fruit tissue culture has been to develop efficient propagation procedures. These investigations have now led us to the point that some stone fruit crops are being commercially propagated in vitro. In addition to the practical thrust of these studies, this research has provided a number of tools that can be of value to the more basic scientist to aid in studies of gene stability, differentiation, and genotype-environment interaction. The stone fruit tissue cultures of the future will continue to be related to propagation, but it is suspected that they will approach the area from a different perspective, i.e., crop improvement. The following paragraphs describe some of the more significant research possibilities for the future.

First, now that there appears to be a system for producing adventitious plants from callus of certain stone fruits using the system of Druart (1980), scientists will want to utilize this adventitious system to study some of the complex interactions of media components, hormones, environmental factors, explant sources on enzyme induction, differentiation, and metabolism. Continued studies with this system may provide information that will help scientists utilize true callus cultures to produce intact plants from single cells.

Second, variability that is encountered in tissue culture is a definite problem to individuals interested in maintaining the genetic integrity of a clone, and future studies may develop methods to minimize or eliminate such variability.

Third, both natural and induced variability in tissue culture can be of value for improvement of particular cultivars through the screening

Table 2. Some *Prunus* Species Grown In Vitro[a]

SPECIES	COMMON NAME	DESCRIPTION
P. accolade	—	Hybrid ornamental cherry
P. amygdalus Batsch (= *P. dulcis*)	Almond	Edible seeds
P. armeniaca L.	Apricot	Edible fruits
P. austera	—	—
P. avium L.	Sweet cherry	Edible fruits, lumber
P. besseyi L.H. Bailey	Sand cherry	Edible fruits, rootstock
P. brigantica	—	—
P. canescens Bois.	—	Rootstock
P. cerasifera J.F. Ehrh.	Cherry plum, myrobalan plum	Ornamental plum, rootstock
P. cerasus L.	Sour cherry	Edible fruits
P. cistena	—	Ornamental flowering cherry
P. dawkyckensis Sealy	—	
P. domestica L.	European plum, prune plum	Edible fruits, dried fruits[b]
P. fruticosa Pall.	European dwarf cherry	—
P. hally jolivette	—	—
P. hillieri	—	—
P. incisa Thunb.	Cherry	—
P. insititia L.	Bullace, Damson plum	Edible fruits, rootstocks
P. kursar	—	—
P. laurocerasus L.	Cherry laurel	Ornamental
P. mahaleb L.	Mahaleb cherry	Ornamental
P. mariana	—	—
P. maritima Marsh	Beach plum	Ornamental, edible fruits
P. myrobalana (= *P. cerasifera*)	—	—
P. nigra Ait.	Canadian plum	—
P. nipponica Matsum	—	—
P. okame	—	—
P. padus L.	Bird cherry	—
P. X pandora (= *P. subhirtella* x *P. yedoensis*)	—	—
P. persica (L.) Batsch	Peach, nectarine	Edible fruits[c]
P. pseudocerasus[d]	—	—
P. rhexii	—	—
P. salicinia Lindl.	Japanese plum	Edible fruits
P. serotina J.F. Ehrh.	Black cherry, wild cherry	Lumber
P. serrula	—	—

Table 2. Cont.

SPECIES	COMMON NAME	DESCRIPTION
P. serrulata Lindl.	Japanese flowering cherry	Ornamental
P. subhirtella Miq.	Higan cherry	Ornamental
P. tenella Batsch	Dwarf Russian almond	Ornamental
P. tomentosa Thunb.	Nanking cherry, Hansen's dwarf cherry	Ornamental, edible fruits
P. yedoensis Matsum	Japanese flowering cherry	Ornamental

[a]Names are based on Bailey and Bailey (1976) and Cummins (1979a,b).

[b]When fruits of this species are dried, the region near the seed does not ferment; therefore this species is used for making prunes.

[c]The nectarine is a "fuzzless" type of peach.

[d]Not *P. pseudocerasus* Lindl., but belonging to section pseudocerasus.

of intraclonal variation, followed by selection for improved genotypes. The system of Hammerschlag (1981c), which showed an in vitro differential sensitivity of peach cells to *Xanthomonas prunui*, could be exploited to select single cells that are resistant to a particular toxin. The cells could then be regenerated into plants.

Fourth, the utilization of anther cultures to produce haploid plants will undoubtedly become more important in the future. Seirlis et al. (1979) have already been able to regenerate some plants from cherry anthers, and other authors have reported haploid cells derived from anther cultures.

Fifth, some stone fruit crops are sexually sterile and do not produce viable offspring due to their ploidy level (e.g., triploidy). It should be possible to isolate plants with doubled chromosome number. Such studies are now underway for the cherry rootstock 'Colt' according to Gayne et al. (1980).

Sixth, meristem cultures have now been utilized to produce virus-free plants. It should be possible to develop methods to store these shoots to facilitate the establishment of germplasm banks for the preservation of species and cultivars.

Finally, propagation of certain difficult-to-propagate or difficult-to-root plants will continue to be of major importance in tissue culture. In some crops, tissue culture will become the principal propagation method.

In conclusion, it is amazing how much progress has been made in the field of stone fruit tissue culture since Tukey's classic embryo culture investigations of 1933. The entire field has now expanded to the point that systems are available to minimize contamination, and propagate and root cultures, as well as to stimulate plants from callus. It is only a matter of time before any stone fruit crop can be grown in vitro. Once these procedures are developed, the scientist should be able to utilize these woody plant cultures to study differentiation, genetic stability, and plant-medium and plant-environment interaction.

REFERENCES

Abou Zeid, A. 1973. Many sided influences on the growth and morphogenesis in the in vitro culture of *Prunus* species embryo axes. I. Myo-inositol as a critical growth factor in the radicular meristem of the embryo axes of dormant cherry, peach and plum embryos. Angew. Bot. 47:227-239.

Alskief, J. and Gautheret, M.R. 1977. Sur le greffage in vitro d'apex sur des plantules décapitées, de Pêcher (*Prunus persica* Batsch.). C. R. Acad. Sci. Ser. D 284:2499-2502.

Bailey, L.H. and Bailey, E.Z. 1976. Hortus Third, Macmillan Pub., New York.

Boxus, P. 1971. La culture de méristèmes de *Prunus*. Note préliminaire relative à d'espèce *P. pandora*. Bull. Rech. Agron. Gembloux 6:3-5.

______ 1975. Meristem culture for the production of virus-free *Prunus*. Acta Hortic. 44:43-46.

______ 1978. The production of fruit and vegetable plants by in vitro culture—actual possibilities and perspectives. In: National Technical Information Service (K.W. Hughes, R. Henke, and M. Constantin, eds.) pp. 44-58. U.S. Department of Energy, Springfield, Virginia.

______ and Quoirin, M. 1972. Culture de méristèmes et d'anthères de diverses espèces de *Prunus*, Station des Cultures Fruitieres et Maraîcheres, Gembloux, C.R. de Récherereches, Années 1971-1972 (19).

______ and Quoirin, M. 1974. La culture de méristèmes apicaux de quelques espèces de *Prunus*. Bull. Soc. Roy. Bot. Belgium 107:91-100.

______ and Quoirin, M. 1977. Nursery behaviour of fruit trees propagated by "in vitro". Acta Hortic. 78:373-379.

Bychenkova, E.A. 1963. An investigation of callus formation in certain trees and shrubs by the method of tissue culture in vitro. Dokl. Akad. Nauk. SSSR Ser. Biol. 151:1077-1080.

Caponetti, J.D., Hall, G.C., and Farmer, R.E., Jr. 1971. In vitro growth of black cherry callus: Effects of medium, environment, and clone. Bot. Gaz. 132:313-318.

Caporali, L. and Weltzien, H.C. 1974. Structural anomalies and growth modifications caused by *Taphrina deformans* (Berk.) Tul. in tissues of *Prunus persica* L. cultivated in vitro. Rev. Gen. Bot. 81:85-133 (transl.).

Chaix, C. 1981. Techniques de production de plantes de merisier (*Prunus avium* L.) par culture in vitro boutrage herbage, boutrage de racines. Mémoire de 3eme année presenté pour l'obtention de diplôme d'ingenieur des techniques forestieres. I.N.R.A. Centre de Recherches Forestieres d'Orleans, France.

Cheng, T.Y. 1978. Clonal propagation of woody plant species through tissue culture techniques. Proc. Int. Plant Prop. Soc. 28:139-155.

Coffin, R., Taper, C.D., and Chong, C. 1976. Sorbitol and sucrose as carbon source for callus culture of some species of the Rosaceae. Can. J. Bot. 54:547-551.

Cummins, J.N. 1979a. Exotic rootstocks for cherries. Fruit Var. J. 33:74-84.

_____ 1979b. Interspecific hybrids as rootstocks for cherries. Fruit Var. J. 33:85-89.

Druart, P. 1980. Plantlet regeneration from root callus of different *Prunus* species. Sci. Hortic. 12:339-342.

Dunstan, D.I. 1981. Transplantation and post-transplantation of micropropagated tree-fruit rootstocks. Intern. Plant Prop. Soc. Combined Proc. 31:39-45.

Durzan, D.J. and Campbell, R.A. 1974. Prospects for the mass production of improved stock of forest trees by cell and tissue culture. Can. J. For. Res. 4:151-174.

Farmer, R.E., Jr. and Besemann, K.D. 1974. Rooting cuttings from physiologically mature black cherry. Silvae Genet. 23:99-134.

Faust, M. and Fogle, H.W. 1980. Potential changes in fruit growing resulting from use of tissue culture. In: Proceedings of the Conference on Nursery Production of Fruit Plants through Tissue Culture: Applications and Feasibility, April 21-22, 1980 (R.H. Zimmerman, ed.) pp. 102-106. USDA Science and Education Administration, Agric. Res. Series Results ARR-NE-11, Beltsville, Maryland.

Feucht, W. 1974. Effects of hormones on growth of *Prunus* tissue cultured in vitro. Proc. XIX Int. Hortic. Congr. 1/A:58 (Abstr.).

_____ 1975. Potential of cell division of stem explants from *Prunus* species with different growth vigor. Gartenbauwissenschaften 40:253-257 (transl.).

_____ and Dausend, B. 1976. Root induction in vitro of easy-to-root *Prunus pseudocerasus* and difficult-to-root *Prunus avium*. Sci. Hortic. 4:49-54.

_____ and Forche, E. 1975. Uber phenolische verbindungen in Prunusarten mit unterschiedlicher Wuchsigkeit. Angrew Bot. 49:15-24.

_____ and Johal, C.S. 1977. Effect of chlorogenic acids on the growth of excised young stem segments of *Prunus avium*. Acta Hortic. 78:109-114.

_____ and Nachit, M. 1977. Flavolans and growth-promoting catechins in young shoot tips of *Prunus* species and hybrids. Physiol. Plant. 40: 230-234.

_____ and Nachit, M. 1978. Flavanol glycosides of different species and hybrids from the *Prunus* section *eucerasus* and the growth promoting activity of quercetin derivatives. Sci. Hortic. 8:51-56.

Fisher, J.B. and Tsai, J.M. 1978. In vitro growth of embryos and callus of coconut palm. In Vitro 14:307-311.

Fridlund, P.R. 1980. Maintenance and distribution of virus-free fruit trees. In: Proceedings of the Conference on Nursery Production of Fruit Plants through Tissue Culture: Applications and Feasibility, April 21-22, 1980 (R.H. Zimmerman, ed.) pp. 86-92. USDA Science and Education Administration, Agric. Res. Series Results ARR-NE-11, Beltsville, Maryland.

Fujii, T. and Nito, N. 1972. Studies on the compatibility of grafting of fruit trees. I. Callus fusion between rootstock and scion. J. Jpn. Soc. Hortic. Sci. 41:1-10.

Gayne, J.A., Jones, O.P., Hopgood, M.E., and Watkins, R. 1980. Mutant production and regeneration from callus in vitro. Annual Report East Malling Research Station for 1980, p. 144.

Ghugale, D., Kulkarni, D., and Narasimhan, R. 1971. Effect of auxin and gibberellic acid on growth and differentiation of *Morus alba* and *Populus nigra* tissues in vitro. Indian J. Exp. Biol. 9:381-384.

Hammerschlag, F. 1980a. Peach micropropagation. In: Proceedings of the Conference on Nursery Production of Fruit Plants through Tissue Culture, April 21-22, 1980 (R.H. Zimmerman, ed.) pp. 48-52. USDA Science and Education Administration, Agric. Res. Series Results ARR-NE-11, Beltsville, Maryland.

______ 1980b. Peach micropropagation. In: Proceedings of the Conference on Nursery Production of Fruit Plants through Tissue Culture, April 21-22, 1980 (R.H. Zimmerman, ed.). USDA Science and Education Administration, Agric. Res. Series Results ARR-NE-11, Beltsville, Maryland.

______ 1981a. Factors affecting the culturing of peach shoots in vitro. HortScience 16:282 (Abstr.).

______ 1981b. Peach micropropagation, present and future. In: Proceedings 40th Annual National Peach Council Convention, pp. 139-143.

______ 1981c. Differential response of peach mesophyll cell membranes to culture filtrates of *Xanthomonas pruni* and *X. pelargonii*. Phytopathology 71:878 (Abstr.).

______ 1981d. In vitro propagation of myrobalan plum (*Prunus cerasifera*). HortScience 16:283 (Abstr.).

______ 1982a. Factors influencing in vitro multiplication and rooting of the plum rootstock myrobalan (*Prunus cerasifera* Ehrh.). J. Am. Soc. Hortic. Sci. 107:44-47.

______ 1982b. Factors affecting establishment and growth of peach shoots in vitro. HortScience 17:85-86.

Harn, L. and Kim, M.Z. 1972. Induction of callus from anthers of *Prunus armeniaca*. Korean J. Breed. 4:49-53.

Hartmann, H.T. and Hansen, C.J. 1958. Effect of season of collecting, indolebutyric acid and preplanting storage treatments on rooting of Marianna plum, peach, and quince hardwood cuttings. Proc. Am. Soc. Hortic. Sci. 71:51-66.

Hartmann, H.T. and Kester, D.E. 1975. Plant Propagation: Principles and Practice, 3rd ed. Prentice-Hall, Englewood Cliffs, New Jersey.

Hedtrich, C.M. 1977. Differentiation of cultivated leaf discs of *Prunus mahaleb*. Acta Hortic. 78:177-183.

Heller, R. 1949. Sur l'emploi de papier filtre sans cendres comme support pour les cultures de tissus végétaux. C. R. Soc. Biol. 143:335-337.

______ 1953. Recherches sur la nutrition minérale des tissues végétaux in vitro. Ann. Sci. Nat. Bot. 14:1-223.

Hesse, C.O. 1975. (The breeding of) peaches. In: Advances in Fruit Breeding (J. Janick and J.N. Moore, eds.) pp. 285-335. Purdue Univ. Press, West Lafayette, Indiana.

Heuser, C.W. 1972. Response of callus cultures of *Prunus persica*, *P. tomentosa*, and *P. besseyi* to cyanide. Can. J. Bot. 50:2149-2152.

Howard, B.H. and Oehl, V.H. 1981. Improved establishment of in vitro-propagated plum micropropagules following treatment with GA_3 or prior chilling. J. Hortic. Sci. 56:1-7.

Ivanicka, J. and Pretova, A. 1980. Embryo culture and micropropagation of cherries in vitro. Sci. Hortic. 12:77-82.

James, D.J. and Thurbon, I.J. 1978. Culture in vitro of M.9 apple: Bacterial contamination. Annual Report East Malling Research Station, pp. 179-180.

Janick, J. 1979. Horticulture's ancient roots. HortScience 14:299-313.

Jones, O.P. 1979. Propagation in vitro of apple trees and other woody fruit plants: Methods and applications. Sci. Hortic. 30:44-48.

______ and Hopgood, M.E. 1979. The successful propagation in vitro of two rootstocks of *Prunus*: The plum rootstock Pixy (*P. insititia*) and the cherry F 12/1 (*P. avium*). J. Hortic. Sci. 54:63-66.

______, Hopgood, M.E., and O'Farrel, D. 1977. Propagation in vitro of M.26 apple rootstocks. J. Hortic. Sci. 52:235-238.

Jordan, M. 1974. Multizellulare pollen bei *Prunus avium* nach in vitro Kultur. Z. Pflanzenzuecht. 71:358-363.

______ 1975. In vitro culture of anthers from *Prunus*, *Pyrus*, and *Ribes*. Planta Med. (suppl.) 1975:59-65.

______ and Feucht, W. 1977. Processes of differentiation and metabolism of phenolics in anthers of *Prunus avium* and *Prunus* x "Pandora" cultivated in vitro. Angew. Bot. 51:69-76.

______, Itturriaga, L., and Feucht, W. 1980. Inhibition of root formation in *Prunus avium* hypocotyls by chlorogenic acid in vitro. Gartenbauwissenschaft 45:15-17.

Kester, D.E., Tabachnik, L., and Negueroles, J. 1977. Use of micropropagation and tissue culture to investigate genetic disorders in almond cultivars. Acta Hortic. 78:95-101.

Kochba, J. and Spiegel-Roy, P. 1977. Cell and tissue culture for breeding and developmental studies of *Citrus*. HortScience 12:110-114.

______, Neumann, H., and Saad, S. 1978. Stimulation of embryogenesis in *Citrus* ovular callus by ABA, ethephon, CCC, and alar and its suppression by GA_3. Z. Pflanzenphysiol. 89:427-432.

Kohlenbach, H.W. and Wernicke, W. 1978. Investigations on the inhibitory effect of agar and the function of active carbon in anther culture. Z. Pflanzenphysiol. 86:463-472.

Lane, W.D. 1979. In vitro propagation of *Spirea bumalda* and *Prunus cistena* from shoot apices. Can. J. Plant Sci. 59:1025-1029.

Legeida, V.S., Levenko, B.A., Berezenko, N.P., Liferova, V.V., and Shchibrya, G.R. 1976. Production and study of callus tissue during culturing of Mazzard cherry and strawberry anthers. Cytol. Genet. 10:10-13.

Linsmaier, E.M. and Skoog, F. 1965. Organic growth factor requirements of tobacco tissue culture. Physiol. Plant. 18:100-127.

Maroti, M. and Levi, E. 1977. Hormonal regulation of the organization from meristem cultures. In: Use of Tissue Cultures in Plant Breeding. Proceedings of the International Symposium 6-11 Sept. 1976,

Olomouc, Czechoslovakia (F.J. Novak, ed.) pp. 337-354. Czechoslovak Academy of Sciences Institute of Experimental Botany, Prague.

Martinez, J., Hugard, J., and Jonard, R. 1979. The different grafting of shoot-tips realized in vitro between peach (*Prunus persica* Batsch), apricot (*Prunus armeniaca* L.), and myrobalan (*Prunus cesarifera* Ehrh.). C. R. Acad. Sci. Ser. D 288:759-762 (Transl.).

Martinez, J., Poessel, L., Hugard, J., and Hugard, R. 1981. L'utilisation du microgreffage in vitro pour l'étude des greffes incompatibles. C. R. Acad. Sci. Ser. D 292:961-964.

Mehra, A. and Mehra, P.N. 1974. Organogenesis and plantlet formation in vitro in almond. Bot. Gaz. 135:61-73.

Michellon, R., Hugard, J., and Jonard, R. 1974. Sur l'isolement de colonies tissulaires de Pêcher (*Prunus persica* Batsch, Cultivars Dixired et Nectared IV) et d'Amandier (*Prunus amygdalus* Stokes, cultivar Ai) a partir d'antherès cultivées in vitro. C. R. Acad. Sci. Ser. D 278:1719-1722.

Miller, C.A., Coston, D.C., Denny, E.G., and Romeo, M.E. 1982. In vitro propagation of 'Nemagard' peach rootstock. HortScience 17:194.

Miller, C.O. and Skoog, F. 1953. Chemical control of bud formation in tobacco stem segments. Am. J. Bot. 40:768-773.

Monsion, M. and Dunez, J. 1971. Young plants of *Prunus mariana* obtained from cuttings cultivated in vitro. C. R. Acad. Sci. Ser. D 272:1861-1864 (transl.).

Mosella, L., and Macheix, J.J. 1979. La microbouturage in vitro du pêcher (*Prunus persica* Batsch). Influence de certains composes phenoliques. C. R. Acad. Sci. Ser. D 289:567-570.

Mosella, L., Riedel, M., Jonard, R., and Signoret, P.A. 1979. Developpement in vitro d'apex de pêcher (*Prunus persica* Batsch). Possibilities d'applications. C. R. Acad. Sci. Ser. D 289:1335-1338.

Mosella, L., Macheix, J.J., and Jonard, R. 1980a. Les conditions du microbouturage in vitro du pêcher (*Prunus persica* Batsch): Influences combinées des substances of croissance et de divers composés phenoliques. Physiol. Veg. 18:597-608.

Mosella, L., Signoret, P.A., and Jonard, R. 1980b. Sur la mise au point de techniques de microgreffage d'apex en vue de l'élimination de deux types de particules virales chez le Pêcher (*Prunus persica* Batsch). C. R. Acad. Sci. Ser. D 290:287-290.

Mouras, A., Salesses, G., and Lutz, A. 1978. Protoplasts use in cytology: Improvement of recent method in order to identify *Nicotiana* and *Prunus* mitotic chromosomes. Caryologia 31:117-127 (Transl.).

Murashige, T. 1974. Plant propagation through tissue culture. Annu. Rev. Plant Physiol. 25:135-166.

______ 1977. Current status of plant cell and organ cultures. HortScience 12:127-130.

______ and Skoog, F. 1962. A revised medium for rapid growth and bioassay with tobacco tissue cultures. Physiol. Plant. 15:473-497.

Navarro, L, Roistacher, C.N., and Murashige, T. 1975. Improvement of shoot-tip grafting in vitro for virus-free *Citrus*. J. Am. Soc. Hortic. Sci. 100:471-479.

Negueroles, J. and Jones, O.P. 1979. Production in vitro of rootstock/scion combinations of *Prunus* cultivars. J. Hortic. Sci. 54:279-281.

Nekrosova, T.V. 1964. The culture of isolated buds of fruit trees. Sov. Plant Physiol. 11:107-113.

Nemeth, G. 1979. Benzyladenine-stimulated rooting in fruit-tree rootstocks cultured in vitro. Z. Pflanzenphysiol. 95:389-396.

Nitsch, J.P. 1965. Culture in vitro de tissus de fruits. III. Mesocarpe et endocarpe de pêche. Bull. Soc. Bot. France 112:22-25.

______ 1972. Haploid plants from pollen. Z. Pflanzenphysiol. 67:3-18.

______ and Nitsch, C. 1969. Haploid plants from pollen grains. Science 163:85-87.

Nozeran, R. and Bancilon, L. 1972. Les cultures in vitro en tant que techniques pour l'approche des problèmes posés par l'amélioration de plantes. Ann. Amelior. Plant. 22:167-185.

Pierik, R.L.M. 1975. Vegetative propagation of horticultural crops in vitro with special attention to shrubs and trees. Acta Hortic. 54:71-82.

Poessel, J.L., Martinez, J., Macheix, J.J., and Jonard, R. 1980. Variations saissonnières de l'aptitude au greffage in vitro d'apex de Pêcher (*Prunus persica* Batsch). Relations avec les teneurs en composés phenoliques endogènes et les activités peroxydasique et polyphenoloxydasique. Physiol. Veg. 18:665-675.

Popov, Y.G., Vysotskii, V.A., and Trushechkin, V.G. 1976. Cultivation of isolated stem apices of sour cherry. Sov. Plant Physiol. 23:435-440.

Quoirin, M. and Lepoivre, P. 1977. Improved media for in vitro culture of *Prunus* sp. Acta Hortic. 78:437-442.

Quoirin, M., Boxus, P., and Gaspar, T. 1974. Root initiation and isoperoxidases of stem tip cuttings from mature *Prunus* plants. Physiol. Veg. 12:165-174.

Quoirin, M., Gaspar, T., and Boxus, P. 1975. Changes in natural growth substances as related to survival and growth of the hybrid ornamental cherry *Prunus accolade* meristems growth in vitro. C. R. Acad. Sci. Ser. D 281:1309-1312 (transl.).

Quoirin, M., Lepoivre, P., and Boxus, P. 1977. Un premier bilan de 10 années de recherches sur les cultures de méristèmes et la multiplication "in vitro" de fruitiers ligneux. C. R. des recherches 1976-1977 de la Station des cultures fruitières et maraîchères, pp. 93-117. Centre Recherches Agronomiques de l'État B-5800 Gembloux (Belgium).

Riffaud, J.L. and Cornu, D. 1981. Use of in vitro cultures to propagate mazzard trees selected in stands. Agronomie 1:633-640 (transl.).

Robitaille, H.A. and Yu, K.S. 1980. Rapid multiplication of peach clones from sprouted nodal cuttings. HortScience 15:579-580.

Rosati, P., Marino, G., and Swierczewski, C. 1980. In vitro propagation of Japanese plum (*Prunus salicina* Lindl. cv. Calita). J. Am. Soc. Hortic. Sci. 105:126-129.

Salesses, G. and Mouras, A. 1977. The use of protoplasts for *Prunus* chromosome study. Ann. Amelior. Plant. 27:363-368 (transl.).

Seirlis, G., Mouras, A., and Salesses, G. 1979. In vitro culture of anthers and organ fragments of *Prunus*. Ann. Amelior. Plant. 29:145-161 (Transl.).

Skirvin, R.M. 1977. Fruit improvement through single-cell culture. Fruit Var. J. 31:82-85.

______ 1978. Natural and induced variation in tissue culture. Euphytica 27:241-266.

______ 1981. (The tissue culture of) fruit crops. In: Cloning Agricultural Plants Via In Vitro Techniques (B.V. Conger, ed.) pp. 51-139. Chemical Rubber Company Press, Boca Raton, Florida.

______ and Chu, M.C. 1977. Tissue culture may revolutionize the production of peach shoots. Ill. Res. 19:18-19.

______ and Larson, D. 1978. Reduction of bacterial contamination in field-grown strawberry shoot tip cultures. HortScience 13:349 (Abstr.)

______, Chu, M.C., and Rukan, H. 1980. Tissue culture of peach, sweet and sour cherry, and apricot shoot tips. Proc. Ill. State Hortic. Soc. 113:30-38.

______, Chu, M.C., and Rukan-Kerns, H. 1982. An improved medium for the in vitro rooting of 'Harbrite' peach. Fruit Var. J. 36:15-18.

Sommer, N.F., Bradley, M.V., and Creasay, M.T. 1962. Peach mesocarp explant enlargement and callus production in vitro. Science 136:264-265.

Stone, O.M. 1963. Factors affecting the growth of carnation shoot apices. Ann. Appl. Biol. 52:199-207.

Tabachnik, L. and Kester, D.E. 1977. Shoot culture for almond-peach hybrid clones in vitro. HortScience 12:545-547.

Tukey, H.B. 1933. Artificial culture of sweet cherry embryos. J. Hered. 24:7-12.

Wetmore, R.h. and Rier, J.P. 1963. Experimental induction of vascular tissues in callus of angiosperms. Am. J. Bot. 50:418-430.

White, P. 1963. The Cultivation of Animal and Plant Cells, 2nd ed. Jacques Cattell Press, Lancaster, Pennsylvania.

Wittenbach, V.A. and Bukovac, 1980. In vitro culture of sour cherry fruits. J. Am. Soc. Hortic. Sci. 105:277-279.

Zwagerman, D. and Zilis, M. 1979. Propagation of *Prunus cerasifera* 'Thundercloud' by tissue culture. HortScience 14:478 (Abstr.).

CHAPTER 17
Strawberry

P. Boxus, C. Damiano, and *E. Brasseur*

HISTORY OF THE CROP

Strawberry was known and cultivated by the Romans before the Christian era. In the Middle Ages it was widely used in Britain. However, only small fruits were available as these were derived from wild strawberry, *Fragaria vesca* L., which was the most prevalent throughout the Northern Hemisphere.

Modern strawberry plants with large fruits are the result of intercrosses of *F. chiloensis* Duch. and *F. virginiana* Duch. These two species were introduced into Europe during the seventeenth and eighteenth centuries. The region of Plougastel, near Brest in Bretagne, France, at that time was the European center of strawberry production, supplying Paris and London. Strawberry production was so important that in 1850, during the peak of the harvest, 20 sailing-ships and a steamboat were required in Brest to load all the freight (Wilhelm, 1974).

Modern strawberries, all octaploids, have been regrouped as *F. ananassa* Duch. The descendants of the original hybrids were very heterogeneous and produced cultivars with either one or two annual blossoms or perpetual blossoms. Strawberries are adapted to many different climates: moderate, mediterranean, subtropical, and even tropical if grown at high altitudes. Most cultivars are propagated vegetatively.

Strawberry production has increased consistently since 1945. However, further development will probably only proceed in less developed countries, where labor is inexpensive. About 75% of the production cost of strawberries is due to labor of which 50% is required for the harvest. This labor cost explains the current interest in mechanical harvesting. It is hoped that within a few years harvesting machines, now in prototype, will be available.

ECONOMIC IMPORTANCE

World production of strawberries is about 1.3 million tons. The main production centers are all located in the Northern Hemisphere. Europe produces 50% of the world's strawberries, and North America produces 30%. Most of the strawberry fields in North America are found in California. Culture is also extensive in Mexico. Asia accounts for 15% of world production, mainly in Japan. World production is summarized in Table 1. The most productive countries in descending order of importance are the United States, Japan, Italy, and Poland.

Table 1. World Strawberry Production (in 1000 tons)[a]

LOCATION	1972	1977
World	1100	1300
Asia	185	199
Israel	5	6.5
Japan	167	175
Turkey	12	17
Europe	545	700
Belgium	36	22
France	62	74
Germany	24	30
Italy	106	152
Netherlands	31	19
Poland	97	150
North America	348	403
Canada	12	20
Mexico	128	86
U.S.A.	208	295
Oceania	7.4	6.4
South America	6.1	8.1

[a]Data taken from Lousteau, 1979.

During the last 15 years, strawberry production has almost doubled from 750,000 tons in 1961 to 1,300,000 tons in 1977. This increase in productivity is due to increased cultivated area and increased yield. In the United States, for example, the yield has increased from 5.8 to 13.1 tons/ha and in Italy, the yield average has increased from 6.5 to 11.6 tons/ha. However, in some countries, such as Poland which produces mostly processing strawberries, the productivity is still poor with only 4-6 tons/ha. The evolution of novel culture methods and the release of superior varieties accounts for these spectacular increases in productivity.

The use of polythene tunnels has become standard in many regions. In France and Italy, 70-80% of strawberries are now grown under protection. Strawberries are grown as an annual crop. The use of cold-stored strawberry plants has gained acceptance, and plastic mulch is frequently utilized. The number of plants per ha has increased dramatically. In California, the average yield is 29 tons/ha, but in some fields with more than 100,000 plants/ha, the average is 80 tons. In Belgium, France, and Italy, the plantation densities are generally from 30,000 to 40,000 plants/ha.

Modern varieties are grown in two distinct climates. They are adapted to conditions of septentrional culture, i.e., short days in autumn and hard winter, or to meridional conditions, i.e., long days in autumn and moderate winters. In the first category are the cultivars Gorella and Redgauntlet. The cultivar Senga Sengana, also septentrional, is still the most important processing strawberry. In the meridional regions the cultivar Tioga is the most important. Two new varieties from California that are widely grown are Tuft, very easy to pick, and Douglas, which has high productivity. The recent creation of day-neutral cultivars will revolutionize present growing techniques.

Strawberry fruit for the processing industry is mainly frozen for use in a range of prepared beverages and desserts. However, the fresh market is still the principal market for the strawberry. In the United States, only 35% of production is allotted for processing. The European market for processing strawberries is almost exclusively supplied by Poland.

STRAWBERRY IMPROVEMENT

Use of Tissue Culture in Breeding Programs

Several tissue culture methods have been applied to strawberry. Rapid clonal multiplication has been widely used (Boxus, 1974). Tissue culture has also been used for preservation of germplasm (Damiano, 1979; Kartha et al., 1980; Mullin and Schlegel, 1976). In addition, efforts have been aimed at production of haplooctoploid and polyhaploid individuals using anther culture (Laneri and Damiano, 1980; Rosati et al., 1975). These efforts, unfortunately, haven't resulted in the chromosome number reduction expected.

Other tissue culture methods, including embryo, ovule, and ovary culture, could be applied for the development of embryos originating from interspecific and intergeneric crosses. The percentage of mature seeds in several crosses has been low and poor seed germination has impaired transfer of desirable characters between species (Scott and Lawrence, 1975).

Sanitary Improvement

As with all the vegetatively propagated plants, such as potatoes, sugarcane, and bulbplants, the strawberry is often infected by virus and mycoplasma diseases. These diseases result in significant drops of yield. Aerts (1974a) listed 54 virus and 8 mycoplasma diseases reported in strawberry plants. Included in this list is the incidence of a complex virus, the strawberry mild yellow edge virus. This virus disease causes a diminution of 20-80% of fruit production depending on the cultivar and a diminution of 50% of runner production on the cultivar Gorella in nursery fields (Aerts, 1974b, 1979). Mullin et al. (1974) noted an increase in productivity of about 15-24% in the cultivar Fresno when freed of mild yellow edge virus. McGrew and Scott (1964) also reported significant improvement in yield, fruit weight, fruit size, and runner production on seven cultivars when freed of latent A virus.

Soil fungi are often more damaging than viruses. At present, commercially applied fungicides are not very effective in reducing these very serious diseases. The "red core" due to *Phytophthora fragariae* or the "wilt diseases" provoked by *Phytophthora cactorum* or *Verticillium* sp. are devastating. Eradication in nursery fields is difficult, even impossible, because most of these diseases do not become evident until after floral induction. Molot et al. (1973) observed in nursery fields symptoms of variable intensity on mother plants, whereas contaminated daughter plants show a normal vegetative growth. These fungal infections are latent.

Other cases of latent infections are well known on the strawberry plant. These include insect parasites which only become evident in certain periods of the year. Tarsonems and nematodes, in particular *Aphelencoides* and *Ditylenchus*, can only be detected, even in the laboratory, in the beginning of the growing season before runners are formed. Currently, different nematocides are sold commercially, but their effectiveness is incomplete. As these products are dangerous to handle, they cannot be used on plants for human consumption. Control is primarily preventive for all of these parasites. It is important to utilize healthy plants in the propagation field, but this is not possible using traditional methods of multiplication. In open field and in screenhouses, it is difficult to control all the parameters involved in the diffusion of pathogens and pests. Therefore, meristem culture and micropropagation are increasingly being utilized for establishment of mother plants.

LITERATURE REVIEW

Meristem Culture

The first meristem cultures of strawberry were initiated in 1962. Belkengren and Miller (1962) eliminated latent A virus from *F. vesca*, and a virus from the garden strawberry (Miller and Belkengren, 1963), using this method. It is therefore not surprising that early tissue culture work in strawberry emphasized virus elimination.

VIRUS ELIMINATION. Posnette and Jha (1960) were successful in eliminating virus using mini-cuttings of plants cured by heat treatment. By propagating the small axillary buds that formed on crown disks sliced from heat-treated plants, some viruses were eliminated. Miller and Belkengren (1963) eliminated a complex of viruses by excision of runner tips that were 3-5 mm in length and transferred them to sterile culture conditions. McGrew (1965) demonstrated that virus eradication depended on the duration of heat-treatment and the dimension of the excised tip. McGrew performed this work on the cultivar Suwannee that was contaminated by the strawberry latent C virus. After 15 days at 35 C, runner tips 2 mm or more in length and grown on culture medium were still infected. However, runner tips from 2 to 6 mm long taken from plants maintained for 25 days under this temperature gave some virus-free plants.

Development of culture media specific for strawberry permitted use of very small explants, often smaller than 1 mm. These smaller explants resulted in more effective elimination of viruses. Some researchers still recommend heat treatment before the removal of the meristem (Hilton et al., 1970; Mullin et al., 1974; Oertel, 1981; Sobczykiewicz, 1979). Currently, a 100% cure rate is obtained if the explants are of sufficiently small size (Adams and Stickels, personal communication; Boxus, 1981; Nishi and Oosawa, 1973).

Recently, McGrew (1980) demonstrated a direct correlation between the size of the meristem and its survival rate in the absence of cytokinin; however, he reported an inverse correlation between the size of the meristem and the rate of virus elimination. While explants of 0.1 mm are not viable, 10% of explants 0.2 mm in length survive and 100% of them produce plants that are free of *Pallidosis* virus. Fifty percent of explants 0.5 mm in length survive but virus elimination is reduced to 95%. Explants of 1.0 mm result in 70% survival but have only 50% virus elimination. Hence, to maximize recovery of the largest number of virus-free plants, meristems 0.5-0.9 mm long are most often used. Using this method, McGrew reported elimination of crinkle, mottle, and veinbanding virus from strawberry plants. Sobczykiewicz (1979) eliminated mottle virus, but the rate of elimination was poor when both mottle virus and crinkle virus simultaneously infected the plant. Sobczykiewicz, as McGrew, did not use cytokinin in the culture medium. Meristem culture was also effective in eradicating heat-resistant strains of crinkle virus (Boxus, 1976). This combination is particularly effective. Moreover, in support of heat therapy, disease reappearance has not been observed even several years after this treatment (Belkengren and Miller, 1962; Boxus et al., 1977; Hilton et al., 1970; Mullin et al., 1974; Oertel, 1981). The major reports of meristem culture for disease eradication are summarized in Table 2. The elimination rate of different viruses, the size of the explants, the type of cytokinin used, and the duration of heat-treatment are listed.

Even if heat pretreatment is not essential for virus elimination, several authors emphasize the importance of the increase in growth rate due to increased temperature. Mullin et al. (1974), Sobczykiewicz (1979), and Vine (1968) report a faster growth of meristems excised from heat-treated plants vs. non-heat-treated plants. The proportion of surviving meristems obtained from heat-treated meristems was higher than that for unheated ones (Table 3). Oertel (1981) reported that heat-treated meristems have a highly improved multiplication rate and also cites a favorable effect on runner production when transferred into the nursery field (146% of untreated control runners).

Table 2. Factors Influencing Virus Elimination

REFERENCE	VIRUS DENOMINATION	EXPLANT SIZE (mm)	CYTO-KININ	HEAT TREATMENT	VIRUS ELIMI-NATION (%)
Adams & Stickels, unpublished	Mottle, mild yellow edge, mottle + crinkle, vein chlorosis	?	BA	None	100 (55/55[a])
Belkengren & Miller, 1962	Latent A	0.5–1.0	CW	38 C, 3–5 days	50 (6/12)
	Latent A		CW	None	100 (1/1)
Boxus, 1981	Crinkle, mild yellow edge, mottle	0.1–0.3	BA	None	100 (144/144)
Hilton et al., 1970	Mild yellow edge	0.5	KIN	None	0
	Mild yellow edge	0.5	KIN	6 weeks	77
	Mild yellow edge	0.7–1.2	KIN	6 weeks	50
McGrew, 1980	Crinkle, mottle, pallidosis, vein banding	1.0	None	None	Good
Miller & Belkengren, 1963	Crinkle	0.5–1.0	CW	None	56 (4/7)
	Vein banding	0.5–1.0	CW	None	20 (2/10)
	Yellows complex	0.5–1.0	CW	None	60 (3/5)
	Yellows complex	0.5–1.0	CW	38 C, 120 hr	33 (1/3)
	Yellows complex	3.0–5.0	CW	None	20 (3/15)
	Yellows complex	3.0–5.0	CW	38 C, 19 hr	100 (1/1)
Mullin et al., 1974	Mild yellow edge	0.3–0.8	KIN	None	25 (2/8)
	Mild yellow edge	0.3–0.8	KIN	36 C, 6 weeks	82 (28/34)
	Mottle	0.3–0.8	KIN	None	Good
	Pallidosis	0.3–0.8	KIN	None	Good
Nishi & Oosawa, 1973	Unspecified	0.2–0.3	BA	None	100
	Unspecified	1.0	BA	None	50
Oertel, 1981	Crinkle	?	BA	10 weeks	Very good
	Mottle	?	BA	10 weeks	Very good

REFERENCE	VIRUS DENOMINATION	EXPLANT SIZE (mm)	CYTO-KININ	HEAT TREATMENT	VIRUS ELIMINATION (%)
Sobczykiewicz, 1979	Latent A	0.4-0.8	None	None	17 (12/69)
	Latent A	0.4-0.8	None	38 C, 5 weeks	71 (39/55)
	Mottle	0.4-0.8	None	None	35 (23/66)
	Mottle	0.4-0.8	None	38 C, 5 weeks	100 (95/95)
	Mottle + crinkle	0.4-0.8	None	None	0 (0/30)
	Mottle + crinkle	0.4-0.8	None	38 C, 5 weeks	17 (3/18)
Vine, 1968	Mild yellow edge	0.8	CW	None	90 (18/20)
	Crinkle	0.8	CW	None	100
	Vein chlorosis	0.8	CW	None	100

[a]Numerator = number of virus-free plants; denominator = number of tested plants.

Table 3. Percentage of Rooted Plantlets Obtained from Meristems Excised on Heat-Treated vs. Nonheat-Treated Mother Plants

REFERENCE	HEAT-TREATED	NONHEAT-TREATED
Mullin et al., 1974	22%	5%
Vine, 1968	50%	28%
Sobczykiewicz, 1979	45%	31%

ELIMINATION OF OTHER PARASITES. Nematodes and tarsonems are external parasites. Both can be eliminated by disinfection of the initial material or by the removal of the external tissue when meristems are excised. The soil fungi, *Phytophthora* or *Verticillium*, are found inside the plant tissues. Molot et al. (1973) stated that *Phytophthora* is located in the young runner tips, most likely in the cells just under the apex. It is necessary to use explants smaller than 0.5 mm to eliminate these fungi. Bacterial colonies are often recovered near the apex along with sporangia or mycelium of *Phytophthora*, resulting in some tissue culture contamination.

CHOICE OF EXPLANTS, DISINFECTION AND TRANSFER. The specific meristem that is used from the mother plant does not influence plant regenerative capability. However, excision is easier from young runner tips than from well developed plants. When meristems are excised from dormant plants or from cold stored plants, a high percentage of the meristems have internal infections (Boxus et al., 1977; Van Hoof, 1974). McGrew (1980) points out that about 20% of explants are contaminated by bacteria. He also notes a variation in the contamination rate associated with the cultivar and the season. Boxus (1981) reported 7% infection based on culture of 3000 meristems.

Disinfection of runner tips is accomplished by submersion for 5-20 min in sodium hypochlorite solution (0.1-5%). Some researchers use 5% calcium hypochlorite. Sometimes, before hypochlorite treatment, a brief dipping in 70-100% ethanol is necessary. Damiano (1980) reports the use of 0.5% merthiolate for sterilization. The use of Difco antibiotic disks has also been reported to result in control of bacterial contamination (Skirvin, 1981).

Dissection of meristems is made under a low-power microscope with fine watchmaker tools and a razor blade. McGrew (1980) describes a method that permits him to easily transfer 20 meristems/hr. He approaches the meristematic area from below by holding the distal end of the runner plantlet and makes thin, transverse slices until a cone including the apical dome and one or two embryonic leaves can be teased out. This permits some control of the size of the explant that is removed.

CHOICE OF CULTURE MEDIA. Knop (1865) or Murashige and Skoog (1962) medium is most often used for strawberry meristems. Van Hoof (1974) considers the mineral solution to be very important. He obtains 82% success with his medium, 10% on WH (White, 1963), 31.5% on that of Vine Medium A, and 78% on the Vine Medium B (Table 4). The presence of BA in the culture medium is essential for success (cf. Table 5).

Varietal differences can be critical in the success rate (Boxus, 1981; Mullin et al., 1974). None of the 253 cultivars cultured at Gembloux has failed to grow in culture. The cultivar Surprise des Halles always gives 85–100% success rate, while the cv. Gorella is the most difficult to culture (15–20% success). There is no difference for European versus Californian cultivars, everbearing versus seasonal, or diploids versus octaploids.

CULTURE CONDITIONS. The first researchers (e.g., Belkengren and Miller, 1962) incubated their cultures in darkness for 8 days, then transferred them to the light. When plantlets reached 1 cm, they were transferred to media with CW. McGrew (1980) and Sobczykiewicz (1979) do not add cytokinin to the culture medium and continue to use the same medium until roots regenerate. According to Sobczykiewicz (1979) 2 or 3 months are needed to obtain rooted plantlets that are 3-5 leaves in size. Almost all the other researchers transfer regenerating shoots onto a rooting medium after 4 weeks in culture. The rooting medium has no CW or BA. However, Adams (1972) transfers onto fresh medium containing 0.44 μM BA for rooting. Also, Boxus et al. (1977) noticed chlorosis followed by a decline in growth rate for meristems maintained over a month in the initial medium. Van Hoof (1974) obtains better rooting by eliminating both BA and adenine.

Duration, intensity, and quality of light are parameters that have not yet been investigated in detail. Sobczykiewicz (1979) and Van Hoof (1974) use 3000 lux, Mullin et al. (1974) use cool white fluorescent lamps with 2000 lux, and Boxus et al. (1977) consider that 1000–1500 lux is sufficient. Belkengren and Miller (1962) maintain cultures under continuous light. Daylength has been varied in different laboratories from 18 hr for Mullin et al. (1974) and 12 hr for Van Hoof (1974) to 16 hr for other workers. The cultivar Ostara grows better and gives a greater average number of plantlets per jar under long daylight photoperiods (Standardi et al., 1978).

The temperature of the growth chamber also influences results. Adams (1972) succeeded at 25 C, but not at 19 C. Boxus et al. (1977) got better results at 24 C while Mullin et al. (1974) maintain 27 C. On the other hand, Van Hoof (1974) uses 20 C and Navatel (1979) and Sobczykiewicz (1979) maintained cultures at 23–25 C. Therefore, the temperature range useful for culture of strawberry meristems appears to be 20–25 C.

LONG TERM STORAGE. Several long term in vitro storage techniques have also been developed. Recently, Kartha et al. (1979) ob-

Table 4. Composition of Media for Meristem Culture and Micropropagation

REFERENCE	PURPOSE	NUTRIENTS Macro	NUTRIENTS Micro	HORMONES (μM)	SUGAR (M)	AGAR (g/l)	pH
Adams, 1972	Meristem culture	MS	MS	BA 0.44 IBA 4.92	Glucose	0	5.2
Belkengren & Miller, 1962	Meristem culture "A"	White	White	CW 10% IBA 2.46	Sucrose .058	7.0	Not specified
Belkengren & Miller, 1962	Meristem culture "B"	White	White	None	Sucrose .058	7.0	Not specified
Boxus, 1974	Meristem culture	Knop	MS	BA 0.44 IBA 4.92 GA 0.29	Glucose 0.22	4.5–7.0	5.6–5.8
Boxus, 1974	Multiplication	Knop	MS	BA 4.4 IBA 4.92 GA 0.29	Glucose 0.22	4.5–7.0	5.6–5.8
Boxus, 1974	Rooting	Knop	MS	IBA 4.92	Glucose 0.22	4.5–7.0	5.6–5.8
McGrew, 1980	Meristem culture	Knop[a]	b	None	Glucose 0.17	8.0	Not specified
Mullin et al., 1974	Meristem culture "A"	Knop	Bertholet	IAA 5.71	Glucose 0.17	0	5.7–5.8
Mullin et al., 1974	Meristem culture "B"	Knop	Bertholet	KIN 0.46 IAA 14.27	Glucose 0.17	0	5.7–5.8
Navatel, 1979	Meristem culture	MS	Heller	BA 0.26 IAA 5.71 GA 0.29	Glucose 0.17	7	5.6
Navatel, 1979	Multiplication	Knop	MS	BA 4.4 IBA 4.92 GA 0.29	Glucose 0.17	7	5.6

Oertel, 1981	Meristem culture	Knop	MS	BA 0.44 IBA 4.92 GA 0.29	Glucose 0.17	7	5.6
Sobczykiewicz, 1979	Meristem culture	Knop	MS	IBA 4.92	Glucose 0.19	7	5.6
Van Hoof, 1974	Meristem culture "A"	[c]	Heller	BA 0.88 IBA 4.92 GA 2.89	Glucose 0.11	8	Not specified
Van Hoof, 1974	Meristem culture	[c]	Heller	BA 0.09 IBA 4.92 GA 0.29	Glucose 0.11 Sucrose 0.02	8	Not specified
Vine, 1968	Meristem culture first method "A"	MS	MS	IBA 4.92 CW 10%	Sucrose .058	0	5.6–5.8
Vine, 1968	Meristem culture first method "B"	Knop	MS	None	Glucose 0.22	0	5.6–5.8
Vine, 1968	Meristem culture second method "A"	White	White	CW 10%	Sucrose .058	0	5.6–5.8
Vine, 1968	Meristem culture second method "B"	White	White	None	Sucrose .058	0	5.6–5.8

[a]Modified Knop (in mM) $Ca(NO_3)_2 \cdot 4H_2O$ = 2.12; $MgSO_4 \cdot 7H_2O$ = 0.5; KH_2PO_4 = 0.92; $(NH_4)_2SO_4$ = 0.95; KCl = 1.67.

[b]Micronutrients McGrew (1980) (in μM) $MnSO_4 \cdot H_2O$ = 473.6; KI = 361.5; $CoCl_2 \cdot 6H_2O$ = 21; $Na_2MoO_4 \cdot H_2O$ + 10.25.

[c]Macronutrients Van Hoof (1980) (in mM) $MgSO_4 \cdot 7H_2O$ = 0.60; Na_2SO_4 = 3.17; KCl = 4.08; $(NH_4)_2SO_4$ = 0.59; $Ca(NO_3)_2$ = 3.41; KNO_3 = 3.68; NaH_2PO_4 = 4.16; $NaNO_3$ = 2.82; NH_4NO_3 = 2.06; $CaCl_2$ = 1.16.

Table 5. Influence of Cytokinins on Meristem Tip Culture

CYTOKININ (μM)	NUMBER EXCISED MERISTEMS	PERCENTAGE OF PLANTLETS	REFERENCE
Benzyladenine			
0.26	—	30-95	Navatel, 1979
0.44	37	72	Adams, 1972
	2928	73	Boxus, 1981
0.88	210	82	Van Hoof, 1974
Kinetin, 0.46[a]	955	12	Mullin et al., 1974
Coconut milk, 10%	46	28	Belkengren & Miller, 1962
	260	35	Vine, 1968
None	1111	34	Sobczykiewicz, 1979

[a]Kinetin is used only after transfer to the second medium, four weeks later.

tained initial results in cryopreservation of strawberry plant meristems. Meristems were removed from plantlets precultured on medium supplemented with 5% dimethylsulfoxide as a cryoprotectant. Ninety-five percent success was obtained when meristems were stored in liquid nitrogen for a week. Samples should be frozen at a cooling velocity of 0.84 degrees/min. More than 55% of the meristems that are stored in liquid nitrogen for 8 weeks regenerate plantlets. Sakai et al. (1978) obtained a whole plant from an excised runner apex frozen to -196 C.

Other techniques requiring less sophisticated equipment have been perfected. Mullin and Schlegel (1976) stored plantlets of 50 cultivars for 6 years on filter paper bridges in small glass test tubes containing 2.5 ml of medium that were sealed with parafilm. Cultures were stored in darkness at 4 C. During the storage, one or two drops of sterile medium was added every 3 months. Addition of a larger quantity of medium killed the plantlets. McGrew (1980) uses a similar method, but the tubes are held in an illuminated incubator at 4 C. Boxus (1974, 1975) maintains plantlets on the basic medium in a culture chamber with very low light intensity. The plantlets are transferred onto a fresh medium after 12-18 months.

TRANSFER TO SOIL. Generally, the transfer of strawberry plantlets to soil is very easy. Success rates are about 95-100% (Boxus, 1979; Damiano, 1977; Navatel, 1979; Sobczykiewicz, 1979). After a rapid washing with tap water to eliminate agar, the plants (Fig. 1a) are set in a peat mixture and are placed under a clear plastic film for 2

Figure 1. Transfer of rooted strawberry plants to soil. (a) Plantlets being removed from a culture jar prior to a rapid wash in tap water. (b) Plants set in a peat mixture under clear plastic film. (c) Plants transferred to the field after 5 or 6 weeks growth in the peat mixture.

weeks to maintain sufficient humidity (Fig. 1b). If the growth temperature is high, the young plantlets grow rapidly. Within 5-6 weeks, the plants have 3-5 trifoliate leaves, and an extensive root system (Fig.

1c). At this time, plants can be transferred to the field. McGrew (1980) estimates that transfer of plants into soil in the greenhouse and subsequent virus indexing requires at least 4 months.

Mullin et al. (1974) report that the young plants are very delicate when first taken out of the tubes and must be carefully manipulated. Molot et al. (1973) report significant losses upon transfer to soil. According to Damiano (1980), the best acclimatization substrate is pure peat, or a 1:1 mixture of sand and peat. The pH of the substrate is important, with optimum results obtained with pH 5.5–7.0.

Micropropagation

Massive production of strawberry plants in vitro was first accomplished in 1974 by Boxus. Within 5 years, European growers adopted this method. The first industrial application of strawberry micropropagation was reported in Germany. In 1976, Reinhold Hummel, a strawberry breeder, created a laboratory exclusively to produce millions of strawberry plants (Merkle, 1981; Riedel, 1977; Westphalen and Billen, 1976). Another, larger production unit will soon be assembled by the cooperatives of North Italy, The Centrale Ortofrutticola of Cesena (Zuccherelli, 1979). At the same time, one of the most important French strawberry nurseries, the Ets J. Marionnet of Soing en Sologne, has announced intentions to set up its own tissue culture laboratory (F.H. 1979).

In addition to these private laboratories, government laboratories are equipped at a semi-industrial level to provide quality plants for nurserymen. In France, the CTIFL (Centre Technique Interprofessionnel des Fruits et Legumes) of Balandran has a laboratory (Navatel, 1979) and in Belgium, the research station for fruit and vegetable crops of Gembloux assumes this role (Boxus, 1977; Dermine, 1977).

The basic principle of this method is extremely simple. The growth of dormant axillary buds is induced by the presence of cytokinin in the culture medium. Elimination of cytokinin permits development of leaves and roots. The factors that optimize this procedure will be examined in detail.

INITIAL MATERIAL. To obtain disease-free material, meristem culture is normally used. However, it is also possible to start with anther culture. Each cultured anther produces a callus capable of regenerating 50-100 plantlets. The mottle, the crinkle, and the mild yellow edge viruses have all been eliminated using anther culture. According to these researchers (Oosawa et al., 1974) this technique is more convenient than the conventional meristem culture methods. As plants are regenerated from callus, the genetic stability of this material should be monitored.

DESCRIPTION OF THE MULTIPLICATION PRINCIPLE. In Belgium (Boxus, 1976; Boxus et al., 1977), France (Navatel, 1979), and Italy (Damiano, 1980) indexed meristem plantlets, stored on a basal medium, are used as starting material for mass propagation (Fig. 2). These

Figure 2. Growth room for strawberry production at Gembloux.

plantlets are placed on a basal medium enriched with 4.4 μM BA after suppression of old leaves and roots. Three or four weeks later 2-3 buds appear at the base of the petioles of the oldest leaves. These young axillary buds grow very quickly and in turn produce new axillary buds. Within a few days, the initial plantlet is transformed into a mass of buds, often without callus. Within 6-8 weeks, the mass of buds grows until it is 3-4 cm in diameter. At this time, the mass may contain 15-25 well-developed buds. These are aseptically separated and transferred to fresh medium. With BA in the culture medium, axillary buds still proliferate, and this method can be repeated indefinitely. In less than a year, millions of buds can be obtained. When this number of plants is reached, one last sterile division is made, but the transfer is made onto a medium without BA. Without BA, the development of new axillary buds stops immediately, and the existing buds grow rapidly producing new leaves which progressively become trifoliated. After 12 days, roots regenerate, so that within 4-6 weeks the young rooted plantlets can be transferred to soil.

Boxus et al. (1977) reported that the 77 cultivars tested react in the same way as wild strawberry. Navatel (1979) does not report a correlation between the bud capacity of a cultivar in vitro and its ability to be propagated in the field.

CULTURE MEDIUM. All production laboratories use the same basal medium: Knop macroelements, MS microelements, MS vitamin mixture, 0.12 M glucose, and 7 g/l agar at pH 5.6. To promote axillary branching, these growth substances are added: 4.4 μM BA, 4.9 μM IBA, and 0.3 μM GA. Roots regenerate on medium containing 4.9 μM IBA.

Few variations have been reported for mineral composition. Zuccherelli (1979) used 9.89 mM KNO_3, four times Knop's concentration. Damiano et al. (1979) improved the growth rate of the cultivar Aliso by adding 4-8 mM KNO_3 to the basal medium. Damiano et al. (1979b) reported that the NH_4^+ ion can be utilized only in the presence of citric or succinic acid.

Several studies have investigated hormonal balance. Waithaka et al. (1980) released axillary dormancy by including a large concentration (50 μM) of KIN in the culture medium. James and Newton (1977) have tested a range of IBA and BA concentrations on the cultivar Gento. They concluded that concentrations of the cytokinin ranging from 0.25 to 2.5 μM, coupled with auxin concentrations ranging from 0.25 to 1.0 μM were most suitable as they resulted in rapid proliferation of adventitious buds. To maintain good quality axillary branching, Damiano (1980a) reduced the BA concentration for some cultivars such as Aliso, Rabunda, and Redgauntlet.

For rooting, addition of 1 or 2 g of activated charcoal per liter of medium promotes elongation of both the shoots and the roots (Damiano, 1978). In Gembloux, we have observed a similar effect when the concentration of activated charcoal is 0.5 g/l. For example, using charcoal, a 30% increase of commercial grade plantlets (Boxus, 1981) was reported for the cultivar Rabunda. The activated charcoal can replace the GA that has been proposed by Navatel (1979). Tests made in Gembloux by Dr. P. Sampet (unpublished) show that 2.88 μM GA not only promotes extension of shoot internodes, but also exerts a depressive effect on root development. The presence of phloroglucinol also reduces root initiation but increases formation of subsequent lateral roots (James, 1979).

STORAGE IN COLD CHAMBER. To distribute production uniformly over the entire year, plantlets are stored in darkness at 1-2 C (Fig. 3). This storage can be made at any stage (Boxus, 1981; Navatel, 1979) and can last for several months. Success of cold storage depends on the cultivar and vigor of the stored material. At the end of storage, plantlets must be cleaned before being transferred to a new medium. For some cultivars, such as Aliso, Gorella or Tioga that have been stored for 16-27 months, 2 or 3 successive transfers are needed before the restoration of a good multiplication rate (Damiano, 1979).

INDUSTRIAL PROCESS OF MASS PROPAGATION

During the past 10 years, 3731 meristems have been cultured in Gembloux from 350 cultivars from America, Asia, and Europe. The rate of success is about 74% with about 7% loss to bacterial contamination. Virus elimination was 100% from contaminated plants received from Asia or Europe. The majority of these 350 cultivars is currently stored in darkness at 2 C, in 25 x 150 mm test tubes. The Kaputs

Figure 3. Storage of plant material at + 2 C.

(Bellco) are sealed using plastic film to avoid dessication and contamination. This germplasm is often used for multiplication.

Since the sale price of a strawberry plant is very low, it is useful to establish standardized propagation methods suitable for industrial processes. Since 1974 we have been focusing on the development of standardized procedures (Boxus, 1977; Damiano, 1980b). All sterile manipulations are made in jars 10 cm in diameter that contain 150 ml of medium. We are able to produce up to 200-500 buds or 45-100 rooting plantlets depending on the cultivar. The culture medium can be prepared in a washing machine that is modified for media preparation. In one hour, 40 liters of culture medium are boiled and dispensed into 250 jars (Fig. 4). Human manipulation is reduced to a minimum by using stainless steel baskets that contain 16 bottles that pass from the washing machine up to the position where jars are filled, into the autoclave, and then into the sterile room. Transfers are made in the sterile room.

Figure 4. A modified washing machine is used to prepare, cook, and dispense 40 liters of medium within one hour.

The same solidified medium is used for all cultivars. The varietal differences, noted by Navatel (1979), are partially compensated by modifiying culture techniques between cultivars. Cultivars with large buds (e.g., Bravo, Pocahontas, Sequoia, Sivetta) do not need particular attention. About 15 explants with 3-5 buds are placed into each jar. One month later the contents of each jar can be divided into five jars. Cultivars with very large buds (e.g., Douglas, Gento, and Tuft) are placed deep into the agar. When subculturing, all traces of callus at the cut end must be removed. Cultivars with small buds (e.g., Ostara, Rabunda, and Redgauntlet) have a high multiplication rate, but many buds are lost during transfer. For these explants it is better to use basal shoots in close contact with the agar surface. Subcultures are not made on a rigid schedule but are transferred approximately every 3 weeks.

Some cultivars, such as Aliso, require optimal culture conditions to develop well. The cultures quickly senesce if the temperature is a little too high. The storage temperature is dependent on the developmental state of material.

Large-scale production must consider the storage capacities allotted to the different cultivars. One worker can produce about 4000 plants per day. The culture room should accommodate up to 8 cycles per year. Each square meter of storage space is capable of producing about 25,000-30,000 plants per year.

For shipping to the nursery, the plantlets are removed from the jars, then cleaned thoroughly with tap water. Plants are sorted, then packed in lightweight plastic boxes (10.5 x 8 x 5 cm) with 200 plants per box. Packaged in this way, the plants can travel great distances with little damage. Small peat pellets that are often used with other seedlings are also a very good and very cheap substrate for the acclimatization. A shipment of 40,000 plantlets weighs only 15 kg. Unfortunately, storage at 2 C in the plastic boxes cannot exceed 4-6 weeks. Plants cannot be stored a long time before the delivery. As packing requires no special qualification, it can be assigned to extra staff; 2000-3000 plantlets can be removed and packed each working day.

PROTOCOL: MERISTEM CULTURE OF STRAWBERRY

1. Meristems from young runner tips are sterilized in 5% sodium hypochlorite for 5-20 min. A brief dip in 70% ethanol is sometimes helpful for difficult material.
2. The meristem tip is now excised using a fine blade under a dissecting microscope. Approach the meristem area from below by holding the distal end of the runner plantlet and make thin, transverse slices until a cone including the apical dome and 1-2 primordial leaves can be teased away.
3. Place meristem onto basal medium (Knop macroelements, MS microelements, MS vitamins, 0.12 M glucose, 7 g/l agar, pH 5.6) containing 4.4 μM BA.
4. Incubate cultures at 20-25 C with 1000-3000 lux light intensity (12-18 hr daylength).

5. Within 3-4 weeks, 2-3 buds will appear at the base of the petioles of the oldest leaves. Many buds will be produced from these axillary buds. When mass contains 15-25 well-developed buds, aseptically separate the buds and place on rooting medium (basal medium without hormones).
6. Root regeneration occurs within 12 days. To transfer rooted regenerated plantlets to soil, wash off all agar from roots and plant in jiffy pots stored under high humidity.

FIELD BEHAVIOR OF MICROPROPAGATED PLANTS

Fruit Production

The first experimental fields to evaluate strawberries propagated in vitro were established between 1974 and 1976. In these field tests, micropropagated plants were not different from mericlone cultivars of the same genotype and disease-free control plants. On the other hand, the micropropagated plants are growing better than commercially grown plants (Boxus, 1981). To take advantage of this improvement, microplants should be used as mother plants. The high price of producing microplants would be offset if used as mother plants in the nursery. These techniques are rapidly being adopted by European nurseries. Today, the majority of production fields have been derived from microplants. Precise statistics are not available on the behavior of these plants, making scientific interpretation of results difficult. We have attempted to analyze the most representative results that are available.

In Belgium, Aerts (1979) noted that in 1977 and 1978, there were no significant differences in either weight or quality between progenies derived from microplants and traditional plants of the cultivar Gorella. Lemaitre (1978) of the Strawberry Research Center at Wepion reported smaller fruit in plants of cvs. Domanil and Gorella derived from microplants than in control plants. A similar observation was made in 1980 in Great Britain for cv. Domanil produced by micropropagation (Walker, personal communication).

In the Netherlands, Dijkstra (1978, 1979) studied the behavior of five Gorella clones produced by micropropagation in vitro versus traditional methods of propagation. Three Dutch clones and two Belgian clones were planted at Wilhelminadorp in August 1978 and 1979. The results of the two tests are summarized in Table 6. In most cases, no significant differences were observed between tissue culture and control plants.

In Italy, in 1979 and 1980, Damiano established different experimental fields in the Po valley for cvs. Belrubi, Gorella, and Pocahontas, and in southern Italy for cv. Aliso. He used fresh and cold stored plants derived from first or second generation microplants and compared these with traditional plants (Tables 7 and 8). In most cases, differences were not significant.

In Switzerland, Bochud et al. (1981) compared the production of the cvs. Gorella and Redgauntlet from the first or second generation in

Table 6. Fruit Production of Five Gorella Clones at Wilhelminadorp, Netherlands[a]

CLONE	WEIGHT PER PLANT[b] (g)		AVERAGE FRUIT WEIGHT[b] (g)	
	1978	1979	1978	1979
Belgium 1, tissue culture	552 a	451 a	13.4 a	12.2 d
Belgium 2, tissue culture	662 bc	471 a	16.8 b	15.5 abc
Netherlands 1, traditional	673 bc	463 a	15.4 b	16.0 ab
Netherlands 2, traditional	611 b	429 a	16.7 b	16.7 a
Netherlands 3, traditional	652 bc	425 a	16.2 b	16.7 a
Netherlands 1, tissue culture	633 bc	487 a	16.3 b	14.4 c
Netherlands 2, tissue culture	619 bc	448 a	16.6 b	15.2 bc
Netherlands 3, tissue culture	695 c	414 a	15.5 b	15.6 abc
Netherlands 1+2+3, traditional	645	439 a	16.1	16.5
Netherlands 1+2+3, tissue culture	649	433 a	16.1	15.1

[a]The experiments were repeated in 1977-1978 and in 1978-1979 with the first generation of three Dutch and two Belgian clones that were propagated by tissue culture or traditional methods (partially in Dijkstra and Van Oosten, 1979).

[b]Values followed by the same letter(s) do not differ significantly.

1981. Gorella gave the same results as traditionally grown plants. On the other hand, tissue culture-derived Redgauntlet plants produced small fruit and a poor yield. In order to interpret the above results, it is important to realize that the Gorella clone used by Lemaitre in Belgium, by Bochud in Switzerland, and by Damiano in Italy is the same as Dijkstra's "Belgium 1 clone." This clone has been propagated at the Gembloux laboratory since 1969, and was used for the first tests performed at Gembloux in 1974. This Gorella clone behaves normally in Bochud's and Damiano's cold stored plant test, where it gives normal production with normal fruit size (Tables 7 and 9). In Italy, this normal production is associated with a normal erect habit in the cold stored outdoor plants. Actually, it is very common, in plants produced by micropropagation, to observe a dwarf habit (Fig. 5), which dis-

Table 7. Fruit Production in Italy, 1979[a]

ORIGIN OF PLANTS	CV. GORELLA UNDER TUNNEL		CV. GORELLA OPEN FIELD		CV. ALISO	
	weight/plant (g)	average fruit weight (g)	weight/plant (g)	average fruit weight (g)	weight/plant (g)	average fruit weight (g)
Fresh plants						
micropropagated	377	10.1	514	8.8[c]	656	10.4
traditional	448	10.6	479[b]	11.1	650	10.4
Cold-storage plants						
micropropagated	495	13.2	512	14.5	—	—
traditional	417	14.7	435	14.7	725	10.0

[a]Weight per plant and average fruit weight of fresh and cold-stored plants that were produced from micropropagated or traditional mother plants (Damiano, 1980a).

[b]Significant difference between micropropagated and traditional plants.

[c]Highly significant difference between micropropagated and traditional plants.

Table 8. Fruit Production in Italy 1980[a]

ORIGIN OF PLANTS	CV. GORELLA[b]		CV. BELRUBI		CV. POCAHONTAS	
	weight/plant (g)	average fruit weight (g)	weight/plant (g)	average fruit weight (g)	weight/plant (g)	average fruit weight (g)
Cold-stored plants						
Traditional	455.7 a	13.5 a	440.7 a	16.5 bc	666.9 b	9.2 bc
First generation of microplants	550.8 b	12.3 a	542.8 b	14.0 a	696.9 b	10.6 d
Fresh plants						
Traditional	455.6 a	13.2 a	369.3 a	14.5 ab	496.3 a	8.0 a
First generation of microplants	548.3 b	12.7 a	416.5 a	16.9 c	708.4 b	10.3 cd
Second generation of microplants	378.4 a	12.3 a	387.0 a	17.1 c	630.6 b	10.0 bd

[a]Weight per plant and average fruit weight of fresh and cold-stored plants.

[b]Values followed by the same letter(s) do not differ significantly.

Table 9. Fruit Production in Switzerland 1981[a]

	CV. GORELLA			CV. REDGAUNTLET		
ORIGIN OF PLANTS	weight/plant (g)	first grade (%)	second grade (%)	weight/plant (g)	first grade (%)	second grade (%)
Cold-stored plants						
Traditional	0.581	87	13	1.010	77	23
First generation of microplants	0.597	97	9	0.649	65	35
Fresh plants						
Traditional	—	—	—	1.152	81	19
Second generation of microplants	—	—	—	0.811	64	36

[a]Bochud et al., 1981.

Figure 5. Dwarf habit of Gorella plants (behind the label), produced directly from micropropagation and grown in the greenhouse.

appears by cold storage or extreme heat treatment. In all the tests using this Gorella clone and fresh plants, size of the fruit is depressed, and the plants are morphologically distinct with dwarf habit, more leaves, but with shorter petioles and reduced leaf surface. Cold storage of the plant at -2 C for several months causes the first generation Gorella plants to grow to normal sizes. The same phenomenon has been observed with Sivetta, a cultivar that was produced in the laboratory and used for early greenhouse production, normal outdoor production, or late production by cold storage at -2 C. Long cold storage has consistently proven beneficial to microplants with improvement in amount and quality of fruit production (Nuyten, 1977).

The results obtained with fresh plants derived from microplants by our group in 1974, and by Dijkstra with four of five Gorella clones are nonetheless encouraging. In fact, in all cases, we were dealing with culture stocks which had not yet been propagated intensively before testing, and these culture stocks had not yet become accustomed to in vitro culture. To completely evaluate this method, it is important to evaluate different cultivars, duration of cold treatment, and number of subcultures. We are studying these variables at Gembloux using the cvs. Domanil, Gorella, and Redgauntlet. Evidence suggests that all three of these variables are important. First, in vitro multiplication of new strawberry hybrids always shows very high production with very large fruit. Lemaitre (1981, personal communication) observed an improvement of yield and fruit quality with 15 cultivars that were subcultured four or five times in the laboratory.

In Great Britain in 1981 we observed a partial recovery of the Domanil clone that had small fruits after a third generation in the field. In Spain, plants were frozen for 30 days during the extremely cold winter of 1981. After this treatment, the cv. Tioga produced by micropropagation had better production than conventional plants. These late strawberries do not flower very much. On the other hand, in hotter regions, fresh plants derived via micropropagation produced using a winter planting system have a tendency to flower profusely and to bear small fruits.

To conclude this section, we should also point out another possible method of modifying the production of plants recovered via micropropagation. It is possible to modify the conditions of in vitro culture by changing the hormonal balance of the solution or by modifying the growth conditions of the plants. It will be very interesting to follow the work performed by Naumann (1979, 1981) on microplants of cv. Senga sengana. Preliminary tests indicate increases in production of 96% over control plants for plantlets raised under specific photoperiods and thermoperiods. It is still too early to make generalizations, but these results are extremely promising.

Runner Production

The improvement in plant production in the nursery through the use of microplants is much more obvious. In fact, only 3 weeks after their transfer to soil, the plantlets produce their first runners when grown under long daylight (14 hr) conditions. These runners are produced in abundance. Damiano (1980) has summarized the number of runners by commercial grade for three cultivars. Data was compiled from a spring planting, harvested the following autumn. Only the Aliso field was irrigated (Table 10). For all three cultivars, the number of runners was increased for tissue culture plants. Lemaitre (1979) obtained a notable improvement for Gorella microplants planted on April 18, 1979, and harvested on August 22 (Table 11). Schwarz et al. (1981) observed an increase of 62% of runnering ability for three cultivars. This tendency to produce large numbers of runners is probably a manifestation of the juvenile state of the plants produced by micropropagation. Other characteristics typical of tissue culture plants, such as the presence of unifoliate leaves, are lost very quickly after planting. Plants of cv. Bordurella, a variety normally without runners, produce 5-7 runners and then stop. This behavior is like a phase change from juvenile to adult phase.

In Europe microplants in the multiplication fields generally have very good health and are frequently extremely homogeneous if the growing conditions are good. Schwarz et al. (1981) also consider that the use of microplants for plant production should result in greater plant uniformity and field survival. Damiano et al. (1980) point out that while the progenies of microplants are not immune to soil parasites, they are certainly not more sensitive to infection than traditionally propagated plants. In France, since the widespread introduction of microplants, wilt symptoms have almost entirely disappeared in

Table 10. Production of Cold-stored Plants in Italy[a]

CULTIVAR AND TYPE	RUNNER PLANTS PER MOTHER PLANT	CROWN DIAMETER ≥ 8 mm		6–7.9 mm		< 6 mm	
		% of total	no. per mother plant	% of total	no. per mother plant	% of total	no. per mother plant
Gorella							
Micropropagated	62.0	71.7	44.4	15.7	9.8	12.6	7.8
Traditional	39.3[b]	79.9	31.4[b]	11.3	4.4[b]	8.8	3.4[b]
Belrubi							
Micropropagated	26.2	88.6	23.3	5.9	1.5	5.6	1.5
Traditional	10.8[b]	83.4	9.0[b]	8.7	0.9[b]	7.9	0.8[b]
Aliso							
Micropropagated	146.3	54.2	79.3	15.3	22.4	30.5	44.6
Traditional	89.5[b]	61.5	55.0	14.8	13.2	23.7	21.2

[a]Field planting dates: cv. Gorella on April 15, cv. Belrubi on May 15, and cv. Aliso on May 1. Only the Aliso field was irrigated.

[b]Highly significant difference between micropropagated and traditional plants from the same cultivar.

Table 11. Production of Fresh Plants in Belgium[a]

MEANS OF PRODUCTION	STOLONS AT CROWN	COMMERCIAL RUNNERS	SMALL RUNNERS
Micropropagated	18	19	14.3
Traditional	7	5.4	4.5

[a]The plants, cv. Gorella, are planted on April 18, the runners are harvested on August 22 (after R. Lemaitre, 1979).

nurseries. It is true, however, that flower supression is a common technique used in nurseries and that this practice could be suppressing disease symptoms. However, Sansavini et al. (1980) have shown disease suppression for "collasso" in cv. Gorella for tissue culture-derived plants.

Genetic Stability

Bunyard has pointed out that the strawberry plant is notable for having a very stable genetic makeup (Schaeffer et al., 1980). This explains why until today among the millions of strawberry plants propagated in vitro only a few mutations have been observed. A green-yellow type variegation has been observed on cvs. Aliso, Belrubi, and Wood (Schaeffer et al., 1980). Schwarz et al. (1981) also noticed this symptom which affects all or part of a bud. In France and in Belgium, this somatic chlorophyll variant has been observed in microplants with a frequency in the range of one in perhaps 10,000 regardless of the cultivar (Fig. 6). This type of variant is also observed using traditional propagation, perhaps with the same frequency. In most cases, the yellow sectors produced by these somatic mutations are limited to two or three leaves and do not affect progenies. However, this depends on the stage at which the mutation occurs during bud development. In any case, the variants are easy to identify and the affected mother plants should be removed.

Other anomalies can occasionally be observed in the microplants after acclimatization to soil. These include various forms of "June yellow," which can appear even in September. This may involve a more or less pronounced yellow mottle, mosaic, or white streaking, which also affects a limited number of leaves. A number of Senga cultivars have proven to be very sensitive to this variation in the fall. However, by spring the symptoms have completely disappeared with the exception of a white streaking in cv. S. Remonta. Cases of white streaking have been reported by Schwarz et al. (1981) on mericlones of cv. Earliglow, at a rate of 8.4%. Scott has also noticed this symptom on cv. Aliso micropropagated in Italy, at a rate of 2% in 1978 and 1.2% in 1979 (Schaeffer et al., 1980). We consider this streaking to be an expression of "June yellow," as described and illustrated by Wills

Figure 6. Somatic chlorophyll mutants of strawberry microclones. These typically appear with a frequency of 1 in 10,000.

(1970). We have also observed this Fugace symptom in several plants derived from traditional multiplication. We have not yet been able to determine whether or not the micropropagation method increases the expression of "June yellow." Under glasshouse conditions, we observed very severe symptoms at Gembloux in the beginning of the season on cv. Tioga. In 1978, on cv. Aliso, we observed yellowing and twisted leaves first in southern France, then in Italy. These leaves are very small and have a necrotic margin. Symptoms first appear in September-October with recovery the next spring. In 1979 and 1980, symptoms were very severe and affected a clone that originated in Gembloux. A new Italian clone propagated since 1980 does not show any yellowing. A similar symptom was observed in the Netherland Limburg in fall 1976 on cv. Redgauntlet, originated from micropropagation.

Sansavini and Gherardi (1980) observed an unusual form of behavior in plants produced by micropropagation. They flower and produce runners at the same time, whereas in traditional plants the generative and vegetative phases occur at different periods. We have already spoken of the unusual appearance of the microplants, with their dwarf habit, numerous small leaves and abundant crowns and flowers, characteristics that appear to be related to the number of subcultures. Some cases of modification of the shape of the fruit or of other fruit malformations have been reported for the cultivars Domanil in England, Hummi grande in Switzerland (Lutz, personal communication), and Redgauntlet in Austria.

The changes that have been reported probably do not reflect genetic modification but represent variants induced by specific physiological conditions during growth of the microplants. A better knowledge of these physiological phenomena will enable us to circumvent their disadvantages. For example, Soede (1981) proposes a decreased salt concentration during micropropagation of *Nephrolepis exalta.* In that way, he completely suppresses the dwarf habit, which is otherwise frequent in some cultivars of Boston fern, after the third subculture. In addition, all production methods are based on the development of axillary buds. The existence of adventitious buds on leaves has been reported by Zimmerman (Schaeffer et al., 1980). Recently at Gembloux we have also observed adventitious buds from leaves, but generally they aborted. Their presence is too rare to explain all the observed deviations.

CONCLUSIONS

The experience acquired during micropropagation of strawberry plants shows that, despite certain difficulties that have been discussed above, this technique permits the production of quality plants. To achieve maximum production, the following must be observed:

1. Begin exclusively with meristems and check the results by indexing.
2. Eliminate all callus and adventitious bud formation.
3. Replace the culture stocks regularly in order to ensure heavy production of large fruit and to avoid the phenomenon of habituation.
4. Perform one, or perhaps two, traditional propagations from microplants, depending on the health condition of the plots and the cultivar used. Everbearing cultivars, for example, should be multiplied only once.

By observing all of these conditions, the speed of the process and the recovery of quality plants will be maximized.

KEY REFERENCES

Darrow, G.M. 1966. The Strawberry: History, Breeding and Physiology. Holt, Rinehart and Winston, New York.

______ 1977. La culture du fraisier. Fruit Belge (numero special) 45: 55-136.

______ 1981. The strawberry, cultivars to marketing. (N.F. Childers, ed.) University of Florida, Horticultural Publications, Gainesville.

REFERENCES

Adams, A.N. 1972. An improved medium for strawberry meristem culture. J. Hortic. Sci. 47:263-264.

Aerts, J. 1974a. Survey of viruses and mycoplasmas in strawberry. Neth. J. Plant Pathol. 80:215-227.

______ 1974b. Verschillen in gevoeligheid van aardbeivarieteiten voor een enkelvoudig virus en virus complex. Mededel. Fakult. Landbouwwetenschappen Gent 39:1203-1215.

______ 1979. Ervaringen met, langs microvermeerdering bekomen, virusvrije aardbeiplanten. Mededel. Fakult. Landbouwwetenschappen Gent 44:981-991.

Belkengren, R.O. and Miller, P.W. 1962. Culture of apical meristems of *Fragaria vesca* strawberry plants as a method of excluding latent A virus. Plant Dis. Rep. 46:119-121.

Bochud, P., Coppey, G., and Berthouzoz, F. 1981. Fraisier 1981, test cultural. Rapport du Centre d'arboriculture et d'horticulture des Fougères, Station fédérale de recherches agronomiques de Changins, pp. 23-30. Conthey, Switzerland.

Boxus, Ph. 1974. The production of strawberry plants by in vitro micropropagation. J. Hortic. Sci. 49:209-210.

______ 1976. Rapid production of virus-free strawberry by in vitro culture. Acta Hortic. 66:35-38.

______ 1977. La micropropagation, procédé industriel de multiplication rapide du fraisier. Fruit Belge 45:120-124.

______ 1981. Commercial production of strawberry plants produced by meristem culture and micropropagation. In: Les Floralies Internationales de Montréal. Colloques Scientifiques, Vol. 15, Perspectives de l'horticulture, arbres fruitiers et petits fruits, pp. 310-348. Collection Etudes et Dossiers de la Documentation Québecoise, Gouvernement du Quebec.

______, Quoirin, M., and Laine, J.M. 1977. Large scale propagation of strawberry plants from tissue culture. In: Applied and Fundamental Aspects of Plant Cell, Tissue and Organ Culture (J. Reinert and Y.P.S. Bajaj, eds.) pp. 130-143. Springer-Verlag, Berlin.

Damiano, C. 1977. La ripresa vegetativa di piantine di fragola provenienti da colture in vitro. Frutticoltura 39:3-7.

______ 1978. Il carbone attivo nella coltura in vitro della fragola. Frutticoltura 40:49-50.

______ 1979. Conservazione in frigorifera di colture in vitro di fragola e ripresa della moltiplicazione. Annual Report, Istituto Sperimentale per la Frutticoltura, Rome, Vol. 10, pp. 53-57.

______ 1980a. Strawberry micropropagation. In: Proceedings of the Conference on Nursery Production of Fruit Plants Through Tissue Culture: Applications and Feasibility, pp. 11-22. USDA-SEA, Agricultural Research Results, ARR-NE-11. Beltsville, Maryland.

______ 1980b. Planning and building a tissue culture laboratory. In: Proceedings of the Conference on Nursery Production of Fruit Plants Through Tissue Culture: Applications and Feasibility, pp. 93-101. USDA-SEA, Agricultural Research Results, ARR-NE-11. Beltsville, Maryland.

______, Laneri, U., and Arias, E. 1979a. Nota preliminare sulla utilizzazione dell' azoto nitrico ed ammoniacale in colture in vitro di fragola. Annual report, Istituto Sperimentale per la Frutticoltura, Rome, Vol. 10, pp. 35-41.

______, Laneri, U., and Arias, E. 1979b. Assunzione dello azoto nella propagazione in vitro della fragola; azione degli ioni nitrico ed ammonio con acido citrico e succinico. Annual report, Istituto Sperimentale per la Frutticoltura, Rome, Vol. 10, pp. 43-52.

______, Faedi, W., and Cobianchi, D. 1980. Prove di stolonizzazione in vivaio e osservazioni agronomiche su piante di fragola ottenute da colture in vitro. Frutticoltura 42:51-60.

Dermine, E. 1977. Nouvelles méthodes pour la production de plants de fraisiers. Fruit Belge 45:115-119.

Dijkstra, J. and van Oosten, A.A. 1979. Culture experiments with strawberries. Annual report, Research Station for Fruit Growing, Wilhelminadorp, The Netherlands, pp. 28-30.

F.H. 1979. Le plant de fraisier multiplié en laboratoire. Semences et Progrès, 20:15-22.

Hilton, R.E., Smith, S.H., Frazier, N.W. and Schlegel, D.E. 1970. Elimination of mild yellow edge virus from fresno and J6 strawberry varieties by a combination of heat treatment and meristem-tip culture. Strawberry News Bulletin 16:1-2.

James, D.J. 1979. The role of auxins and phloroglucinol in adventitious root formation in *Rubus* and *Fragaria* grown in vitro. J. Hortic. Sci. 54:273-277.

______ and Newton, B. 1977. Auxin-cytokinin interaction in the in vitro micropropagation of strawberry plants. Acta Hortic. 78:321-331.

Kartha, K.K., Leung, N.L., and Pahl, K. 1980. Cryopreservation of strawberry meristems and mass propagation of plantlets. J. Am. Soc. Hortic. Sci. 105:481-484.

Knop, W. 1865. Quantitative untersuchungen uber die Ernahrungsprozesse der Pflanzen. Landwirtsch. Vers. Stn. 7:93-107.

Laneri, U. and Damiano, C. 1980. Strawberry anther culture. In: Application de la Culture In Vitro à l'Amélioration des Plantes Potagères (C. Dore, ed.) pp. 150-155. Station de Genetique et d'Amelioration des Plantes, INRA, Versailles, France.

Lemaitre, R. 1979. Rapport annuel, Centre d'études pour la récupération des énergies résiduelles, Gembloux, p. 52-56.

Lousteau, P. 1979. La fraise, panorama économique 1979. CTIFL Documents 62:81-107.

McGrew, J.R. 1965. Eradication of latent C virus in the Suwannee variety of strawberry by heat plus excised runner tip culture. Phytopathology 55:480-481.

______ 1980. Meristem culture for production of virus-free strawberries. In: Proceedings on the Conference on Nursery Production of Fruit Plants Through Tissue Culture: Applications and Feasibility, pp. 80-85. USDA-SEA, Agricultural Research Results, ARR-NE-11. Beltsville, Maryland.

______ and Scott, D.H. 1964. The effect of strawberry latent A virus on growth and fruiting of seven varieties of strawberry. Plant Dis. Rep. 48:929-932.

Merckle, S. 1981. Erdbeervermehrung durch Meristemkultur. Erfahrungen aus der Praxis. Ostbau 6:95-96.

Miller, R.W. and Belkengren, R.O. 1963. Elimination of yellow edge, crinkle, and vein banding viruses and certain other virus complexes from strawberries by excision and culturing of apical meristems. Plant Dis. Rep. 47:298-300.

Molot, P.M., Nourisseau, J.G., Leroux, J.P., Navatel, J.C., and Germain, P. 1973. Le *Phytophthora cactorum* sur fraisier. Connaissances actuelles. Mise au point des méthodes de lutte. CTIFL Documents 40:1-12.

Mullin, R.H. and Schlegel, D.E. 1976. Cold storage maintenance of strawberry meristem plantlets. HortScience 11:100-101.

Mullin, R.H., Smith, S.H., Frazier, N.W., Schlegel, D.E., and McCall, S.R. 1974. Meristem culture frees strawberries of mild yellow edge, pallidosis and mottle disease. Phytopathology 64:1425-1429.

Murashige, T. 1977. Current status of plant cell and organ cultures. HortScience 12:127-130.

______ and Skoog, F. 1962. A revised medium for rapid growth and bioassays with tobacco tissue cultures. Physiol. Plant. 15:473-497.

Naumann, W.D. and Synowsky, B. 1979. Wuchs und Ertragsleitung von Erdbeerpflanzen nach Mikrovermehrung. Erwerbsobstbau 21:142-146.

______ 1981. Die Ertragsleistung mikrovermehrter Erdberrpflanzen und ihrer Nachzuchten. Obstbau 6:90-94.

Nishi, S. and Oosawa, K. 1973. Mass production method of virus free strawberry plants through meristem callus. Jpn. Agric. Res. Q. 7: 189-194.

Nuyten, H. 1977. Meristeemcultuur . . . de oplossing? Groenten en Fruit, 33, Kerstnummer, p. 77.

Oertel, Cl. 1981. Zur Effektivität der Wärmebehandlung bei der Virusfreimachung von Zierpflanzen und Erdbeeren. Archiv. Gartenbau, Berlin 29:217-231.

Oosawa, K., Toka, M., and Nishi, S. 1974. Studies on the anther culture of vegetable crops. II. Breeding of a great quantity of virus free plants by means of strawberry anther culture method. Bull. Veg. Ornam. Res. Station, Japan, A1:41-57.

Posnette, A.F. and Jha, A. 1960. The use of cuttings and heat treatment to obtain virus free strawberry plants. Annual Report, East Malling Research Station, p. 98.

Riedel, R. 1977. Die Meristemvermehrung von Erdbeeren. Obstbau 2: 211-213.

Rosati, P., Devreux, M., and Laneri, U. 1975. Anther culture of strawberry. HortScience 10:119-120.

Sakai, A., Yamakawa, M., Sakata, D. Harada, T., and Yakuwa, T. 1978. Development of a whole plant from an excised strawberry runner apex frozen to -196 C. Low Temp. Sci. Ser. B. 36:31-38.

Sansavini, S. and Gherardi, G. 1980. Selezione clonale e stabilità genetica di fragole micropropagate. Frutticoltura 42:39-46.

Schaeffer, G.W., Damiano, C., Scott, D.H., McGrew, J.R., Krul, W.R., and Zimmerman, R.H. 1980. Transcription of panel discussion on the genetic stability of tissue culture propagated plants. In: Proceedings of the Conference on Nursery Production of Fruit Plants Through Tissue Culture: Applications and Feasibility, pp. 64-79. USDA-SEA, Agricultural Research Results, ARR-NE-11. Beltsville, Maryland.

Scott, D.H. and Lawrence, F.J. 1975. Strawberries. In: Advances in Fruit Breeding (J. Janick and J.N. Moore, eds.) p. 80. Purdue Univ. Press, West Lafayette, Indiana.

Skirvin, R.M. 1981. Fruit crops. In: Cloning Agricultural Plants via in vitro Techniques (B.V. Conger, ed.) pp. 51-139. CRC Press, Boca Raton, Florida.

Sobczykiewicz, D. 1979. Heat treatment and meristem culture for the production of virus-free strawberry plants. Acta Hortic. 95:79-82.

Soede, A.C. 1981. Vermeerdering van *Nephrolepis* in weefselkweek nu betrouwbaarder. Vakblad voor de Bloemisterij 51:48-51.

Standardi, A., Boxus, Ph., and Druart, Ph. 1978. Preliminary research into the effect of light on the development of axillary buds and on the rooting of plantlets cultivated in vitro. In: Round-table Conference, In vitro Multiplication of Woody Species (Ph. Boxus, ed.) pp. 269-282. Centre de Recherches Agronomiques de l'Etat, Gembloux, Belgioque.

Swartz, H.J., Galetta, G.J., and Zimmerman, R.H. 1981. Field performance and phenotypic stability of tissue culture-propagated strawberries. J. Am. Soc. Hortic. Sci. 106:667-673.

Van Hoof, P. 1974. Méthode pratique de culture de méristèmes de fraisiers. Bull. Soc. Roy. Bot. Belg. 107:5-8.

Vine, J.J. 1968. Improved culture of apical tissues for production of virus free strawberries. J. Hortic. Sci. 43:293-297.

Waithaka, K., Hildebrandt, A.C., and Dana, M.N. 1980. Hormonal control of strawberry axillary bud development in vitro. J. Am. Soc. Hortic. Sci. 105:428-430.

Westphalen, H.J. and Billen, W. 1976. Erzeugung von Erdbeerpflanzen in grossen Mengen durch Sprossspitzenkultur. Erwerbsobstbau 18:49-50.

Wilhelm, S. 1974. The garden strawberry: A study of its origin. Am. Sci. 62:264-271.

Wills, A.B. 1970. June yellows of strawberry. In: Virus Diseases of Small Fruits and Grapevines (N.W. Frazier, ed.) pp. 55-56. Univ. of California, Division of Agricultural Sciences, Berkeley.

White, P.R. 1963. The cultivation of animal and plant cells. Ronald Press, New York.

Zuccherelli, G. 1979. Tecniche di propagazione industriale delle piantine di fragola da coltura in vitro per il vivaio. In: Societa Orticola Italiane Incontro Nazionale Sulla Coltura della Fragola (S. Sansavini, ed.) pp. 169-178. S.O.I., Ferrara.

SECTION VIII
Fiber and Wood

CHAPTER 18
Cotton

H. J. Price and *R. H. Smith*

HISTORY AND GEOGRAPHICAL DISTRIBUTION

The genus *Gossypium* L. consists of 39 species. All but six are diploid (n = 13), concentrated in Africa, Central and South America, and Australia (Fryxell, 1980). Its species occupy mainly arid to semiarid regions of the tropics and subtropics with only two wild species extending into temperate regions. Cytologically, the species are divided into six genome groupings (A,B,C,D,E,F), with American D genomes well differentiated from the African (A,B,E,F) and Australian (C) genomes (Beasley, 1942; Phillips, 1962, 1966, 1976; Wilson et al., 1968).

Four species are cultivated; *G. herbaceum* L., *G. arboreum* L., *G. hirsutum* L., and *G. barbadense* L. *Gossypium herbaceum* (A genome) occurs in Africa, including Ethiopia, and through the eastern Mediterranean region as far east as India. *Gossypium arboreum* (A genome) ranges generally to the east of *G. herbaceum* from India to Burma, Indonesia, and China (Fryxell, 1980).

The New World tetraploid species possess both the A and D genomes (Beasley, 1940; Phillips, 1966; Wilson et al., 1968). The tetraploid apparently originated in South America without the influence of man and in preagricultural times (Fryxell, 1980). How the A genome species migrated to South America from Africa has been the subject of much speculation which will not be discussed here. However, an allotetraploid occurred by chromosome doubling following hybridization of species similar to *G. raimondii* Ulbr., D genome, and *G. herbaceum*, A genome (Gerstel, 1953; Phillips, 1962; Cherry et al., 1970). From this, adaptive radiation resulted in a highly variable complex of six species (Fryxell, 1980). Man has domesticated two of these, *G. hirsutum* and *G. barbadense*.

The earliest records of man's use of cotton in the Old World are from the Mohenjo-Daro archeological finds from Pakistan, which date to about 3000 B.C. (Hutchinson, 1962). By 1500 B.C. cotton was apparently used as a textile for weaving in India and between 400 B.C. and 100 A.D. was grown as a crop around the Persian and Arabian Gulfs, Palestine and Egypt (Berger, 1969). Cotton was a field crop in China by 1100 A.D. (Hutchinson, 1962), and was introduced into southern Europe in the ninth and tenth centuries A.D. (Berger, 1969).

In the New World *G. hirsutum* was apparently fully domesticated by 3500 B.C. in the Tehuacan Valley of Mexico (Smith and Stephens, 1971). *Gossypium barbadense* dates back to 2500 B.C. from coastal Peru (Stephens and Moseley, 1973). Cotton was used by man extensively in pre-Columbian Central and South America, and southwestern North America. When Columbus discovered America in 1492 he found cotton growing in the Bahama Islands. By 1616 colonists were growing cotton along the James River in Virginia.

Upland cotton (*G. hirsutum*) made its appearance into the southeastern United States by the middle of the eighteenth century. Extensive cultivation of cotton in the New World followed the invention of the cotton gin in 1793. Both *G. hirsutum* and *G. barbadense* have been cultivated in the United States. Cultivation of *G. barbadense* (Sea Island) was essentially abandoned in 1920, due to its late maturity which made it susceptible to attack by the boll weevil. However, *G. barbadense* (American-Egyptian) of the Pima varieties currently is grown on a limited scale in the southwestern United States. The agronomically superior tetraploid cotton varieties are replacing the diploid cultivars of the Old World. Although these are primarily varieties of upland cotton, some areas grow *G. barbadense* in abundance, e.g., Egypt.

The extensive cultivation of cotton since the 1700s has resulted in selection pressures dictated by man's requirements of the crop, e.g., increased yield and quality of fibers. Specific breeding problems will be discussed later. However, the general effects of domestication have been neatly summarized and discussed by Fryxell (1980). The following traits were apparently influenced by man's selection and are important to the domestication of cotton: (a) ease of germination or loss of seed dormancy, (b) reduction of plant size to a more disciplined growth habit, (c) increased productivity by increases in both size and number of fruits per plant and by increases in the amount of fiber on each seed, and (d) development of the annual habit.

ECONOMIC IMPORTANCE

Cotton is a multipurpose crop that supplies five basic products: lint, oil, meal, seed hulls, and linters. The lint is the most important product of the cotton plant and provides much of the high-quality fiber for the textile industry. The most important by-product of the seed is oil, which is used primarily for shortening, cooking oil, salad dressing, mayonnaise, and margarine. The meal from glandless seeds or from seeds with gossypol extracted is used as a protein supplement for livestock. Glandless cotton seeds have been recently used for human consumption, e.g., a popped seed snack called TAMU nuts, a flour used for commercial production of a high-protein bread, and as a meat extender. Cotton seed hulls are used for cattle feed and as a soil covering called mulch. Linters, which are short fibers hanging to the seed after ginning, are removed at the oil mill to become a valuable source of cellulose for plastics, synthetic fibers, explosives, and batting for cushions. The National Cotton Council of America claims over 100,000 uses for cotton, which makes it a truly remarkable resource.

In the 1980/81 year, 11,125,000 bales (480 lb/bale) of fiber were produced in the United States from 4,350,000 tons of harvested seed. Processing 4,230,000 tons of crushed seed produced 1,363,000,000 lb of oil, 1,929,000 tons of cake and meal, 1,100,000 bales of linters, and 1,135,000 tons of hulls (Booker et al., 1981). The business revenue stimulated by cotton in the nation's economy is estimated at greater than $41 billion. United States cotton exports are valued in excess of $1 billion annually, thus making an important contribution to the balance of trade.

The United States is a major cotton producing country accounting for about 20% of the world's production. Table 1 presents USDA estimates for the 1980/81 and 1981/82 U.S. lint production. Estimates for other major cotton growing countries are also in Table 1. In the United States alone, over 14 million acres are planted in cotton (Booker et al., 1981). Cotton is the most important industrial crop and the nation's fourth leading crop.

The geographic limitations of cotton cultivation are climatically determined by the length of the growing season, range in temperatures, and the amount of sunlight. Cotton is native to the tropics and subtropics, but the annual habit selected during domestication has allowed it to be grown in temperate regions with long warm summers, e.g., the southern United States. A detailed analysis of the requirements for cotton cultivation is presented by Berger (1969). Basically, cotton requires a long frost-free period of about 180-200 days and 4-5 months of uniformly high temperatures with predominantly sunny days. At least 175-200 mm precipitation are required during the growing season and minimally 500 mm annually. Some cotton production occurs on irrigated lands, e.g., in the western and southwestern United States. Cotton is grown to 30° N in the United States, 47° N in the Ukraine and 42° N in Manchuria. In the southern hemisphere it is cultivated to 32° S in South America and Australia and to 30° S in Africa. Cotton is grown in a region stretching across the lower half of the United States from the Carolinas to California.

The upland cotton varieties that account individually for more than 1% of the 1981 U.S. acreage are presented in Table 2. Deltapine and Stoneville are the leading upland cotton varieties planted and account for about 16% of the national acreage. Acala, Lankart, Paymaster, Tamcot, and GSA account for 13, 10, 9, 9, and 7% respectively. Pima S-5 accounts for 99% of *G. barbadense* acreage, and Pima P-34-8P

Table 1. USDA Estimates of U.S. and Foreign Cotton Production

COUNTRY OR REGION	PRODUCTION[a] 1980–81	1981–82
ASIA AND OCEANIA[b]		
Australia	440	465
China (Peoples Republic)	12,430	13,000
India	6,100	6,300
Indonesia	10	15
Republic of Korea	14	15
Pakistan	3,259	3,400
Syria	540	565
Thailand	275	300
Turkey	2,218	2,295
EUROPE[b]		
Greece	535	500
Italy	3	—
Spain	225	250
USSR	14,300	13,500
AFRICA[b]		
Egypt	2,442	2,200
Sudan	407	440
NORTH AND SOUTH AMERICA[b]		
Argentina	385	620
Brazil	2,750	2,760
Guatemala	580	540
Mexico	1,595	1,600
Nicaragua	350	420
Paraguay	480	475
FOREIGN TOTAL	54,326	54,789
USA[c]		
Southeast	496	618
Mid-south	2,433	3,400
Southwest	3,521	5,820
West	4,675	4,951
USA TOTAL	11,125	14,789

[a] 480 lb. bales x 1000.

[b] From Cotton Economic Review, National Cotton Council, July 30, 1981 (Vol. 12, No. 6) and Aug 31, 1981 (Vol. 12, No. 7).

[c] From Booker et al. (1981).

Table 2. Cotton Varieties by State Comprising the 1981 United States Crop[a]

VARIETY[b]	STATE[c]															
	NC	SC	GA	AL	MS	TN	MO	AR	LA	OK	TX	NM	AZ	CA	All other[d]	Total
Acala 1517-75	—	—	—	—	—	—	—	—	—	—	*[e]	32	2	—	—	1
Acala SJ-2	—	—	—	—	—	—	—	—	—	—	—	—	—	74	9	10
Acala SH-5	—	—	—	—	—	—	—	—	—	—	—	*	—	18	—	2
Blightmaster A-5	—	—	—	—	—	—	—	—	—	—	2	—	—	—	—	1
Cascot B-2	—	—	—	—	—	—	—	—	—	3	1	*	—	—	—	1
Cascot L-7	—	—	—	—	—	*	*	*	—	1	2	—	1	—	—	1
Coker 304	17	24	23	—	—	1	1	—	—	—	*	*	—	—	—	1
Coker 310	47	22	19	3	*	*	6	*	—	—	*	—	*	—	—	1
Coker 315	6	36	13	1	—	—	—	—	—	—	*	—	*	—	—	1
Deltapine 41	—	—	2	14	12	13	12	14	23	1	*	*	4	—	10	4
Deltapine 55	—	—	7	7	18	8	16	4	16	—	*	—	6	*	42	4
Deltapine 61	—	—	2	12	9	2	6	15	11	2	*	—	48	7	18	6
Deltapine 70	—	—	—	—	—	—	—	—	—	—	*	—	22	1	—	1
Deltapine SR-4	—	—	—	—	—	—	—	—	—	*	1	*	—	—	—	1
Dunn 119	—	—	—	—	—	—	—	—	—	—	2	10	—	—	—	1
Dunn 120	—	—	—	—	—	—	—	—	—	—	2	1	—	—	—	1
Dunn 219	—	—	—	—	—	—	—	—	—	—	1	1	—	—	—	1
GP 3774	—	—	—	—	—	—	—	—	—	*	2	—	—	—	—	1
GSA 71	—	—	—	—	—	—	—	—	—	1	11	2	—	—	—	6
GSA 75	—	—	—	—	—	—	—	—	—	*	2	—	—	—	—	1
Lankart 57	—	—	—	—	—	—	—	—	—	44	5	—	—	—	—	4
Lankart 611	—	—	—	—	—	—	—	—	—	16	7	—	—	—	—	4
Lankart LX-571	—	—	—	—	—	—	—	—	—	10	3	—	—	—	—	2
Lockett 77	—	—	—	—	—	—	—	—	—	1	3	*	—	—	—	2
McNair 220	24	8	15	2	*	1	—	—	—	—	1	1	*	—	—	1

Table 2. Cont.

VARIETY[b]	STATE[c]															
	NC	SC	GA	AL	MS	TN	MO	AR	LA	OK	TX	NM	AZ	CA	All other[d]	Total
Paymaster 18	—	—	—	—	—	—	—	—	—	—	1	—	—	—	—	1
Paymaster 266	—	—	—	—	—	—	—	—	—	—	3	2	—	—	—	2
Paymaster 303	—	—	—	—	—	—	—	—	—	*	9	3	—	—	—	5
Paymaster 792	—	—	—	—	—	—	—	—	—	*	2	1	—	—	—	1
Quapaw	—	—	—	—	—	—	—	—	—	1	4	2	—	—	—	2
Stoneville 213	—	—	1	9	15	15	24	29	12	2	2	*	1	*	—	5
Stoneville 825	—	—	13	45	39	51	34	31	33	2	1	1	8	*	—	11
Stripper 31	—	—	—	—	—	—	—	—	—	—	4	—	—	—	—	1
Tamcot CAMD-E	—	—	—	—	—	—	*	—	—	*	2	*	—	—	—	1
Tamcot SP-21S	—	—	—	—	—	—	—	—	—	*	2	4	—	—	—	1
Tamcot SP-37H	—	—	—	—	—	—	—	—	—	*	1	—	—	—	—	1
Tamcot SP-21	—	—	—	—	—	—	—	—	—	*	8	11	—	—	—	4
Tamcot SP-37	—	—	—	—	—	—	—	—	—	1	4	—	—	—	—	2
Westburn M	—	—	—	—	—	—	—	—	—	12	*	—	—	—	—	1
Western SP-44	—	—	—	—	—	—	—	—	—	—	1	—	—	—	—	1
All other	2	2	1	2	*	—	*	*	—	3	2	1	4	—	—	2
Total	96	92	96	95	93	91	99	93	95	100	89	74	96	100	79	100

[a] Data summarized from "Cotton varieties planted 1981 crop," USDA Marketing Service Cotton Division, Memphis (August, 1981).

[b] Varieties comprising less than 1% of the total U.S. acreage are not included.

[c] All other includes Florida and Nevada.

[d] Percent by state.

[e] Less than 0.5%.

accounts for 1%. Less than 5% (60,000 acres) of the total U.S. acreage is in Pima cotton, and this predominantly is restricted to Arizona, New Mexico, and Texas.

Other leading cotton producing countries are the Soviet Union, the Peoples Republic of China, India, Brazil, Egypt, Pakistan, and Turkey. The cotton species grown in these and other countries are presented in Table 3.

KEY BREEDING AND CROP IMPROVEMENT PROBLEMS

Cotton has some special breeding requirements. First, a premium is placed on fiber quality, i.e., length, strength, and uniformity, that is dictated by the modern ginning and spinning technology. Secondly, there is a need for a uniformly short-stature, early, determinate flowering plant that is suitable for once-over machine harvesting in areas where mechanical agriculture is practiced. Late season insect buildups and damage from fall rains have made it more economical to grow short-season varieties. Breeding for characteristics such as small bract, frego bract, and minimum exposure boll can reduce dust during processing. Increased herbicide resistance is sought in addition to resistance to insect, fungal, nematode, and bacterial pests.

Over 100 insect species damage cotton in the cotton belt (Berger, 1969). Two-thirds of the loss due to insects is by the boll weevil, cotton bollworm, and tobacco bollworm. Controlling insect pests is critical and chemical pesticides are extensively used. Varieties that have increased insect resistance or escape, therefore, require fewer insecticide treatments.

Several bacterial and fungal parasites infect cotton and contribute to reduced yield or crop loss. Common diseases are seedling disease, fusarium wilt, verticillium wilt, bacterial blight, *Ascochyta* blight, boll rot, cotton root rot, and cotton rust. Obviously, genes providing increased resistance to any of these diseases are important to cotton breeding.

Several approaches are currently being used by cotton breeders to develop improved varieties. One of these is the "multiple-adversity resistance" (MAR) program (Bird, 1973, 1975, 1981). The MAR breeding places emphasis on selection for resistance to several pathogens, nematodes, insects, stresses, morphological traits, and fiber characteristics in the same breeding line. Bird (1981) has shown that certain genetic interrelationships exist, and selection for key characters, i.e., seed coat resistance to mold, slow rate of germination after 8 days at 13.3 C, and cotyledon resistance to races of the bacterial blight pathogen, can result in simultaneous improvement in resistance to other major diseases, higher yield potential, and earliness. This approach led to improved cotton varieties, e.g., those of the Tamcot series, which currently account for 17% of the cotton acreage in Texas (Table 2). Other researchers have been involved with transferring desirable traits of other species into upland cotton (Meyer, 1974). Genes for bacterial blight resistance and fiber length and strength have been transferred from *G. barbadense*. The complex of wild species provides a potential

Table 3. *Gossypium* Species Grown in Major Cotton Producing Countries

COUNTRY	*GOSSYPIUM* SPECIES[a]	TOTAL WORLD PRODUCTION (%)[b]
USA	*hirsutum*	22.5
USSR	*hirsutum*	20.1
China	*hirsutum*	15.7
India	*arboreum*	8.7
	herbaceum	
	hirsutum	
	barbadense	
Pakistan	*hirsutum*	4.6
	arboreum	
Brazil	*hirsutum* (S and NE)	4.1
	barbadense (NE)	
Turkey	*hirsutum*	3.6
	herbaceum	
Egypt	*barbadense*	3.4
Mexico	*hirsutum*	2.4
Guatemaula	*hirsutum*	1.04
Argentina	*hirsutum*	1.0
Syria	*hirsutum*	0.94
Sudan	*barbadense*	0.93
	hirsutum	
Iran	*hirsutum*	0.78
	herbaceum	
Nicaragua	*hirsutum*	0.78
Colombia	*hirsutum*	0.77
Greece	*hirsutum*	0.7
El Salvador	*hirsutum*	0.5
Israel	*hirsutum*	0.5
Peru	*barbadense*	0.5
	hirsutum	

[a]Data from Berger, 1969.

[b]Data from FAO Crop Production Handbook, 1979.

rich source of genes for disease and insect resistance, fiber quality, and morphological traits.

Genetic manipulations to transfer genes between species are not always easy, due to the cytogenetic differentiation among the species, various reproductive isolation barriers, and ploidy level differences. Genes have been transferred by producing synthetic AD amphidiploids and crossing these with upland cotton. Also wild species have been crossed to upland cotton. The resulting triploid is made hexaploid by experimentally doubling the chromosome number. Selection for the desired trait and agronomic properties can be exercised, and the chromosome number reverts to tetraploid following recurrent backcrossing to *G. hirsutum* (Meyer, 1974).

Over the last 20 years, considerable interest has been directed toward the possible commercial production of hybrid cotton. Hybrid vigor has been demonstrated in some crosses (Wilson et al., 1968; Pathak and Kumar, 1976), and male sterile cytoplasms have been developed (Meyer, 1974). However, the cotton pollen is not windblown and insects are necessary for pollen transport in the hybrid seed production process. The lack of adequate pollinators, or their loss from insecticides, is a major problem. Other problems with potential production of hybrid cottons are the increased cost of hybrid seed, yield losses due to incomplete restoration of fertility in the hybrid (Fisher, 1980), and the general low degree of heterosis (Pathak and Kumar, 1976).

The high nutritional quality of cottonseed protein could make it an important human food. However, the cottons contain gossypol and related terpenoids that are concentrated in glands and are toxic to nonruminant animals. The chemical removal is costly and reduces protein quality. Genes for glandless cotton exist and have been incorporated into some commercially grown varieties. The glandless seeds are free of gossypol and are suitable for human consumption.

There is concern that glandless cotton may be more susceptible to insect damage (Bell and Stipanovic, 1979; Fryxell, 1980). Bell (1974) and Bell and Stipanovic (1979) consider the terpenoids to be major antibiotics essential to the cotton plant's disease resistance. A major genetic manipulation would be to transfer the "glanded-plant glandless-seed" trait found in wild Australian *Gossypium* (Fryxell, 1980) into upland cotton. Then selection for the quantity and composition of terpenoids to obtain potential increased disease resistance (Bell and Stipanovic, 1979) could occur without affecting the terpenoid levels in the glandless seed.

REVIEW OF THE LITERATURE

Six categories of plant tissue culture have been reported for cotton: callus, ovule-embryo, shoot tip, anther, and protoplast culture, as well as somatic embryogenesis. The intent of these investigations was varied. Workers involved with callus cultures ultimately wanted to initiate suspension cultures, investigate protoplast isolation and fusion to form somatic hybrids with subsequent plant regeneration. Ovule cultures were examined to gain basic information on fiber development, and ultimately, a better understanding of fiber yield and quality. Embryo cultures have been studied to examine embryo rescue techniques to obtain hybrids from embryos which normally abort. Anther culture was pursued to produce haploid and/or homozygous diploids for breeding purposes and mutagenesis studies. In this review each of these tissue culture techniques will be examined to determine the status of its potential application in cotton improvement.

Callus

Gossypium species from which callus has been initiated are listed in Table 4. Callus initiated from funicular tissue explants or the micropylar region of cultured ovules from *G. hirsutum* was first reported by

Table 4. Callus Cultures of *Gossypium* Species

SPECIES	MEDIUM	GROWTH REGULATORS (μM)		EXPLANT	TEMPERATURE, ILLUMINATION	REFERENCE
		1° Medium	2° Medium			
G. hirsutum	WH			Ovule	30 C, dark	Beasley, 1971
G. hirsutum Auburn 56	SH	2.3 2,4-D, 10.7 pCPA, 0.5 KIN		Mesocotyl	27±2 C, continuous illumination	Schenk & Hildebrandt, 1972
G. hirsutum Deltapine 16	LS	0.5 2,4-D		Leaf disc	30 C, 9000 lux	Davis et al., 1974
G. hirsutum (Deltapine 16), *G. barbadense* (Pima S-4)	LS	5.1 IAA, 4.8 BA		Peeled stem	28-30 C, dim fluorescent	Sandstedt, 1975
G. arboreum	MS	11.0 IAA, 4.6 KIN	11.0 NAA, 2.2-4.4 BA, or 5.4 NAA, 5-10 2iP	Hypocotyl	29±1 C, 15:9, 130 $\mu Em^{-2}/s$	Smith et al., 1977
G. barbadense	LS	27.0 NAA, 4.4 BA	11.0 NAA, 0.44 BA	Anther	29 C, 18:6, 44 $\mu Em^{-2}/s$	Barrow et al., 1978
G. hirsutum	Gresshoff & Doy, 1972; DBM II	27.0 NAA, 4.4 BA	5.4 NAA, 0.4 BA	Anther	29 C, dark	Barrow et al., 1978
G. hirsutum, *G. arboreum*	MS	5.4 NAA, 4.6 KIN		Hypocotyl, stem segments	25±2 C, diffuse light	Rani & Bhojwani, 1976

SPECIES	MEDIUM	GROWTH REGULATORS (μM)		EXPLANT	TEMPERATURE, ILLUMINATION	REFERENCE
		1° Medium	2° Medium			
G. hirsutum Hancock	Beasley & Ting, 1973	9.9 CEPA		Ovule	30 C, dark	Hsu & Stewart, 1976
	Beasley & Ting, 1973	9.9 CEPA, 0.06 GA_3		Ovule	30 C, dark	Stewart & Hsu, 1977
G. barbadense Pima S-4 & S-5	1° Beasley & Ting, 1973; 2° LS	91.3 NAA, 50 mg/l	11.0 NAA, 0.44 BA	Cotyledon	29 C, 18:6, 2000 lux	Katterman et al., 1977
G. anomalum	MS	11.0 IAA, 4.6 KIN	11.0 IAA, 4.6 KIN or 5.7 IAA 49.0 2iP	Hypocotyl	29±1 C, 15:9, 130 $\mu Em^{-2}/s$	Price et al., 1977
G. arboreum	MS	11.0 IAA, 4.6 KIN	11.0 IAA, 2.2–4.4 BA or 5.7 NAA, 25–49 2iP	Hypocotyl	29±1 C, 15:9, 130 $\mu Em^{-2}/s$	Price et al., 1977
G. armourianum	MS	11.0 IAA, 4.6 KIN	29.0 IAA, 4.9 2iP	Hypocotyl	29±1 C, 15:9, 130 $\mu Em^{-2}/s$	Price et al., 1977

Table 4. Cont.

SPECIES	MEDIUM	GROWTH REGULATORS (μM)		EXPLANT	TEMPERATURE, ILLUMINATION	REFERENCE
		1° Medium	2° Medium			
G. hirsutum	MS	11.0 IAA, 4.6 KIN	29.0 IAA, 4.9 2iP or 0.6 NAA, 4.9 2iP	Hypocotyl	29±1 C, 15:9, 130 $\mu Em^{-2}/s$	Price et al., 1977
G. klotzschianum	MS	11.0 IAA, 4.6 KIN	0.6 NAA, 4.9 2iP or 5.4 NAA, 49.0 2iP	Hypocotyl	29±1 C, 15:9, 130 $\mu Em^{-2}/s$	Price et al., 1977
G. raimondii	MS	11.0 IAA, 4.6 KIN	0.6 NAA, 4.9 2iP	Hypocotyl	29±1 C, 15:9, 130 $\mu Em^{-2}/s$	Price et al., 1977

Beasley (1971). This was the beginning of several studies by Beasley to determine the physiological role of growth regulators on fiber development in fertilized (Beasley and Ting, 1974) and unfertilized (Beasley et al., 1974) cotton ovules in culture.

Friable callus from the micropylar region of *G. hirsutum* ovules was reported by Hsu and Stewart (1976) in studies on growth and development of the ovule. Ethylene, released in the culture medium by (2-chloroethyl) phosphonic acid (CEPA), stimulated cell division and expansion in developing callus. Further studies by Stewart and Hsu (1977) reported callus growth to be greatly stimulated by the combination of CEPA and GA_3.

Schenk and Hildebrandt (1972) were successful in establishment of friable callus cultures from *G. hirsutum* mesocotyl explants. Similarly, suspension cultures have been established from callus originating from leaf explants (Davis et al., 1974). Rani and Bhojwani (1976) obtained slow-growing friable callus from *G. hirsutum* and *G. arboreum*.

A strong antioxidant, dithiothreitol (DTT), and high concentrations of NAA were used by Katterman et al. (1976) to initiate a compact callus from *G. barbadense* cotyledons. Other antioxidants, ascorbic acid or polyvinylpyrollidone, were not as effective in preventing polyphenolic oxidation of the callus. Subculture of this callus involved changing the carbohydrate from sucrose to glucose and removing the DTT. These workers observed root formation from subcultured callus and also reported regeneration of one plant.

Smith et al. (1977) defined conditions for initiation and subculture of hypocotyl-derived callus of *G. arboreum*. The combination of glucose, high light intensity, and high temperature resulted in optimal callus growth without browning. These workers also regenerated one plant from cotyledon-derived callus.

Initiation and subculture of callus from five species were defined by Price et al. (1977) and Price and Smith (1977) (Table 5). With glucose as a carbon source and high light intensity, subculture of callus required mixtures of different growth regulators at different concentrations. Inhibition of browning or coloring of the medium was attributed to the glucose and high light intensity.

Difficulties with callus induction, browning of the callus, and subculture generally have been overcome with *Gossypium* species (Smith et al., 1977; Price et al., 1977; Price and Smith, 1977). Phenolic oxidation can be avoided by using glucose, high light intensity, and 29-30 C culture conditions. All attempts to regenerate plants from callus cultures have been unsuccessful. The two instances of plant regeneration mentioned above may have originated directly from cotyledon tissue. The further development of a viable cell culture methodology for *Gossypium* is dependent on development of repeatable, high frequency shoot proliferation from callus or somatic embryogenesis.

Ovule-Embryo Culture

In vitro studies of cotton ovule culture have served two purposes. First, they have been valuable in providing information on the physiology of fiber development and, second, in obtaining interspecific

Table 5. Callus Induction Medium for a Wide Range of *Gossypium* Species

COMPONENTS	CONCENTRATION
MS salts	
Myo-inositol	555.0 μM
Thiamine·HCl	1.2 μM
IAA	11.0 μM
KIN	4.6 μM
Glucose	166.0 mM
Agar	6 g/l
pH	5.7

hybrids. In many instances, hybrids fail to develop in vivo due to endosperm and embryo incompatibility or breakdown of endosperm development. Weaver (1957) observed that interspecific crosses between *G. hirsutum* and *G. arboreum* abort. Cotton embryos excised 20-27 days after flowering were grown in vitro by Lofland (1950); however, embryos excised earlier than 15 days did not undergo further development. Cells of embryos weighing 37 mg fail to divide or elongate in vitro even with IAA and GA_3 treatments (Dure and Jensen, 1957).

Hybrid embryos of *G. arboreum* x *G. hirsutum* excised 20 days after pollination were successfully cultured to maturity by Weaver (1958). Mauney et al. (1967) cultured young embryos (12-14 days old) at the heart stage of development with up to 75% survival on the Mauney et al. (1975) medium (Table 6). The cultures were grown at 30 C under continuous 100 ft-c illumination. The addition of ammonium or calcium malate to the culture medium at 4 mg/ml was necessary for rapid growth and enhanced viability of the embryos. An analysis of the liquid endosperm from 12 to 14 day-old cotton ovules from intact plants has revealed an endogenous malic acid concentration in excess of 7 mg/ml. Eid et al. (1973) cultured fertilized cotton ovules of *G. hirsutum* on MS medium without growth regulators. Normal seedlings were obtained on ovule cultures at 5-10 days post anthesis.

Valuable information concerning the in vitro culture of cotton ovules (Beasley, 1973; Beasley and Ting, 1973) was later used in studies of in ovule embryo culture by Stewart and Hsu (1977, 1978) to obtain interspecific hybrids. Stewart and Hsu (1977) cultured ovules in the dark at 30 ± 2 C on liquid medium for germination (Table 7). After germination, balanced shoot and root growth of the embryos occurred following transfer of the ovules to a secondary medium (Table 8) at 30 C with 16 hr light at 58 $\mu Em^{-2}s^{-1}$ (Stewart and Hsu, 1977).

Shoot-Tip Culture

Shoot tip cultures were initiated by Chappell and Mauney (1976) to determine the nutritional requirements for normal growth and develop-

Table 6. Medium for Heart Stage Embryo Culture of Cotton[a]

COMPONENT	CONCENTRATION
$Ca(NO_3)_2 \cdot 4H_2O$	3.05 mM
NH_4NO_3	8.24 mM
KCl	4.36 mM
KNO_3	3.96 mM
$MgSO_4 \cdot 7H_2O$	1.46 mM
KH_2PO_4	1.37 mM
$MnSO_4 \cdot 4H_2O$	0.15 mM
H_3BO_3	0.13 mM
$ZnSO_4 \cdot 7H_2O$	0.05 mM
$Fe_2(SO_4)_3$	0.03 mM
KI	0.02 mM
Glycine	0.13 mM
Pyridoxine·HCl	0.012 mM
Nicotinic acid	0.02 mM
Thiamine	1.5 µM
Adenine sulfate	0.17 mM
KIN	0.47 µM
CH	250 mg/l
CW	15% v/v
Sucrose	0.058 M
Ammonium malate	0.026 M

[a]Mauney et al., 1976.

ment of apical meristems of *G. hirsutum.* They were unsuccessful in achieving normal growth and development, but did observe new leaf primordia initiation. Roots did not develop. GA_3 at 1.4–2.9 µM in a simple salt medium seemed to stimulate growth, but coconut milk, auxins, and vitamins did not appear to have an effect on meristematic development.

Cultured axillary buds from mature *G. hirsutum* TM-1 developed into shoots at a frequency of 20–40% (Smith, unpublished) on a medium containing MS salts, 1.2 µM thiamine·HCl, 4.6 µM KIN, 5.7 µM IAA, 83 mM glucose, and activated charcoal. The subculture of shoots onto the same medium with 11.0 µM NAA resulted in low frequency root formation. Further work is in progress to enhance the frequency of shoot and root development.

Anther Culture

A mixture of haploid and diploid callus resulted from anther cultures of *G. barbadense* and *G. hirsutum* (Barrow et al., 1978). Liquid cultures originating from these cultures had the diploid chromosome complement. Culture medium for callus induction is summarized in Table 4. A change from sucrose to glucose was made for the secon-

Table 7. Medium upon Which Ovules[a] Were Cultured for Development of Mature Embryos of *G. hirsutum* L. cv. Hancock[b]

COMPONENT	CONCENTRATION
KH_2PO_4	2.0 mM
H_3BO_3	0.1 mM
$Na_2MoO_4 \cdot 2H_2O$	1.0 µM
$CaCl_2 \cdot 2H_2O$	3.0 mM
KI	0.05 mM
$CoCl_2 \cdot 6H_2O$	0.1 µM
$MgSO_4 \cdot 7H_2O$	2.0 mM
$MnSO_4 \cdot H_2O$	0.1 mM
$ZnSO_4 \cdot 7H_2O$	0.03 mM
$CuSO_4 \cdot 5H_2O$	0.1 µM
KNO_3	50.0 mM
$FeSO_4 \cdot 7H_2O$	0.03 mM
Na_2EDTA	0.03 mM
NH_4NO_3	15.0 mM
Nicotinic acid	4.0 µM
Pyridoxine·HCl	4.0 µM
Thiamine·HCl	4.0 µM
Myo-inositol	1.0 mM
Sucrose	0.058 M
IAA	5.1 µM
GA_3	0.6 µM
KIN	0.47 µM

[a]Two days after fertilization.

[b]Stewart and Hsu, 1977.

dary medium. Roots differentiated from callus when the light intensity was increased from 26 to 56 to 150 $\mu Em^{-2}s^{-1}$.

Somatic Embryogenesis

Price et al. (1979) reported somatic embryogenesis in suspension cultures of *G. klotzschianum* after 3-4 weeks of culture in a medium containing glutamine (optimally, 10-15 mM). Embryogenesis occurred only after the callus was precultured on a medium containing 49 µM 2iP. Asparagine was much less effective than glutamine in promoting embryo development. The presence of 2,4-D in the medium resulted in increased vigor of the suspension cultures and subsequently in a higher frequency of embryo formation. Intact plants have not been recovered.

Protoplasts

Gossypium hirsutum protoplasts were isolated from hypocotyl-derived callus by Bhojwani et al. (1977). The enzyme solution contained 4%

Table 8. Medium for Balanced Root and Shoot Growth from in vitro Germinated Embryos[a]

COMPONENT	CONCENTRATION
KNO_3	5.0 mM
NH_4NO_3	3.0 mM
$MgSO_4 \cdot 7H_2O$	2.0 mM
$CaCl_2 \cdot 2H_2O$	1.2 mM
KH_2PO_4	0.2 mM
FeEDTA	1.1 mM
H_3BO_3	1.0 mM
$MnSO_4 \cdot 4H_2O$	1.0 mM
$ZnSO_4 \cdot 7H_2O$	1.0 mM
KI	1.0 mM
$NaMoO_4 \cdot 2H_2O$	1.0 mM
$CuSO_4 \cdot 5H_2O$	1.0 mM
$CoCl_2 \cdot 6H_2O$	1.0 mM
Nicotinic acid	4.0 µM
Pyridoxine·HCl	4.0 µM
Thiamine·HCl	4.0 µM
Sucrose	0.058 M
Agar	10 g/l
pH	5.5

[a]Stewart and Hsu, 1977.

Meicelase, 0.25% Driselase, 0.4% Macerozyme, 11% mannitol, and inorganic salts at pH 5.6. Incubation was in the dark for 3 hr at 31 C. Callus older than 18 days did not yield viable protoplasts; however, callus 6-9 days old resulted in a 15% yield of protoplasts. Protoplasts were cultured in darkness at 27-29 C on B-5 salts and vitamins with 0.55 mM inositol, 27.7 mM glucose, 29.2 mM sucrose, 0.6 M mannitol, 2.3 µM 2,4-D, 1.1 µM NAA, and 0.9 µM BA at pH 5.6. The addition of 3.12 mM NH_4NO_3 and 5.31 mM $CaCl_2$ enhanced survival. Culture in the light resulted in budding and at temperatures below 25 C no divisions were observed. Only 1.2% of the protoplasts divided to form colonies. Colonies of 25-30 cells were visible at 5 weeks.

Cotyledon tissue from 10 to 12 day seedlings of *G. hirsutum* was used by Khasanov and Butenko (1979) for protoplast isolation. Yields of 68.5% protoplasts were reported using 1.5% cellulase, 0.5% pectinase, and 0.5 M mannitol, incubated at 18-21 C for 16-18 hr. Protoplasts were plated on an MS medium with 2.3 µM KIN, 4.5 µM 2,4-D, sucrose, glucose, and 16.7 mM galactose. Galactose was necessary for cell wall formation which was observed at three days. Macroscopic callus was not regenerated.

Isolation and culture of *G. klotzschianum* protoplasts from hypocotyl-derived callus yielded 20.5% protoplasts (Finer and Smith, 1982). The enzyme solution contained 1% Cellulysin, 0.5% Macerase, and 0.7 M mannitol at pH 4.7. Incubation was in the dark with gentle agitation (40 rpm) for 14 hr. Following washing, the protoplasts were suspended in an MS medium with 0.55 mM inositol, 1.2 μM thiamine·HCl, 83.1 mM glucose, 0.6 M mannitol, and 3 g/l agar, pH 5.7. After incubation for four days in the dark, protoplasts were placed under low light (50 $\mu Em^{-2}/s$) at 16:8 photoperiod, 29 C. Macroscopic colonies were formed within 2 weeks.

As the preceding review indicates, much progress remains to be accomplished before plant tissue culture methods can be applied to enhance breeding programs in cotton. The induction of callus from a wide range of *Gossypium* species is feasible as is the isolation of protoplasts; however, plant regeneration is not currently possible and the reestablishment of a vigorous callus after protoplast isolation has not been accomplished.

PROTOCOLS

The well-estabished protocols for callus induction and subculture, ovule-embryo culture, somatic embryogenesis, and protoplast isolation are presented below. Other aspects of cotton tissue culture are not included, since more experimentation is necessary before they are at the protocol stage of development.

Callus

Callus from a wide range of *Gossypium* species can be obtained by the following procedure:

1. Surface-sterilize delinted seeds in 1% sodium hypochlorite with one drop Tween 20/100 ml, rinse, dip in 95% alcohol, and flame.
2. Aseptically germinate in sterile vermiculite moistened with distilled water.
3. At 5-14 days, depending on the genotype, excise hypocotyl, split and culture cut surface on agar. Medium for callus induction is outlined in Table 5. Culture at 130 $\mu Em^{-2}/s$, 29 ± 2 C, 15:9 photoperiod.
4. Subculture callus following 1-2 months of growth. The subculture media varies according to the genotype. Table 4 indicates growth regulator changes for subcultures of individual species.

Ovule-Embryo Culture

Heart stage embryos can be cultured to maturity by the following procedure (Mauney et al., 1967):

1. Harvest fruits 12-14 days after pollination.
2. Surface sterilize.
3. Excise embryos using a sterile dissecting needle and scalpel. Embryos at the heart stage are approximately 0.2-0.3 mm long.

4. Culture on medium in Table 6 at 30 C with continuous 100 ft-c cool-white fluorescent light.

A procedure for in ovule embryo culture and seedling development is as follows (Stewart and Hsu, 1977):

1. Harvest fruits 2 days after pollination.
2. Surface-sterilize by dipping in ethanol and flame.
3. Excise ovules and float on the medium in Table 7. Culture in the dark at 30 ± 2 C for about 10 weeks.
4. Viable plants with balanced root and shoot growth are obtained by subculturing ovules to the agar-solidified medium in Table 8 at 30 C with a 16 hr photoperiod of about 58 $\mu Em^{-2}/s$.

Somatic Embryogenesis (Price and Smith, 1979; Finer et al., 1983)

1. Initiate callus cultures from hypocotyl segments of aseptically germinated seeds on MS salts, 1.1 µM thiamine•HCl, 0.055 mM myo-inositol, 0.088 M glucose, 11 µM IAA, 4.6 µM KIN, pH 5.7, 0.6% agar at 29 ± 1 C, 16-hr photoperiod, 130 $\mu Em^{-2}/s$.

[For enhanced embryo yield, initiate callus from previously obtained asexual embryo on the same medium as in step 2.]

2. Callus develops in 4-7 weeks and is subcultured bimonthly onto above medium modified by replacing IAA and KIN with 5.4 µM NAA and 49.0 µM 2iP. Reduce light intensity to 30 $\mu Em^{-2}/s$.
3. Initiate suspension cultures with 2-3 g of callus following 2 weeks of subculture into 50 ml of liquid medium containing Gamborg B5 salts and vitamins, 15 mM glutamine, 0.45 µM 2,4-D, 0.058 M sucrose, pH 5.7 at 150 rpm, 16-hr photoperiod, 60 $\mu Em^{-2}/s$.
4. After 2 weeks filter cultures through 6-8 layers of cheesecloth and resuspend cell aggregates retained on cheesecloth into fresh suspension medium without 2,4-D.
5. Somatic embryos are apparent after 3-4 weeks.

Isolation of Protoplasts from Callus

1. Initiate callus cultures from hypocotyl segments of aseptically germinated seeds on MS salts, 1.1 µM thiamine•HCl, 0.55 mM myo-inositol, 0.088 M glucose, 11.0 µM IAA, 4.6 µM KIN, pH 5.7, 0.6% agar at 29 ± 1 C, 16-hr photoperiod, 30 $\mu Em^{-2}/s$.
2. Callus develops in 4-7 weeks. Subculture callus onto above medium modified by replacing IAA and KIN with 5.7 µM NAA and 49.0 µM 2iP.
3. Use callus 2 weeks after subculture and isolate protoplasts in the enzyme mixture containing 1% cellulysin, 0.5% macerase, 0.7 M mannitol, pH 4.7 at 40 rpm, 14 hr in dark.

4. Wash protoplasts 3 times by centrifuging at 100 x g for 5 min. After each centrifugation, decant off supernatant and discard.
5. Suspend ca. 10^4-10^5 protoplasts in 1 ml of the agar-solidified medium and pour into 6-cm petri dishes while warm.
6. Cell divisions are apparent after 3 days and macroscopic colonies are formed by 2 weeks.

FUTURE PROSPECTS

The application of tissue culture techniques to cotton breeding unfortunately will be minimal in the immediate future. The most promising techniques for immediate use are ovule and bud culture. Stewart and Hsu (1978) foresee ovule culture as a way of obtaining interspecies crosses that are either not possible or difficult by conventional hybridization methods. If the interspecific hybrids obtained are triploid, or sterile due to lack of chromosome homology, then colchicine doubling of the chromosome number is needed to restore fertility. Colchicine treatment may be easier to administer and control and be more effective if done in vitro on cultured bud explants.

The key to any major application of tissue culture in breeding programs is the achievement of high frequency plant regeneration from callus or suspension cultures. Plant regeneration from tissue cultures has been achieved only with one cotton species, *G. klotzschianum*; however, the protocol is unreliable. The fact that low frequency organogenesis has been partially achieved with one species provides optimism that further research will result in more reliable regeneration protocols. Once high frequency and vigorous plant regeneration can be obtained, then there can be straightforward uses for tissue culture in plant breeding.

In a plant breeding program, it would be desirable to test the performance of particular genotypes. The direct cloning of single plants from segregating populations would permit early evaluation of selections under different environmental conditions. This in turn would allow plants that perform well in different environments to be used in breeding programs. Also, the time needed between the development of new varieties and the seed increases for commercial production could be shortened significantly if cotton plants could be cloned in vitro.

The development of *Gossypium* protoplast technology has exciting potential uses for genetic manipulation. First, somatic cell hybridization may permit the production of previously unobtainable hybrids, which may be used to introgress genes from wild *Gossypium* species into upland cotton. New nuclear-cytoplasm combinations could provide male-sterile cytoplasms for hybrid cotton studies and/or increase the cytoplasmic diversity among cultivated varieties.

Anther culture has a promising use for induced mutagenesis studies in cotton. Mutagenizing microspores or haploid tissues, with subsequent in vitro regeneration of plants would permit the direct recovery of mutant phenotypes, thereby reducing the time and space requirements of conventional induced mutation screening programs. Mutants such as "glanded plant-glandless seeds," or the removal of the cyclopropene ring from malvalic acid potentially could be induced in A and D genome plants and then introgressed into *G. hirsutum* by conventional techniques. The removal of the cyclopropene ring in malvalic acid may be desirable since cyclopropenoid fatty acids are known to exert toxic or adverse biological effects when ingested by a variety of animals (Phelps et al., 1965; Mickelsen and Yang, 1966; Scarpelli, 1974).

Work needs to be done with cotton tissue culture before it can be used as a technique in plant breeding. Most importantly, the techniques need to be developed for the wild as well as the cultivated species. It is necessary to consider the genus *Gossypium* as a resource, and not just the cultivated species and their varieties.

ACKNOWLEDGMENTS

Technical article from the Texas Agricultural Experiment Station. We thank Drs. K. El-Zik, C. Nessler, and J.D. Smith for constructive comments concerning the manuscript and Dr. L. Bird for valuable conversations pertaining to cotton breeding. We appreciate the numerous publications and pamphlets of the National Cotton Council that were provided by Dr. J. Wyrick.

KEY REFERENCES

Price, H.J. and Smith, R.H. 1979. Somatic embryogenesis in suspension cultures of *Gossypium koltzschianum* Anderss. Planta 145:305-307.

______, and Grumbles, R.M. 1977. Callus cultures of six species of cotton (*Gossypium* L.) on defined media. Plant Sci. Lett. 8:115-119.

Smith, R.H., Price, H.J., and Thaxton, J.B. 1977. Defined conditions for the initiation and growth of cotton callus in vitro. I. *Gossypium arboreum*. In Vitro 13:329-334.

Stewart, J.McD. and Hsu, C.L. 1977. In-ovulo embryo culture and seedling development of cotton (*Gossypium hirsutum* L.). Planta 137: 113-117.

REFERENCES

Barrow, J.B., Katterman, F., and Williams, D. 1978. Haploid and diploid callus from cotton anthers. Crop Sci. 18:619-622.

Beasley, C.A. 1971. In vitro culture of fertilized cotton ovules. Bioscience 21:906-907.

______ 1973. Hormonal regulation of growth in unfertilized cotton ovules. Science 179:1003-1005.

______ and Ting, I.P. 1973. The effects of plant growth substances on in vitro fiber development from fertilized cotton ovules. Am. J. Bot. 60:130-139.

______ and Ting, I.P. 1974. Effects of plant growth substances on in vitro fiber development from unfertilized cotton ovules. Am. J. Bot. 61:188-194.

______, Ting, I.P., Linkins, A.E., and Birnbaum, E.H. 1974. Cotton ovule culture: A review of progress and a review of potential. In: Tissue Culture and Plant Science (H.E. Street, ed.) pp. 169-192. Academic Press, New York.

Beasley, J.O. 1940. The origin of American tetraploid *Gossypium* species. Am. Nat. 74:285-286.

______ 1942. Meiotic chromosome behavior in species, species hybrids, haploids and induced polyploids of *Gossypium*. Genetics 27:25-54.

Bell, A.A. 1974. Biochemical bases of resistance of plants to pathogens. In: Proceedings of the Summer Institute on Biological Control of Plant Insects and Disease (F.G. Maxwell and F.A. Harris, eds.) pp. 403-462. Univ. Press of Mississippi, Jackson.

______ and Stipanovic, R.D. 1979. Biochemistry of disease and pest resistance in cotton. Mycopathologia 65:91-106.

Berger, J. 1969. The World's Major Fibre Crops, Their Cultivation and Manuring. Conzett and Huber, Zurich.

Bird, L.S. 1973. Interrelationships of resistance and escape from multi-diseases and other adversities. In: Proceedings Beltwide Cotton Production Research Conference, pp. 92-97. National Cotton Council, Memphis.

______ 1975. Genetic improvement of cotton for multi-adversity resistance. In: Proceedings Beltwide Cotton Production Research Conference, pp. 150-152. National Cotton Council, Memphis.

______ 1982. Multi-adversity (diseases, insects and stresses) resistance (MAR) in cotton. Plant Dis. Rep. 66:173-176.

Bhojwani, S.E., Power, J.B., and Laking, E.C. 1977. Isolation, culture and division of cotton callus protoplasts. Plant Sci. Lett. 8:85-89.

Booker, G., Hunnings, A., Howell, J., Crawford, W., Duren, R., and Bowling, A. 1981. The Economic Outlook for U.S. Cotton 1981. National Cotton Council of America, Memphis.

Chappel, E.J. and Mauney, J.R. 1967. Culture of the apical meristem of *Gossypium hirsutum* in vitro. Phyton 24:93-100.

Cherry, J.P., Katterman, F.R.H., and Endrizzi, J.E. 1970. Comparative studies of seed proteins of species of *Gossypium* by gel electrophoresis. Evolution 22:431-447.

Davis, D.G., Dusbabek, K.E., and Hoerauf, R.A. 1974. In vitro culture of callus tissues and cell suspensions from okra (*Hibiscus esculentus* L.) and cotton (*Gossypium hirsutum* L.). In Vitro 9:395-398.

Dure, L.S. and Jensen, W.A. 1957. The influence of gibberellic acid and indoleacetic acid on cotton embryos cultured in vitro. Bot. Gaz. 117:254-261.

Eid, A.A., DeLanghe, E., and Waterkeyn, L. 1973. In vitro culture of fertilized cotton ovules. I. The growth of cotton embryos. Cellule 69:361-371.

Finer, J.J. and Smith, R.H. 1982. Isolation and culture of protoplasts from cotton (*Gossypium klotzschianum* Anderss.) callus cultures. Plant Sci. Lett. 26:147-151.

Finer, J.J., Reilley, A.A., and Smith, R.H. 1983. Stimulation of somatic embryogenesis in cotton. TCA Report 17(5):8,17.
Fisher, W.D. 1980. Hybrid cotton update. In: Summary Proceedings Western Cotton Production Conference (Fresno, California) p. 25. National Cotton Council, Memphis.
Fryxell, P.A. 1980. The Natural History of the Cotton Tribe. Texas A&M Press, College Station, Texas.
Gerstel, D.U. 1953. Chromosomal translocations in interspecific hybrids of the genus *Gossypium*. Evolution 7:234-244.
Hsu, C.L. and Stewart, J.M. 1976. Callus induction by (2-chloroethyl) phosphonic acid on cultured cotton ovules. Physiol. Plant. 36:150-153.
Hutchinson, J. 1962. The history and relationship of the world's cotton. Endeavour 21:5-15.
Katterman, F.R.H., Williams, D.M., and Clay, W.F. 1977. The influence of a strong reducing agent upon the initiation of callus from the germinating seedlings of *Gossypium barbadense*. Physiol. Plant. 40:98-101.
Khasanov, M.M. and Butenko, R.G. 1979. Cultivation of isolated protoplasts from cotyledons of cotton (*Gossypium hirsutum*). Fiziol. Rast. 26:103-108.
Lofland, H.B. 1950. In vitro culture of the cotton embryo. Bot. Gaz. 111:307-311.
Mauney, J.R., Chappell, J., and Ward, B.J. 1967. Effects of malic acid salts on growth of young cotton embryos in vitro. Bot. Gaz. 128: 198-200.
Meyer, V.G. 1974. Interspecific cotton breeding. Econ. Bot. 28:56-60.
Mickelsen, O. and Yang, M.G. 1965. Naturally occurring toxicants in foods. Fed. Proc. 25:104-123.
Pathak, R.S. and Kumar, P. 1976. A study of heterosis in upland cotton (*Gossypium hirsutum* L.). Theor. Appl. Genet. 47:45-49.
Phelps, R.A., Shenstone, F.S., Kemmerer, A.R., and Evans, R.J. 1965. A review of cyclopropenoid compounds: Biological effects of some derivatives. Poult. Sci. 44:358-394.
Phillips, L.L. 1962. Segregation in new alloploids of *Gossypium*. IV. Segregation in New World x Asiatic and New World x wild American hexaploids. Am. J. Bot. 49:51-57.
______ 1966. The cytology and phylogenetics of the diploid species of *Gossypium*. Am. J. Bot. 53:328-335.
______ 1976. Cotton Gossypium (Malvaceae). In: Evolution of Crop Plants (N.W. Simmond, ed.) pp. 196-200. Longman, New York.
Price, H.J. and Smith, R.H. 1977. Tissue culture of *Gossypium* species and its potential in cotton genetics and crop improvement. In: Beltwide Cotton Production Research Conference Proceedings, pp. 51-55. National Cotton Council, Memphis.
______ 1979. Somatic embryogenesis in suspension cultures of *Gossypium klotzschianum* Anderss. Planta 145:305-307.
______ 1979. Somatic embryogenesis from *Gossypium* suspension cultures. In: Beltwide Cotton Production Research Conference Proceedings, pp. 36-37. National Cotton Council, Memphis.

______, and Grumbles, R.M. 1977. Callus cultures of six species of cotton (*Gossypium* L.) on defined media. Plant Sci. Lett. 8:115-119.

Rani, A. and Bhojwani, S.S. 1976. Establishment of tissue cultures of cotton. Plant Sci. Lett. 7:163-169.

Sandstedt, R. 1975. Habituated callus cultures from two cotton species. In: Beltwide Cotton Production Research Conference Proceedings, pp. 52-53. National Cotton Council, Memphis.

Scarpelli, D.G. 1974. Mitogenic activity of sterculic acid, a cyclopropenoid fatty acid. Science 185:958-959.

Schenk, R.U. and Hildebrandt, A.C. 1972. Medium and techniques for induction and growth of monocotyledonous and dicotyledonous plant cell cultures. Can. J. Bot. 50:199-204.

Smith, C.E. and Stephens, S.G. 1977. Critical identification of Mexican archeological cotton remains. Econ. Bot. 25:160-168.

Smith, R.H., Price, H.J., and Thaxton, J.G. 1977. Defined conditions for the initiation and growth of cotton callus in vitro. I. *Gossypium arboreum*. In Vitro 13:329-334.

Stephens, S.G. and Moseley, M.E. 1973. Cotton remains from archeological sites in central coastal Peru. Science 180:186-188.

Stewart, J.M. and Hsu, C.L. 1977. In-ovulo embryo culture and seedling development of cotton (*Gossypium hirsutum* L.). Planta 137:113-117.

______ 1978. Hybridization of diploid and tetraploid cottons through in-ovulo embryo culture. J. Hered. 69:404-408.

Weaver, J.B. 1957. Embryological studies following interspecific crosses in *Gossypium*. I. *G. hirsutum* x *G. arboreum*. Am. J. Bot. 44:209-214.

______ 1958. Embryological studies following interspecific crosses in *Gossypium*. II. *G. arboreum* x *G. hirsutum*. Am. J. Bot. 45:10-16.

Wilson, F.D., Lee, J.A., and Bridge, R.R. 1968. Genetics and Cytology of Cotton 1956-67. South. Coop. Ser. Bull. 139.

CHAPTER 19
Hardwoods

H. E. Sommer and *H. Y. Wetzstein*

The terms hardwoods, broadleaf trees, and deciduous trees are relatively misleading and imprecise, commonly used to denote woody, dicotyledonous trees as opposed to pines and other gymnosperms. "Hardwood" is a misnomer in that the term has nothing to do with the physical hardness of the wood. Within this group woods range in physical strength all the way from balsa to oak. The "broadleaf" designation for dicotyledonous trees is also misleading in that exceptions arise. Gingko might appear to be a "broadleaf tree," but it is a gymnosperm. "Deciduous trees" are not necessarily dicotyledonous. Larches are "deciduous" but again are gymnosperms. Others might point out that hardwoods is a forestry term and does not include domesticated dicots such as apple, peach, pecan, and walnut. Yet pecan and walnuts are also among the more valuable species of forest trees. In addition, categorizing trees as hardwoods and softwoods (conifers) leaves out the aborescent monocots.

"Hardwoods" is also a term for a type of forest and associated vegetation. We may have a pure hardwood forest with gymnosperms making up a relatively minor portion of the forest, or a mixed hardwood coniferous forest, with gymnosperms a considerable part of the forest. These forests may have a closed canopy or consist of scattered trees and thickets.

The exact number of dicotyledonous tree species that exist seems to be unknown. However, of Hutchinson's 82 orders of dicots, he classifies 54 in his group Lignosae, and only 28 in the Herbaceae. Of these only a very small fraction have been studied in tissue culture, and a still smaller number have differentiated plantlets and even plantlets of fewer species have been field tested. In this paper we will concentrate on those species of woody dicots for which information on their culture is available. Some may be shrubs rather than trees, while others may be more often seen as ornamentals, or in use for food production rather than forest trees. However, all are woody dicots.

HISTORY OF THE CROP

The history of the use of hardwoods by man is parallel to the history of humanity. The use of carbon 14 analysis of charcoal and of tree ring analysis for dating of sites by archaeologists is well known. However, we do not know much of the forestry practices of the early civilizations. In the area around the Mediterranean Sea, however, wood must have been abundant in at least some areas. The Mediterranean Basin forests were used to build the walls of Athens, and the fleets of the Phoenicians, Persians, Greeks, Cathaginians, Syrians, and Romans. However, Fernow (1913) indicates that possible wood scarcity first developed in more populated areas in the eleventh century B.C. At that time building timbers for temples at Tyre and Sidon had to be brought from Mount Lebanon 50-75 miles away. By 333 B.C. the south slopes of the Lebanese mountain forests were largely denuded. During Roman rule, wood shortages did not seem to be a problem of much concern due to the ease with which wood could be brought to market by water even from as far away as India. Major supplies of wood came from areas bordering the Black Sea, northwestern Italy, and Morocco. However, through the actions of man and an overpopulation of goats and sheep eating any natural regeneration of the forest, marginal forests have been eliminated and many formerly fully stocked forests were reduced to a sparse stocking in the Mediterranean area. Indeed the famous and great cedar forests of Lebanon have vanished leaving a few scattered trees. Over all, it appears that forests were treated by these earlier cultures more as a resource to be mined, than as a crop. One exception was the practice of coppicing oak, chestnut, and willows to grow vineyard stakes.

The history of the forest has many similarities in the evolution of its use and in the eventual development of forest management systems. Fernow (1913) considers Germany the Fatherland of Forestry and gives a detailed account of the German forests and treatment under the numerous states that developed.

With the dissolution of the feudal system, Germany was composed of many independent principalities, some cities, and no developed transportation system. Each unit had to provide its own wood needs from the local forest. The result was a great variety of regulations, varying from possibly none to fairly extensive.

By 1800 forestry and forest sciences had started to become an area of systematic as well as empirical studies. The first school of forestry was started in 1763 in Wernigerode by V. Zanthier. And starting about 1770 forestry courses were offered at most German universities. The defining of proper forestry practices began in the 19th century and continues today. Thus hardwoods have come under management as a crop, though usually not to the extent conifers have, since over the course of several centuries the forests had gone from abundance to a resource requiring careful management.

ECONOMIC IMPORTANCE

Hardwoods have a worldwide distribution limited primarily by temperature and water. The major hardwood forests (Edlin, 1976) with few exceptions are found in the Northern Hemisphere. In North America the hardwood forest includes areas from the east coast to the prairies west of the Mississippi, from the coniferous forests of Canada and to the deserts and higher elevations of the southwest. There is also a much smaller area in middle California. In the Eastern Hemisphere the broadleaf forest starts in Ireland and west through northern Spain and continental Europe and Turkey to the Ural mountains. Only a small area of southern Sweden on the Scandinavian peninsula has a broadleaf forest, and in USSR hardwoods are generally south of Riga, Moscow, and Ufa. The hardwood forest follows a narrow strip south of the Siberian coniferous forest to Manchuria, Korea, Japan, and eastern China to about Guangzhou. Small areas exist in the southern Himalayan region. Other areas where temperate hardwood forests exist include the extreme south of Chile and Argentina, the south island of New Zealand, the southeast Australian coast, and Tasmania. In tropical areas there are extensive hardwood forests from southern Mexico to limits of the Amazon basin. To the south of the Amazon base is the broadleaf savanna the "Cerado of Brasil" extending to 20-22° south latitude. In Africa we find a savanna with broadleaf trees over much of the areas south of the Sahara and west of the horn of Africa except for the equatorial forest of the Congo basin and the Gold Coast and monsoon forest of south central Africa. Likewise equator forests and monsoon forests are found in northeastern India, the Malaysian Peninsula, Indonesia, Indochina, New Guinea, and northern and southwestern Australia. While broadleaf trees may predominate, conifers will be found intermixed and in some of these areas predominate. In many of the areas, particularly in the case of savannas and Mediterranean forest, the canopy is not closed. Likewise while an area may be described here as forested, in many cases the majority of the forest has been removed or otherwise altered by man (Edlin, 1976).

The last *World Forest Inventory* was published by the United Nations in 1963 (FAO, 1963). Table 1 is based on that data. No real monetary value can be placed on the products of the hardwood forest. Despite the age of the data in the report a rough idea of the potential economic importance of the hardwood resource can be discerned. One thing that should be understood about these tables is that no uniform system was used in collecting much of the data. Some are based on statistical sampling, while others are only unofficial estimates. Also definitions of the main headings even varied from country to country. Depending upon which part of the report is used, 29-34% of the land

Table 1. World Hardwood Resources[a]

LOCATION	TOTAL FOREST AREA (10^6 ha)	AREA IN HARDWOOD (10^6 ha)	AREA IN MIXED FOREST (10^6 ha)	TOTAL MIXED AND HARDWOOD FOREST (10^6 ha)	REMOVAL OF HARDWOODS[b]		
					Indus-trial	Fuel	Total
United States	293	109	—	109	59	30	90
Canada	713	30	58	88	5	34	39
USSR	910	175	—	175	30	35	66
Australia	207	204	—	204	0.9	5	14
New Zealand	6	0.6	1.3	2	0.2	0.06	0.3
Brazil	335	328	1	329	7.6	88	96
South Africa	4	0.7	—	0.7	3	0.6	3
Europe	138	52	13	65	52	83	135
Asia	500	222	12	234	61	212	273
Latin America (minus Brazil)	566	161	16	177	12	79	91
Africa (minus S.A.)	696	399	—	399	17	171	188
Pacific Islands (minus Aust. & N.Z.)	40	14	1.4	16	0.3	4	4
WORLD	441	169	103	180	256	744	1000

[a] Based on World Forest Inventory, 1963, FAO.

[b] Millions of cubic meters of round wood.

area of the earth is forests. Of this about 40% is some type of hardwood or mixed forest. When removals and their uses are checked, three patterns are quite evident. In the United States, Australia, New Zealand, and South Africa, most of the removals are classified as going for industrial uses such as sawlogs and pulp. In Canada, USSR, and Europe (and possibly parts of Asia) a much smaller portion of the hardwood removals appear to go to industrial uses, which probably are the result of their large conifer resources and the preferences of conifers for industrial uses. For instance, in Canada about three times as many cubic meters roundwood of conifers are harvested for industrial uses as hardwoods per year. The other pattern found is where often more than half of the forest is classified as hardwoods yet only a small fraction, often less than 10% of the annual removal, is used for industrial wood, while the majority is used for fuel wood. On a worldwide basis three times as much hardwood is consumed for fuel as for industrial uses. More likely than not because fuel wood is usually consumed locally, this is greatly underestimated.

The industrial uses of the hardwoods are numerous. Primary manufacturing includes the products of saw mills, planing mills, veneer mills, plywood mills, pulp, paper and paperboard mills, particle board manufacture, and gum and wood chemicals recovery. Secondary manufacturing includes millwork and prefabricated wood products, wooden containers, furniture, paper and paperboard products, fibers, plastics, textiles, wood preserving and other wood products. The 1972 census of manufacturers for the United States includes 90 types of firms engaged in primary or secondary wood manufacturing. The use of wood for fuel has increased recently in some areas. Often the increasing price of oil is given sole blame for this situation. The practicality of using wood as opposed to alternate fuels is governed primarily by the relative cost, availability, and trade balance.

On a nationwide basis wood is not a major energy source (except possibly in terms of barrels of oil saved), contributing only 1.5 quads of the energy used annually in the United States. A well organized and planned biomass energy program might raise this to 5 quads (USDA Forest Service, 1980). Even if all the timber cut in one year in the United States (about 200 million tons) were burned to generate electric power, it would be about a tenth of energy needed just to provide electric power for the nation (Youngs, 1978).

There are other uses for wood. For instance, 1 cord (2-1/2 tons) of wood will produce when pyrolyzed 50 bushels of charcoal, 11,500 cubic feet of fuelgas, 25 gallons of tar, 10 gallons of wood alcohol, and 250 pounds of nitrate of lime. An average size tree (3700 lb.) is equivalent to 3.4 barrels of oil in BTUs. Wood from this standpoint may thus have an important energy role in industrialization.

This role of hardwood plantations, particularly *Eucalyptus*, in industrial development can be seen in the case of the iron industry in Brasil (Ayling and Martins, 1981). In one day of operation, a steel mill consumes about 5000 m^3 of charcoal. The primary source of carbon for the iron industry is charcoal.

Charcoal is still being made from wood from the natural forest. Since 1966 the Brazilian government has provided fiscal incentives for reforestation. During ten years of incentives, 1.5 million hectares of eucalyptus were established. The vast majority of these were on a short rotation. Yields of 20-40 m^3/ha per year are potentially possible in the Cerrado area.

Thus in this case hardwood plantations are essential for the development of a nation's steel and future chemicals industry. All too few developing nations have shown the wisdom of including reforestation and high yielding plantations in their development plans. The only other example that comes readily to mind is South Africa, which has gone from a wood importing nation to one that has been nearly self-sufficient since 1945.

Many third world countries do not have the forest developmental potential of Brasil and have adopted the attitude that the forest is a barrier to feeding people. However, these countries could in fact derive much benefit from development of hardwood plantations. Approximately 1.5 billion people obtain 90% of their energy needs from wood, and another billion obtain 50% of their energy from wood (NAS, 1980). At the present net rate of removal, 250 million people will be without wood in 20 years. A good example of the extent of the problem is that all accessible trees near roads within 40 km of Ouagadougou, Upper Volta have been harvested for fuel. No reforestation has taken place, and the stripped area has expanded (NAS, 1980). Obviously this can not go on indefinitely. If these countries are to survive, reforestation must take place. Also since in many of these areas the locals will cut even a good sized seedling for wood, new plantations must be protected.

Regardless of the limitations and potentials of wood as an energy supply, history does indicate that the forest can not be left to itself to regrow. However, with current effective artificial regeneration, a harvest rotation of 25 years or 4 rotations per century is feasible. If left to itself the forest may not reach maturation for 30-35 years. Thus one rotation per century is lost. With even more advanced applications of forestry research, the rotation can be lowered to approximately 20 years.

Trees also contribute important aesthetic value and therefore have value as ornamentals. The genetic makeup of many ornamental cultivars is highly heterozygous and the unique characteristics are not sexually transmitted. This has led to the extensive use of vegetative means of propagation such as cutting, grafting, and budding. Trees are often selected for their beauty and variation of form and color. Cultivars have been developed for many landscape needs. Ornamentals have been selected for variations in foliage color during the growing season and for fall color. Examples would include *Prunus cerasifera*, the purple leaf plum, and a large number of *Acer* species which can range from true greens to rusts and vibrant reds in the new foliage. Much interest also exists in variegated foliage forms. For example, Japanese maple cultivars range widely in the degree and location of leaf variegation. *Acer palmatum* 'Kagiri nishiki' has white and rose pink coloration on the leaf margins. 'Nishiki gasane' leaves are deep green and flecked with gold. 'Orido nishiki' leaves have irregular and varied variegation with some leaves being almost entirely white. Other ornamentals with selected leaf variegation include *Ilex* and *Pyracantha*.

Fall color may be a selection factor as seen in the autumn foliage colors of sweetgum (*Liquidambar*), red maple and sugar maple (*Acer rubrum*, *A. saccharum*), American beech (*Fagus grandifolia*) and dogwood (*Cornus florida*).

Flowering characteristics have also held much horticultural interest. For example, there are nearly 200 *Prunus* species with very diverse flower forms and colors. Hybrids can be single or double flowering types. The Sargent cherry, *P. sargentii*, has single, deep pink flowers. *Prunus serrulata*, the Japanese flowering cherry, has either single or double flowers with white to deep pink flowers. Other ornamentals developed for flower production include the crabapples (*Malus* spp.), almonds (*Amygdalus* spp.), flowering dogwood (*Cornus florida*), *Magnolia* spp., and redbud (*Cercis*). Some woody ornamentals are valued for fruit development including the hollies (*Ilex* spp.) and firethorn (*Pyracantha*).

Ornamentals are also valued for their size, form, and growth rate, whether for shade (as with maples, oaks, sycamores), as screens or specimen plantings. Dwarf plants have had increasing interest in landscape plantings.

BREEDING AND TREE IMPROVEMENT PROBLEMS

Wright (1976) and others have written extensively on tree genetics and its application to forest tree improvement and have discussed the problems involved. The problems can be divided into three general areas: the lack of basic genetic information, the need to obtain gains in productivity in which quantitative genetic aspects rather than specific single gene traits are important, and the need for a long term commitment to tree research. Selections in the field based on a particular phenotype tell little about behavior in a breeding program. The only reliable way to determine if the tree will have offspring which will give greater growth than average is to include them in a progeny test with other offspring of other trees under similar conditions. Only by use of this indirect evidence can a tree's potential for inclusion in a tree improvement program be determined. This isn't so bad except that if scions are removed from the parent, grafted and placed in a breeding orchard it may be 5 years before they start yielding seed. Once the progeny are planted, it may be 15-25 years before a useful evaluation of their growth is obtained. Hormone treatments to promote early flowering and improved statistical procedures to allow earlier evaluation of progeny are under development in some cases, mostly in the case of conifers. It would seem vegetative methods could be used to propagate hardwoods to insure that certain traits are retained. However, with rare exceptions (such as some poplar and willow), forest trees are propagated from seed.

REVIEW OF THE LITERATURE

There are several methods and approaches used to propagate hardwood species in culture. Table 2 lists numerous species of woody dicots that have been propagated in vitro. Most of these have been

Table 2. Some Woody Dicotyledonous Forest Tree Species from which Plantlets Have Been Regenerated from Tissue or Organ Culture

SPECIES	TREE AGE[a]	EXPLANT SOURCE	REFERENCE
Acacia koa	J	Shoot tip	Skolmen & Mapes, 1976, 1978
Acer rubrum	J	Shoot tip, axillary buds	Brown, 1980, 1981
	-	Shoot tip, node sections	Welsh et al., 1979
Actinidia chinensis	-	Stem, root segments	Harada, 1975
A. chinensis var. *chinensis* var. *hispida*	M	Stem segments	Kwei et al., 1978
Albizzia lebbeck	J	Hypocotyl	Gharyal & Maheshwari, 1981
Alnus glutinosa	J	Axillary buds	Hosier et al., 1981
	J	Shoot tip	Brown, 1980, 1981
A. rubra	J	Shoot tip, buds, hypocotyls	Brown, 1980, 1981
Amelanchier laevis	-	Shoot tip	Behrouz & Lineberger, 1981a,b
	-	Shoot tip	Lineberger, 1981
Atriplex canescens	J	Shoot tip	Wochok & Sluis, 1980
Betula papyrifera	J	Buds	McCown, 1979
B. pendula	J	Stem	Winten & Huhtinen, 1976
	J	Stem	Huhtinen & Yahyaoglu, 1974
	-	Stem	Huhtinen, 1976
	M	Catkins	Srivastava & Steinhauer, 1981
B. platyphylla 'Schezuanica'	J	Buds	McCown, 1979
	J	Internodes from shoot tip cultures	McCulloch & McCown, 1979
	J,M	Shoot tips	McCown & Amos, 1979; Amos & McCown, 1979
Brachycome sp.	-	Leaf, stem, cotyledons	Turiansky et al., 1981
Broussonetia kazinoki	J	Hypocotyl	Oka & Ohyama, 1974; Ohyama & Oka, 1980

SPECIES	TREE AGE[a]	EXPLANT SOURCE	REFERENCE
Camellia	J	Shoot tip in vitro grafting	Creze, 1980
Carya illinoensis	J	Axillary buds	Knox & Smith, 1981
Castanea dentata	J	Embryo culture	McPheeters et al., 1980
C. sativa	J	Embryonic axes	Vieitez & Vieitez, 1980a
	J	Lateral buds	Vieitez & Vieitez, 1980b
Catalpa bignonioides	J	Shoot tip	Brown, 1980, 1981
Cinchona ledgerana	J	Shoot tip	Hunter, 1979
Corylus avellana	J	Embryos	Radojevic et al., 1975
	J	Embryonic axes	Jarvis et al., 1978
Diospyros kaki	J	Hypocotyl	Yokoyama & Takeuchi, 1976
Eucalyptus alba	J	Hypocotyl	Kitahara & Caldas, 1975
E. bancrofti	J	Shoot tip	Cresswell & de Fossard, 1974
E. camaldulensis	J	Shoots	Hartney & Barker, 1980
	–	Node	Goncalves et al., 1979
E. citriodora	J	Hypocotyl, cotyledons	Sita, 1979
	M	Terminal buds	Gupta et al., 1981
	–	Lignotubers	Aneja & Atal, 1969
E. curtisii	J	Shoots	Hartney & Barker, 1980
E. dalrympleana	J	Apex, axillary buds	Durand & Boudet, 1979
	J	Nodes	Durand-Cresswell & Nitsch, 1977
E. deglupta	J	Nodes, shoot tip	Cresswell & de Fossard, 1974
E. ficifolia	J,M	Nodes	de Fossard et al., 1977, 1978a; Barker et al., 1977
	J	Nodes	de Fossard & Bourne, 1976; de Fossard, 1978; Gorst & de Fossard, 1980
	J	Shoots	Hartney & Barker, 1980

Table 2. Cont.

SPECIES	TREE AGE[a]	EXPLANT SOURCE	REFERENCE
Eucalyptus globulus	J	Shoots	Hartney & Barker, 1980
E. grandis	J	Nodes	Barker et al., 1977; de Fossard et al., 1974, 1977; Cresswell & Nitsch, 1975; Durand-Cresswell & Nitsch, 1977
	J	Shoots	Hartney & Barker, 1980
	J	Nodes, shoot tip	Cresswell & de Fossard, 1974
	-	Nodes	Goncalves et al., 1979
E. gunnii	J	Apex, axillary buds	Durand & Boudet, 1979
E. nitens	J	Shoots	Hartney & Barker, 1980
E. obtusifolia	J	Shoots	Hartney & Barker, 1980
E. orcades	J	Shoots	Hartney & Barker, 1980
E. pauciflora	J	Shoots	Hartney & Barker, 1980
	J	Apex, axillary buds	Durand & Boudet, 1979
E. regnans	J	Shoots	Hartney & Barker, 1980
	J	Nodes	Barker et al., 1977
E. robusta	J	Shoot tip	Brown, 1980, 1981
	-	Nodes	Goncalves et al., 1979
Fagus silvatica	-	Axillary buds	Chalupa, 1979
Ficus benjamina	-	Shoot tip	Makino et al., 1977
F. decora 'Elastica'	-	Shoot tip	Makino et al., 1977
F. lyrata	-	Leaves	Deberg & de Wael, 1977
F. pandurata	-	Shoot tip	Makino et al., 1977
Gleditsia triocanthos var. *inermis*	J	Shoot tip	Brown, 1980
G. triocanthos	-	Shoots	Rozozinska, 1968
	J	Cotyledons	Rozozinska, 1969

SPECIES	TREE AGE[a]	EXPLANT SOURCE	REFERENCE
Hamamelia virginiana	J	Embryos	Zilis & Meyer, 1977
H. vernalis	J	Embryos	Zilis & Meyer, 1977
Hedera helix	J,M	Callus from stems	Banks, 1979
Ilex aquifolium	J	Cotyledon	Hu & Sussex, 1971; Hu, 1976; Hu et al., 1978
I. cornuta	J	Cotyledon	Hu, 1976
I. opaca	J	Cotyledon	Hu, 1976
Kalmia latifolia	J	Shoot tip	Lloyd & McCown, 1980
Liquidambar styraciflua	J	Shoot tip, hypocotyls	Sommer & Brown, 1980; Sommer, 1981
Liriodendron tulipifera	M	Terminal buds	Brown, 1980, 1981
Moghania macrophylla	J	Seed	Nag & Ahmad, 1978
Morus sp.	J	Hypocotyl	Ohyama, 1970
Paulownia taiwaniana	-	Ovules	Fu, 1978
	J	Stem	Fan & Hu, 1976
P. tomentosa	J	Hypocotyls, cotyledons	Marcotrigiano & Stimart, 1981
	J	Shoot tip, axillary buds, petiole, leaf base	Brown, 1980, 1981
	M	Unfertilized ovules	Radojevic, 1979
Populus alba	-	Axillary buds	Chalupa, 1979
	J	Twig	Chalupa, 1975
	-	Axillary buds, apical buds	Christie, 1978
P. alba x *glandulosa*	-	Axillary buds, apical buds	Christie, 1978
P. alba x *tremula*	-	Axillary buds, apical buds	Christie, 1978
Populus x *canadensis*	J	Stem tip	Berbee et al., 1972
P. canescens	J	Cambium, shoot tip	Chalupa, 1974
	-	Axillary buds, apical buds	Christie, 1978

Table 2. Cont.

SPECIES	TREE AGE[a]	EXPLANT SOURCE	REFERENCE
Populus euramericana	-	Shoot tip, callus	Lester & Berbee, 1977
P. euramericana 'Robusta'	J	Cambium, shoot tip	Chalupa, 1974
	-	Axillary buds	Chalupa, 1979
	J	Twig	Chalupa, 1975
	-		Dujickova et al., 1977
	-		Riou et al., 1975
Populus 'Flevo' (*P. deltoides* x *P. nigra)*	-	Axillary buds	Whitehead & Giles, 1976
P. nigra 'Italica'	M	Twigs	Brand & Venverloo, 1973; Venverloo, 1973
	-	Axillary buds	Whitehead & Giles, 1976
P. nigra 'Typica'	J	Cambium, shoot tip	Chalupa, 1974
	-	Axillary buds	Chalupa, 1979
P. nigra	J	Twig	Chalupa, 1975
P. tremula	J		Winton, 1971
	-	Axillary buds	Chalupa, 1979
	J	Twig	Chalupa, 1975
	J	Cambium, shoot tip	Chalupa, 1974
	-	Axillary buds, apical buds	Christie, 1978
P. tremuloides	-	Axillary buds	Christie, 1978
	-	Cambium	Wolter, 1968
	-	Root sprouts	Winton, 1968, 1970
	J	Stem tip of root sucker	Mathes, 1964
P. tristis x *P. balsamifera* 'Tristis #1'	-	Shoot apex	Thompson & Gordon, 1977
P. yunnanensis	-	Axillary buds	Whitehead & Giles, 1976

SPECIES	TREE AGE[a]	EXPLANT SOURCE	REFERENCE
Prunus amygdalus	J	Stem	Mehra & Mehra, 1974
	J	Epicotyl	Williams & Ryugo, 1980
P. amygdalus and hybrids	-	Shoot tip, dormant shoot buds	Tabachink & Kester, 1977
	J	Shoot tip	Davis et al., 1981
P. avium	M	Buds	Riffaud & Cornu, 1981
Quercus robur	-	Axillary bud	Chalupa, 1979
Rhododendron canescens	-	Shoot tip	Hannapel et al., 1981
R. catawbiense 'Nora Zembla'	M	Florets	Meyer, 1981a,b
R. catawbiense 'Roseum Elegans'	M	Florets	Meyer, 1981b
R. prunifolium	-	Shoot tip	Hannapel et al., 1981
Rhododendron spp.	-	Shoot tip, axillary buds	Economou et al., 1981
Robinia pseudoacacia	-	Roots	Seeliger, 1959a,b
Salix aquatica	-	Lateral buds	Garton et al., 1981
S. babylonica	J	Axillary buds	Letouze, 1977
S. interior	-	Lateral buds	Garton et al., 1981
S. purpurea	-	Lateral buds	Garton et al., 1981
S. viminalis	-	Nodes	Garton et al., 1981
Santalum acuminatum	J	Nodes & internodes, apex, cotyledons, hypocotyl, roots	Barlass et al., 1980
S. album	J	Endosperm	Sita et al., 1980a
	-	Shoots	Sita et al., 1980b
	J	Hypocotyls	Rao & Bapat, 1978; Bapat & Rao, 1979
	J	Embryo	Rao, 1965; Rao & Rangaswamy, 1971

Table 2. Cont.

SPECIES	TREE AGE[a]	EXPLANT SOURCE	REFERENCE
Santalum album	M	Shoot segments, shoot tip	Sita et al., 1979
S. lanceolatum	J	Nodes & internodes	Barlass et al., 1980
Sapium sebiferum	J	Hypocotyl	Venketeswaran & Gandi, 1981
Simmondsia chinensis	-	Shoot tip	Madani et al., 1979
Tectona grandis	M,J	Shoot tip, stem, cotyledons, axillary buds	Gupta et al., 1980
Tupidanthus calyptratus	-	Shoot tip	Matsuyama & Murashige, 1977
Ulmus americana	J	Hypocotyls	Durzan & Lopushanski, 1975
U. campestris	J	Twigs	Chalupa, 1975
	-	Axillary buds	Chalupa, 1979
U. effusa	-	Axillary buds	Chalupa, 1979
U. scabra	-	Axillary buds	Chalupa, 1979
Vaccinium angustifolium	J	Hypocotyls	Nickerson, 1978
	M	Buds	Frett & Smagula, 1981a,b
V. corymbosum	-	Shoot tip	Zimmerman & Broome, 1980

[a]M = mature; J = juvenile.

done in the last 5 years, juvenile explants have been generally used, and the source of explants varies. We will consider the propagation methodologies in three major categories: *Populus*, *Eucalyptus*, and other woody dicots.

However, first we should recall that Gautheret (1940) reported the formation of adventitious buds in cambial cultures of *Ulmus campestris*. Later, Jacquiot (1949, 1951, 1955) reported further on this phenomena in *U. campestris* and several other species, and in some instances obtained roots, but he never observed simultaneous root and shoot development.

Populus

Later reports of plant regeneration (Mathes, 1964; Wolter, 1968; Winton, 1968), while of historical importance, left some doubt as to the viability and degree of abnormality to be found in these plantlets. Later Winton (1970, 1971) demonstrated that normal trees could be obtained, though only at low frequencies. Since these initial results were published, numerous studies on the tissue culture of *Populus* have been reported. Some researchers investigated organogenesis and its control in vitro by hormones (Venverloo, 1973) and other media components (Saito, 1980a,b). Other studies have been more concerned with the origin and development of organs in culture (Brand and Venverloo, 1973; Bawa and Stettler, 1972).

In attempts to use tissue culture to propagate *Populus* two strategies have been used: callus culture and micropropagation. The use of callus culture for propagation has been investigated in detail by Berbee and coworkers (Berbee et al., 1972, 1976; Lester and Berbee, 1977). Their primary objective was to attempt to cure poplar decline in selected Aigeiros poplar clones by the combined use of tissue culture and heat treatment. The general approach was to obtain callus from shoot tips, regenerate shoots, and finally root the regenerated shoots. The plantlets were then established in mother plant blocks in the nursery. There were several practical problems. No one series of culture media was successful for the culture of all clones, and as few as a third of the calluses produced shoots. There were significant differences in the height of some subclones from the same parent as well as between subclones derived from callus from the same shoot tip, and subclones of the same callus. Growth rate differences were apparent even when the subclones and parents were repropagated from the same length cuttings, and were thus not due to secondary effects of culture environment. Chromosome counts were made on the root tips of some subclones. Many had chromosome numbers within the normal range; however, 8 of 46 subclones ranged anywhere from haploid to tetraploid and included some aneuploids. None of these plants was indexed for virus; thus we may have variability from both chromosomal aberration and degree of infection by the virus.

Chalupa (1974, 1975) has likewise produced *Populus* plantlets from callus. Up to 86% of his calli differentiated shoots. Rather than root the adventitious shoots on agar, he used either sterile perlite-sand or perlite-sand-peat mixtures as rooting substrate to produce better root development. The plantlets grew to trees in pots. Later Chalupa (1977) compared callus derived plantlets with the original parent trees. The growth rate of plantlets from the same mother tree was similar, as were color, form and size of leaves, and the photosynthetic rate. The chromosome number of the plantlets was the same as for the mother tree. Thus in this case the variability Berbee found for callus-regenerated plantlets was not found.

Riou et al. (1975) obtained friable callus of *Populus euramericana* cv. robusta, and place it into liquid suspension culture. The free cells of the suspension were obtained by filtration and centrifugation, and plated into agar in petri dishes. The colonies that developed were transferred to tubes for shoot differentiation, then the shoots were excised and rooted. Each step required a different medium. One can only wonder if each colony represented a distinct clone and also if all the shoots from a callus were identical. It is unfortunate no data were given on this point, as the use of suspension culture could allow rapid growth of large amounts of cells for organogenesis.

Since the results of Berbee leave some doubt as to the usefulness of callus culture methods to multiply selected genotypes, attempts have been made to adopt micropropagation to the multiplication of tree species. Whitehead and Giles (1976) investigated the application of rapid micropropagation to the propagation of *Populus nigra* 'Italica', *P.* 'Flevo' (*P. deltoides* x *P. nigra*), and *P. yunnonensis*. Axillary buds were removed, surface sterilized, and placed on a bud break medium. After bud break and shoot elongation, the shoots were cut in 0.5-cm sections and recultured. Existing axillary buds grew out and adventitious buds formed also. By this method 120-220 shoots were formed from each original bud. The shoots could be rooted in vitro or in vivo. In 3 months the plantlets were 1-1.5 m tall. Thus this method has the potential to produce large numbers of shoots in a short time. Whether the presence of adventitious buds will add variability to the progeny was not investigated. Christie (1978) has done further development of this system with the goal of producing 20,000 plantlets of rust resistant poplars for plantings. He has modified the media so either adventitious shoots are produced or enhanced axillary branching of shoots is stimulated. The adventitious shoots are small and hard to handle. The enhanced axillary branching can be continued at 4-6 week intervals by subculturing. Their reported multiplication rate per bud after establishment of the culture line is 10 shoots/culture per month. Good root development was obtained on 80% of the shoots. A hardening off procedure has been developed and 1.5-2.0 m growth made in 3 months. Chalupa (1979) applied similar methods to not only poplars but several other forest trees.

It would appear several species of *Populus* can be successfully propagated using micropropagation.

Eucalyptus

One of the most frequently cited papers on the tissue culture of eucalyptus is Aneja and Atal (1969), although the description of their medium is incomplete and the morphological evidence is poor. In addition, there are discrepancies between a subsequent description of the medium (Lee and de Fossard, 1974) and the original.

Cresswell and de Fossard (1974) attempted to propagate *E. grandis* using node cuttings from seedlings in organ cultures. Axillary buds enlarged and roots developed in some; however, the main contribution of this paper is an anatomical study of root development from the nodal cuttings. Cresswell and Nitch (1975) continued studies on node cultures from *E. grandis* branches above node 14, the highest node from which rooted cuttings can be obtained. Using a 2 hr soak, presumably to remove inhibitors, they were able to obtain axillary bud development and root initiation from about 30% of the nodes. de Fossard and Bourne (1976) worked on extending this technique to adult trees of *E. ficifolia.* Some nodal cultures produced multiple buds; however, rooting was not reported for cultures established from adult tissue. Further extension of this work was done (Barker et al., 1977) using an improved medium that gave more vigorous growth of nodes from adult trees and multiple bud formation. Shoots could be obtained from buds after repeated subculture, to a point where all adult tissue had been removed. In addition, *E. ficifolia, E. grandis, E. regnans,* and *E. polybractea* were cultured without successful propagation from adult sources. As this work continued (de Fossard, 1978; de Fossard et al., 1978), considerable progress was made in multiplication of shoots and apparently also in rooting of seedling material of *E. ficifolia.* Unfortunately, detailed data were not given. The multiplication of adult shoots was apparently improved, however rooting of adult shoots appears to be accompanied by the degradation of the shoot. Thus the mass propagation of *E. ficifolia* starting from nodes from adult trees still needs additional work.

Other groups have worked on the propagation of *Eucalyptus.* Kitahara and Caldas (1975) obtained shoots from *E. alba*; however, root development was poor (Goncalves et al., 1979). Attempts have been made to propagate *E. citriodora* from seedling parts (Sita, 1979; Sita and Vaidyanathan, 1979). Though small shoots were obtained, only 3 plantlets were recovered. No differentiation was obtained from mature plants. Hartney and Barker (1980) have reported a method for obtaining multiple shoots from seedlings of 10 species of eucalyptus and rooting from 5 of these. Some of these had been planted and were 2 years old. More recently, there has been a report that 20-year-old trees of *E. citriodora* can be propagated by tissue culture (Gupta et al., 1981). Terminal and axillary buds from 20-year-old trees were used as explants, and under specified conditions 30% did produce 5-8 shoots. These shoots could be subcultured. After four subcultures, 35-40% could be rooted. Field survival was 40-50%. A total of 22 plants were raised.

M. Boulay (personal communication) with Association Forêt Cellulose (AFOCEL) states that they are developing a method to propagate a species of Eucalyptus in culture. Their criteria for a successful method is to produce 100,000 plantlets ready for the field. At present, they have only 40,000 shoots, representing 108 clones, and thus do not recommend the use of their procedure for commercial propagation.

Miscellaneous Woody Dicots

Many other woody dicots have responded with morphogenesis in tissue culture. In some instances somatic embryos and plantlets have been obtained. Table 2 lists many examples for which plantlets of hardwoods have been obtained. Numerous horticultural tree crops have also been propagated in vitro. For this presentation a few examples have been chosen to demonstrate different methods and to include some species of general interest. Callus culture has been used for both birch (Huhtinen, 1976) and elm (Durzan and Lopushanki, 1975). Young seedlings were used as the explant source. The birch had been selected for early flowering at one year of age and at this stage had only reproductive buds. After culture, vegetative buds were found on the second growth phase. These results open up interesting questions concerning the phase change from juvenile to mature and the possibility of reversion of tissue to the juvenile state by using tissue culture. Callus had also led to embryogenesis, which will be covered separately.

Micropropagation has been applied to multiplication and propagation of numerous woody plants in the last few years. Axillary or terminal buds are used as the functional explant, even in the case of cultures of *Castanea sativa* (Vieitez and Vieitez, 1980a) which was started from an embryonic axis. Axillary buds, which differentiate as the axis develops into a plantlet, are most often used for the multiplication step. The use of axillary buds is important to maintaining genetic stability. A good example of micropropagation is the system for *Betula platyphylla* (McCown and Amos, 1979) since propagated shoots have since been rooted in a potting mix, hardened off, and outplanted. First year results demonstrated that the tissue cultured birch plantlets went dormant sooner in the field, thus exhibiting less growth than the median for seedlings from the same seed lot. However, the tissue cultured plantlets were more uniform in height. The work with *Tectona grandis* (Gupta et al., 1980) is particularly interesting in the fact that 100-year-old trees could be used as the explant source. The rate of multiplication was not as high as for the seedling explants. One interesting observation from the work with the older material from teak is that the percent of rooting increases from first to second subculture from 10 to 60%. This may represent a case of rejuvenation or a selection of rootability.

These few examples of woody dicots that have been propagated in tissue culture were chosen to illustrate some of the methods used. A more complete list is found in Table 2.

Embryogenesis

The production of somatic embryos in tissue culture has several advantages over organogenesis as a practical means of propagation. Growth, multiplication, and even differentiation often can be accomplished in suspension culture. Or alternatively the suspension may be plated out and differentiation accomplished on agar. Regardless of the exact method, suspension culture allows rapid multiplication from stock cultures and reduces the labor per plantlet, since each flask has the potential of producing thousands of plantlets. It has even been proposed that the embryos can be encapsulated and handled as seeds. Throughout the tissue culture literature, there are many claims of embryogenesis in tissue cultures, some of which are of doubtful authenticity. Haccius (1978) has presented a definition of the embryo (zygotic or nonzygotic) as a new individual arising from a single cell and having no vascular connection with maternal tissues. Single cell origin is difficult to prove. However, anatomically the most characteristic feature of an embryo, by her criteria, is the presence of a discrete radicular end (vascularly closed from the parent tissue). This is not as hard to establish. We should probably add the subsequent growth of the embryo to form a functional plantlet as an additional criteria.

The earliest report of embryogenesis in culture of a forest tree (citrus is omitted from this review) is in *Santalum album* L. (Rao, 1965). The embryos were "easily separable from the parent." Embryos had an easily distinguished root, but a very poorly developed fasciated embryonic shoot. The media used had undefined constituents, and no functional plants developed.

Hu and Sussex (1972) reported on the development of somatic embryos on the cotyledon of *Ilex aquifolium*. The conditions were somewhat unusual in that embryos developed on the cotyledons of zygotic embryos that failed to develop shoots and normal cotyledons in culture on basal LS medium. Callus formation was not apparent. Anatomical studies showed no vascular connections to the parent tissue. Anatomically the embryos appeared normal and under specific conditions could be germinated. In later studies (Hu et al., 1978) some of the embryos were bicolor corresponding to two color sectors of the background cotyledonary tissue. This suggests these embryos may not be of single cell origin.

Another very valuable set of studies on nonzygotic embryogenesis in woody plants are those of Radojevic et al., 1975, 1979; Vijicic et al., 1976; Radojevic, 1979; Vijicic, 1979. While previous papers included light and scanning electron microscopy these studies include transmission EM. In the case of *Corylus avellana* and *Paulownia tomentosa*, embryogenic callus was obtained from zygotic embryos, or from unfertilized ovules in the case of *P. tomentosa*. In the presence of 2,4-D, plantlet formation was inhibited, more so for *C. avellana* than *P. tomentosa*. During the transmission electron micrograph studies of embryogenic tissue of both species, they found an orderly regular alignment of membrane-bound ribosomes. Such an arrangement has been rarely reported in plant cells. In those cases in which it has been reported, i.e., following fertilization in fern eggs and active secretory glands, it is associated with reorganization of the growth pattern of the cells or high synthetic capability. These endoplasmic reticulum (ER) patterns are not found in plantlets, but histochemical studies show the meristematic cells which embryoids may differentiate from are continually synthesizing starch and are rich in sugars, lipids, and protein. It is likely these regular arrangements of ER are associated with the competence of the cells for embryogenesis.

Not all problems of somatic embryogenesis are unambiguous. Both the juvenile and mature forms of *Hedera helix* can be grown in culture (Banks, 1979). The cultures were grown in the dark for 3 months then transferred to the light. Juvenile callus formed shoots while mature callus on White's medium with activated charcoal formed embryos. Mature callus without charcoal did not differentiate. The plantlets that developed showed juvenile character.

Anther Culture

For several years anther culture of hardwood species was attempted unsuccessfully. A few reports (Table 3), such as those of Zenkteler et al. (1975), showed initial production of haploid callus, while others, such as those of Sato (1974), reported differentiation of callus derived plantlets from anther cultures. These plantlets proved to be diploid. From 1975, reports began to appear on haploids from anther cultures of populus. The Tree Improvement Laboratory at Heilungkiang Institute of Forestry (1975) reported haploid plantlets of *Populus ussuriensis* and *Populus simonii* x *Populus nigra*. Wang et al. (1975) reported haploids from anther cultures of *P. nigra*. In the former instance, the plantlets were obtained from callus. For the hybrid, 33.8% of the calluses were from pollen. Six plantlets were verified as haploid from chromosome counts. Later Zhu et al. (1980) cultured 14 species of populus and obtained 900 plantlets, most of which were haploid.

Since mosaic plants can arise from callus, direct androgenesis from microspores is preferable. Notable in this respect are two reports. Radojevic (1978) cultured anthers of *Aesculus hippocastanum* and was able to demonstrate with the aid of light and electron microscopy the plantlets and embryoids obtained came from microspores. Karyotypes of metaphase figures were haploid. Later (Radojevic et al., 1980), an ultrastructural study of the developing embroys was published. In addition, she has obtained pollen plantlets from *Chimonanthus praecox* (Radojevic, 1980). We might also mention that androgenesis direct from microspores has been achieved from *Hevea brasiliensis* (Chen et al., 1978).

Protoplasts

Development of protoplast technology with hardwood species has lagged behind that of herbaceous species. Most authors have used leaf mesophyll as a source of tissue (Table 4). Wakasa (1973) was able to isolate some protoplast from *Prunus* leaves. Saito (1976) reported what appears to be better yields from the leaves of *Paulownia fortunei* and *Populus euramericana*. Later, Saito (1980a,b) extended this work and was able to obtain 1:1 fusion in the presence of PEG of different strains of *Paulownia* and *Populus* and of *Populus* with *Paulownia*. Redenbaugh et al. (1980, 1981) have worked extensively on anther culture of *Ulmus* including the isolation of protoplasts from pollen mother cells, tetrads, and microspores for possible use in the production of elm hybrids. While the gametophytic protoplasts did not develop cell walls,

Table 3. Some Woody Dicotyledonous Forest Species from which Plantlets Have Been Generated from Anther Culture

SPECIES	REFERENCE
Aesculus hippocastanum L.	Radojevic, 1978
Betula pendula Roth	Huhtinen, 1978
Hevea brasiliensis Nuell.-Arg.	Chen et al., 1978
Populus nigra L.	Wang et al., 1975
Populus ussuriensis Komar	Anonymous, 1975
Populus (hybrids)	Lu et al., 1978
Populus (species and hybrids)	Zhu et al., 1980
Populus (hybrids)	Anonymous, 1975

somatic protoplasts of *Ulmus pumila* did develop cell walls. The only case to date of regeneration of protoplasts from a tree has been in the case of *Datura* where a herbaceous species has been fused with a tree species and hybrid plantlets regenerated (Schieder, 1980).

STATE OF THE ART AND FUTURE PROSPECTS

From Table 2 it is obvious that the culture types used for propagation of hardwood trees are not much different from those used for other plant species. The fundamental difference is in most cases only juvenile tissues have been used or given successful results. Success with mature tissue is rare. In some cases where we were in doubt as to degree of the maturity of the explant source, no classification was made. Otherwise the cultures can be handled as any other dicot culture. The protocols that are needed are those for hardening-off the plantlets, readily controlled embryogenesis, and androgenesis, and further investigations on protoplast technology.

The future prospects for tissue culture clones of hardwoods are bright as a research tool, but otherwise are limited by economics. For horticultural uses, a tissue culture propagated tree, particularly of a patentable selection or cultivar, probably could be priced competitive to that of one produced by conventional means. In forestry, however, nursery produced hardwood seedlings sell for a few cents each, while current estimates for tissue culture plantlets seem to range from 20-30 cents each. So while we do not expect to see mass plantings of tissue culture hardwoods, we do expect them to make substantial contributions to understanding the genetics and physiology of dicot trees.

ACKNOWLEDGMENTS

This research was funded in part by State and Hatch funds allocated to the Georgia Agricultural Experiment Station.

Table 4. Protoplasts from Hardwoods

HARDWOOD SPECIES	REFERENCE
Betula spp.	McCulloch & McCown, 1980; Smith & McCown, 1981
Corylus avellana	Radojevic & Kavoor, 1978
Datura candida	Schieder, 1980
D. sanguinea	Schieder, 1980
Paulownia fortunei	Saito, 1976
P. taiwaniana	Saito, 1980a,b
Populus euramericana	Saito, 1976, 1980a,b
Rhododendron	Smith & McCown, 1981
Ulmus americana	Redenbaugh et al., 1981; Lange & Karnosky, 1980
U. parvifolia	Redenbaugh et al., 1981; Lange & Karnosky, 1980
U. pumila	Redenbaugh et al., 1981; Lange & Karnosky, 1980

REFERENCES

Amos, R. and McCown, B.H. 1979. Initial trials with commercial micropropagation of birch selections. Proc. Int. Plant Prop. Soc. 29:387-393.

Aneja, S. and Atal, C.K. 1969. Plantlet formation in tissue cultures from lignotubers of *Eucalyptus citriodora* Hook. Curr. Sci. 38:69.

Anonymous, Tree Improvement Laboratory, Heilungkiang Institute of Forestry. 1975. Induction of haploid poplar plants from anther culture in vitro. Sci. Sin. 18:769-777.

Arola, R.A. and Miyata, E.S. 1981. Harvesting wood for energy. USDA Forest Service Research Paper NC-200.

Ayling, R.D. and Martins, P.J. 1981. The growing of eucalypts on short rotation in Brasil. For. Chron., Feb.

Banks, M.S. 1979. Plant regeneration from callus from two growth phases of English Ivy, *Hedera helix* L. Z. Pflanzenphysiol. 92:349-353.

Banks, P.F. and Metelerkamp, H.R.R. 1981. A comparison between the cost of growing *Eucalyptus grandis* for fuel-wood and the cost of coal in Zimbabwe Rhodesia. S. Afr. For. J. 117:16-18.

Bapat, V.A. and Rao, P.S. 1979. Somatic embryogenesis and plantlet formation in tissue cultures of sandalwood (*Santalum album* L.). Ann. Bot. 44:629-630.

Barlass, M., Grant, W.J.R., and Skene, K.G.M. 1980. Shoot regeneration in vitro from native Australian fruit-bearing trees—Quandong and plum bush. Aust. J. Bot. 28:405-409.

Barker, P.K., de Fossard, R.A., and Bourne, R.A. 1977. Progress toward clonal propagation of *Eucalyptus* species by tissue culture techniques. Int. Plant Prop. Soc. Proc. 27:546-556.

Behrouz, M. and Lineberger, R.D. 1981a. Influence of light quality on in vitro shoot multiplication of *Amelanchier laevis*. HortScience 16: 406.

______ 1981b. Interactive influences of auxins and cytokinins on proliferation of cultured *Amelanchier* shoot tips. HortScience 16:453.

Berbee, F.M., Berbee, J.B., and Hilderbrandt, A.C. 1972. Induction of callus and trees from stem tip cultures of hybrid poplar. In Vitro 7: 629.

Brand, R. and Venverloo, C.J. 1973. The formation of adventitious organs. II. The origin of buds formed on young adventitious roots of *Populus nigra* L. 'Italica.' Acta Bot. Neerl. 22:399-406.

Brown, C.L. 1981. Application of tissue culture technology to production of woody biomass. Internation Energy Agency Report NE 1981: 18. National Swedish Board for Energy Source Development, Stockholm.

Chalupa, V. 1974. Control of root and shoot formation and production of trees from poplar callus. Biol. Plant. 16:316-320.

______ 1975. Induction of organogenesis in forest tree cultures. Commun. Inst. For. Czech. 9:39-50.

______ 1979. In vitro propagation of some broad-leaved forest trees. Commun. Inst. For. Czech. 11:159-170.

Chen, C., Chen, F., Chien, C., Wang, J., Chang, H., Hsu, H., Ou, Y., Ho, T., Lu, T. 1978. Obtaining pollen plants of *Hevea brasiliensis* Muell.-Arg. Proceedings of the Symposium on Plant Tissue Culture. Science Press, Peking. pp. 11-12.

Christie, C.B. 1978. Rapid propagation of aspens and silver poplar using tissue culture techniques. Int. Plant Prop. Soc. Proc. 28:256-260.

Cresswell, R.J. and de Fossard, R.A. 1974. Organ culture of *Eucalyptus grandis*. Aust. For. 37:55-69.

Cresswell, R. and Nitsch, C. 1975. Organ culture of *Eucalyptus grandis* L. Planta 125:87-90.

Creze, J. 1980. The graft of *Camellia* meristem in vitro. Camellia Rev. 42:11-12.

Davis, M., Liu, L., Kester, D.E., and Papadatas, A. 1981. Micropropagation of almond and almond-hybrid genotypes. HortScience 16:417.

Deberg, P. and de Wael, J. 1977. Mass propagation of *Ficus lyrata*. Acta Hortic. 78:361-364.

Dickerman, M.D., Duncan, D.P., Gallegos, C.M., and Clark, F.B. 1981. Forestry today in China: Report of a month's tour by a team of American foresters. J. For. 79:65-80.

Dujickova, M., Chalupa, V., and Fikacova, J. 1977. Rust a morfogenese kalusovych tkani *Populus euramericana* (Dode) Guinier cv. Robusta (The growth and morphogenesis of tissue culture of *Populus euramericana* (Dode) Guinier cv. Robusta). In: Use of Tissue Cutlures in Plant Breeding. Proceedings of the International Symposium, 6-11 Sept., 1976, Olomouc, Czechoslovakia (F.J. Novak, ed.) pp. 81-89. Czechoslovak Academy of Sciences, Institute of Experimental Botany, Prague.

Durand, R. and Boudet, A.M. 1979. Le bouturage "in vitro" de l'eucalyptus. AFOCEL Etudes et Recherches n° 12—6/79, Micropropagation D'Arbres Forestiers, pp. 57-66. Paris.

Durand-Cresswell, R. and Nitsch, C. 1977. Factors influencing the regeneration of *Eucalyptus grandis* by organ culture. Acta Hortic. 78:149-155.

Durzan, D.J. and Lopushanski, S.M. 1975. Propagation of American elm via cell suspension cultures. Can. J. For. Res. 5:273-277.

Economou, A.S., Read, P.E., and Pellett, H.M. 1981. Micropropagation of hardy deciduous azaleas. HortScience 16:452.

Edin, H.L. 1976. Trees and Man. Columbia Univ. Press, New York.

Fan, M.-L. and Hu, T.-W. 1976. Plantlets from *Paulownia* tissue cultures. Bulletin 286, Taiwan Forestry Research Institute, Taipei.

Fernow, B.E. 1913. A Brief History of Forestry in Europe, the United States and Other Countries. University Press, Toronto.

Food and Agriculture Organization of the United Nations. 1963. World Forest Inventory. FAO, Rome.

Forest Service. 1978. Forest Statistics of the U.S., 1977. USDA Forest Service.

______ 1980. A National Energy Program for Forestry. Misc. Pub. No. 1394, USDA Forestry Service.

de Fossard, R.A. 1978. Tissue culture propagation of *Eucalyptus ficifolia* F. Muell. Proceedings of Symposium on Plant Tissue Culture (May 25-30, 1978, Peking) pp. 425-438. Science Press, Peking.

______ and Bourne, R.A. 1976. Vegetative propagation of *Eucalyptus ficifolia* F. Muell. by nodal culture in vitro. Int. Plant Prop. Soc. Proc. 26:373-378.

______, Nitsch, C., Cresswell, R.J., and Lee, E.C.M. 1974. Tissue and organ culture of *Eucalyptus*. N. Z. J. For. Sci. 4:267-278.

______, Barker, P.K., and Bourne, R.A. 1977. The organ culture of nodes of four species of *Eucalyptus*. Acta Hortic. 78:157-165.

______, Bennett, M.T., Gorst, J.R., and Bourne, R.A. 1978. Tissue culture propagation of *Eucalyptus ficifolia* F. Muell. Int. Plant Prop. Soc. Proc. 28:427-435.

Frett, J.J. and Smagula, J.M. 1981a. Effect of stock plant pretreatment on lowbush blueberry explant shoot multiplication. HortScience 16:418.

______ 1981b. The effect of 2iP, IAA, and explant bud position of lowbush blueberry shoots produced in vitro. HortScience 16:459.

Fu, M.L. 1978. Plantlets from *Paulownia* tissue culture. In: Fourth International Congress Plant Tissue and Cell Culture (T.A. Thorpe, ed.) Abstract 1736.

Garton, S., Read, P.E., and Farnham, R.S. 1981. Micropropagation of *Salix* spp. HortScience 16:405.

Gautheret, R. 1940. Nouvelles recherches sur le bourgeonnement du tissu cambial d'*Ulmus campestris* cultivé in vitro. CR Acad. Sci. Paris 210:744-746.

Gharyal, P.K. and Maheshwari, S.C. 1981. In vitro differentiation of somatic embryoids in a leguminous tree: *Albizzia lebbeck* L. Naturwissenschaften 68:379-380.

Goncalves, A.N., Marchado, M.A., Caldas, L.S., Sharp, W.R., and Mello, H.A. 1979. Tissue culture of *Eucalyptus*. In: Plant Cell and Tissue

Culture: Principles and Applications (W.R. Sharp, P.O. Larsen, E.F. Paddock, and V. Raghavan, eds.) pp. 509-526. Ohio State Univ. Press, Columbus.

Gorst, J.R. and de Fossard, R.A. 1980. Riboflavin and root morphogenesis in *Eucalyptus*. In: Plant Cell Cultures: Results and Perspectives (B. Sala, B. Parisi, R. Cella, and O. Ciferri, eds.) pp. 271-275. Elsevier/North Holland, Amsterdam, The Netherlands.

Gupta, P.K., Nadgir, A.L., Mascarenhas, A.F., and Jagannathan, V. 1980. Tissue culture of forest trees: Clonal multiplication of *Tectona grandis* L. (Teak) by tissue culture. Plant Sci. Lett. 17:259-268.

Gupta, P.K., Mascarenhas, A.F., and Jagannathan, V. 1981. Tissue culture of forest trees: Clonal propagation of mature trees of *Eucalyptus citridora* Hook, by tissue culture. Plant Sci. Lett. 20:195-201.

Hannapel, D.J., Dirr, M.A., and Sommer, H.E. 1981. Micropropagation of native *Rhododendron* species. HortScience 16:452.

Harada, H. 1975. In vitro organ culture of *Actinidia chinensis* Planch. as a technique for vegetative multiplication. J. Hortic. Sci. 50:81-83.

Hartney, V.J. and Barker, P.K. 1980. The vegetative propagation of eucalypts by tissue culture. IUFRO Symposium and Workshop on Genetic Improvement and Productivity of Fast-growing Tree Species, August 25-30, Aguas de Sao Pedro, Sao Paulo, Brazil.

Hosier, M., Garton, S., Read, P.E., and Farnham, R.S. 1981. In vitro propagation of *Alnus glutinosa*. HortScience 16:453.

Hu, C.Y. 1976. In vitro embryoids production on a tree angiosperm, *Ilex aquifolium* L. In: Proceedings Fourth North American Forest Biology Workshop, 1976 (H.E. Wilcox and A.F. Hamer, eds.). School of Continuing Education, College of Environmental Science and Forestry, Syracuse, New York.

______ and Sussex, I.M. 1971. In vitro development of embryoids on cotyledons of *Ilex aquifolium*. Phytomorphology 21:103-107.

______, Ochs, J.D., and Mancini, F.M. 1978. Further observations on *Ilex* embryoid production. Z. Pflanzenphysiol. 89:41-49.

Huhtinen, O. 1976. Early flowering of birch and its maintenance in plants regenerated through tissue culture. Acta Hortic. 56:243.

______ 1978. Callus and plantlet regeneration from anther culture of *Betula pendula* Roth. In: Fourth International Congress Plant Tissue and Cell Culture (T.A. Thorpe, ed.) Calgary, Canada.

______ and Yahyaglu, Z. 1974. Das frühe Blüher von aus Kalluskulturen herangezogenen Pflänzohen bei der Birke (*Betula pendula* Roth.). Silvae Genet. 23:32-34.

Hunter, C.S. 1979. In vitro culture of *Cinchona ledgériana* L. J. Hortic. Sci. 54:111-114.

Jacquiot, C. 1949. Observations sur la néoformation de bourgeons chez le tissu cambial d'*Ulmus campestris* cultivé in vitro. CR Acad. Sci. Paris 229:529-530.

______ 1951. Action du mesoinositol et de l' adenine sur la formation de bourgeons par le tissu cambial d'*Ulmus campestris* cultivé in vitro. CR Acad. Sci. Paris 233:815-817.

______ 1955. Formation d'organes par le tissu cambial d'*Ulmus campestris* L. et de *Betula verrucosa* Gaertn. cultivés in vitro. CR Acad. Sci. Paris 240:557-558.

Jarvis, B.C., Wilson, D.A., Fowler, M.W. 1978. Growth of isolated embryonic axis from dormant seeds of hazel (*Corylus avellana* L.). New Phytol. 80:117-123.

Kitahara, E.H. and Caldas, L.S. 1975. Shoot and root formation in hypocotyl callus cultures of *Eucalyptus*. For. Sci. 21:242-243.

Knox, C.A. and Smith, R.H. 1981. Progress in tissue culture methods for production of 'Riverside' stocks. Pecan Q. 15:27-31.

Kwei, Y., An, H.-H., Tsai, T.-R., and Wang, J.-R. 1978. Induction of callus and plantlet from stem segment in Chinese gooseberry. In: Proceedings of Symposium on Plant Tissue Culture (May 25-30, Peking) p. 526. Science Press, Peking.

Lange, D.D. and Karnosky, D.F. 1981. Techniques for high-frequency isolation of elm protoplasts. In: Forest Genetics: 25 Years of Progress. Proceedings of the Twenty-seventh Northeastern Forest Tree Improvement Conference, Burlington, Vermont, July 29-31, 1980 (D.H. DeHayes, ed.) pp. 213-222.

Lester, D.T. and Berbee, J.G. 1977. Within-clone variation among black poplar trees derived from callus culture. For. Sci. 23:122-131.

Letouze, R. 1977. Croissance du bourgeon axillaire d'une bouture de saule (*Salix babylonica* L.) en culture in vitro. Physiol. Veg. 12:397-412.

Li, H.L. 1963. The origin and cultivation of shade and ornamental trees. Univ. Pennsylvania Press, Philadelphia.

Lineberger, R.D. 1981. Tissue culture approaches commercial production. Am. Nurserymen 154:14-15, 69-70.

Lloyd, G.B. and McCown, B.H. 1980. Commercially feasible micropropagation of mountain laurel, *Kalmia latifolia*, by use of shoot-tip culture. Proc. Int. Plant Prop. Soc. 30:421-427.

Lu, C.H., Chang, H., and Liu, Y. 1978. Induction and cultivation of pollen plants from poplar pollen. In: Proceedings Symposium on Plant Tissue Culture (May 25-30, Peking) p. 242. Science Press, Peking (Abstr.).

Madani, A., Lee, C.W., and Hogan, L. 1979. In vitro culture and rooting of *Simmondsia chinensis* shoot tips. Plant Physiol. Suppl. 63:138 (Abstr.).

Makino, R.K., Nakano, R.T., Makino, P.J., and Murashige, T. 1977. Rapid cloning of Ficus cultivars through application of in vitro methodology. In Vitro 13:169.

Marcotrigiano, M. and Stimart, D.P. 1981. In vitro organogenesis and shoot proliferation of *Paulownia tomentosa* Steud. (Empress tree). HortScience 16:405.

Mathes, M.C. 1964. The in vitro formation of plantlets from aspen tissue. Φyton 21:137-141.

Matsuyama, J. and Murashige, T. 1977. Propagation of *Tupidanthus calyptratus* through tissue culture. In Vitro 13:169.

McCown, B.H. 1979. Rapid reproduction of *Betula* clones through shoot tip cultures. HortScience 14:431.

______ and Amos, R. 1979. Initial trials with commercial micropropagation of birch selections. Proc. Int. Plant Prop. Soc. 29:387-393.

McCulloch, S. and McCown, B.H. 1980. Isolation and plating of protoplasts of *Betula* spp. HortScience 15:433.

McPheeters, K., Skirvin, R.H., and Bly-Monnen, C.A. 1980. Culture of chestnut (*Castanea* spp.) in vitro. HortScience 15:328 (Abstr.).

Mehra, A. and Mehra, P.N. 1974. Organogenesis and plantlet formation in vitro in almond. Bot. Gaz. 135:61-73.

Meyer, M.M., Jr. 1981a. Tissue culture propagation of *Rhododendron* plants using florets as explants. In Vitro 17:228-229.

______ 1981b. In vitro propagation of rhododendron from flower buds. HortScience 16:452.

Nag, K.K. and Ahmad, J. 1978. Organogenesis in tissue cultures of a lac host *Moghania macrophylla*. Curr. Sci. 47:238-239.

Nickerson, N.L. 1978. In vitro shoot formation in lowbush blueberry seedling explants. HortScience 13:698.

Ohyama, K. 1970. Tissue culture in mulberry tree. Jpn. Agric. Res. Q. 5:30-34.

______ and Oka, S. 1980. Bud and root formation in hypocotyl segments of *Broussonetia kazinoki* Sieb. in vitro. Plant Cell Physiol. 21: 487-492.

Oka, S. and Ohyama, K. 1974. Studies on the in vitro culture of excised buds in mulberry tree. J. Sericul. Sci. Jpn. 43:230-235.

Radojevic, L. 1978. In vitro induction of androgenic plantlets in *Aesculus hippocastanum*. Protoplasma 96:369-374.

______ 1979. Somatic embryos and plantlets from callus cultures of *Paulownia tomentosa* Steud. Z. Pflanzenphysiol. 91:57-62.

______ 1980. Haploid embryos, plantlets and callus formation in woody species. In: The Plant Genome, Proceedings of the Fourth John Innes Symposium, September 1979 (D.R. Davies and D.A. Hopwood, eds.) p. 259. The John Innes Charity, Norwich, Connecticut.

______ and Kovoor, A. 1978. Characterization and estimation of newly synthesized DNA in higher plant protoplasts during the initial period of culture. J. Exp. Bot. 29:963-968.

______, Vujicic, R., and Neskovic, M. 1975. Embryogenesis in tissue culture of *Corylus avellana* L. Z. Pflanzenphysiol. 77:33-41.

______, Zylberberg, L., and Kovoor, J. 1980. Etude Ultrastructureale des Embryons Androgenetiques d'*Aesculus hippocastanum* L. Z. Pflanzenphysiol. 98:255-261.

Rao, P.S. 1965. In vitro induction of embryonal proliferation in *Santalum album* L. Phytomorphology 15:175-179.

______ and Bapat, V.A. 1978. Vegetative propagation of sandalwood plants through tissue culture. Can. J. Bot. 56:1153-1156.

______ and Rangaswamy, N.S. 1971. Morphogenetic studies in tissue cultures of the parasite *Santalum album* L. Bipl. Plant. 13:200-206.

Redenbaugh, K., Karnosky, D.F., and Westfall, R.D. 1981. Protoplast isolation and fusion in three *Ulmus* species. Can. J. Bot. 59:1436-1443.

Riffaud, J.-L. and Cornu, D. 1981. Utilisation de la culture in vitro pour la multiplication de merisiers adultes (*Prunus avium* L.) selectionnes en forêt. Agronomie 1:633-640.

Riou, A., Harada, H., and Taris, B. 1975. Production des plantes entieres à patir des cellules separées de cals de Populus. C.R. Acad. Sci. Paris 280D:2657-2659.

Rogozinska, J.H. 1968. The influence of growth substances on the organogenesis of honey locust shoots. Acta Soc. Bot. Pol. 37:485-491 (in Polish).

______ 1969. Organogenetic differences in honey locust cotyledons. Bull. Acad. Pol. Sci. 17:721-725.

Saito, A. 1976. Isolation of protoplasts from mesophyll cells of *Paulownia fortunei* Hemsl. and *Populus euramericana* cv. I-45/51. J. Jpn. For. Soc. 58:301-305.

______ 1980a. Isolation of protoplasts from mesophyll cells of *Paulownia* and *Populus*. Bulletin of the Forestry and Forest Products Research Institute No. 309, pp. 1-6.

______ 1980b. Fusion of protoplasts isolated from somatic cells of tree species. Bulletin of the Forestry and Forest Products Research Institute No. 309, pp. 7-12.

Schieder, O. 1980. Somatic hybrids between a herbaceous and two tree *Datura* species. Z. Pflanzenphysiol. 98:119-127.

Seeliger, I. 1959a. Uber die Bildung wrzelburtiger Sprosse und das Wachstum isolierter Wurzeln der Robinie (*Robinia pseudocacia* L.). Flora 148:218-254.

______ 1959b. Zur Entwicklungsgeschichte und Anatomie wurzelburtiger Sprosse an isolierten Robinienwurzel. Flora 148:119-124.

Sita, G.L. 1979. Morphogenesis and plant regeneration from cotyledonary cultures of *Eucalyptus*. Plant Sci. Lett. 14:63-68.

______ and Vaidyanathan, C.S. 1979. Rapid multiplication of *Eucalyptus* by multiple shoot production. Curr. Sci. 48:350-352.

______, Raghava Ram, N.V., and Vaidyanathan, C.S. 1979. Differentiation of embryoids and plantlets from shoot callus of sandalwood. Plant Sci. Lett. 15:265-270.

______, Raghava Ram, N.V., and Vaidyanathan, C.S. 1980a. Triploid plants from endosperm cultures of sandalwood by experimental embryogenesis. Plant Sci. Lett. 20:63-69.

______, Shobha, J., and Vaidyanathan, C.S. 1980b. Regeneration of whole plants by embryogenesis from cell suspension cultures of sandalwood. Curr. Sci. 49:196-198.

Skolmen, R.G. and Mapes, M.O. 1976. *Acacia koa* Gray plantlets from somatic callus tissue. J. Hered. 67:114-115.

______ 1978. After care procedures required for field survival of tissue culture propagated *Acacia koa*. Proc. Int. Plant Prop. Soc. 28: 156-164.

Smith, M.A. and McCown, B.H. 1981. The effect of tissue source and preconditioning on protoplast release and viability from *Betula* and *Rhododendron* germplasm. HortScience 16:453.

Sommer, H.E. 1981. Propagation of sweetgum by tissue culture. In: Proceedings Sixteenth Southern Forest Tree Improvement Committee, pp. 184-188.

______ and Brown, C.L. 1980. Embryogenesis in tissue cultures of sweetgum. For. Sci. 26:257-260.

Srivastava, P.S. and Steinhauer, A. 1981. Regeneration of birch plants from catkin tissue. Plant Sci. Lett. 22:379-386.

Tabachnik, L. and Kester, D.E. 1977. Shoot culture for almond and almond-peach hybrid clones in vitro. HortScience 12:545–547.
Thompson, D.G. and Gordon, J.C. 1977. Propagation of poplars by shoot apex culture and nutrient film technique. In: Tappi Conference Papers, Forest Biology and Wood Chemistry Conference, pp. 77-82. TAPPI, Atlanta.
Turiansky, G.W., Chesney, R.H., and Kirby, E.G. 1981. In vitro growth and antibiotic sensitivity of *Brachycome*. Plant Physiol. Suppl. 67: 117.
Venketeswaran, S. and Gandhi, V. 1981. Tissue culture and plantlet regeneration of Chinese tallow tree, *Sapium sebiferum* L. In Vitro 17:218-219.
Venverloo, C.J. 1973. The formation of adventitious organs. I. Cytokinin-induced formation of leaves and shoots in callus cultures of *Populus nigra* L. 'Italica.' Acta Bot. Neerl. 22:390–398.
Vieitez, A.M. and Vieitez, E. 1980a. Plantlet formation from embryonic tissue of chestnut grown in vitro. Physiol. Plant. 50:127-130.
Vieitez, A.M. and Vieitez, M.L. 1980b. Culture of chestnut shoots from buds in vitro. J. Hortic. Sci. 55:83-84.
Wang, C., Sun, C., and Chu, Z. 1975. The induction of *Populus* pollen-plants. Acta Bot. Sinica 17:56-59.
Welsh, K., Sink, K.C., and Davidson, H. 1979. Progress on in vitro propagation of red maple. Proc. Int. Plant Prop. Soc. 29:382-386.
Whitehead, H.C.M. and Giles, K.L. 1976. Rapid propagation of poplars by tissue culture methods. Proc. Int. Plant Prop. Soc. 26:340-343.
Williams, K.M. and Ryugo, K. 1980. Root initiation and formation on excised epicotyls of almonds, *Prunus amygdalus*. HortScience 15:432.
Winters, R.W. 1974. The Forest and Man. Vantage Press, New York.
Winton, L.L. 1968. Plantlets from aspen tissue cultures. Science 160: 1234-1235.
______ 1970. Shoot and tree production from aspen tissue culture. Am. J. Bot. 57:904-909.
______ 1971. Tissue culture propagation of European aspen. For. Sci. 17:348-350.
______ and Huhtinen, O. 1976. Tissue culture of trees. In: Modern Methods in Forest Genetics (J.P. Miksche, ed.) pp. 241-264. Springer-Verlag, New York.
Wochok, Z.S. and Slvis, C.J. 1980. Gibberellic acid promotes *Atriplex* shoot multiplication and elongation. Plant Sci. Lett. 17:363-369.
Wolter, K.E. 1968. Root and shoot initiation in aspen callus cultures. Nature 219:509-510.
Wright, J.W. 1976. Introduction to Forest Genetics. Academic Press, New York.
Yokoyama, T. and Takeuchi, M. 1976. Organ and plantlet formation from callus in Japanese persimmon (*Diospyros kaki*). Phytomorphology 26:273-275.
Youngs, R.L. 1978. Meeting the energy demand through efficient use of wood. In: Proceeding of the Western Forestry Conference of the Western Forestry and Conservation Association 69:1-6.
Zerbe, J.I. 1978. Impacts of energy developments on utilization of timber in the northwest. In: Proceedings of the Northwest Private Forestry Forum, Portland, Oregon.

Zhu, X., Wang, R., and Liang, Y. 1980. Induction of poplar pollen plantlets. Sci. Silv. Sin. 16:190-197.

Zilis, M.R. and Meyer, M.M. 1977. Embryoid production in the woody plants, *Hamamelis virginiana* and *H. vernalis*. HortScience 12:389.

Zimmerman, R.H. and Broome, O.C. 1980. Blueberry micropropagation. In: Proceedings Conference on Nursery Production of Fruit Plants through Tissue Culture: Applications and Feasibility (Apr. 21-22, 1980) pp. 44-47. USDA Agricultural Research Service, Beltsville, Maryland.

SECTION IX
Extractable Products

CHAPTER 20
Cacao

P. Dublin

HISTORY AND ECONOMIC IMPORTANCE

The seeds or beans of *Theobroma cacao* provide the raw material for the manufacture of cocoa and chocolate and for the extraction of cocoa butter. The wet beans newly removed from the fruits are first fermented in fermentation boxes, then spread on a drying floor or tray and exposed to the sun. This period of drying may take one to two weeks, depending on the weather. Chocolate is then made from these dried beans, which are roasted, ground, and mixed with sugar.

Humans started using cacao beans a long time ago, and it is thought that Mayan farmers were the first to domesticate cacao trees. In the fourteenth century, cacao trees belonging to the Criollo group were cultivated in Mexico to make a local beverage called cacahuatl. The cacahuatl, made from ground, cured cacao beans, maize flour, and peppers in hot water, was very popular among the Aztec nobility. Presumably, the peppers and maize flour of the cacahuatl were later replaced with milk and sugar to make a drink that was pleasant to the European palate, which was called cacao.

During the fourteenth and fifteenth centuries, Mexico was the only producer of cacao, but by the end of the sixteenth century cacao was cultivated in the tropical regions of South America, Central America, Trinidad, Jamaica, and the Caribbean Islands. In 1585, cacao beans and a locally manufactured paste of cacao were exported to Spain for the first time. In Spain, cacao became very popular; its use was then extended to Italy and France. During the seventeenth century the

Mexican Criollo cacao types were introduced into Venezuela, which then became a major producer, exporting 4000 tons annually by 1750.

A century later, Ecuador took the lead in production, and by 1850 was exporting 40,000 tons annually. The cacao exported was called Arriba and was produced from a local cacao type, the "Cacao National."

In Africa, cacao was first introduced into the Fernando Poo and Sao Thome regions. It is thought that cacao was first introduced in 1857 in Ghana, from Surinam, by Swiss missionaries. From Ghana, cacao was introduced into Nigeria, Cameroon, and the Ivory Coast, and in the early twentieth century production of cacao in Africa increased considerably.

In 1830 the total world crop of 10,000 tons was produced by the tropical countries of America (Venezuela, Ecuador, Haiti, and Brazil). In 1900 the world production was 115,000 tons, of which 50% was from South America (Ecuador, Brazil, and Venezuela), 28% from Central America (Trinidad and St. Domingo), 17% from Africa (Ghana), and 4% from Asia and Oceania (Indonesia, Ceylon, and the Philippines). In 1964 the world production was 1,528,000 tons, of which 13% was from South America (Brazil), 6% from Central America, 78% from Africa (Ghana, Nigeria, Cameroon, and the Ivory Coast), and 2% from Asia and Oceania.

Cacao has become very important for the economy of all of these African countries. Most of the cacao beans produced by these countries are consumed by Western Europe, the United States, and Canada (Tables 1 and 2).

REPRODUCTION AND MORPHOLOGY

The genus *Theobroma* is a member of the family Sterculiaceae and is related to the genera *Herrania, Abroma,* and *Bytneria.* According to Cuatrecasas (1964), the most widely accepted authority, the genus *Theobroma* can be divided into six sections containing 22 species. *Theobroma cacao* is the only cultivated species of this genus. Some of the wild species of *Theobroma* are used for local beverages in different countries in the Americas (e.g., Mexico, Colombia, and Brazil).

The origin of *T. cacao* is probably the upper reaches of the Amazon River. This hypothesis was suggested by Cheesman (1944) based upon material collected by Pound (1938) along the Amazon and its tributaries.

Cultivated cacao has a great deal of variability in color, size, and shape of flowers, pods, and seeds. According to Cheesman (1944) and other authors, cultivated cacao should be divided into two major populations: Criollo and Forastero. The well-known cultivar Amelonado that is cultivated in West Africa and Brazil is a Forastero type. The Trinitario type is derived from spontaneous hybrids between Forastero (the Guyana Amelonado) and the Venezuelan Criollo population (Toxopeus, 1969).

Table 1. World Production of Cacao Beans[a]

COUNTRY	1971	1975	1980
AFRICA			
Cameroon	123	96	117
Ghana	470	396	280
Ivory Coast	226	227	390
Nigeria	257	215	150
Total	1172	994	996
AMERICAS			
Brazil	164	251	325
Ecuador	65	60	325
Total	372	448	553
WORLD TOTAL	1589	1502	1629

[a]Thousands of tons.

Table 2. World Per Capita Cocoa Consumption[a]

COUNTRY	1975	1979
WESTERN EUROPE		
Austria	2.48	3.01
Belgium	2.49	3.04
Germany	2.37	2.57
Switzerland	3.23	3.82
United Kingdom	1.84	1.65
France	1.82	1.93
EASTERN EUROPE		
Germany	1.64	1.39
Hungary	1.33	1.30
NORTH AMERICA		
Canada	1.57	1.44
United States	1.73	1.42
AUSTRALIA	1.49	1.58

[a]Kilograms.

Theobroma cacao L. is a tree of small size (4-15 m). It has a dimorphic growth habit with two kinds of branches—the orthotropic shoot or "chupon" which has a spiral leaf arrangement (phyllotactic index 3/8), and the plagiotropic branches or "fan" branches which have an alternate leaf arrangement (phyllotaxic index 1/2). The plagiotropic branches appear at the end of the stem when the seedling is about 2 years old. They are formed in whorls of three to six (normally five) twigs. These whorls are called jorquettes. The leaves are 15-50 cm long and 4-15 cm broad with an acute or subacute apex. Cacao has a cauliflorous growth habit with flowers and fruits formed on the stem and on the major branches. The inflorescences are formed in old leaf axils that have produced flowers for several years. These axils become thickened tubecles called flower-cushions (Fig. 1).

The flower of cacao is complete, hermaphroditic, and made up of 5 sepals, 5 petals, 5 anthers, 5 staminodes, a pistil, and a superior ovary. Flower size depends on the cultivar, as flowers of Trinitario and Amazon cocoa are generally bigger than those from West African types, such as Amelonado. The ovary contains 30-60 ovules arranged in 5 rows. The base of the petal is cup shaped, and the anther is located in this cup so that the pollen sacs, each containing only about 1000 pollen grains, are hidden.

Mature seeds of cacao can germinate immediately after being removed from the ripe pod and have epigeal germination. The duration of viability of cacao seeds is very short, and this characteristic makes exchange of germplasm between distant countries quite difficult. The cacao seeds maintained in the ripe pod remain viable for a maximum of 3 weeks. Methods of storage to preserve viability of seeds removed from pods have been investigated but with little success (Swarbrick, 1965). Seeds covered with charcoal powder and wrapped in a plastic bag maintain viability for 2 weeks or more.

After germination, the cotyledons open and the plumule starts to grow. The first phase of growth of the young seedlings ends with the hardening-off of the first leaves. This first phase of growth is followed by a visible pause, then growth resumes. Subsequent growth pulses appear at approximately 6-week intervals, as internodes become longer, and the leaves are well spaced in a spiral arrangement. Depending on the ecological conditions and the genotype, the young seedling initiates its next growth phase by forming its first jorquette between its second and fourth year. Vigorous Amazonian hybrids cultivated in good ecological conditions often form a jorquette when they are only 18 months old. The formation of a jorquette is due to the simultaneous growth of five axial buds on the terminal end of the plant. These five buds give five to six branches with a plagiotropic growth habit, which collectively is called a fan. After a few months, a new "chupon" (orthotropic bud) may start growing just below the joint of the first jorquette, and when it has attained a certain length, a new jorquette is formed. This process may be repeated several times. In *T. cacao*, the foliage of an adult tree is made up of plagiotropic branches that grow in pulses alternating with periods of latent meristems.

Inflorescences are formed on the stem, and the major plagiotropic branches form between the second and fourth year. They arise where

Figure 1. Cacao tree originating from a seedling. *T. cacao* has a cauliflorous habit with flowers and fruits formed on the stem and on the major branches.

old axil leaves were located. Opening of flowers takes place in the early morning. The sepals of the mature floral bud start to split in the afternoon and continue to open through the night so that the flower is completely opened between 5 and 7 a.m. Dehiscence of the anthers occurs immediately after the opening of the flower. To be successful, pollination must be achieved the same day the flower opens. Failure of fertilization at this time results in abscission of the flower, within 48 hr following the flower opening. Cacao shows a periodicity of flowering throughout the year, and noticeable differences can be detected between genotypes. A single cacao tree can produce 20,000–100,000 flowers annually, but only a very low percentage (ca. 1-5%) of these flowers become fruit. The main pollination agent is a small midge (*Forcypomia* sp.) that is ubiquitous in the cacao-growing world. Aphids and ants also effect pollination.

After fertilization the zygote starts dividing within 40-50 days, and during this first month after fertilization the young pod grows slowly. After this period of slow growth the young pod grows more rapidly and reaches a maximum size at about 75 days. About 140 days after pollination the development of the embryo is complete, and the pod begins to ripen. The pod can be harvested about 150 days after pollination.

Only a limited number of young pods form mature fruit. The name Cherelle is used to describe a young pod that is 5-20 cm long. If Cherelles fail, wilt, and shrivel on the tree and the failure is not caused by a disease, this is called Cherelle wilt. Cherelle wilt is a physiological fruit-thinning mechanism.

All traditional methods of vegetative propagation—cutting, budding, and marcotting—can be successfully applied to cacao. However, only cuttings are used on a large scale for vegetative reproduction. Cuttings originating from orthotropic shoots are more desirable than those from plagiotropic stems (Fig. 2). Pyke (1933) in Trinidad was the first to successfully utilize cuttings. Cuttings to be rooted generally consist of two to three internodes of fan shoots that have been recently hardened-off. As no rooting can be obtained on cuttings without leaves, two to three half-leaves are usually retained. Rooting is induced in propagules when the atmosphere is very moist and cool. Treatment of the basal part of the cuttings with auxins facilitates root induction. Methods have been developed for large-scale production of cuttings, and some centers for commercial cuttings have been able to produce thousands of cuttings annually.

PESTS AND DISEASES

Cultivated cacao is subject to a number of pests and diseases which are responsible for loss of fruits and for death of cacao trees in most of the cacao-growing countries.

Capsids are a major pest of cacao in West Africa and South America. The two species of capsids that cause the most damage are *Sahlbergella singularis* and *Distantiella theobromae*. Capsids attack the pods and young shoots and suck the sap through a feeding puncture. The damage inflicted on young branches may kill off the young pulses causing a die-back. There are some other pests of minor importance, including thrips that attack young leaves and cause defoliation of trees and borers that bore holes in the stems.

Figure 2. Cacao trees originating from plagiotropic cutting are low spreading without main trunk and need frequent pruning to be shaped.

Among the fungi that are responsible for cacao diseases, *Phytophthora palmivora* is certainly the most important and most widespread in cacao-growing countries. *Phytophthora palmivora* causes the black pod disease and can also affect stems and leaves. Pod rot is by far the most important damage caused by *P. palmivora*, which attacks pods at all ages and may in some countries (e.g., Cameroon) cause losses and affect up to 50% of the pods.

The witch broom disease is caused by the fungus *Marasmius perniciosus*, indigenous to South America. This fungus probably originated in the upper Amazon Valley, from which it has spread to the surrounding cacao-growing countries such as Ecuador, Colombia, Peru, and Venezuela. It is so called because the most obvious symptom of the disease is hypertrophied shoots, which are much thicker than healthy shoots, resulting in many lateral shoots with underdeveloped leaves and giving the appearance of a broom. *Marasmius periciosus* can also infect cushion flowers, flowers, and pods. Only young developing tissue is attacked by this fungus.

The monilia disease caused by the fungus *Monilia roreri* has resulted in losses of pods in Colombia, Peru, and Venezuela. Apart from these important diseases of cacao, there are other minor diseases, such as root diseases caused by different fungi belonging to the group of *Armillaria*, *Rosellenia*, and *Fomes*. The importance of each disease varies depending on soil, cultural, and climatic conditions.

The first known virus disease in cacao was the swollen shoot disease. This disease was responsible for the death of thousands of cacao trees in the West African cacao-growing countries of Ghana and Togo. It is so called because the most obvious symptom is a swelling of shoots and young chupon in infected trees. The virus vectors are mealybugs. Two other viruses are also important to mention—cacao mottle leaf virus and cacao necrosis virus.

GENETICS AND BREEDING

Very little is known about the genetics and heritability of useful characteristics such as yield factors, vigor, and disease resistance in cacao. The genetics of sexual incompatibility and some morphological characteristics such as axil spot and bean color are best known.

Axil spot is a red anthocyanin coloration at the junction of the petiole with the main axis. The color intensity varies from petioles that are completely dark red to those with a spot of red at the petiole junction. This axil spot character, used for the detection of haploids (Dublin, 1973a), is controlled by two complementary genes (Harland and Frecheville, 1927).

The bean color varies from white to purple and includes various shades of purple. According to Wellensieck (1932), the purple color is due to the action of a dominant allele and white results from the action of a recessive allele. Similarly, both the number and size of seeds have high heritability.

Pound (1932) was the first to discover the existence of incompatibility in cacao. The Trinidad trees tested by Pound were classified based on their setting capacity as self incompatible (SI) and self compatible (SC). Pound found that pollen from SC trees is effective on any stigma, whereas pollen from SI trees only caused setting on SC trees. Self incompatible trees have been found in several other countries including Java (Ostendorf, 1948), Colombia (Naundorf and Villamil, 1950), and Ghana (Posnette, 1945a). In contradiction with the early findings of Pound, successful crosses between two SI trees have been reported by several authors. Muntzing (1947) successfully crossed two SI trees in Ecuador. Posnette (1945a), working with a small population of Amazon trees introduced from the upper reaches of the Amazon, found all tested trees were SI but cross compatible to some extent.

Cacao provides an unique example of incompatibility, where the diploid tissue of the style does not prevent fertilization. All incompatibility reactions take place in the embryo sac. The incompatibility mechanism in cacao is based on the genetic control of the success or failure of syngamy (Knight and Rogers, 1953, 1955; Cope, 1962).

The emergence of SC types from SI members of the primordial population of cacao may have occurred somewhere in the lower Amazon. All material collected in and near the origin of cacao is SI. Self compatible types, though, have been found in the Colombian valley of Cauca, in Central America, in the lower Amazon Valley, and in Venezuela. The vast population of trees in West Africa, which provides the bulk of the world cacao, is uniformly SC and is reputed to have initiated from a small population of cacao collected near the mouth of the Amazon.

Genetic improvement of cultivated cacao is mainly based upon exploitation of the heterosis that occurs in hybrids between upper Amazon types and the Amelonado or Trinitario genotypes. Hybrid vigor was first reported in cultivated cacao many years ago. In Indonesia, the Djati Roenggo hybrids, which were famous for vigor and production during the early 1900s, were actually derived from chance hybrids of a Forastero type introduced from Venezuela and the local Java Criollo.

The cacao trees from the upper Amazon have greater vigor than existing varieties and offer unique sources of disease resistance. Although the Amazon hybrids require different management methods than traditional cultivated varieties, the improved early vigor, ease of establishment, and precocity are so great that hybrids with Amazon types are being used increasingly in most cacao-growing countries. Since several authors have demonstrated the existence of hybrid vigor in cacao, the genetic improvement of this plant has been essentially based on the utilization of group heterosis, which occurs when the Amazon parent is combined with an Amelonado or Trinitario parent. These Amazonian hybrids have several advantages over local cultivars (vigor, earliness, disease resistance) but are highly heterogeneous. This heterogeneity of hybrids is a direct consequence of the heterozygosity of the upper Amazon parent. Obtaining a homozygous, self-fertile Amazon parent should permit elimination of this heterogeneity and yield homogeneous hybrids of greater vigor. Hence, a great deal of emphasis has recently been placed on the production of haploid plants.

The first known haploid seedlings of cacao ($n = x = 10$) were obtained by Dublin (1972). These first haploid cacao trees were obtained following the dissociation of embryos in polyembryonic seeds and were verified by counting chromosomes of young leaves. It was found that the mean proportion of polyembryos in cultivated cacao trees was 1.55

and that 2.7% of polyembryonic seeds produced haploids. In cultivated cacao trees the rate of haploids is associated with formation of twins. Depending on the lot of seeds, this rate varies between 0 and 9%.

Haploid plantlets have also been obtained by germinating flat beans under controlled environmental conditions. Under ordinary conditions, these beans have a low germination rate so that the recovery rate of haploid embryos has been very low (Dublin, 1973b). Haploids have also been recovered by delayed pollination of SI Amazon types with the aid of pollen carrying a genetic marker, axil spot (Dublin, 1973a). Haploids have also been obtained by screening seedlings derived from monoembryonic seeds. The haploid rate using this method varies greatly from one group of cacao trees to another. Within the Amazon group the rates have sometimes reached 1% of certain genotypes. Haploid seedlings derived from monoembryonic seeds developed more rapidly and tolerated colchicine treatment better than haploids derived from polyembryos or flat beans.

The phenotype of haploid cacao trees is unique. The haploid cacao grows slowly with small irregular-shaped leaves. The erect haploid leaf is supported by a short erect pedicel that is distinct from the decumbent limb of the diploid leaf (Fig. 3). The haploid leaf has a distinct puckered aspect of the limb due to a series of invaginations. Normally, the haploid cacao tree is incapable of living beyond 5-12 months. By grafting haploid shoots onto a diploid rootstock of *T. cacao*, it is possible to maintain the haploid state of the cacao tree for several years.

Chromosome doubling is achieved by depositing a small ball of cotton-wool soaked in an aqueous solution of colchicine onto the terminal meristem of a 4-week-old plant (Dublin, 1974). If successful, it is possible to observe an increase in the size of the bud and deeper anthocyanin coloration 15 days following treatment with colchicine (Fig. 4). Homozygous diploid cacao trees derived from haploids develop vegetatively, flower, and set fruit normally. These homozygous trees, derived from diploidized haploids, have been used for crossing with Amelonado parents in the genetic improvement of cacao trees in the Ivory Coast.

Interspecific and Intergeneric Hybridization

Many of the wild species of *Theobroma* or *Herrania*, another member of Sterculiaceae, have desirable characters that would be worth transferring into *T. cacao*. These include thick pods in *T. bicolor*, resistance to black pod and viral diseases in *T. grandiflora* (Martinson, 1966), and high butter fat content (50-60%) in *T. grandiflora*.

Posnette (1945b) was the first to suggest transfer of desirable characters from wild species of *Theobroma* into the cultivated varieties and was the first to work on interspecific hybridization between *T. cacao* and related species.

The results and conclusions of several interspecific crosses between *T. cacao* and related species of the genus *Theobroma* and *Herrania* are similar (Table 3). The percentage of flower set is low (Williams, 1975),

Figure 3. *T. cacao*, haploid plantlet (n = x = 10), the erect haploid leaves supported by a short pedicel contrast with the classical decumbent limb of the diploid leaves.

and only a small amount of fruit is obtained (Jacob and Opeke, 1971). Interspecific crosses in *T. cacao* generally produce only a few hybrid seeds that are capable of germination. The growth of hybrid seedlings from *T. cacao* x *T. grandiflora* can be improved by grafting the hybrid

Figure 4. *T. cacao* haploid plantlet fifteen days after a colchicine treatment of the terminal bud. An elongation of the stipules and hairs, a deeper coloration of the young leaves, an increase of the diameter of the stem are the first signs of a successful diploidization.

plant on rootstock of either *T. cacao* or *T. grandiflora* (Martinson, 1966).

LITERATURE REVIEW

To date there have been a limited number of studies on tissue culture of *Theobroma cacao.* In general, all attempts to initiate callus of cacao were successful as callus was rapidly obtained from various organs or explants on a wide range of culture media. On the contrary, all attempts to regenerate plantlets from cacao callus have failed.

Archibald (1954) was the first to investigate tissue culture of cacao. He obtained callus from explants of bark or stem on culture media of Gautheret or White without any growth regulators. For sustained subculture of the callus derived from cambial tissue the nutritional requirements were more complex than those of the original explants. White's medium supported only very slow growth, while spectacular increases in the growth rate were obtained by adding CW to it. The primary callus produced from cultured explants was compact, white and granular, but during subculture on medium supplemented with CW, the tissue became colorless. Archibald observed considerable differences between tissues cultured on solid media and those cultured on filter paper soaked in liquid medium. On agar medium the typical white cal-

Table 3. Flower Pollination, Flower Set, and Fruit Harvested from Interspecific Hybridization (from Jacob and Opeke, 1971).

INTERSPECIFIC COMBINATION	FLOWERS POLLINATED (No.)	FLOWERS SET		FRUITS HARVESTED (No.)
		No.	Percent	
T. cacao × *T. bicolor*	491	8	1.6	2[a]
T. cacao × *T. microcarpa*	14	0	0	0
T. cacao × *T. speciosa*	69	0	0	0
T. cacao × *T. grandiflora*	93	35	37.6	3[b]
T. cacao × *Herrania* sp.	436	27	6.2	6[c]
Herrania sp. × *T. cacao*	295	2	0.7	0

[a]Pods contained only normal cacao beans.

[b]Pods contained only flat beans.

[c]Pods contained various proportions of flat beans and normal cacao beans.

lus mass was gradually overgrown by disorganized pale brown, loosely granular tissue. Using liquid medium, growth was restricted to where the culture was in contact with the soaked filter paper, so that the original tissue was raised from the surface of the paper by the newly formed tissue. Callus was also initiated and subcultured from explants of plants infected with swollen shoot virus (Archibald, 1954).

The second attempt to culture cacao in vitro was reported by Ibanez (1964). Using a modified medium of Rudolph and Cox (1943), investigations were reported on the action of different sugars (sucrose, dextrose, maltose, lactose, and sorbose) on the respiration rate of cotyledon-free mature cacao embryos under sterile conditons.

Among the different media tested, the largest proportion of explants with maximum callus development was obtained with MS. Tests with media supplemented with various concentrations of sucrose, auxins, cytokinins, and vitamins suggest that the best callus growth occurred on MS media supplemented with 11.0 μM IAA, 0.47 μM KIN, and twice the normal concentration of MS vitamins.

Improvement of callus growth was also obtained by varying the concentration of IAA or NAA. Callus growth increased for 7 weeks. After 7 weeks, growth declined and senescence occurred. Consequently, an 8-week interval was used for subculture.

A large range of media supplemented with extracts derived from leaves, pod walls, and young seeds were tested in attempts to regenerate plantlets from callus. With the exception of periodic root initiation, no organogenesis was obtained.

Orchard et al. (1979) and Miller (personal communication) also examined in vitro culture of apical buds of cacao for vegetative propagation. The shoot apex in most plants has alternate periods of growth and dormancy. In temperate species, one cycle of growth and dormancy extends over a year, while in cacao the cycle is repeated several times a year. Orchard et al. (1979) initiated tissue cultures of apical buds of cacao to establish suitable material to investigate the endogenous control of intermittent bud growth. Excised terminal buds from young Amazon seedlings that were raised in a glasshouse were sterilized in 0.1% mercuric chloride and were placed on LS medium (Linsmaier and Skoog, 1965) supplemented with different growth substances (KIN, ZEA, IAA, IBA, and GA). Both solid and agitated liquid media were tested.

Some growth was observed on both agar and liquid medium but the degree of response varied with the stage of bud development and with hormone treatment. The response on agar medium was limited to bud swelling, whereas in some liquid medium treatments bud swelling was followed by leaf and stem elongation. Breakage of dormancy as manifested by bud swelling followed by stipule opening was promoted by both KIN and GA. No intact plants were recovered from cultured dormant buds.

Comparable and more encouraging results were obtained by Miller (personal communication), who cultured axillary buds and nodes containing axillary buds to maintain organized growth in cacao with as little callus formation as possible. Any fragment of cacao stem placed in vitro on a media with or without growth regulators results in formation of callus. This tendency in cacao to produce callus in vitro may be very inconvenient when the objective is to initiate and maintain organized growth.

Miller cultured explants on MS medium supplemented with several growth regulators (IBA, IAA, NAA, 6BA, 2iP, and GA). In addition, effects of media were tested by using SH, B5, and LS media. With a combination of IBA and KIN on MS medium, bud dormancy was broken and in some cases was followed by leaf expansion and shoot elongation. In most cases, though, only dormancy breakage was observed and eventually the buds died. In addition, Miller has suggested that losses through contamination can be reduced by using plants that are kept in a glasshouse and are not watered overhead.

Both Esan (1975) and Pence et al. (1979) were able to obtain somatic embryos in vitro from cultured cotyledon and hypocotyl tissues of very young cacao embryos. Esan (1975), in attempts to develop a method for production of cacao plantlets in vitro, used numerous explants including ovules from fruit 6-8 weeks old, immature embryos from 90-day-old fruit, and the embryo axis (axes of cacao bean) from mature unripe pods and anthers. The basal medium used for culture of somatic cacao explants was composed of inorganic salts of MS, with WH vitamins and CH. Cacao anthers were cultured on Nitsch medium.

Esan (1975) reported that some growth additives can trigger a growth response in specific parts of the embryo. For example, tryptone in a range of 100-1000 mg/l enhances growth of the cotyledon more than other parts of the embryo. Addition of NAA to the basal culture medium promoted direct somatic embryogenesis rather than an increase of root growth. The ensuing adventive embryos that were spherical or bell-shaped developed through a budding process. Most of these somatic embryos were derived from the hypocotyl portion of the seedling embryo, while some were derived from the adaxial portion of the cotyledon. The initiation of these adventive embryos is not preceded by callus formation.

Pence et al. (1979) initiated tissue cultures of cacao in order to establish the necessary conditions for cacao regeneration in vitro. Various explants including leaves, pericarp, ovules, immature embryos, cotyledons from mature embryos, and the axis of mature embryos were cultured on different culture media to determine morphogenetic potential. The basal medium contained inorganic salts of MS: myo-inositol, 0.55 mM; nicotinic acid, 4.1 μM; pyridoxine HCl, 2.4 μM; thiamine HCl, 0.03 μM; glycine, 0.03 mM; CH, 1000 mg/l; sucrose, 0.088 M; Bacto agar, 10,000 mg/l; and CW, 100 ml/l. Three cytokinins--KIN (0-43 μM), 2iP (0-98 μM), and BA (0-88 μM)—and three auxins—IAA (0-110 μM), NAA (0-110 μM), and 2,4-D (0-90 μM)—were tested in combinations. The cultures were incubated at a temperature of 26 C in both light and dark.

Callus was obtained with all explants used and on practically all media tested. On media supplemented with NAA, roots were obtained from leaf explants, from immature ovules, and mature cotyledon explants in both light and dark. Immature sexual embryos cultivated in dark or in light on basal medium supplemented with both NAA and CW produced adventive embryos, which proliferated by budding from the cotyledon of the immature sexual embryo (Pence et al., 1980). When transferred from a solid medium to an agitated liquid medium, these adventive embryos developed roots and primary leaf growth occurred for some of them (Pence et al., 1981). Further development of complete normal plantlets was not obtained. In some treatments, up to 80% of the cotyledons of the sexual embryos initiated asexual embryos (Table 4).

Table 4. Proliferation of Asexual Embryos from Immature Cotyledons of Cacao[a] (after Pence et al., 1979)

EXPLANT SOURCE	EMBRYO LENGTH (mm)	SEXUAL EMBRYOS CULTURED (No.)	CULTURED EMBRYOS FORMING ASEXUAL EMBRYOS (%)
Amelonado	2-4	25	80
UF 221 (Costa Rica)	2-4	16	81
41 R (Mexico)	2-4	7	86
SIAL 93 (Brazil)	10-15	19	47
UF 11 (Costa Rica)	10-15	32	38

[a]After 18 days, cultured in a basal medium with CW and NAA.

Jalal and Collin (1977, 1979) suggested using callus of cacao to investigate the biosynthesis of polyphenols and flavor compounds in cacao. The polyphenols of the cacao bean have long been regarded as important components of flavor in the roasted fermented cotyledons of cacao. Jalal and Collin (1977) developed tissue cultures of cacao to test their value to study the biosynthesis of cacao polyphenols. Although some polyphenols were found both in cacao callus and in tissue of the intact plant, most of the polyphenols discovered in the callus were not detected in the plant.

In order to test the value of using cacao tissue culture as a source of cacao flavor, Jalal and Collin (1979) developed methods for the subculture of callus and suspension cultures of cacao derived from cotyledon and stem explants (Tables 5 and 6). The production of specific components of cacao flavor such as the flavonoids and the methylated purines were examined in both types of tissues. Vigorous growth was obtained on MS medium supplemented with 4.9 μM IBA and 0.23 μM ZEA (Jalal and Collin, 1979).

Table 5. Effect of Light on Mean Callus Initiation from Hypocotyl Explants of Amazon and Amelonado Seedlings Inoculated onto White's Medium + CW (from Hall and Collin, 1975)

	FRESH WEIGHT OF CALLUS AS PERCENT OF EXPLANT FRESH WEIGHT		
CULTIVAR	Dark	Intermittent Light	16-hr Light
Amazon	45.3	54.8	49.4
Amelonado	65.2	77.1	37.0

Table 6. The Effect of Temperature on Mean Callus Initiation from Hypocotyl Explants of Amelonado Cacao Seedlings Inoculated onto White's Medium + CW (from Hall and Collin, 1975)

TEMPERATURE (C)	NO. OF DAYS FOR CALLUS TO APPEAR	FRESH WEIGHT OF CALLUS AS PERCENT OF EXPLANT
25.0	11.2	67.5
27.5	9.8	80.2
30.0	13.4	77.0
32.5	10.7	67.0
35.0	14.0	35.8

During callus initiation, Hall and Collin (1975) observed that explants from Amazon seedlings develop callus earlier than explants from Amelonado seedlings (Table 7). This observation suggests a correlation for Amazon plants between growth in vitro and superior vigor in vivo. If such a correlation could be quantified and extrapolated to all cacao varieties, conditions of callus growth could be used as a test of vigor to permit selection of plants at an early stage.

Callus tissues of cacao have been used for investigations in plant pathology by Prior (1977). The basidiomycete *Oncobasidium theobromae* (Talbot and Keane, 1971) is a pathogen that causes a vascular disease in cacao in New Guinea. The development of a rapid technique for screening resistance requires a regular supply of basidiospores of this fungus. As the fungus appears to flourish only in association with a living host, live callus tissue was proposed as a suitable growth medium for this fungus. Callus tissue cultures of cacao as a living support for culture of *O. theobromae* have been established by Prior (1977) on MS medium supplemented with 2,4-D and KIN. The callus was obtained from somatic tissue of cacao anthers or from fragments of the axis of the mature embryo. Callus was incubated in 25 C with intermittent light for 12 hr every day. Cultures were subcultured every 4-5 weeks. As previously reported, Prior (1977) observed that the callus growth rate was variable between cacao clones. Mycelium of *O. theobroma* grew vigorously on cacao callus and microscopic examination of infected callus demonstrated that the mycelial growth was intercellular. Living cacao callus therefore can be useful for growth of the fungi responsible for cacao diseases, particularly those fungi whose nutritional requirements are not known yet.

PROTOCOLS FOR CALLUS CULTURE IN *THEOBROMA CACAO*

In cacao, callus can be induced from all sorts of explants with a wide variety of media and under different physiological conditions of temperature, light intensity, and light period.

The major difficulty in obtaining explants from trees grown in the field is sterilization of the plant material. Regardless of the procedure used to sterilize these fragments of stem, the percentage of infected explants is always over 90%. By using stem fragments from plants growing in a greenhouse, contamination can be significantly reduced.

Better results can be obtained by using other sources of explants such a hypocotyl fragments from young sterile seedlings, ovary walls of floral buds, pericarp fragments of mature and immature pods, and cotyledon fragments of mature and immature embryos.

Whole Mature Pods as Explants

Establishment of a sterile culture of young cacao seedlings is very easy when initiated from freshly harvested mature pods.

Table 7. Callus Growth from Explants of Hypocotyl, Root, Stem, and Cotyledon of Amazon and Amelonado Seedlings on Different Nutrient Medium after 10 Weeks (from Hall and Collin, 1975)

MEDIUM	CULTIVAR	TOTAL NO. EXPLANTS	MEAN NO. DAYS FOR CALLUS TO APPEAR	PERCENT NO. TOTAL EXPLANTS IN EACH RATING[a]				
				0	1	2	3	4
White's	Amazon	184	17	16.0	11.0	30.0	18.0	25.0
	Amelonado	48	20	22.0	14.0	19.0	23.0	22.0
Wood &	Amazon	249	15	25.0	7.0	14.0	21.0	33.0
Braun	Amelonado	*[b]	*	*	*	*	*	*
Heller's	Amazon	170	40	30.0	25.0	25.0	15.0	5.0
	Amelonado	48	60	25.0	64.0	10.0	*	*
MS	Amazon	226	9	16.0	3.0	5.0	12.0	64.0
	Amelonado	85	21	18.0	14.0	6.0	18.0	44.0

[a]Rating system for the extent of callus growth on each explant:
0 = no callus
1 = callus just appearing on cut surfaces
2 = callus covering cut surfaces and as nodules on root explants
3 = callus on cut surfaces and appearing on undamaged surfaces
4 = callus completely covering explant.

[b]All treatments contaminated.

1. After cutting the stalk end of the pod, which generally contains bacterial and fungal spores, the pod is washed with tap water, dried with filter paper, flamed with alcohol, and dipped in a 10% calcium hypochlorite solution for an hour or more. The waxed film of the pod prevents the hypochlorite from penetrating inside the pericarp of the pod.
2. The pod is cut open in a laminar air-flow cabinet. The testa of the extracted seeds are peeled off and these seeds are dipped in an 8% calcium hypochlorite solution for 20 min, and rinsed three times with sterile water. Due to the coating that protects the cotyledons and the radical end, the embryo remains undamaged by the calcium hypochlorite solution.
3. The seeds can then be placed in test tubes (200–250 mm long) filled with 20 ml of the following medium: half-strength MS, 0.058–0.088 M sucrose, 7 g/l agar, and 5–6 pH.
4. When placed in an incubation room of 27 C, germination occurs and in 4–5 weeks, 95% of the seed gives rise to sterile seedlings that are long enough to be sectioned into explants. The nutritional medium helps to detect seeds that are contaminated with bacteria or fungi.

Pericarp of Mature and Immature Pods as Explants

1. The procedure of sterilization of the whole pod is identical to that of the previous protocol.
2. The pericarp is cut into large fragments in a laminar air-flow cabinet. After removing the epidermis and the soft part of the endocarp, the pericarp is cut into small fragments of 1 cm.
3. The segments are place in test tubes containing MS inorganic elements, vitamins (12 µM calcium panthotenate, 0.55 mM myo-inositol, 8.1 µM nicotinic acid, 3.0 µM thiamine HCl, 5.0 µM pyridoxine HCl, 0.04 µM biotin), 0.088 M sucrose, 2.3-4.5 µM 2,4-D, 4.4 µM BA or 4.6 µM ZEA.
4. Incubate in the dark at 25-27 C. In these conditions callus can be obtained after 4-6 weeks of culture.
5. Callus can then be maintained by subculturing in the dark or in light on callus initiation medium.

There are several other combinations of media capable of inducing callus on various explants in *T. cacao.*

FUTURE PROSPECTS

The preceding discussion demonstrates that very few tissue culture investigations have been reported for *Theobroma cacao.* The results of tissue culture research in cacao indicate that callus can be obtained very easily on many media from any vegetative or flowering part of the plant such as leaf, stem, ovule, or pericarp. In cacao, callus cultures can be established and maintained for years by periodic transfer to fresh media.

As reported by Prior (1977), callus of cacao has also been used as a living substrate for the culture of fungi responsible for diseases of cacao. This method can facilitate research programs directed toward identification of resistant varieties. Particularly exciting is the recent report of basidiospore production by *Oncobasidium* grown in culture with cacao callus (Prior, 1982).

The propensity of any tissue of cacao to produce callus is sometimes an obstacle particularly when in vitro development of an organized structure (bud, meristem) is desired. Attempts by Miller (personal communication) and Orchard et al. (1979) to culture terminal and axillary buds of cacao and to develop in vitro procedures for vegetative propagation of cacao were unsuccessful. Only swelling of buds followed by elongation of leaves was obtained.

Esan (1975) and Pence et al. (1979) obtained somatic embryos through budding of cotyledon and hypocotyl tissue of young sexual embryos. Attempts to develop these adventitious embryos into complete plants with normal leaves and roots were unsuccessful. Even if further development of the adventitious embryos could be obtained their utilization in cloning of cacao would be limited, as cultures were originated from immature embryos whose genotype is unknown. On the other hand, somatic embryogenesis from immature sexual embryos is helpful in achieving interspecific hybridization in cacao. In cacao, young hybrid embryos generally are produced in limited number, aborted at early stages of development, and fail to develop into normal seeds.

Numerous attempts to obtain haploid plantlets in cacao through in vitro anther and ovule culture have been reported. In all cases, callus formation was only reported from the diploid tissue of the anthers or ovule. In some cases, root formation was obtained. As in other tropical crops newly established cultures of cacao frequently become contaminated. The epidermis of the vegetative parts of cacao is covered with glandular hairs that hinder disinfection. This is particularly true when the explants come from adult plants. Disinfection is particularly difficult in field-grown material.

Although most researchers who investigate tissue culture of cacao have failed to use tissue culture for crop improvement, these techniques could be very useful for genetic improvement of cultivated cacao. Practical use of tissue culture might be very useful in cacao cultivation, such as vegetative propagation, interspecific hybridization, production of homozygous plants, recovery of virus-free plants, establishment of a micro-collection of useful genotypes, understanding physiological development, and biosynthesis of cacao flavor.

Somatic embryogenesis, though reported many years ago, is only now being used for vegetative propagation for tropical crops such as oil palm (Paranjothy, Vol. 3, Chap. 22). In the near future, embryogenesis could also be used for vegetative propagation of coconut and date palm (Tisserat, Vol. 2, Chap. 18).

In comparison to in vitro propagation by cuttings, somatic embryogenesis when perfected can provide a greater rate of multiplication. In coffee, for example, small fragments of leaf, stem, or ovules can develop into thousands of embryos that give rise to normal plants (Sondahl et al., Vol. 3, Chap 21).

Cultivated cacao trees are all affected by a vegetative dimorphism and have both plagiotropic and orthotropic shoots. All foliage in a mature cacao tree is formed on plagiotropic branches, and from such a tree only plagiotropic cuttings can be obtained. Cacao trees originating from plagiotropic cuttings are low spreading with many branches and require frequent pruning to be shaped. Potential production from plagiotropic shoots is inferior to trees originated from an orthotropic bud. In cacao, orthotropic shoots may develop spontaneously in plagiotropic trees originated from plagiotropic cuttings. Generally these spontaneous orthotropic branches occur on badly damaged trees of cacao and should produce only orthotropic plantlets.

Because of the rapid expansion of newly developed crop varieties, the genetic variability of many crops is diminishing rapidly. Hence, preservation of genetic resources is now a daily concern for both geneticists and plant breeders. Using shoot cultures, plantlets can be maintained in vitro to establish microcollections of cacao trees freed of viruses, fungi, or insects. If these collections are established in multiple locations, tissue cultures could facilitate exchange of valuable genetic material necessary for genetic improvement of cacao.

In cacao, identification of spontaneous haploids and their possible use in genetic improvement of cacao has been reported by Dublin (1978).

The rate of spontaneous haploidy in cacao varies from zero, for some genotypes of Amelonado or Trinitario group, to 1% for some genotypes of the Amazonian group (Table 7). Induced haploidy through anther or ovary culture may provide a larger number of homozygous plants to the plant breeder than spontaneous haploidy.

Interspecific crosses and transfer of disease resistance genes from wild species into cultivated cacao could be very important for development of new cultivars resistant to cacao diseases. All attempts to date have failed due to precocious abortion of the hybrid embryo. Culture of immature embryos of cacao should help to overcome barriers to interspecific hybridization in cacao. Somatic embryogenesis from cultured immature hybrid embryos would increase the number of hybrids in experimental research for the transfer of useful wild genes into cultivated cacao trees.

All of these examples illustrate the importance of the technique of plant regeneration of cacao in cell culture through organogenesis or embryogenesis. Cacao, a very important economic crop for tropical countries, still represents a large and untouched field for investigations on the practical use of cell culture techniques.

KEY REFERENCES

Dublin, P. 1978. Diploidized haploids and production of fertile homozygous genotypes in cultivated cocoa trees (*Theobroma cacao*). Cafe Cacao Thé 22:275-289.

Orchard, J.E., Collin, H.A., and Hardwick, K. 1979. Culture of shoot apices of *Theobroma cacao*. Physiol. Plant. 47:207-210.

Pence, V.C., Hasegawa, P.M., and Janick, J. 1980. Initiation and development of asexual embryos of *Theobroma cacao* in vitro. Z. Pflanzenphysiol. 98:1-4.

REFERENCES

Archibald, J.F. 1954. Culture "in vitro" of cambial tissue of cocoa. Nature London 173:351-352.

Cheesman, E.E. 1944. Notes on the nomenclature, classification and possible relationship of cacao populations. Trop. Agric. Trinidad 21: 144-159.

Cope, F.W. 1962. The mechanism of pollen incompatibility in *Theobroma cacao* L. Heredity 17:157-182.

Cuatrecasas, J. 1964. Cacao and its allies a txonomic revision of the genus *Theobroma*. Contrib. U.S. Nat. Herb. 35:379-614.

Dublin, P. 1972. Polyembryony and haploidy in *Theobroma cacao*. Cafe Cacao Thé 16:295-311.

______ 1973a. On the use of a genetic marker in the search for haploids in the cacao tree (*Theobroma cacao* L.). Cafe Cacao Thé 17: 205-210.

______ 1973b. Hybridisation intérspecifique et matromorphie chez *Theobroma cacao*. Rap. Ann. IFCC Côte d'Ivoire 2:73-76.

______ 1974. Haploids of *Theobroma cacao* L. Diploidization and production of homozygous individuals. Cafe Cacao Thé 18:83-96.
Esan, E.B. 1975. Tissue culture studies on cacao (*Theobroma cacao* L.). In: Proceedings, Fifth International Conference on Cacao Research (Ibadan, Nigeria) pp. 116-124.
Hall, T.R.H. and Collin, H.A. 1975. Initiation and growth of tissue cultures of *Theobroma cacao*. Ann. Bot. 39:555-570.
Harland, S.C. and Frecheville, G.E. 1927. Natural crossing and the genetic of axis spot in cacao. Genetica 9:279-288.
Ibanez, M.L. 1964. The cultivation of cacao embryos in sterile culture. Trop. Agric. Trinidad 41:325-328.
Jacob, V.K. and Opeke, L.K. 1971. Interspecific hybridization in *Theobroma*. In: Proceedings Third International Conference on Cacao Research (Accra Ghana, 1969) pp. 552-555.
Jalal, M.A.F. and Collin, H.A. 1977. Polyphenols of mature plant, seedling and tissue cultures of *Theobroma cacao*. Phytochemistry 16: 1377-1380.
______ 1979. Secondary metabolism in tissue culture of *Theobroma cacao*. New Phytol. 83:343-349.
Knight, R. and Rogers, H.H. 1953. Sterility in *Theobroma cacao* L. Nature London 172:164.
______ 1955. Incompatibility in *Theobroma cacao*. Heredity 9:69-77.
Linsmaier, E.M. and Skoog, F. 1965. Organic growth factor requirements of tobacco tissue cultures. Physiol. Plant. 18:100-127.
Martinson, V.A. 1966. Hybridization of cacao and *Theobroma grandiflora*. Heredity 57:134-136.
Muntzing, A. 1947. Some observations on the pollination and fruit setting in Ecuador cacao. Hereditas 33:397-404.
Murashige, T. and Skoog, F. 1962. A revised medium for rapid growth and bioassays with tobacco tissue culture. Physiol. Plant. 15:437-497.
Naundorf, G. and Villamil, G.F. 1950. Contribucion al estudio de la fisiologia del cacao (*Theobroma cacao* L.) II. Palmira Est. Agric. Colombia 3:87-90.
Ostendorf, F.W. 1948. Fertility in cacao. Chron. Nat. 104:101-105.
Pence, V.C., Hasegawa, P.M., and Janick, J. 1979. Asexual embryogenesis in *Theobroma cacao* L. J. Am. Soc. Hortic. Sci. 104:145-148.
______ 1981. Sucrose-mediated regulation of fatty acid composition in asexual embryos of *Theobroma cacao*. Physiol. Plant. 53:378-384.
Posnette, A.F. 1945a. Incompatibility in Amazon Cacao. Trop. Agric. Trinidad 22:184-187.
______ 1945b. Interspecific pollination in *Theobroma cacao*. Trop. Agric. Trinidad 22:188-190.
Pound, F.J. 1932. Criteria and methods of selection in cacao. Annu. Rep. Cacao Res. Trinidad 1:10-24.
______ 1938. Cacao and witch broom disease (*Marasmius perniciousus*) of South America, with notes on other species of *Theobroma*: Report on a visit to Ecuador, the Amazon valley and Colombia—April 1937-April 1938, p. 58. Port of Spain Yuille's printer.
Prior, C. 1977. Growth of *Oncobasidium theobromae*, Talbot and Keane in dual culture with callus tissue of *Theobroma cacao* L. J. Gen. Microbiol. 99:219-222.

______ 1982. Basidiospore production by *Oncobasidium theobromae* in dual culture with cocoa callus tissue. Trans. Br. Mycol. Soc. 78:571-574.

Pyke, E.E. 1933. The vegetative propagation of cacao II. Softwood cuttings. In: Second Annual Report on Cacao Research (1932) Trinidad, pp. 3-9.

Rudolph, L.F. and Cox, L.C. 1943. Factors influencing the germination of iris seed and the relation of inhibiting substances to embryo dormancy. Proc. Am. Soc. Hortic. Sci. 43:284-300.

Swarbrick, J.T. 1965. Storage of cocoa seeds. Exp. Agric. 1:201-207.

Talbot, P.H.B. and Keane, P.J. 1971. *Oncobasidium*, a new genus of tulasnelloid fungi. Aust. J. Bot. 19:203-206.

Toxopeus, H. 1969. Cacao (*Theobroma cacao* L.). In: Outlines of Perennial Crop Breeding in the Tropics (F.P. Ferwerda and F. Wit, eds.) pp. 79-109. Wageningen, The Netherlands.

Wellensieck, S.J. 1932. Observations in floral biology of cacao. Arch. Koffie Cult. Med. Indie 6:87-101.

Williams, J.A. 1975. Interspecific crosses between *Theobroma* and *Herrania* species. In: Proceedings, Fifth International Conference on Cocoa Research (Ibadan, Nigeria) pp. 1-10.

CHAPTER 21
Coffee

M. R. Sondahl, T. Nakamura, H. P. Medina-Filho, A. Carvalho, L. C. Fazuoli, and *W. M. Costa*

ECONOMIC IMPORTANCE

Coffee is one of the most important agricultural products in the international market and many countries are involved in its production, trade, or consumption. The beverage is popular and well accepted due to its stimulating effects and organoleptic qualities. The world production of coffee in 1977-1978 and in 1978-1979 was 69.2 and 76.9 million 60-kg bags, respectively. In 1977, about 75% of the coffee produced was sold in the international market generating 12 billion U.S. dollars. There are several countries that depend heavily upon the revenue of coffee exportation in order to buy modern goods and advanced technology including Burundi (94%), Ethiopia (75%), Rwanda (45%), Colombia (63%), El Salvador (62%), Ivory Coast (47%), Haiti (73%), Kenya (42%), Tanzania (41%), and Costa Rica (40%). The ten leading coffee producers in the world in the period 1970-1979 have been Brazil, Colombia, Ivory Coast, Mexico, Indonesia, Ethiopia, El Salvador, Guatemala, Uganda, and India (Table 1).

Brazil has been the leading coffee producer since the 1800s. This country is a 20-million-bag coffee producer (1940-1950 and 1970-1980) with the exception of the late 1950s when the newly established fields in Parana State contributed significantly to the doubling of Brazilian production. The cultivated coffee areas in Brazil shifted from equatorial regions toward the south, reaching Parana State (ca. 23° S) during the early 1950s. The high fertility of the uncultivated red soil triggered this southern shift of the coffee fields, despite the high risk of

Table 1. World Production of Green Coffee[a]

COUNTRY	1970/71-1977-78[b]	1978/79	1979/80	PERCENT OF TOTAL 1970-1979
Brazil	17.5	20.0	22.5	25.7
Colombia	9.2	11.3	11.5	14.0
Ivory Coast	4.4	4.7	4.8	6.2
Mexico	3.7	3.8	3.8	5.1
Indonesia	2.9	4.7	4.2	5.2
Ethiopia	2.7	3.0	3.0	3.9
El Salvador	2.6	3.0	3.0	3.8
Guatemala	2.3	2.0	2.2	3.1
Uganda	2.3	2.6	2.7	3.3
India	1.8	1.9	1.8	2.5
Other	19.6	19.9	20.8	27.0
Total[c]	69.2	76.9	80.3	

[a]Anonymous (1979).

[b]Millions of 60-kg bags.

[c]In 1977 total coffee exportation was ca. 12 billion U.S. dollars.

frost. Such periodic frosts have caused sharp fluctuations in coffee production and prices in the international market. For example, following the 1975 frost in Brazil, coffee prices increased fourfold, causing a tremendous instability in the world coffee market. The Brazilian goal is to stabilize its yearly production through the establishment of new planting areas with highly productive cultivars. Presently, new coffee plantations (ca. 1.4 million trees) are being established in lower latitudes (21-20° S) to avoid frequent frost damage.

There are two widely cultivated coffee species: *Coffea arabica* L. and *C. canephora* Pierre (Robusta coffee). Quality beverage is produced from *C. arabica* which is cultivated at higher altitudes. This species represents 70% of the commercial coffee of the world and about 99% of Latin American production. *Coffea canephora* is usually grown in tropical areas at lower altitudes. Eighty percent of the African production is of this type. On a very reduced scale, *C. liberica* is grown in Liberia, Surinam, and Malaysia, *C. racemosa* in Mozambique, and *C. dewevrei* in Ivory Coast and Zaire. These species produce beans of lower quality that are only acceptable in the local market.

A well established coffee plantation in Brazil consists of about 1250 plants/ha with an average production of 25 bags/ha each weighing 60 kg. Considering that the value of coffee in the international market in April, 1982, a total of $3896 (U.S. dollars)/ha would be generated by such a coffee field. The gross revenue of a Brazilian coffee farmer is 56% of this total. The remaining 44% represents Federal and State taxes. The net profit of a coffee farm depends on the productivity of the plantation and its direct and indirect costs. Taking into account the prices of April, 1982, a farmer's estimated net profit would be $444.80/ha.

New farming techniques and more productive cultivars have been introduced as a consequence of a long-term commitment to coffee research. It has been estimated that in Brazil a 27% return per year has resulted from the coffee research effort over the past 35 years (Fonseca, 1976; Fonseca et al., 1977).

HISTORY OF THE CROP

Based on nearly 150 relevant papers, Wellman (1961) has reviewed the picturesque history of the migration of coffee throughout the world. The classic writings from the Egyptians, Greeks, and Romans contain no information about coffee. As far as one can tell, Arab and Persian invaders introduced *C. arabica* from the Ethiopian highlands into Yemen. From these early plantations, seven seeds reached India and then Ceylon. The Dutch secured a few plants from Ceylon (presently Sri Lanka) and brought them to Java isle (Indonesia) from which a single plant was shipped to the Botanical Garden of Amsterdam. The best plant from its first progeny was sent as a gift to King Louis XIV of France, who cultivated it in a specially built glasshouse in his celebrated Chateau Marly in Paris. Progenies of these trees were used to establish the coffee plantations in the Americas. It is interesting to note that the major introductions (Yemen, India, Ceylon, Amsterdam, and Paris) involved very few plants, thus restricting the variability of the first coffee plants introduced into America to an extremely narrow genetic basis. Attempts to improve the primitive arabica introduced in Brazil were unsuccessful. Progress was achieved in later years with germplasm introduced from the islands of Reunion and Sumatra. Several mutations and new introductions from other countries increased the genetic diversity of coffee in Latin America. Intensive exchange of genetic materials has resulted in further dispersion of coffee among these American countries.

BREEDING AND CROP IMPROVEMENT

Breeding for high yields and vigorous plant growth has been successfully applied to coffee. The coffee breeding program in Campinas began in 1933 and since that time hundreds of thousands of individual plants have been investigated. Individual plant selection followed by progeny evaluation, hybridization, and pedigree selection are the most usual methods employed. In addition, backcross methods have been used to transfer specific traits such as short internodes and disease and pest resistance to *C. arabica* from other cultivars or related species. Individual plant selection can be performed either in commercial fields or in experimental plots planted specifically for breeding purposes. Using this process, some outstanding *C. arabica* cultivars have been developed such as Bourbon Vermelho, Bourbon Amarelo, and Mundo

Novo (Carvalho et al., 1952). The progress of the breeding program can be seen in Table 2 where the yield of those cultivars is compared with cv. Arabica (Carvalho et al., 1964).

Table 2. Average Yields of *C. arabica* cultivars Bourbon Vermelho (BV), Bourbon Amarelo (BA), and Mundo Novo (MN) in Relation to Arabica (A)[a]

LOCATION (Sao Paulo, Brazil)	PERCENT YIELD OF CULTIVAR			
	A	BV	BA	MN
Campinas	100	164	201	251
Mococa	100	189	240	316
Jau	100	180	246	332
Average	100	178	229	300

[a]Data from yield trials carried out in different locations during 15-25 years.

Leaf rust and coffee berry disease are the most destructive coffee diseases, and breeding programs are being carried out to develop material with high yield associated with resistance and good cup quality. Leaf rust disease (*Hemileia vastatrix*) has a long history of devastation in the Old World. The first record of attack on *C. arabica* dates back to 1868 in Ceylon where coffee plantations were abandoned and replaced by tea. After 1868 the disease gradually spread to many coffee areas in the tropics (Chaves et al., 1970), reaching Brazil in 1970. From the northeastern coast of Brazil it spread rapidly and at present it is found in Argentina, Paraguay, Bolivia, Peru, Nicaragua, El Salvador, Guatemala, Mexico, and Colombia. Favorable climatic conditions and the high genetic uniformity of Latin American plantations are the major factors contributing to the development of this epidemic. All commercial coffee varieties are homozygous for the gene SH5 which makes them susceptible to race II of leaf rust.

Coffee berry disease (*Colletotrichum coffeanum*) occurs mainly in Africa and results in significant yield reduction. The fruits are susceptible at various stages of development requiring several chemical sprayings during the season to keep them on the trees. Resistance to this disease is mainly found in the Rume Sudan cultivar of *C. arabica*, in some accessions from Ethiopia, and in the Timor hybrid.

Three other coffee diseases deserve to be mentioned: bacterial blight caused by *Pseudomonas garceae*, "llaga macana" caused by *Ceratocystis fimbriata*, and tracheomycosis caused by *Gibberella xylarioides*. In addition, nematodes attack coffee all over the world. In Sao Paulo State, several species are found that attack *C. arabica*. The root-knot nematodes *Meloidogyne exigua*, *M. incognita*, and *M. coffeicola*, and the root-lesion nematodes *Pratylenchus brachyurus* and *P. coffeae* are the most important species. *Meloidogyne exigua* and *M. incognita* have wide geographic distribution. The broad range of host plants and the severe destructive effects in the root cortex make *M. incognita* the most important species of root-knot nematode. The susceptibility of commercial *C. arabica* cultivars to nematodes has prompted the development of breeding programs aimed at establishing resistance or tolerance of coffee plants to nematodes.

Despite the fact that coffee is one of the major agricultural products in the international trade market and has economic importance for many tropical countries, coffee plantations have not yet benefited from technological developments applied to other cash crops. It is expected that new research accomplishments pertaining to coffee will give rise to higher and more stable world production and consequently more stable prices.

REVIEW OF THE LITERATURE

Many aspects of in vitro techniques can be applied to the improvement of cultivated coffee. Pioneer work with coffee tissues was published by Staritsky (1970). He was successful in inducing somatic embryos and plantlets from orthotropic shoots of *C. canephora.* Colonna (1972) established cultures of embryos of *C. canephora* and *C. dewevrei.* Callus from endosperm tissues of *C. arabica* was induced by Keller et al. (1972) with the objective of studying caffeine synthesis. Sharp et al. (1973) cultured somatic and haploid tissues of *C. arabica* obtaining callus growth (petioles, leaves, green fruits), proembryo formation (anthers), and shoot development (orthotropic shoots). Monaco et al. (1977) were able to induce abundant friable callus from perisperm tissues of *C. arabica* and *C. stenophylla.* Perisperm proliferated rapidly in the absence of auxin, suggesting that it was autonomous for auxin. With the purpose of producing coffee aroma from suspension cultures, Townsley (1974) established liquid cultures of coffee cells from friable callus derived from orthotropic shoots of *C. arabica* cv. El Salvador. These same suspension cultures were used to analyze caffeine and chlorogenic acid contents (Buckland and Townsley, 1975) and to compare unsaponifiable lipids in green beans with those in cell suspensions (van der Voort and Townsley, 1975). Anthers were placed in culture from several *Coffea* species; however, rapid and friable callus development was observed only with *C. liberica* (Monaco et al., 1977). Herman and Hass (1975) reported the presence of organoids from leaf explants of *C. arabica*; the normal plantlets were rooted and transferred to soil after 7 months. Solid cultures from stem explants of *C. arabica* cv. Bourbon Vermelho were established for analysis of caffeine production and possible commercial utilization of in vitro synthesis of such alkaloids (Frishchknecht et al., 1977). The distinction of high and low frequencies of somatic embryo induction from cultured mature leaf explants of *C. arabica* cv. Bourbon was described by Sondahl and Sharp (1977a, 1979). A histological study demonstrated that somatic embryos from coffee leaf explants were derived from mesophyll cells (Sondahl et al., 1979b). Low frequency somatic embryos (LFSE) were present as early as 70 days whereas embryogenic tissue and high frequency somatic embryos (HFSE) were detected only 90-120 days of secondary culture (Sondahl et al., 1979b). HFSEs were derived from a distinct cellular phenotype (small and round-shaped, ca. 20 µm diameter) in contrast with callus cells (long rods, ca. 150 µm long) (Sondahl et al., 1979a). Mature leaf cultures of *C. canephora*, *C. congensis*, *C. dewevrei* cv. Excelsa, and *C. arabica* cvs. Mundo Novo, Catuai, Laurina, and Purpurascens have also yielded HFSE (Sondahl and Sharp, 1977b, 1979).

Protoplasts have been isolated from leaf-derived callus tissues of *C. arabica* cv. Bourbon and protoplast-derived callus was obtained in about 30% of the cultures (Sondahl et al., 1980). Small colonies from coffee protoplasts isolated from young leaves were observed by Orozco and Schieder (1982). Similarly to previous work, these cultures did not survive the first subculture in semisolid medium. The histogenesis of callus induction and early embryo formation from *C. canephora* stem explants was described by Nassuth et al. (1980). It was observed that the parenchymatous cells of the cortex contribute to formation of callus tissues. Embryolike structures were present in the callus zone after 14 days of culture. Similar to LFSE described for leaf cultures of *C. arabica*, somatic embryos from leaf explants of Arabusta (F_1 hybrid of *C. arabica* x *C. canephora*, 4x) were described on medium rich in cytokinin and auxin-free after 90 days of culture (Dublin, 1981). Somatic embryos were also described from *C. canephora* egg cells (35 days following fertilization) after 90 days of primary culture (Lanaud, 1981). Cultures of nodal explants of *C. arabica* led to the development of 2.2 plantlets from arrested buds of leaf axils (Custer et al., 1980). Nodal cultures of *C. arabica* cv. Mundo Novo and two interspecific hybrids (*C. canephora* x *C. eugenioides; C. congensis* x *C. eugenioides*) gave rise to 4.5 plants per node following 120 days of culture (Sondahl and Nakamura, 1980; Nakamura and Sondahl, 1981; Sondahl, 1982). Shoot development has also been obtained from arrested buds of Arabusta as a means of vegetative propagation (Dublin, 1980). Apical meristems of *C. arabica* were cultured with the goal of coffee germplasm preservation (Kartha et al., 1981).

In summary, plant regeneration with coffee tissues has been accomplished from solid cultures of orthotropic shoots and mature leaves covering four species of the genus *Coffea* and five *C. arabica* cultivars. Direct (LFSE) and indirect (HFSE) somatic embryogenesis are frequently associated with coffee tissues. Shoot development from arrested orthotropic buds and apical meristems has been described. Finally, limited callus proliferation from isolated protoplasts has been observed.

Leaf Culture

Mature leaves from orthotropic or plagiotropic branches of greenhouse plants are surface sterilized in 1.6% sodium hypochlorite (30% commercial bleaching agent) for 30 min and rinsed 3 times in sterile double-distilled water. If the coffee plants are growing in the field, the surface sterilization is much more difficult. The following procedure has given some degree of success for field material: 2.6% sodium hypochlorite for 30 min, rinse in sterile water, incubate in sealed petri dishes overnight followed by another exposure to 2.6% sodium hypochlorite for 30 min, then rinse 3 times with sterile water. It has been found that 70% ethanol and solutions of $HgCl_2$ are toxic to coffee leaves; on the other hand, solutions of sodium or calcium hypochlorite are effective and nontoxic. Before immersing the leaves in the sterilizing solution, it is advisable to wash them by hand with 1% detergent solution, and rinse in distilled water. If ozone water is available, a 5-min immersion is recommended. The addition of antibiotics and systemic fungicides in the saline-sugar medium used for the 3-day preculture period has been advantageous in reducing bacterial and fungal contaminations.

Leaf explants of about 7 mm^2 are cut, excluding the midvein, margins, and apical and basal portions of the leaf blade. The elimination of the midvein also excludes the domatia, consequently lowering the number of contaminated leaf explants. Domatia are deep pores (ca. 0.18 mm diameter) located in the acute angle formed by the midvein with secondary veins on the abaxial side of the coffee leaf. Leaf sections are usually cut on top of sterile filter paper (or paper towel) that is changed frequently to avoid cross contamination. All sides of a leaf explant must be cut since callus proliferation occurs only from cut edges. Histological studies have demonstrated that the callus tissue originates from mesophyll cells of the leaf explant (Sondahl et al., 1979b).

The leaf explants are placed on 20 x 100-mm petri dishes containing the solidified saline-agar medium (half-strength MS salts with 0.06 M sucrose). The preculture period of the leaf explants can be in the dark or in the light (no difference has been detected between these two conditions) for about 72 hr. This preculture incubation period has been found to be very useful for the selection of viable explants and the elimination of contaminated leaf pieces. The abaxial surface of the leaf explants, which is clearly distinguished a dull pale green coloration (in contrast to the shiny dark green coloration of the adaxial surface) is always placed upwards. French square bottles of about 50 ml capacity containing 10 ml autoclaved basal medium made up of MS inorganic salts (Murashige and Skoog, 1962), 30 µM thiamine-HCl, 210 µM L-cysteine, 550 µM meso-inositol, 117 mM sucrose, and 8 g/l Difco Bacto agar are used. In primary culture, an "induction medium" containing a combination of KIN (20 µM) and 2,4-D (5 µM) is used, and the bottles are incubated in the dark at 25 ± 1 C for 45-50 days. The composition of this "induction medium" was found to be ideal for HFSE induction of *C. arabica* cv. Bourbon (Sondahl and Sharp, 1977a).

Secondary cultures are established under conditions of a 12-hr light period at 24-28 C by subculturing 45- to 50-day-old tissues onto a "conditioning medium" containing half-strength MS organic salts (except KNO_3, which is added at 2x concentrations), sucrose (58.4 mM), KIN (2.5 µM), and NAA (0.5 µM). Following transfer to the conditioning medium, the massive parenchymatous type of callus growth ceases and the tissues slowly turn brown. Two sequences of morphogenetic differentiation have been characterized in secondary cultures of leaf explants in *Coffea*: low frequency somatic embryo (LFSE or direct embryogenesis) and high frequency somatic embryo (HFSE or indirect embryogenesis). Adopting the standard culture protocol described for *C. arabica* cv. Bourbon, LFSE is observed after 13-15 weeks and HFSE after 16-19 weeks of secondary culture. LFSE appeared 3-6 weeks before the visible cluster of HFSE in more than 5000 cultured flasks (Sondahl and Sharp, 1979; Sondahl et al., 1979a,b).

The appearance of isolated somatic embryos developing into normal green plantlets in numbers ranging from 1 to 10 per culture is typical of LFSE. The occurrence of HFSE follows a unique developmental sequence: a white friable tissue containing globular structures defined as *embryogenic tissue* develops from the nonproliferating brown callus cell mass; the globular structures appear to develop synchronously for a period of 4-6 weeks. The embryogenic tissue gives rise to somatic embryos and finally to plantlets, but this latter developmental process lacks the synchrony of the earlier stage. The amount of this embryogenic tissue varies, but on the average about 100 somatic embryos develop from such clusters of embryogenic tissue. To speed up development and increase the percentage of fully developed plantlets, it is advisable to isolate the embryogenic tissues and grow them under light conditions at 26 C in 5-10 ml of liquid basal medium devoid of growth regulators for 4-6 weeks. After this period, the torpedo-shaped somatic embryos and young plantlets can be plated onto saline-agar medium containing 0.015-0.03 M sucrose in the presence of light. Individual plantlets bearing a developed top root are removed from the agar medium, gently washed, and immediately transferred to small pots inside a humid chamber. After a hardening period of 1 month, they can be exposed to normal atmospheric humidity and transferred to a greenhouse. Another way of hardening the plantlets is the use of approximately 100 ml of saline-agar medium without sucrose in 250 Erlenmeyer flasks closed with cotton plugs and paper. The Erlenmeyer flasks are exposed to sunlight in a shaded portion of the greenhouse (usually protected with plastic screens that filter about 60-70% of the sun's rays). After about 2 months, the plantlets have good leaf and root development and can be transferred to small pots containing a light soil or promix.

This procedure has been applied to four different species of coffee and to many *C. arabica* cultivars. Presently somatic embryos from leaf tissues of some morphological coffee mutants have been obtained in order to evaluate the genotype stability of this methodology. Differentiation has occurred in the following single gene mutants: Erecta (Er), Purpurascens (pr), Angustifolia (ag), Nanna (na), San Ramon (Sr), and Volutifolia (vf). Presently these cultures have embryogenic tissues and somatic embryos at different stages of development, as summarized in Table 3. The genetic analysis of plants derived from the leaf explant culture of these mutants will provide new insights on the feasibility of somatic embryo production for vegetative propagation of plants. Moreover, it will possibly permit a further distinction between organogenesis and embryogenesis in regard to genetic stability and cytological origin.

The culture of cotyledonary leaves of coffee has not been described in the literature. The feasibility of this explant source was tested

Table 3. Leaf Culture and Somatic Embryo Regeneration from Six Morphological Mutants of *Coffea arabica*

MUTANTS	TOTAL NO. FLASKS	REGENERATION FREQUENCY[a] (%)	TOTAL NO. SOMATIC EMBRYOS
Erecta (Er)	58	19	62
Purpurascens (pr)	8	17	16
Angustifolia (ag)	26	38	14
Nanna (na)	10	50	9
San Ramon (Sr)	14	7	8
Volutifolia (vf)	28	25	7

[a]Including flasks with embryogenic tissues and embryos.

with 3-month-old seedlings of *C. arabica* cv. Mundo Novo using the same methodology described for mature leaves, except the concentration of 2,4-D during primary culture (conditioning medium) was varied. The data recorded after 2 months of primary culture are summarized in Table 4. It is interesting that there was little contamination recorded among all 550 cotyledonary segments cultivated, and that the lowest concentration of 2,4-D (1 μM) also induced the differentiation of adventitious roots. A concentration of 5 μM 2,4-D, which is the same concentration used for mature leaves, was adopted for future cultures of cotyledonary leaf explants. Embryogenic tissue was obtained profusely after 4 months on secondary culture (inducing medium).

Culture of Hypocotyl Explants

In a manner similar to the culture of cotyledonary tissues, hypocotyl explants were tested for their ability to produce callus and differentiate. Explants were excised from 3-month-old seedlings of *C. arabica* cv. Mundo Novo using the same protocol described for the culture of mature leaves of coffee. Table 5 presents the data for callus growth following 2 months of primary culture. Callus proliferation from hypocotyl tissues was similar to callus production from leaf explants. However, 1 μM 2,4-D did not induce root differentiation as it did with cotyledonary leaf explants. Embryogenic tissues were recorded following secondary culture on inducing medium.

Embryo Culture

Coffee seeds are surface sterilized before embryo excision. Colonna et al. (1971) used 95% ethanol for 1 min under vacuum followed by 5 min in 0.7% hypochlorite solution. Before the embryos were removed,

Table 4. Callus Induction from Cotyledonary Leaf Explants of *Coffea arabica* cv. Mundo Novo[a]

2,4-D (μM)	TOTAL NO. EX-PLANTS	CONTAMINATION		CALLUS GROWTH INDEX	ROOT GROWTH NO.
		Medium	Explant		
1	189	16	20	1.86	159
5	189	20	26	1.88	-
10	187	10	9	1.85	-

[a]After 2 months on primary culture.

Table 5. Callus Induction from Hypocotyl Explants from Three-Month-Old Seedlings of *Coffea arabica* cv. Mundo Novo[a]

2,4-D (μM)	TOTAL NO. EX-PLANTS	CONTAMINATION		OXIDATION	CALLUS GROWTH INDEX
		Medium	Explant		
1	53	9	22	4	1.8
5	58	8	21	1	2.0
10	55	9	24	-	1.4

[a]After 2-month culture on primary medium.

the sterilized seeds were soaked in sterile distilled water for 36-48 hr. In our laboratory a 75% solution of commercial bleach has been used with success (ca. 3.9% NaOCl) for 30 min with continuous agitation (rotatory shaker, 150 rpm), followed by 3 rinses of sterile water. The sterile seeds are incubated 3-4 days on saline sucrose agar plates before excision. Soaking the seeds in sterile water has given negative results because of cross contamination in the liquid environment. Both $HgCl_2$ and ethanol are deleterious to the seeds. Before the sterilization procedure mentioned above, the parchment is removed from the seeds. The silver skin is almost completely eliminated with the agitation in the shaker. An alternative method to obtain sterile coffee embryos consists of soaking seeds (without parchment and silver skin) in water for 12 hr, excising the embryos, and sterilizing them (1% NaOCl for 6 min). During culture of immature coffee embryos, the young berry is surface sterilized with 70% alcohol for 3 min and followed by 3.9% NaOCl for 30 min. After rinsing with sterile water, the immature embryos are excised under sterile conditions.

Solid and liquid Heller's salt solution plus sucrose (58.4 mM), meso-inositol (1.1 mM), thiamine (3.0 μM), cysteine (6.5 μM), calcium pantothenate (1.0 μM), nicotinic acid (8.1 μM), adenine (7.4 μM), pyridoxine (3.9 μM), and glycine (4.0 μM), has been used by Colonna (1972) with mature embryos of *C. canephora* cv. Robusta and *C. dewevrei* cv. Excelsa and cv. Neo-anddiana. Subculturing was necessary to provide for continued development. The liquid medium provided longer root growth, but development was more regular in agar medium. The first pair of leaves in both varieties developed on the 45th day of culture and on the 75th day a second pair of leaves appeared in *C. dewevrei* cv. Excelsa. The addition of 2 μM IAA to the culture medium promoted a 17% increase in elongation, while 10% CW in addition to the IAA produced a 38% increase in elongation at the 40th day of culture. The cultures were maintained under a 12-hr photoperiod at 28 C day and 21 C night temperatures.

Coffea arabica cv. Mundo Novo and cv. Catuai mature embryos were cultivated in half-strength B5 inorganic salts, sucrose (58.4 mM), nicotinic acid (30 μM), thiamine (30 μM), meso-inositol (550 μM), pyridoxine (15 μM), BA (1 μM), IAA (0-20 μM), and agar (8 g/l) at pH 5.5 in the presence of light, and at a temperature of 24-28 C. Evaluations made after 60 and 90 days indicated that 5-10 μM IAA was excellent for induction of roots but the subsequent development of roots was better in media containing 0-0.25 μM IAA. Good hypocotyl and leaf development were obtained using IAA concentrations in the range of 0-2.5 μM. Considering these results, the medium described above has been adopted for *C. arabica* with 2.5 μM IAA for initial culture of the excised embryos. Since IAA is naturally degraded by the IAA oxidase system, a reaction enhanced by light, its concentration will be lowered as the culture ages. After 50 to 60 days, the culture bottles were exposed to light with higher intensity or even transferred to the greenhouse in the shade (70-80% shade). Immature embryos from *C. arabica* x *C. canephora* crosses have been successfully raised with this protocol (Fig. 1).

Apical Meristem Culture

The development of a protocol for meristem culture is a prerequisite for the future cryopreservation of coffee germplasm. Kartha et al. (1981) isolated 0.3-mm apical meristems from sterile seedlings of *C. arabica* cv. Caturra Rojo and cv. Catuai. Culture media consisting of MS salts and B5 vitamins, supplemented with benzyladenine or ZEA (5-10 μM) and NAA (1 μM), induced multiple shoots, whereas lower concentrations of cytokinins produced single shoots. Rooting occurred only in media with half-strength MS salts devoid of sucrose in the presence of 1 μM IBA.

In our laboratory apical meristems of 3-month-old *C. arabica* cv. Mundo Novo have been excised and plated on saline-sucrose agar plates. After 5 days of preculture of 192 meristems, the meristems were transferred to primary culture after discarding the contaminants (32%) and oxidation (4%). The culture media consisted of MS salts (0.1x micro solution), pyridoxine (15 μM), nicotinic acid (15 μM), thiamine (30 μM), inositol (550 μM), cysteine (500 μM), sucrose (0.087 M), IAA (0.5

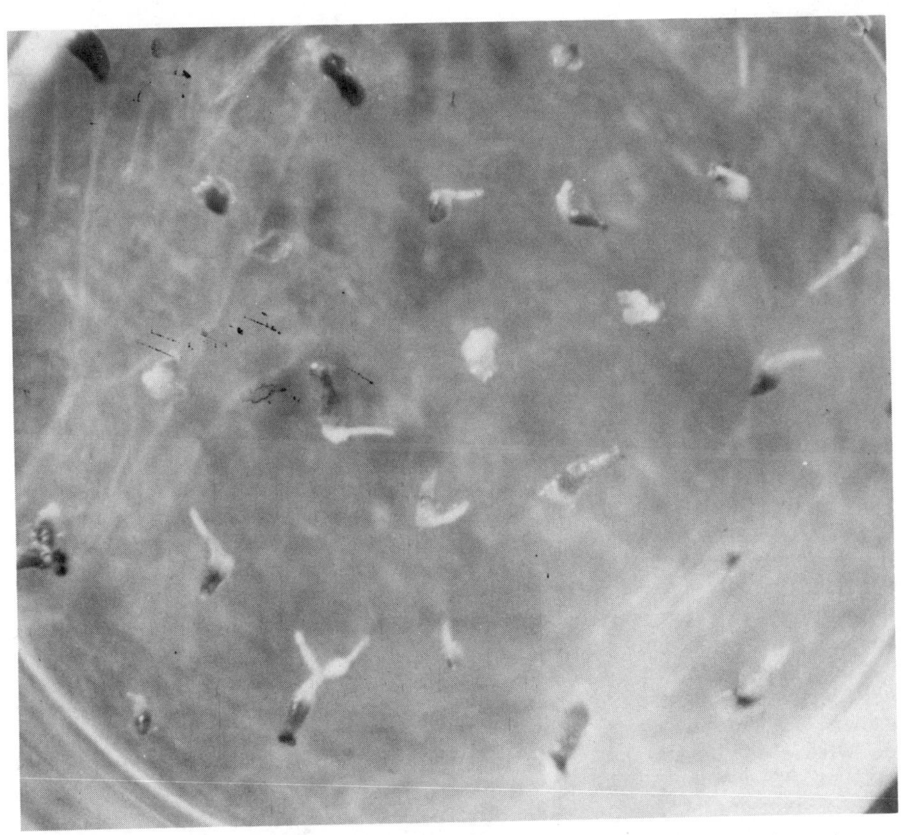

Figure 1. Early stages of immature embryo culture derived from interspecific cross between *Coffea arabica* and *C. canephora*.

µM), and various concentrations of BA (Table 6). After 10 weeks of culture, the meristems cultured on media containing 5 µM BA were the most vigorous. The rooting medium adopted consisted of half-strength MS salts (0.1x microelements), sucrose (0.058 M), vitamins, IBA (5–10 µM), and NAA (5–10 µM). In all cases, only long and single roots were obtained. In order to form secondary roots and root hairs, the plantlets were subcultured on sand:vermiculite (2:3) sterile substrate. Apical meristems of *C. arabica* cv. Catuai were cultivated with the same protocol described for cv. Mundo Novo with similar results.

Axillary Bud Culture

Characteristically, coffee plants have multiple arrested orthotropic buds and two plagiotropic buds at each stem node. The plagiotropic buds differentiate only after the 10th–11th node of a developing seedling, whereas the orthotropic buds are present in the first node (cotyledonary node). The removal of the apical meristem results in the

Table 6. Apical Meristem Culture of Three-Month-Old Seedlings of *Coffea arabica* cv. Mundo Novo[a]

BA (µM)	AVERAGE HEIGHT (cm)	AVERAGE NO. LEAF PAIRS	AVERAGE LEAF AREA (cm^2)
1	1.6	4.6	3.9
5	2.0	4.8	5.2
10	1.5	4.2	4.0
25	1.7	4.6	3.3
50	1.6	4.2	2.1

[a]After 60 days in primary culture.

development of two orthotropic shoots, one at each leaf axil, at the most apical node. At each leaf axil of plagiotropic branches, there are 4 serially arrested buds (Fig. 2). These buds will differentiate into flower buds under proper environmental conditions. Sometimes, the uppermost buds of that series at one or both sides of a node will develop as a vegetative bud instead of a floral bud, leading to a secondary plagiotropic branch.

The presence of these arrested buds has been explored as a means of vegetative propagation of coffee plants. Nodal explants of aseptically grown *C. arabica* plants have been cultured on MS medium supplemented with BA (44 µM) and IAA (0.6 µM), under 16-hr photoperiod (2000 lux) at 25 ± 0.5 C (Custer et al., 1980). Shoot development occurs on the average of 2.2 per node after 2-5 weeks. The use of NAA (1.1 µM) in complete darkness encourages good rooting of these young coffee shoots. It is recommended that at least 3-month-old in vitro plants be used for excision of nodal sections and that the leaves be retained during culturing. Similar techniques have been used by Dublin (1980) with *Arabusta* plants. Shoot development was observed in medium supplemented with malt extract (400 mg/l) and BA (4.4 µM).

PLAGIOTROPIC BUDS. Arrested plagiotropic buds were excised before flower induction and placed on the following medium: MS salts (0.25x microelements), pyridoxine (15 µM), nicotinic acid (15 µM), thiamine (30 µM), inositol (550 µM), cystein (500 µM), sucrose (0.087 M), IAA (0.5 µM), BA (25 µM), streptomycin (1 g/l), and agar (8 g/l). Plagiotropic branches were washed with detergent and tap water and incubated for 48 hr in a humid chamber. After this incubation period, the plant material is surface sterilized with 50-75% commercial bleach solution for 30 min and rinsed 3 times with sterile water. Table 7 summarizes the culture of plagiotrophic buds of four *C. arabica* cultivars.

The incidence of contamination for the establishment of these nodal cultures was still very high despite the use of high concentrations of

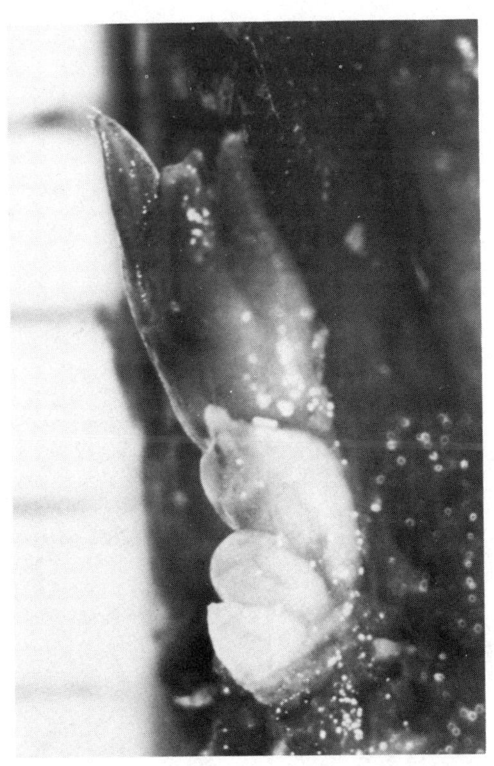

Figure 2. Axillary serial buds from a plagiotropic node of *Coffea arabica* cv. Catuai.

hypochlorite and antibiotics. On the other hand, the degree of oxidation was low. 'Mundo Novo' buds were slower in development (0.40 leaf pairs/bud) in comparison with 3 cultivars (1.0 leaf pairs/bud). Following 50–60 days on primary culture, the plagiotropic buds were transferred to the following secondary medium for further development: MS salts (0.25x), cysteine (500 μM), pyridoxine (15 μM), nicotinic acid (15 μM), thiamine (30 μM), inositol (550 μM), activated charcoal (5 g/l), sucrose (0.03 M), and agar (8 g/l).

Presently, all developing plagiotropic buds grow upward but so far the differentiation of lateral branches has not been recorded. It is interesting to note that rooting of plagiotropic branches leads to plagiotropic growth, i.e., they do not grow upward (orthotropically). The ability to change this pattern of growth by early excision and medium manipulation, thus recovering orthotropic shoots instead of plagiotropic shoots, would be very valuable.

ORTHOTROPIC BUDS. Nodal cultures were established from orthotropic branches of coffee. The donor plants were derived from the greenhouse (ca. 10 months old), nursery (1-2 years old), or field plots

Table 7. Development of Arrested Plagiotropic Buds of Four *C. arabica* Cultivars

CULTIVAR	AGE (days)	NO. INOC. BUDS	NO. VIABLE BUDS	CONTAMINATION (%)	OXIDATION (%)	NO. LEAF PAIR	LEAF PAIR/BUD
B. Amarelo	50	125	80	25.6	10.4	77	0.96
M. Novo	46	510	174	73.0	-	70	0.40
Catuai	41	538	167	67.3	0.3	167	1.00
B. Vermelho	30	255	68	73.3	-	68	1.00

(15-20 years old). Green orthotropic shoots containing 4-6 nodes were excised from the donor plants. After washing with 1% detergent and sterile water, the tissues were surface sterilized with 2.6% sodium hypochloride for 30 min under continuous agitation (150 rpm). Shoots excised from the nursery or field-growing plants were allowed to incubate in a humid chamber overnight and submitted to a second sterilization as described above. The apical meristems were eliminated from all orthotropic shoots. The uppermost node beneath the apical meristem was coded No. 1. Whenever possible, the nodes were cultivated with the attached leaf pair trimmed to half to a third of the original size.

The primary medium consisted of a modified B5 medium containing pyridoxine (15 μM), nicotinic acid (15 μM), thiamine (30 μM), inositol (550 μM), cysteine (500 μM), sucrose (87 mM), BA (25-50 μM), IAA (10 μM), activated charcoal (2.5 g/l), PVP-40 (1 g/l), and agar (7 g/l). Round screw cap bottles of 150-ml capacity charged with 25 ml were used. The bottles were inoculated with individual nodes and sealed with a 16.5-μm PVC film. This film has the following permeability characteristics: CO_2 (10 cm^3/cm^2 per 24 hr); oxygen (2 cm^3/cm^2 per 24 hr); nitrogen (0.4 cm^3/cm^2 per 24 hr); and water vapor (10 mg/cm^2 per 24 hr).

The nodal cultures were maintained in a growth room under 500 lux illumination with a 14-hr photoperiod. Contamination frequencies were very high during primary culture. Uppermost nodes from small plants presented 60% contamination whereas the lower nodes (4th-6th) reached 90-100% levels. Average data from 824 nodal explants from 4 different cultivars of field-growing coffee plants demonstrate frequencies of 43% contamination and 42% oxidation (Table 8).

Table 8. Recovery of Nodal Cultures from Selected Field-Growing Coffee Plants[a]

CULTIVAR	NO. OF NODES	CONTAMINATION (%)	OXIDATION (%)	FREQUENCY BUDS/NODE
Mundo Novo	225	53	42	1.6
Catuai	104	41	47	1.7
Icatu[b]	287	40	30	1.5
Robusta Conilon	208	37	50	1.6

[a]After 30-60 days of primary culture.

[b]Similar results for nodal cultures of Catimor and Timor hybrid.

The development of the arrested orthotropic buds from nodal cultures becomes visible during the third week of culture (Fig. 3a). The number of developed buds per node varies from 1 to 6 and is partially controlled by the level of cytokinin. Higher cytokinin levels induce the development of higher numbers of buds, but they may cause adverse

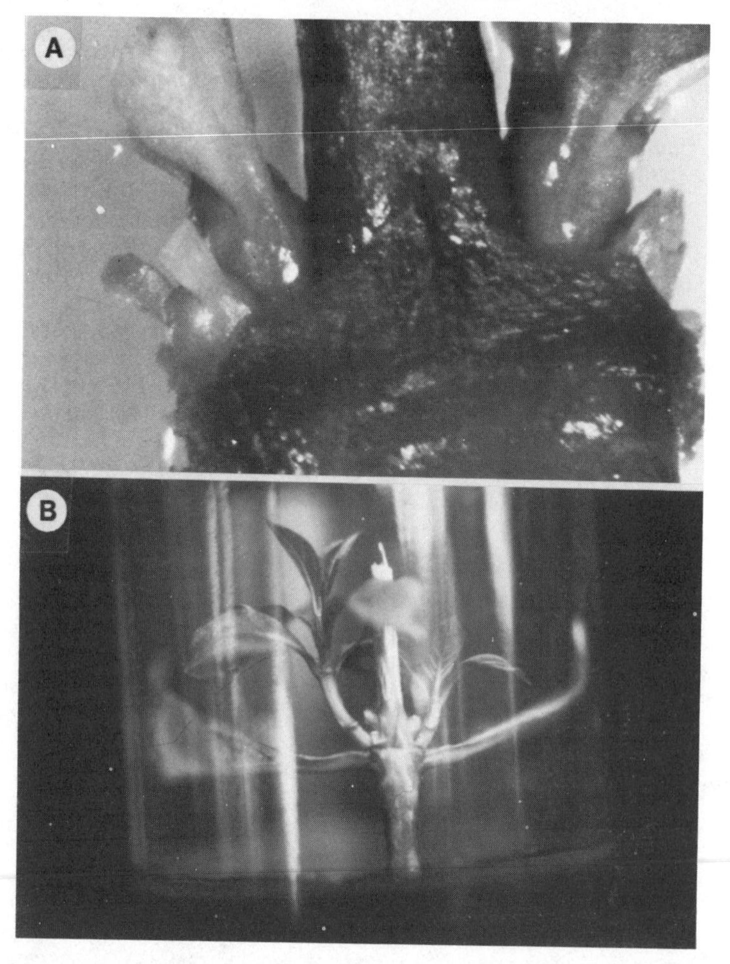

Figure 3. Nodal cultures of orthotropic shoots of *Coffea arabica* cv. Mundo Novo. (A) Multiple bud development after 18 days of culture in basic medium containing high cytokinin. (B) Orthotropic shoot development from arrested axillary buds following 2 months of culture.

effects on the subsequent growth of the derived shoots and lower the rooting frequencies. Pertaining to the cytokinin source, BA has proven more effective than KIN at 25 μM (2.1 versus 1.2 developed bud/node) and 50 μM (2.3 versus 1.4 bud/node), respectively.

The effects of GA on the elongation of shoot internodes were tested. Unsatisfactory results were found with the addition of GA in the solid medium sterilized either by autoclave or membrance filters. Effective treatments consisted of 25-50 ppm of GA applied directly to the developing shoots at the time of excision for rooting.

When the developed shoots have 3 nodes in length (ca. 60 days of primary culture), they can be excised and rooted (Fig. 3b). The best rooting frequencies (60%) were found with shoots treated with IBA (10 µM). A double-layer medium can be used for rooting, containing basic medium supplemented with IBA (10 µM) in the top layer and the basic medium with added charcoal (2.5 g/l) in the bottom (Fig. 4a). Another rooting technique consists of treating the undeveloped shoots with a liquid auxin solution for 10 days and then transferring the shoots to a basic solid medium or on a paper bridge (Fig. 4b). Shoots treated with a talc preparation of the auxin also gave high frequencies of rooting (60%) on the basic solid medium.

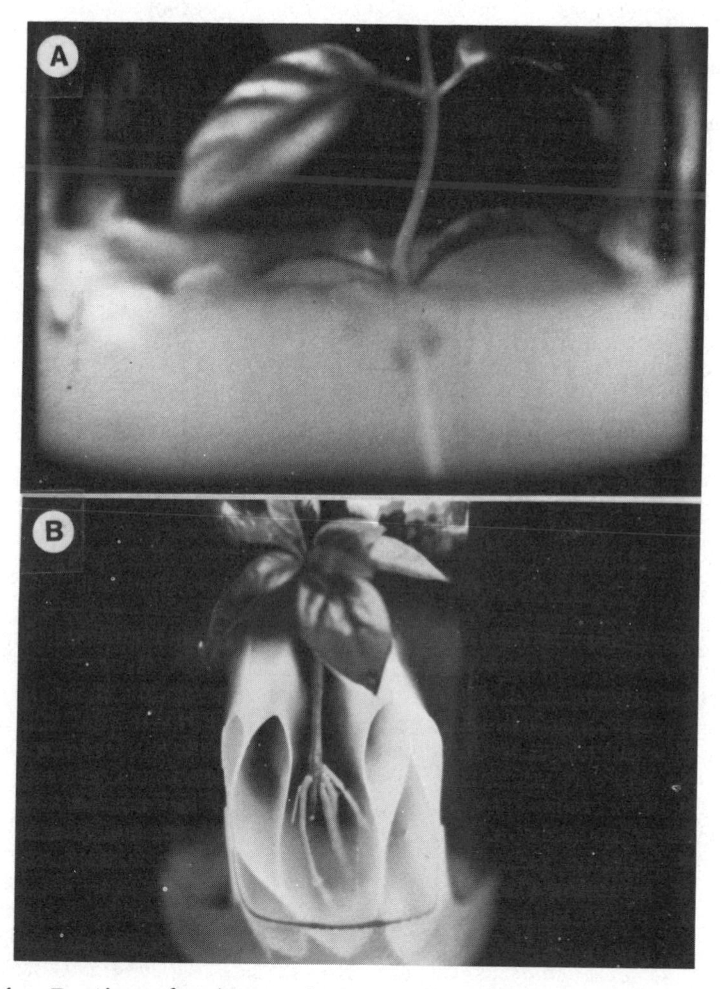

Figure 4. Rooting of primary shoots derived from arrested orthotropic buds. (A) Double-layer medium containing auxin in the top layer and basic medium with added charcoal in the bottom layer. (B) Root development on filter paper soaked with basic medium. Shoot base was treated with an auxin solution for 10 days before transfer to filter paper.

After excision of the first developed shoots, the nodal explants are transferred to fresh primary medium in order to recover the remaining arrested orthotropic buds. A nodal explant may survive 2-3 passages and so permit the recovery of most of the arrested buds. In several instances, 9-10 developing buds were recovered after the third bimonthly transfer (6 months of continuous culture). The developed buds derived from the original nodal explants were called primary buds or shoots. The buds developed from the nodes of primary shoots were designated secondary buds or shoots.

In one experiment of nodal culture from 1-year-old coffee plants (*C. arabica* cv. Mundo Novo), the frequencies of primary and secondary bud development per node were recorded using 6 months of continuous culture with bimonthly transfers (Table 9). The frequency of 1.33 developed buds/node was observed after the first 60 days of culture. At the end of the secondary culture period (120 days), an accumulated data of 4.5 buds/node were recorded. A final accumulated frequency of 7.5 buds/node was found at the end of the third culture period (180 days).

These data demonstrate the potential to explore the presence of arrested orthotropic buds for vegetative propagation of superior coffee plants by nodal culture technique (Sondahl and Nakamura, 1980; Nakamura and Sondahl, 1981). A 45% recovery (85 buds/19 nodes) of all arrested buds was recorded at the end of the second culture period (120 days). Taking into account the experimental data presented in Table 9, a theoretical estimation of the multiplication rate can be made as follows:

1. First growth period (6 months)—26 shoots with 3 nodes derived from nodal cultures of one orthotropic branch containing 6 nodes.
2. Second growth period (3 months)—104 shoots with 3 nodes.
3. Third growth period (3 months)—415 shoots with 3 nodes.

This protocol permits the recovery of 415 shoots/6-node branch after 12 months of successive culture. Since a mature coffee plant has several orthotropic branches that can be used for this vegetative purpose, a higher number of plants can be obtained by this technique within the time frame of 1 year. In conclusion, the final number of propagated shoots will vary according to the number of orthotropic branches utilized from each donor plant, the losses due to contamination and oxidation, and the efficiency of rooting.

Nodal explants of two sterile interspecific hybrids (*C. congensis* x *C. eugenioides* and *C. canephora* x *C. eugenioides*) were cultured with the techniques just described. The development of arrested orthotropic buds was similar to other coffee material. However, direct differentiation of somatic embryos on the cut surface of both orthotropic and plagiotropic internodal segments of the hybrid *C. congensis* x *C. eugenioides* was recorded. These embryos apparently arose from the cortex with an average frequency of 8 somatic embryos per segment after 44 days of culture on primary medium. The shoots developed from nodal cultures of the other hybrid (*C. canephora* x *C. eugenioides*) were sectioned and cultured on the same primary medium in order to

Table 9. Orthotropic Bud Development from Nodal Culture of *C. arabica* cv. Mundo Novo[a]

NODE NO.[b]	CULTIVATED NODE	BUDS/NODE RECOVERED	PRIMARY BUDS[c]	SECONDARY BUDS[d]	TOTAL BUDS
1	9	5	32	23	55
2	5	7	24	19	43
3	3	9	17	11	28
4	2	7	12	5	17
5	-	-	-	-	-
6	-	-	-	-	-
Total	19	28	85	58	143
Average no. shoots/node			4.5		7.5

[a]After 6 months of continuous culture with bimonthly transfers.

[b]The uppermost node beneath the apical meristem was coded No. 1.

[c]Accumulated data at the secondary culture (120 day).

[d]Accumulated data at the tertiary culture (180 day).

induce bud development. It is interesting that a great deal of embryogenic tissue (Sondahl et al., 1979,a,b) differentiated from the cut ends of the leaf pair attached to each node after about 60 days of culture. This embryogenic tissue was transferred to the conditioning medium (Sondahl and Sharp, 1977, 1979; Sondahl et al., 1980) where the somatic embryos completed their development. These results stress once again the differential response of genotypes in the presence of a common growth medium. Presently, more than 1000 nodal cultures have been successfully established and direct embryogenesis has been observed just with *C. congensis* x *C. eugeioides* hybrid tissues.

PROTOCOLS

Somatic Embryogenesis from Leaf Tissues of *Coffea arabica*

1. Surface sterilize leaves in 1.6% sodium hypochlorite (30% commercial bleach) for 30 min. Rinse 3 times in distilled water.
2. Cut leaf explants (7 mm^2) avoiding midvein and margins. The explants are submitted to a preincubation period upon a saline-sugar agar plate for 36–48 hr in darkness prior to inoculation onto induction medium. Preincubation medium: half-strength MS inorgan-

ic salts, thiamine (30 μM), cysteine (210 μM), inositol (550 μM), sucrose (87 mM), and agar (8 g/l).

3. Transfer sterile leaf explants with normal coloration to induction medium (primary culture). Basal medium: MS inorganic salts, thiamine (30 μM), cysteine (210 μM), inositol (550 μM), sucrose (117 mM), and agar (8 g/l). Growth regulators: KIN (20 μM) and 2,4-D (5 μM). Incubate in darkness for 7-8 weeks.
4. Transfer primary cultures to conditioning medium (secondary culture): basal medium with half-strength MS salts (except KNO_3 which is added 2x), KIN (2.5 μM), and NAA (0.5 μM). The culture bottles are kept under light for 3-4 months. The original callus develops a brown color after transfer to the conditioning medium.
5. Within 13 to 16 weeks after subculture to conditioning medium, the embryogenic tissue (white-cream color, friable) emerges from the periphery of the original callus mass. After 3 to 6 more weeks, clusters of globular shaped embryos develop.
6. This protocol allows the recovery of somatic embryos through indirect embryogenesis (HFSE) and direct embryogenesis (LFSE). The LFSE route can be preferentially obtained with a modified induction medium: basic medium supplemented with KIN or BA (10-20 μM) and 2,4-D (0.5 μM).
7. Somatic embryos are also obtained with this same protocol from cotyledon and hypocotyl tissues.

Embryo Culture

1. Coffee berries are surface sterilized by 70% alcohol for 5-10 min and 3.9% sodium hypochlorite (75% commercial bleach) for 30 min with agitation followed by 3 rinses with sterile distilled water. Individual seeds can be sterilized in the same way after removal of the parchment and silver skin.
2. Embryos are dissected by gentle incisions on the seed tip, convex side.
3. Culture medium: half-strength B5 inorganic salts, nicotinic acid (30 μM), pyridoxine (15 μM), thiamine (30 μM), meso-inositol (550 μM), sucrose (58.4 mM), BA (1 μM), IAA (2.5 μM), and agar (7 g/l).
4. Growth conditions: low intensity light (500 lux), 16-hr photoperiod.
5. After 50-60 days, the cotyledonary leaves will be developed. The embryos are subcultured to fresh medium without growth regulators and transferred to higher light intensity (3000 lux) or to 80% shaded sunlight.

Apical Meristem Culture

1. Green orthotropic shoots are washed in 1% detergent solution with continuous agitation (150 rpm) for 5-10 min and rinsed with sterile water.
2. Surface sterilization is accomplished by 2.6% sodium hypochlorite for 30 min with agitation (150 rpm) followed by 3 rinses with

sterile water. If orthotropic shoots are taken from field grown plants, the tissues are incubated overnight in a humid chamber and submitted to a second sterilization. Sterile seedlings can be used as a source of apical meristems.

3. Small apical meristems (ca. 0.3 mm) are isolated and immediately transferred to agar medium. Kartha's medium: MS salts, B5 vitamins, BA or ZEA (5-10 μM), and NAA (1 μM). Sondahl's medium: MS salts with 0.1x micronutrients, pyridoxine (15 μM), nicotinic acid (15 μM), thiamine (30 μM), meso-inositol (550 μM), cysteine (500 μM), sucrose (87 mM), BA (5 μM), IAA (0.5 μM), and agar (8 g/l).
4. Growth conditions: as described in steps 4 and 5 in the protocol for embryo culture.
5. Root in IBA (10 μM).

Nodal Culture

1. Apical portions of orthotropic shoots bearing 4-6 green nodes are washed in 1% detergent solution with agitation (150 rpm) and rinsed with sterile water 3 times.
2. Surface sterilization as described in step 2 of protocol for apical meristem culture.
3. Culture individual nodes with adjacent pair of leaves. Usually half to two-thirds of leaf blade area is cut to facilitate inoculation and culture.
4. Culture medium: B5 salts, pyridoxine (15 μM), nicotinic acid (15 μM), thiamine (30 μM), meso-inositol (550 μM), cysteine (500 μM), sucrose (87 mM), activated charcoal (2.5 g/l), PVP-40 (1 g/l), BA (25-50 μM), IAA (10 μM), and agar (7 g/l).
5. Growth condition: as described in step 4 of protocol for embryo culture.
6. Every 60 days, excise the primary shoots and transfer the nodal explant to fresh medium.
7. New nodal explants are obtained from in vitro developed shoots.
8. Rooting media: (a) double layer flasks containing basic medium supplemented with IBA (10 μM) in the top layer and basic medium with added charcoal (2.5 g/l) in the bottom layer; (b) liquid basic medium with IBA (10 μM) for 10 days followed by paper bridge culture in liquid basic medium ; (c) dip the cut end of shoots in a talc preparation containing IBA (10 μM) and culture in basic solid medium.
9. Shoots can be treated (dip) with GA (25-50 ppm) just before inoculation in the rooting medium.

FUTURE PROSPECTS

It has been well recognized that in vitro techniques are valuable tools for the improvement of crop plants (for review see Evans et al.,

1984; Vol. 2 of this series). Although conventional breeding techniques have been successfully applied to coffee improvement, there will be situations where tissue culture techniques may have unique applications. There is a need for high quality coffee in the world market. On the whole, coffee producing countries have been emphasizing yield increase and not quality. Coffee breeding for cup quality has been hindered by the lack of defined chemical parameters for selections as well as long-term support for multidisciplinary research. Correlations of chemical parameters associated with superior beverages between green and roasted coffee and between leaves and roasted characteristics are desirable. A modern improvement program for coffee would take advantage of standard breeding methods and in vitro techniques. Somaclonal variation and gametoclonal variation would be valuable for both arabica and robusta coffee. Large-scale clonal propagation would be important for robusta coffee.

Vegetative propagation by means of axillary bud culture has been successful with coffee (Dublin, 1980; Sondahl and Nakamura, 1980). Theoretical calculations with model cultures of *C. arabica* suggest a production capability of about 415 shoots per orthotropic branch after 12 months of continuous culture. This technique can be very useful to propagate (a) sterile interspecific hybrids (triploids); (b) segregating hybrids for multiple field plot studies and selection; (c) commercial production of superior mother plants (polygenic traits); (d) rapid increase of resistant genotypes to withstand the rapid evolution of physiological races of leaf rust; and (e) establishment of field plots for seed production of selected plants. In the case of *C. canephora*, rooting orthotropic shoots is very feasible using conventional nursery methods. However, as with *C. arabica*, axillary bud culture can be applied to this allogamous species for (a) propagation of breeding lines at several phases of a breeding program; (b) establishment of fields for seed production; and (c) commercial release of superior mother trees to establish new plantations.

Coffea arabica cultivars were derived from a very narrow genetic reservoir. Although hybridization has been the primary method of developing new breeding lines, induced mutations could offer an opportunity to increase the range of genetic variability for the breeder. However, the data currently available indicate that cuttings, germinating seeds, and pollen grains of *C. arabica* are quite resistant to ionizing radiatins such as gamma- and X-rays (2-100 krad) and to chemical mutagens. Although a very large number of samples have been irradiated, mutants have been isolated only occasionally. Treatment of seeds with recognized powerful mutagens, e.g., EMS or sodium azide, have failed to produce a single distinct mutation despite examination of several thousands of plants. Tissue culture somaclonal genetic variability and subsequent plant regeneration would therefore offer an excellent opportunity for developing unique types of genetic variability for the selection of new breeding lines. Somaclonal genetic variability is also likely to occur among plants regenerated from cell or tissue culture (Larkin and Scowcroft, 1981). Furthermore, since somaclonal variants are derived from spontaneous mutations (nuclear or cytoplasmic), it will be possible to release new varieties in shorter periods of time provided that the cultures were derived from tissues of superior commercial genotypes.

Coffea stenophylla (2n = 22) is immune to the attack of leaf miner (*Perileucoptera coffeela*). Attempts to cross this diploid species with the cultivated *C. arabica* have failed due to the abortion of hybrid triploid embryos. Up to now, the doubling of the chromosome number of *C. stenophylla* to facilitate the desirable cross at the 4n level has been hindered since colchicine treatments were ineffective—yielding only diploids or useless periclinal chimeras. Culture of immature interspecific embryos would make it possible to obtain viable hybrids between these species, as well as to assist in the transfer of genes in subsequent backcross generations. Similarly, crosses between other species of the tertiary gene pool of *C. arabica* that possess desirable characteristics could open new horizons for the genetic improvement of *C. arabica.* This would not only allow exploitation of the variability present in those species per se but would also provide an opportunity for the expression of novel variation generated by wide hybridizations and introduction of alien cytoplasm. Protoplast fusion followed by plant regeneration is also a valid alternative for overcoming genetic barriers. This technique would broaden even more the perspective for utilizing wide relatives in the improvement of *C. arabica.*

Haploid plants obtained by macro- or microspore culture might be used to produce commercial hybrids in the self-compatible species *C. canephora* (Sondahl et al., 1981). Microspores of mother plants with superior general combining ability would be used for the production of haploid plants. After chromosome doubling, four homozygous diploid clones with the best combining ability would be interplanted to produce F_1 hybrid seed or to produce double cross hybrids. Gametoclonal variation from F_2 anthers would permit the access of a broader variability for selection and breeding.

Plant regeneration from cultures of mature leaves of four species of the genus *Coffea* and five *C. arabica* cultivars has been achieved (Sondahl and Sharp, 1979). The culture media necessary for induction of somatic embryogenesis in high frequencies and for the culture of coffee protoplasts to yield callus have also been described (Sondahl et al., 1980).

From the topics discussed above and in the preceding sections it is obvious that from the breeder's perspective there are several areas that need to be further exploited using in vitro techniques. Major research emphasis must be placed on the development of appropriate methodology in coffee for (a) culture of immature interspecific hybrid embryos, (b) protoplast fusion and subsequent plant regeneration, (c) organelle and chromosome transfer, (d) in vitro chromosome doubling, (e) production of haploids via ovary or anther culture, and (e) cell suspension cultures, mutant selection, and plant regeneration. Since the necessary basic work and the protocols for dependable coffee tissue culture have already been established (Sondahl and Sharp, 1979; Sondahl et al., 1981), it is likely that in the near future the potential for the genetic improvement of coffee will be substantially broadened by the application of the new technologies of tissue culture. It is imperative that these improved techniques are fully integrated with conventional methods of breeding in order to guarantee the release of new, improved coffee varieties.

KEY REFERENCES

Sondahl, M.R. and Sharp, W.R. 1977a. High frequency induction of somatic embryos in cultured leaf explants of *Coffea arabica* L. Z. Pflanzenphysiol. 81:395-408.

Sondahl, M.R. and Sharp, W.R. 1979. Research in *Coffea* spp. and applications of tissue culture methods. In: Plant Cell and Tissue Culture: Principles and Applications (W.R. Sharp, P.O. Larsen, E.F. Paddock, and V. Raghavan, eds.) pp. 527-584. Ohio State Univ. Press, Columbus.

Sondahl, M.R., Salisbury, J.L., and Sharp, W.R. 1979a. SEM characterization of embryogenic tissue and globular embryos during high frequency somatic embryogenesis in coffee callus cells. Z. Pflanzenphysiol. 94:185-188.

Sondahl, M.R., Spahlinger, D.A., and Sharp, W.R. 1979b. A histological study of high frequency and low frequency induction of somatic embryos in cultured leaf explants of *Coffea arabica* L. Z. Pflanzenphysiol. 94:101-108.

Sondahl, M.R., Caldas, L.S., Maraffa, S.B., and Sharp, W.R. 1980. The physiology of in vitro asexual embryogenesis. In: Horticultural Reviews (J. Janick, ed.) pp. 268-310. AVI Pub., Westport, Connecticutt.

Sondahl, M.R., Monaco, L.C., and Sharp, W.R. 1981. In vitro methods applied to coffee. In: Tissue Culture and its Application in Agriculture (T. Thorpe, ed.) pp. 325-358. Academic Press, New York.

Staritsky, G. 1970. Embryoid formation in callus cultures of coffee. Acta Bot. Neerl. 19:509-514.

REFERENCES

Anonymous. 1979. Coffee economics. Kenya Coffee 44:521.

Buckland, E. and Townsley, P.M. 1975. Coffee cell suspension cultures. Caffeine and chlorogenic acid content. J. Inst. Can. Sci. Technol. Aliment 8:164-165.

Carvalho, A., Krug, C.A., Mendes, J.E.T., Antunes, H., Moraes, H., Aloisi, J., Moraes, M.V., and Rocha, T.R. 1952. Melhoramento do cafeeiro IV: Cafe Mundo Novo. Bragantia 12:97-129.

Carvalho, A., Monaco, L.C., and Campana, M.P. 1964. Melhoramento do cafeeiro XXVII: Ensaio de seleçoes regionais de Jau. Bragantia 23: 129-142.

Chaves, G.M., Cruz Filho, J., Carvalho, M.G., Matsuaka, K., Coelho, D.A., Shimoya, C. 1970. A ferrugem do cafeeiro (*Hemileia vastatrix* Berk. and Br.). Revisao de literatura com observacoes e comentarios sobre a enfermidade no Brasil. Seiva 75.

Colonna, J.P. 1972. Contribution à l'étude de la culture in vitro d'embryos de cafeiers. Action de la cafeine. Cafe Cacao The 16: 193-203.

Custer, J.B.M., Van Ee, G., and Buijs, L.C. 1980. Clonal propagation on *Coffea arabica* L. by nodal culture. In: Proc. IX International Colloquium of Coffee. Assoc. Scient. Intern. Coffee, London.

Dublin, P. 1980. Multiplication végétative in vitro de l'Arabusta. Cafe Cacao The 24:281-290.

______ 1981. Embryogénèse somatique directe sur fragments de feuilles de cafeier Arabusta. Cafe Cacao The 25:237-241.

Fonseca, M.A.S. 1976. Retorno social aos investimentos empesquisa na cultura de cafe. M.S. Thesis, ESALQ, Universidade de Sao Paulo, Brazil.

______, Araujo, P.F.C. and Pedroso, I.A. 1977. Retorno social aos investimentos em pesquisa na cultura do cafe. In: Proceedings V, Congresso Brasileiro Pesquisas Cafeeiras, pp. 111-112, Guarapari, ES.

Frischknecht, P.M., Baumann, T.W., and Wanner, H. 1977. Tissue culture of *Coffea arabica*. Growth and caffeine formation. Planta Med. 31:344-350.

Herman, F.R.P. and Haas, G.J. 1975. Clonal propagation of *Coffea arabica* L. from callus culture. Hortic. Sci. 10:588-589.

Kartha, K.K., Mronginski, L.A., Pahl, K., and Leung, N.L. 1981. Germplasm preservation of coffee (*Coffea arabica*) by in vitro culture of shoot apical meristems. Plant Sci. Lett. 22:301-307.

Keller, H., Wanner, H., and Baumann, T.W. 1972. Caffeine synthesis in fruits and tissue cultures of *Coffea arabica*. Planta 108:339-350.

Krug, C.A. and Carvalho, A. 1939. Genetic proof of the existence of coffee endosperm. Nature 144:515.

Lanaud, C. 1981. Production de plantules de *C. canephora* par embryogénèse somatique relisée à partir de culture in vitro d'ovules. Cafe Cacao The 25:231-235.

Larkin, P.J. and Scowcroft, W.R. 1981. Somaclonal variation: A novel source of variability from cell cultures for plant improvement. Theor. Appl. Genet. 60:197-214.

Monaco, L.C., Sondahl, M.R., Carvalho, A., Crocomo, O.J., and Sharp, W.R. 1977. Applications of tissue cultures in the improvement of coffee. In: Applied and Fundamental Aspects of Plant Cell, Tissue, and Organ Culture (J. Reinert and Y.P.S. Bajaj, eds.) pp. 109-129. Springer-Verlag, Berlin.

Murashige, T. and Skoog, F. 1962. A revised medium for rapid growth and bioassays with tobacco tissue cultures. Physiol. Plant. 15:473-497.

Nakamura, T. and Sondahl, M.R. 1981. Multiplicacao in vitro de gemas ortotropicas em *Coffea* spp. Proceedings 9 Congresso Brasileiro Pesquisa Cafeiras, pp. 162-163. Sao Lourenco, Brazil.

Nassuth, A., Wormer, T.M., Bouman, F., and Staritsky, G. 1980. The histogenesis of callus in *Coffea canephora* stem explants and the discovery of early embryoid initiation. Acta Bot. Neerl. 29:49-54.

Orozco, F.J. and Schieder, D. 1982. Aislamento y cultivo de protoplastos a partir de hojas de cafe. Cenicafe 33:129-136.

Sharp, W.R., Caldas, L.S., Crocomo, O.J., Monaco, L.C., and Carvalho, A. 1973. Production of *Coffea arabica* callus of three ploidy levels and subsequent morphogenesis. Phyton 31:67-74.

Sondahl, M.R. 1982. Tissue culture of morphological mutants of coffee. In: Proceedings 5th International Congress Plant Tissue and Cell Culture, Japan, pp. 417-418.

______ and Nakamura, T. 1980. Propagacao vegetativa in vitro de *Coffea* spp. In: Proceedings 8 Congresso Brasileiro Pesquisa Cafeeiras, Campos Jordao, p. 129.

______ and Sharp, W.R. 1977b. Interactions of cytokinin and auxin growth and embryogenesis in cultured leaf explants of *Coffea*. In: International Conference on Regulation of Developmental Processes of Plants, Halle, p. 180.

Townsley, P.M. 1974. Production of coffee from plant cell suspension cultures. Can. Inst. Food Sci. Technol. J. 7:79-81.

van der Voort, F. and Townsley, P.M. 1975. A comparison of the unsaponificable lipids isolated from coffee cell cultures and from green coffee beans. J. Inst. Can. Sci. Technol. Aliment. 8:199-201.

Wellman, F.L. 1961. Coffee: Botany, Cultivation and Utilization. Interscience, New York.

CHAPTER 22

Oil Palm

K. Paranjothy

Oil palm (*Elaeis guineensis* Jacq.) is grown commercially in Africa, South America, Southeast Asia, and the South Pacific, generally within 10° of the equator. The oil palm has been an important source of edible oil in Africa since historical times. It was introduced in 1848 into the Far East. Oil palm is now a major plantation crop in this region, with Malaysia and Indonesia accounting for almost 80% of world production in 1980.

Palm oil accounts for about 15% of the world's total production of vegetable oils. Two distinct types of oil are obtained from oil palm fruits: palm kernel oil, which closely resembles coconut oil, is obtained from the kernel (or seed), while palm oil is obtained from the fleshy mesocarp. Both oils have a wide variety of uses in the food and detergent industry. The residue remaining after extraction of kernel oil is an important source of animal feed.

Oil palm is a perennial crop, having an economically productive life span of 20-25 years. It is the highest yielding oil crop, averaging 4-6 tons of oil/hectare per year. Present-day planting materials are largely, if not entirely, of a hybrid variety known as tenera. The fruits of the tenera hybrids are characterized by a thin endocarp. The character for endocarp (or shell) thickness is determined by a single gene. The tenera hybrid is obtained from crosses between the dura (homozygous for thick shell) and pisifera (homozygous for absence of shell) types.

Oil palm is propagated by seed, conventional horticultural methods being inapplicable to this crop. There is much variation in yield between individual palms within a stand, reflecting to some degree the genetic heterogeneity of an outbreeding species. There is thus an obvious potential for tissue culture methods of vegetative propagation of oil palm.

LITERATURE REVIEW

Much information pertaining to culture of palms, especially that of oil palm, remains unpublished. This is no doubt due to the commercial potential that advances in tissue culture of palms over the past decade have generated. Indeed at the present time, plans for commercial scale production and sale of clonal plants are well advanced (Lioret and Ollagnier, 1981; Ahee et al., 1981; Pannetier et al., 1981; Noiret, 1981; Corley et al., 1981).

An attempt has been made to collate as much published information as possible in this review in an effort to elicit an understanding of the essential features of tissue culture of oil palm. Several stages are recognizable in the production of plantlets from callus cultures in vitro (Lioret, 1981; Fig. 1), and some of these, along with other aspects of in vitro culture of palms, are discussed below.

Plant Material

Aseptically grown seedlings, developed from seed or embryo cultures, have sometimes been used as a source of explants. Their use offers the convenience that methods for establishment of callus can be investigated without the loss of cultures through microbial contamination or the inhibitory or toxic effects of sterilants on explants. It would be advantageous, however, for purposes of clonal propagation and crop improvement to use material from proven and therefore mature palms. Roots can be easily dug out from the soil without destruction of the palm. In spite of the reported difficulty of surface sterilization (Smith and Thomas, 1973), these appear to have been used successfully as starting materials for clonal multiplication of oil palm (Corley et al., 1979) and date palm (Ammar and Benbadis, 1977; Tisserat, 1979). Young inflorescence and leaf tissues close to the apex are sterile and therefore convenient tissues for culture, though their removal may involve destruction of the apex. These have been used successfully for multiplication of oil palm (Corley et al., 1979). Rabechault and Martin (1976) describe callus initiation from young leaf tissue of oil palm obtained without damage to the apex. Several hundred such explants can be obtained from a single tree, between 20 and 69% of the explants yielding callus (Lioret, 1981). In date palm, callus-yielding tissues have been obtained from offshoots which provide a convenient alternative to the mature apex (Tisserat, 1979).

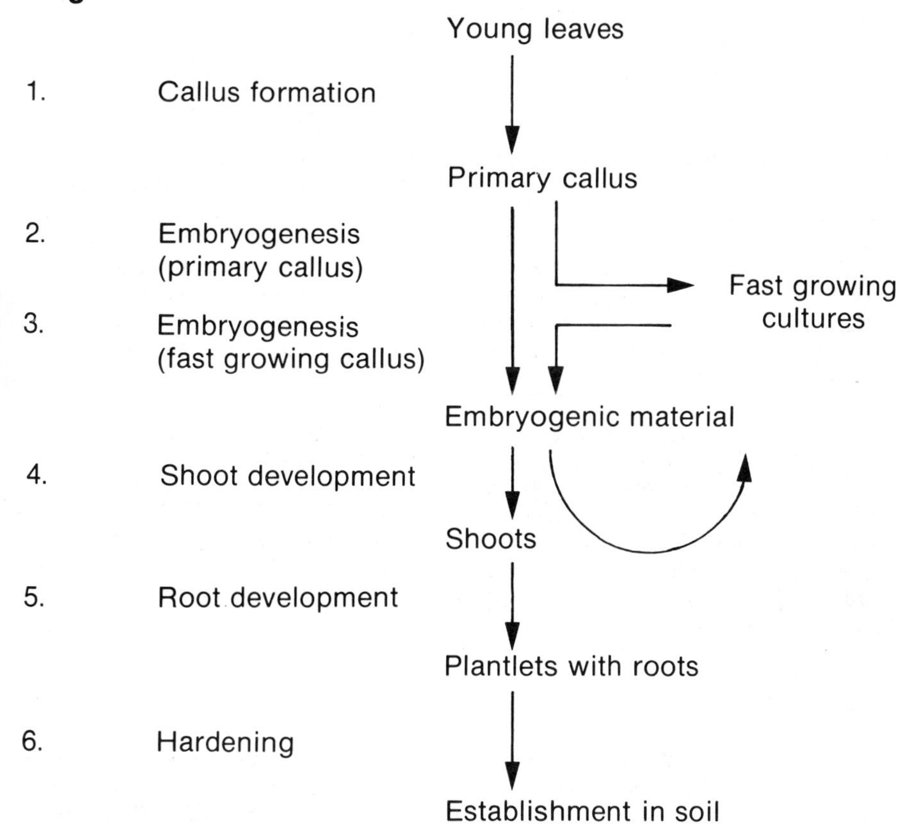

Figure 1. Stages in production of plantlets from cell cultures (from Lioret, 1981).

Callus Initiation

Anatomical studies of callus initiation are insufficient to ascertain the kinds of cells that normally give rise to callus in palm explants. Nevertheless, it appears that callus is generated mostly from meristematic sites or undifferentiated cells. Thus, Guzman et al. (1978) found that callus arose along procambial strands in germinating coconut embryos. Smith and Thomas (1973) found that callus initiation in roots occurred at sites from which laterals would normally have developed, and Martin et al. (1972) found that callus usually originated at the tips of oil palm roots.

Browning of explants is evidently a frequently encountered difficulty in palm in vitro culture studies. Browning is generally believed to be a response to wounding. In palm cultures browning appears to be promoted by cytokinins and agar (Rabechault and Martin, 1976; Martin and Rabechault, 1978). The inclusion of activated charcoal appears to be necessary or advantageous for callus initiation in date and other palms (Tisserat, 1979; Reynolds and Murashige, 1979). It is noteworthy that in both of these studies cytokinins were incorporated in the callus initiation medium. In several other studies where cytokinins were used at lower concentrations or not used at all, callus was successfully initiated without the use of activated charcoal (Rabechault et al., 1970; Martin et al., 1972; Smith and Thomas, 1973; Rabechault and Martin, 1976). An auxin is clearly essential for callus initiation (Smith and Thomas, 1973). In their requirements for growth substances palm cultures seem to have some similarity with cereals, the latter evidently not requiring cytokinins for callus initiation (Yeoman and Forche, 1979). Callus growth appears to be influenced to some extent by the particular auxin used. Thus callus grown on media containing NAA was found to be whiter in appearance, besides having a propensity to form roots, than callus grown on 2,4-D (Smith and Thomas, 1973).

The mineral nutrient requirements for callus initiation seem to have been satisfied in many studies by standard media formulations such as that of Murashige and Skoog. Coconut explants were reported to require higher levels of iodine than that found in MS medium (Eeuwens, 1976), but oil palm does not appear to have a requirement for iodine (Smith and Thomas, 1973).

Differentiation

Differentiation of somatic embryos and of roots and shoots has been observed in callus cultures of oil palm, date palm, and some other palms of less economic importance (Rabechault et al., 1972; Jones, 1974; Tisserat, 1979; Reynolds and Murashige, 1979).

Embryogenesis appears to have taken place as a result of a reduction or omission of auxin in some studies (Reynolds and Murashige, 1979; Tisserat, 1979). In the view of the results of Reynolds and Murashige (1979), the conditions under which asexual embryogenesis is obtained in palm cultures are the same as that for other plants. That is, callus is first obtained on a medium rich in auxin. Callus is then recultured to another medium lacking auxin to achieve embryogenesis. Cytokinins have been reported to promote embryogenesis (Rabechault and Martin, 1976). Light, osmolarity, and interim culture in liquid media are other factors implicated in embryogenesis in oil palm cell cultures (Rabechault et al., 1972). Jones (1974) indicated that embryogenesis in oil palm cell cultures appeared to be more a function of the starting inoculum than cultural conditions.

Rhizogenesis in callus, organ explants, and seedlings appears to be promoted by NAA (Sajise and Guzman, 1972; Smith and Thomas, 1973; Guzman et al., 1975; Eeuwens, 1978). High sugar levels also appear to promote root initiation (Guzman et al., 1975; Eeuwens, 1978).

Street (1979) discussed evidence for induction of embryogenesis in callus cultures of several plants and concluded that it is very difficult from available evidence to justify claims that variation of culture medium specifically induces morphogenetic events in cell cultures. The induction of embryogenesis is a poorly understood phenomenon in cell cultures and palm cell cultures are certainly no exception. Work aimed at elucidating the factors controlling embryogenesis in these cultures would obviously be valuable.

Plantlet Development

Little published information exists on nutrient conditions required for development of plantlets from somatic embryos. Martin and Rabechault (1978) reported that transfer of oil palm embryos to basal nutrient media containing 0.03 M sucrose led to shoot development. Root development was subsequently induced by treatment with NAA. Reynolds and Murashige (1979) obtained plantlets when date palm and other palm embryos were transferred to nutrient media lacking growth substances. Tisserat (1984) obtained plantlets when date-palm embryos were transferred to media lacking 2,4-D but containing 15 μM N^6-γ-γ dimethylallyaminopurine and 0.3% activated charcoal. Transfer of somatic embryos to fresh media for plantlet development has in some cases led to multiple plantlet development (Ahee et al., 1981).

Genetic Stability

The genetic stability of callus cultures is obviously one important factor that will determine the successful utilization of cell cultures for clonal propagation of oil palm.

Chromosome counts of root tip cells of oil palm plantlets produced from callus cultures have been reported to be normal and variability in growth parameters in oil palm clonal populations has been found to be less than that in seedling populations (Corley et al., 1979). Gross abnormalities that could be attributed to genetic or cytological changes were not evident in a population of about 13,000 plants representing several clones that had been produced from cell cultures. One clone, however, showed four abnormal plants, representing about 1.8% of the plants of the particular clone (Corley et al., 1981).

Oil palm callus is reported to be slow growing (Smith and Thomas, 1973), and this may partly account for its stability. The origin of callus from meristematic cells, which generally do not contain polyploid elements, and the absence of a requirement for cytokinins, known sometimes to induce chromosomal aberrations for callus initiation, are other factors which possibly favor stability. Polyploid and aneuploid cells have been recorded in a rapidly growing callus line of oil palm (Smith and Thomas, 1973), but cells of this line did not produce embryos (Corley et al., 1979). It is likely that cells with chromosomal aberrations rarely give rise to embryos.

Nevertheless, cytological studies of palm cell cultures are few and even these are briefly reported. It would seem useful, particularly in view of the commercial potential of palm cell cultures, to build up an understanding of their chromosomal cytology in relation to such factors as clonal origin, growth rates, capacity for differentiation, and media components, especially growth substances. The value of these cytological studies would, however, be limited by their failure to detect changes at the molecular level.

Zygotic Embryo Cultures

Besides fundamental studies, palm embryos are of interest as they are used as a convenient source of explant material for callus initiation. Also, aseptic seedlings developed from excised embryos cultured in vitro have been used as explant material. In coconut, embryo culture appears to be potentially useful for propagation of the nongerminating makapuno mutant. It has been suggested that collecting coconut embryos instead of seednuts will facilitate cheap exchange of germplasm on a large scale, overcoming at the same time restrictions on international transport of plant material (Guzman, 1969). The value of such aseptic germplasm transfer, however, would be doubtful if viral diseases were involved.

One suggestion evident in connection with culture of palm embryos is that toxic substances present in the medium or alternately toxic substances leached into the medium inhibit the germination and growth of the embryos. Some palm embryos including coconut embryos developed into seedlings only in media containing activated charcoal (Wang and Huang, 1976). Improved root development was evident in makapuno coconut embryos cultured in charcoal containing media (Guzman and Manuel, 1975). In addition, culture of these embryos in liquid media, in which toxic substances would be better dispersed, was found to promote root development (Guzman, 1969). There is surprisingly no report on the use of activated charcoal for culture of oil palm embryos, especially since the pattern of germination on media not supplemented with growth substances in one investigation included lack of root growth (Jones and Dethan, 1973).

Hodel (1977) was able to obtain plantlets without the use of activated charcoal from embryos of *Pritchardia kaalae* and *Veitchia joannis* cultured in Vacin and Went's medium containing 0.058 M sucrose.

PROTOCOLS

Callus Cultures

ESTABLISHMENT OF IN VITRO CULTURES. Exposed plant parts such as roots and leaves must be surface sterilized before culture. It is often difficult to secure satisfactory asepsis without sterilants causing injury to explant material and it would be useful, therefore, to survey some of the procedures that have been used with palm materials.

Surface sterilization of roots of oil palm is reported to be difficult (Smith and Thomas, 1973). Of a variety of sterilants tested, complete asepsis was obtained by treatment with 3% peracetic acid for 45 min, but none of the roots formed callus. Treatment with 0.1% mercuric chloride for 20 min resulted in approximately 50% sterility, and less than 20% of aseptic roots obtained by this treatment formed callus. Ong (1975) obtained callus from oil palm seedling roots that had been surface-sterilized by treatment with 95% alcohol for 2 min followed with 0.1% mercuric chloride for 2 min and finally 10% sodium hypochlorite for 20 min.

Rabechault and Martin (1976) were successful in obtaining callus from segments of oil palm leaf laminae that had been sterilized by treatment with 0.1% mercuric chloride for 20 min.

Tisserat (1979) reported that disinfection of date palm explants from offshoots (and root tips) was difficult. After surface sterilization in 1% sodium hypochlorite for 15 min the explants were washed and dipped in 1% sodium hypochlorite for 5 sec before culture on nutrient media, without rinsing. This procedure was found to be effective in producing contaminant-free cultures and bud and shoot tip explants subjected to this final dip in bleach solution were reported to grow without inhibition.

Well concealed parts, such as the stem apex, young inflorescence, or seed embryo, do not normally require treatment with sterilants. Thus, uncontaminated young coconut inflorescence tissue was obtained without the use of chemical sterilants by careful separation of the inner and outer spathes under aseptic conditions (Eeuwens, 1976). Precautions, however, are often necessary to prevent entry of microorganisms into the unexposed parts during sampling. As a precaution, Staritsky (1970) treated excised oil palm "cabbages" (comprising the stem apex and surrounding young tissues) with a saturated solution of calcium hypochlorite for 1-2 hr before excision of explants. He noted that materials from vigorously growing palms only were not heavily contaminated. Rabechault et al. (1972) disinfected oil palm "cabbages" in 0.1% mercuric chloride for 20 min. Eeuwens (1976) obtained young stem and leaf explants of coconut from a 4 x 4 x 4 cm block of "cabbage" that had been sterilized in a 2% solution of calcium hypochlorite for 15 min. Cork borers were used for excision of explants from the surface sterilized "cabbages" (Rabechault et al., 1972; Eeuwens, 1976).

A variety of explants have been used for callus initiation, including embryos, roots, young leaves, stem apex tissues, and inflorescences. Rabechault et al. (1970) reported the formation of nodular cell colonies on seed embryos cultured in a liquid medium containing Heller's salts, 1 μM 2,4-D, and 1 μM KIN. The nodules regenerated new nodular cell colonies when transferred to media containing 1 μM IAA.

Following experiments which involved callus generation from several hundred embryos, Smith and Thomas (1973) concluded that the composition of the mineral nutrients was not important for generation of callus. No differences in callus growth were observed when MS mineral formulation was compared with their own formulations, "P.C.l." and "P.C.m." Growth of callus was observed over a wide pH range, from 3.7 to 7.3, with no apparent optimum. Sucrose and glucose supported growth at concentrations from 1 to 5%. The presence of an auxin was found to be essential for culture of oil palm cells, with initiation of callus requiring a more 'vigorous' auxin than its subsequent maintenance. Thus, 2,4-D and 2,4,5-T were found to be more effective than NAA for callus initiation.

Of several cytokinins tested only KIN at 0.5 or 5 μM showed occasional and even then slight effects on callus growth. No response to GA was evident at concentrations in the range of 3-150 μM. As a generally satisfactory medium the MS mineral salts (without potassium iodide), and 0.1% casein hydrolysate, 5% CW, 0.03 M sucrose, and 0.055 M glucose were adopted, auxins and other substances being added as required.

Martin et al. (1972) reported initiation of callus from the tips of roots excised from seedlings that had been germinated in vitro. Callus was formed when roots were cultured in rotating vessels containing a liquid medium consisting of MS salts, 0.058 M sucrose, and 2 μM 2,4-D, or in agar media of similar composition containing NAA instead of 2,4-D.

Smith and Thomas (1973) found that aseptic roots obtained from cultured embryos or seeds formed callus within three weeks of culture. Callus was formed on secondary or tertiary roots at positions from which laterals would have developed. Callus also developed at the root tips but more slowly. When agar media containing MS salts, 0.1% CH, and 5% CW were used, a suitable concentration of auxin for callus initiation was 25 μM of either NAA or 2,4-D. Lower concentrations were used for subculture of callus. KIN at 2.5 μM was inhibitory to growth, while at 0.05 μM no stimulation of growth was observed.

The most promising tissues from the apex appeared to be those closest to the apex, such as the white bases of petioles, the potential for callus proliferation decreasing in tissues excised away from the apex (Rabechault et al., 1972). Best growth was seen in liquid followed by semisolid and solid media. Agar appeared to give rise to browning. A high concentration of minerals (>1800 mg/l) was seen as one of the requirements for callus formation. Cytokinins appeared to cause browning and these were therefore used at low concentrations (0.1 μM). BA was found to cause less browning than KIN. Thus, as a standard procedure, Rabechault et al. (1972) initiated callus in liquid media containing salts of Knop, Heller, or MS (x 2/5), 0.058 M sucrose, trace elements, 2 μM 2,4-D, and 0.1 μM KIN or BA.

Experiments of Smith and Thomas (1973) with explants from the apical region indicated differences in media requirement for callus initiation and maintenance. Twelve of 13 callus lines obtained were initiated on P.C.m. medium containing 0.1% soya peptone and 2,4,5-T. The 13 callus lines gave rise to 55 cultures, 16 on the same medium and 55 on P.C.I. medium containing 5% CW and 0.1 mM NAA. Callus initiation did not appear to depend on source of plants or time of the year but appeared to be positively influenced by vigor of the plant at time of isolation. The base of the spear leaf was a more rewarding source than the central region.

Rabechault and Martin (1976) reported initiation of callus from young leaf tissue. The latter was obtained without damage to the apical growing region. Squares of leaf laminae were cultured in the dark in MS medium containing 0.058 M sucrose and either 2,4-D or 2,4,5-T at a level between 10 and 50 μM. The pH was adjusted to a low level of 4.5 before autoclaving. Callus formed along the veins after 15-60 days.

Smith and Thomas (1973) found that explants from male and female inflorescences showed some development, usually of their normal structures but no callus was obtained. From the results of their investigations they concluded that early inflorescences (spadices < 20 cm long) were more likely to develop callus than more mature material and that a high auxin concentration would be required to disrupt normal development.

Staritsky (1970) cultured oil palm shoot apices in Miller's nutrient medium containing 0.088 M sucrose, 0.5 μM KIN, and 0.25 μM NAA or IAA and found regeneration of lanceolate leaves that resembled the juvenile leaves of young seedlings. The shoot apex explants also initiated roots and nodules of callus.

MULTIPLICATION OF PROPAGULES. Nodular cell colonies that had been obtained from embryos cultured on 2,4-D- and KIN-supplemented media were reported to develop roots when transferred to media containing low concentrations of IAA (Rabechault et al., 1970). Shoot development, however, was not obtained. Callus derived from roots has also been reported to regenerate roots when transferred to low auxin media (Martin et al., 1972).

Rabechault et al. (1972) obtained root and shoot organogenesis in callus derived from explants of apical tissue. This was achieved by a series of transfers, the first transfer being a liquid medium with mineral salts increased to 3000 mg/l (Miller's B and MS medium being suitable) and 2,4-D replaced with 1 μM NAA. The second transfer was to a similar liquid medium but with sucrose increased from 0.058 to 0.12 M. Finally, the cultures were transferred to media with reduced concentration of mineral salts (1800 mg/l), 0.058 M sucrose, and 0.1 μM IAA.

Jones (1974) found that callus obtained from apical explants as well as roots regenerated a mass of fibrous roots and pneumathodes in media containing low concentrations of NAA. Occasionally parts of the roots greened but no shoots developed in these cultures. Development of embryos was sometimes seen in callus derived from seedling root and leaf base tissues. Details of media used for callus initiation or differentiation were not described but formation of embryos was noted to be more a function of the origin of the explant than of the subsequent cultural conditions. The embryos developed initially as white structures with a well defined epidermis. The structures elongated and acquired polarity and showed other embryo-like features such as accumulation of storage protein and triglycerides in their cells. Development of plantlets from embryos was observed. It was noted that the embryos did not respond to exogenously applied IAA, unlike seed embryos (Jones and Dethan, 1973).

Rabechault and Martin (1976) found that callus derived from leaf explants obtained from mature palms formed embryos when the callus was incubated under light in media containing 2.5 μM NAA and 10 μM of KIN, BA, or 2iP. The formation of embryos was also believed to be aided by impoverishment of the medium. Callus transferred onto media without auxin did not produce embryos. Embryo formation was stimulated by cytokinins and amongst the cytokinins tested only ZEA was effective below 0.25 μM in stimulating embryogenesis.

ESTABLISHMENT OF PLANTLETS IN SOIL. Plantlets produced from cell cultures of oil palm (Rabechault and Martin, 1976; Corley et al., 1976, 1979) have been successfully established in soil. Various procedures have been studied in attempts to improve the survival of oil palm plantlets on transfer to soil (Anonymous, 1978). Reduction of moisture loss by leaf clipping or application of anti-transpirent films were of no beneficial effect when the plantlets were kept under conditions of high humidity in a plastic tent. Steam sterilization of soil appeared to be of doubtful value. Plantlets growing in steam sterilized soil had higher nitrogen concentrations but appeared to accumulate toxic concentrations of manganese. Reasonable establishment in nonsterile soil could be achieved if the plants had been adequately hardened before planting from culture. Tisserat (1984) describes in detail a procedure for establishment of in vitro developed date palm plantlets in soil. The initial size of the plantlets was a critical factor in their survival. A minimum height of between 10 and 12 cm appeared to be necessary for maximum survival on transplantation. Plantlets were transplanted into a 1:1 mixture of peat moss and vermiculite. The plantlets were covered with plastic lids to maintain high humidity. For the first 2 months in soil, plantlets were sprayed with 0.5% benomyl and watered with 1/4 strength Hoagland's solution once a week.

Embryo Cultures

Hussey (1958) cultured excised oil palm seed embryos in White's medium containing 0.058 M sucrose and found poor and unbalanced growth, as the embryos produced only shoot or root growth. A series of detailed studies by Rabechault and co-workers has since led to much useful information. Growth of embryos in vitro was found to be dependent on both the size and age of source seeds. Embryos from large seeds (> 100 mg) were found to grow more vigorously than embryos from smaller seeds and inferior growth was generally seen with embryos obtained from seeds more than 6 months old (Rabechault and Ahee, 1966). Rabechault et al. (1968) found that embryos obtained from seeds which had been stored for a year developed satisfactorily in vitro if the moisture content of the seeds was raised to 20-22% before removal of the embryo. Storage of rehydrated seeds for 10-15 days further improved subsequent growth of the embryos in vitro, indicating that rehydration of embryos within seeds was superior to their rehydration on nutrient media. Rabechault et al. (1969) found that growth in vitro of embryos was also related to the dormancy of the source seeds, the effects of dormancy and dehydration being independent of one another.

Light was found to stimulate growth of the haustorium and root system but seedling shoots in light were shorter and had fewer leaves (Rabechault, 1962). The severed haustorium was found to increase markedly in diameter and to green and accumulate starch on exposure to light, but embryos did not develop unless a small portion of the haustorium remained attached (Rabechault and Cas, 1974).

Sucrose appeared to be the best source of carbon and energy, vigorous shoot and root growth being obtained by culture of embryos in media containing 0.088 M sucrose (Buffard-Morel, 1968). Partial replacement of sucrose with mannitol did not lead to reduction in growth, indicating that osmotic effects of sucrose in the medium were also important (Rabechault et al., 1974). Embryos also appeared to grow better in media with high overall salt concentration (Rabechault et al., 1970). Thus, the poor growth seen in media containing salts of Randolph and Cox, Rijven, or White was improved when the overall salt concentration of these media was increased, without change in ionic balance, to that of Heller's medium, in which good growth of embryos had been obtained.

Rabechault et al. (1972) obtained better growth of embryos in liquid rather than on agar media. Progressive stimulation of growth was seen when liquid medium was supplemented with 5, 10, 15, 20, and 25% CW. Root initiation was earlier and more frequent in liquid media containing 5% CW. This effect of CW was further increased by incubation of cultures in light. Rabechault et al. (1973) found that growth of embryos from fresh seeds was vigorous and their growth was not stimulated by CW. Embryos from seeds 45-60 days old responded to CW but embryos from seeds 70 days old with a low moisture content were no longer responsive. When these seeds were rehydrated to a moisture content of 20.1%, a marked improvement of growth was obtained with increasing quantities of CW.

GA stimulated shoot growth in proportion to the concentration used, but root formation and growth were not affected (Bouvinet and Rabechault, 1965). IAA, NAA, 2,4-D, and KIN had an increasingly inhibitory effect on embryos at concentrations from 0.01 to 100 μM. NAA and 2,4-D caused an increase in the diameter of the haustorium while KIN promoted growth in length of the haustorium. Browning of the tissues and the medium increased with concentration of growth substances, KIN causing the most intense browning (Rabechault et al., 1976).

Jones (1974) found a typical pattern of shoot growth without root development when embryos were cultured on media that were not supplemented with growth substances. Balanced shoot and root development was obtained by supplementing media with low concentrations of IAA and KIN, the optimal concentrations being 2.5 μM and 0.5 μM, respectively. Significantly better results were obtained with 0.088 M sucrose than with 0.073 M, while 0.15 M inhibited growth. In confirmation of earlier findings of Rabechault et al. (1970), better development of embryos was seen in media formulations containing high overall salt concentrations. Murashige and Skoog's formulation was hence adopted as a basal medium. The addition of 200 mg/l of CH resulted in good shoot growth and was raised to 1000 mg/l as a standard component. Ferulic acid promoted root emergence in White's medium but much less effectively in embryos cultured in MS medium. Other phenolic compounds such as caffeic and sinapic acids (at 20 μM) also promoted rooting in White's medium.

FUTURE PROSPECTS

The use of in vitro methods in conjunction with cryopreservation for storage of germplasm has obvious attractions for palms, particularly since they are not amenable to multiplication by conventional propagation methods. Tisserat et al. (1981) have demonstrated the feasibility of using cryogenic methods for storage of date palm callus and have also been successful in regenerating plantlets from revived cultures. Similar methods could perhaps be utilized for storage of palm embryos.

One aspect of in vitro investigations, which has potential in crop improvement but which appears not to have attracted significant research attention in palms, is that of production of haploids by culture of anthers. It is quite likely that generation of haploids will be difficult in palms that are outbreeders. Nevertheless, anther culture research may be worthwhile if there is scope for beneficial utilization of haploids in breeding or in the promising and developing field of genetic engineering.

REFERENCES

Ahee, J., Artuuis, P., Cas, G., Duval, Y., Guenin, G., Hanower, J., Hanower, P., Lievoux, D., Lioret, C., Malaurie, B., Pannetier, C., Raillot, D., Varechon, C., and Zuckerman, L. 1981. La multiplication végétative in vitro du palmier à huile par embryogénèse somatique. Oleagineux 36:113-115.

Ammar, S. and Benbadis, 1977. Multiplication végétative du palmier-dattier (*Phoenix dactylifera* L.) par la culture de tissus de jeunes plantes issués de semis. C.R. Acad. Sci. Paris 284:1789-1792.

Anonymous. 1978. In Progress Report, 1978-1979. Bakasawit Clonal Oil Palm Reserach Unit, Malaysia.

Bouvinet, J. and Rabechault, H. 1965. Recherches sur la culture in vitro des embryons de palmier à huile (*Elaeis guineensis* Jacq.). II. Effets de l'acide gibberellique. Oleagineux 20:79-87.

Buffard-Morel, J. 1968. Recherches sur la culture in vitro des embryons de palmier à huile (*Elaeis guineensis* Jacq. var. dura). Effets du glucose, du levulose, du maltose et du saccharose. Oleagineux 23: 707-711.

Corley, R.H.V., Barrett, J.N., and Jones, L.H. 1976. Vegetative propagation of oil palm via tissue culture. In: Malaysian International Agriculture Oil Palm Conference, Incorporated Society of Planters, Kuala Lumpur, Malaysia.

Corley, R.H.V., Wooi, K.C., and Wong, C.Y. 1979. Progress with vegetative propagation of oil palm. Planter, Kuala Lumpur 55:377-380.

Corley, R.H.V., Wong, C.Y., Wooi, K.C., and Jones, L.H. 1981. Early results from the first oil palm clone trials. In: The Oil Palm in Agriculture in the Eighties, Vol. 1, pp. 173-196. Incorporated Society of Planters, Kuala Lumpur, Malaysia.

Eeuwens, C.J. 1976. Mineral requirements for growth and callus initiation of tissue explants excised from mature coconut palms (*Cocos nucifera*) and cultured in vitro. Physiol. Plant. 36:23-28.

______ 1978. Effects of organic nutrients and hormones on growth and development of tissue explants from coconut (*Cocos nucifera*) and date (*Phoenix dactylifera*) palms cultured in vitro. Physiol. Plant. 42:173-178.

Guzman, E.V. 1969. The growth and development of coconut 'Makapuno' embryo in vitro. I. The induction of rooting. Philipp. Agric. 53:65-78.

______ and Manuel, G.C. 1975. Improved root growth in embryo and seedling cultures of coconut 'Makapuno' by the incorporation of charcoal in the growth medium. Fourth Session of the FAO Technical Working Party on Coconut Production, Protection and Processing, Kingston, Jamaica, 14-25 Sept. 1975. FAO, Kingston, Jamaica.

______, Manuel, G.C., and Del Rosario, A.G. 1975. The growth and morphogenesis of coconut Makapuno embryos in vitro: The enhancement of callus growth of 2,4-D through a modification of the sugar, calcium and potassium content of the medium. Proceedings of the Second Regular Session Symp. Biol. Res., S.U. (April 4-5, 1975) pp. 126-130. Silliman University, Dumaguete City, Philippines.

______, del Rosario, A.G., and Ubalde, E.M. 1978. Proliferative growths and organogenesis in coconut embryo and tissue cultures. Philipp. J. Coconut Stud. 3:1-10.

Hodel, D. 1977. Notes on embryo culture of palms. Principes 21:103-108.

Hussey, G. 1958. An analysis of the factors controlling the germination of the seed of the oil palm, *Elaeis guineensis* Jacq. Ann. Bot. 22:259-284.

Jones, L.H. 1974. Propagation of clonal oil palms by tissue culture. Oil Palm News 17:1-8.

______ and Dethan, S.K. 1973. Establishment of oil palm plants from aseptically grown excised embryos. Eucarpia Congr. Assoc. Eur. Amelior. Plant.

Lioret, C. 1981. Vegetative propagation of oil palm by somatic embryogenesis. In: The Oil Palm in Agriculture in the Eighties, Vol. 1, pp. 163-172. Incorporated Society of Planters, Kuala Lumpur, Malaysia.

______ and Ollagnier, M. 1981. La culture in vitro de tissus chez le palmier à huile. Oleagineux 36:111.

Martin, J.P. and Rabechault, H. 1978. Procede de multiplication végétative de végétaux et plants ainsi obenus. Brevet D'invention No. 76, 28361, Institut National de la Propriété Industrielle, Paris.

Martin, J.P., Cas, S., and Rabechault, H. 1972. Note préliminaire sur la culture in vitro de racines de palmier à huile. Oleagineux 27: 303-305.

Noiret, J.M. 1981. Application de la culture in vitro a l'amélioration et à la production de material clonal chez le palmier à huile. Oleagineux 36:123-124.

Ong, H.T. 1975. Callus formation from roots of the oil palm (*Elaeis guineensis* Jacq.). Proceedings of the National Plant Tissue Culture Symposium, Kuala Lumpur, pp. 26-31. Rubber Research Institute of Malaysia, Kuala Lumpur.

Pannetier, C., Arthuis, P., and Lievoux, D. 1981. Néoformation de jeunes plantes d'*Elaeis guineensis* à partir de cals primaires obtenus sur fragments foliaires cultivés in vitro. Oleagineux 36:119-120.

Rabechault, H. 1962. Recherches sur la culture 'in vitro' des embryons de palmier à huile *Elaeis guineensis* Jacq. I. Effets de l'acide B-indole-acétique. Oleagineux 17:757-765.

_____ and Ahee, J. 1966. Recherches sur la culture in vitro des embryons de palmier à huile (*Elaeis guineensis* Jacq.). III. Effets de la grosseur et de l'age des graines. Oleagineux 21:729-734.

_____ and Cas, S. 1974. Recherches sur la culture in vitro des embryons de palmier à huile (*Elaeis guineensis* Jacq. var. dura Becc.). X. Culture de segments d'embryons. Oleagineux 29:73-78.

_____ and Martin, J.P. 1976. Multiplication végétative du palmier à huile (*Elaeis guineensis* Jacq.) à l'aide de cultures de tissus foliaries. C.R. Acad. Sci. Paris 238:1735-1737.

_____, Ahee, J., and Guenin, G. 1968. Recherches sur la culture in vitro des embryons de palmier à huile (*Elaeis guineensis* Jacq.). IV. Effets de la teneur en eau des noix et de la durée de leur stockage. Oleagineux 23:233-237.

_____, Guenin, G., and Ahee, J. 1969. Recherches sur la culture in vitro des embryons de palmier à huile (*Elaeis guineensis* Jacq. var. dura Becc.). VI. Effets de la deshydratation naturelle et d'une rehydration de noix dormantes et non dormantes. Oleagineux 24:263-268.

_____, Ahee, J., and Guenin, G. 1970. Colonies cellulaires et formes embryoids obtenues in vitro à partir de cultures d'embryons de palmier à huile (*Elaeis guineensis* Jacq. var. dura Becc.). C. R. Acad. Sci. Paris 270:3067-3070.

_____, Ahee, J., and Guenin, G. 1972. Recherches sur la culture in vitro des embryons de palmier à huile (*Elaeis guineensis* Jacq. var. dura Becc.). VIII. Action du lait de coco autoclave en presence ou non de gelose et de lumière et en raison de l'age des graines. Oleagineux 27:249-254.

_____, Guenin, G., and Ahee, J. 1973. Recherches sur la culture in vitro des embryons de palmier à huile (*Elaeis guineensis* Jacq. var. dura Becc.). IX. Activation de la sensibilité au lait de coco par une rehydratation des graines. Oleagineux 28:333-336.

___, Buffard-Morel, J., and Varechon, C. 1974. Recherches sur la culture in vitro des embryons de palmier à huile (*Elaeis guineensis* Jacq.). XI. Effets de la pression osmotique sur la croissance et le développement et sur l'absorption des sucres. Oleagineux 29:351-356.

_____, Ahee, J., and Guenin, G. 1976. Recherches sur la culture in vitro des embryons de palmier à huile (*Elaeis guineensis* Jacq.). XII. Effets de substances de croissance à des doses supraoptimales. Relation avec le brunissement des tissus. Oleagineux 31:159-163.

Reynolds, J.F. and Murashige, T. 1979. Asexual embryogenesis in callus cultures of palms. In Vitro 15:383-387.

Sajise, J.U. and Guzman, E.V. 1972. Formation of adventitious roots in coconut makapuno seedlings grown in medium supplemented with naphthalene acetic acid. Kalikasan Philipp. J. Biol. 1:197-206.

Smith, W.K. and Thomas, J.A. 1973. The isolation and in vitro cultivation of cells of *Elaeis guineensis*. Oleagineaux 28:123-127.

Staritsky, G. 1970. Tissue culture of the oil palm (*Elaeis guineensis* Jacq.) as tool for its vegetative propagation. Euphytica 19:288-292.

Street, H.E. 1979. Embryogenesis and chemically induced organogenesis. In: Plant Cell and Tissue Culture: Principles and Applications (W.R. Sharp, P.O. Larsen, E.F. Paddock, and V. Raghavan, eds.) pp. 123-125. Ohio State University Press, Columbus.

Tisserat, B. 1979. Tissue culture of the date palm. J. Hered. 70:221-222.

______ 1984. Date palm. In: Handbook of Plant Cell Culture, Vol. 2 (W.R. Sharp, D.A. Evans, P.V. Ammirato, and Y. Yamada, eds.) pp. 505-545. Macmillan, New York.

______, Ulrich, J.M., and Finkle, B.J. 1981. Cryogenic preservation and regeneration of date palm tissue. HortScience 16:47-48.

Yeomann, M.M. and Forche, E. 1980. Cell proliferation and growth in callus cultures. In: Perspectives in Plant Cell and Tissue Culture (I.K. Vasil, ed.) pp. 1-24, Suppl. 11A, International Review of Cytology. Academic Press, New York.

Wang, P.J. and Huang, L.C. 1976. Beneficial effects of activated charcoal on plant tissue and organ cultures. In Vitro 12:260-262.

Species Index

Subject Index